T0135818

Adaptive Wavelet Methods for Variational Formulations of Nonlinear Elliptic PDEs on Tensor-Product Domains

Inaugural-Dissertation

zur

Erlangung des Doktorgrades

der Mathematisch-Naturwissenschaftlichen Fakultät

der Universität zu Köln

vorgelegt von

Roland Pabel

aus Köln

Köln, 2015

Berichterstatter/in:	Prof. Dr. Angela Kunoth
	Prof. Dr. Gregor Gassner
	Prof. Dr. Helmut Harbrecht (Universität Basel)
Tag der mündlichen Prüfung:	18. Mai 2015

Bibliographic information published by the Deutsche Nationalbibliothek

The Deutsche Nationalbibliothek lists this publication in the Deutsche Nationalbibliografie; detailed bibliographic data are available on the Internet at http://dnb.d-nb.de .

ISBN 978-3-8325-4102-6

Logos Verlag Berlin GmbH
Comeniushof, Gubener Str. 47,
10243 Berlin
Tel.: +49 (0)30 42 85 10 90
Fax: +49 (0)30 42 85 10 92
INTERNET: http://www.logos-verlag.de

Danksagung

Nach vielen Jahren intensiver Arbeit liegt sie nun vor Ihnen: meine Dissertation. Damit ist es an der Zeit, mich bei denjenigen zu bedanken, die mich in dieser Phase meiner akademischen Laufbahn begleitet haben.

An erster Stelle danke ich Frau Prof. Dr. Angela Kunoth für die Möglichkeit, überhaupt promovieren zu können. Bereits im Kindergarten hatte ich mir die Promotion in Mathematik als Ziel gesetzt, welches ich nun drei Jahrzehnte später erreicht habe. Es war ein herausfordernder, aber auch ungemein lohnender Weg (der durch ganz Nordrhein-Westfalen führte), den ich ohne Frau Kunoths Unterstützung nun nicht hätte beenden können. Ohne ihre fachliche Hilfe, bei gleichzeitiger Gewährung umfassender Freiheiten in der Forschung, würde diese Arbeit nicht existieren und ich könnte sie nicht als mein Werk ansehen. Ich möchte mich auch für die zeitweise Anstellung im Projekt C1 des TransRegio 32 „Patterns in Soil-Vegetation-Atmosphere Systems: Monitoring, Modelling and Data Assimilation“ bedanken.

Den weiteren Gutachtern, Prof. Dr. Gregor Gassner und Prof. Dr. Helmut Harbrecht, sowie den weiteren Mitgliedern der Prüfungskommission, Prof. Dr. Yaping Shao und Dr. Roman Wienands, danke ich für ihren Arbeitseinsatz, welcher, geschuldet dem Thema und Umfang dieser Arbeit, nicht unterschätzt werden sollte.

Aufgrund der zwei Rufe und anschließenden Universitätswechsel von Frau Prof. Dr. Kunoth während meines Promotionsstudiums kann ich allen Kollegen und Studenten der Universitäten Bonn, Paderborn und Köln nur zusammenfassend für die gute Zusammenarbeit danken. Besonders möchte ich hier nur meinen Bürokollegen und Mitdoktoranden Christian Mollet erwähnen, an dem scheinbar ein guter Hutmacher verloren gegangen ist. Der Arbeitsalltag an der Universität Paderborn hätte ohne ihn sehr viel weniger Spaß gemacht.

Des Weiteren möchte ich Frau Nurhan Sulak-Klute in Paderborn dafür danken, dass sie immer ein offenes Ohr (und Kekse) für uns Mitarbeiter und Studenten hatte. Frau Anke Müller danke ich für ihre unermüdliche (und am Ende erfolgreiche) Suche nach einem geeigneten Farbdrucker am Morgen der offiziellen Abgabe dieser Arbeit.

Meinen Eltern Rainer und Walburga danke ich ganz herzlich für die Unterstützung, insbesondere die finanzielle, der letzten zwei Jahre, ohne die ich diese Arbeit in dieser Form nicht hätte vollenden können. Meiner langjährige Partnerin Stephanie bitte ich um Entschuldigung für den Stress der letzten Jahre und hoffe auf viele weitere schöne gemeinsame Jahre mit ihr. Außerdem danke ich ihr noch für die Durchsicht meiner Diplomarbeit, in deren Vorwort ich sie dafür zu erwähnen vergessen hatte. Julian und Julia danke ich dafür, dass sie Probevorträge über sich haben ergehen lassen.

Allen Freunden und Weggefährten danke ich für ihre Hilfe und einfach dafür, da gewesen zu sein.

Last, but definitely not least, some rightfully deserved praise for Dr. Panagiotis Kantartzis:

Ξεχωριστές ευχαριστίες χρωστώ στο φίλο μου Τάκη, τον Δρ Παναγιώτη Κανταρτζή, που επί μήνες διάβαζε τη διατριβή μου και πρότεινε άπειρες βελτιώσεις. Από την Πολίνα και τα παιδιά ζητώ κατανόηση που τους στέρησα για τόσο καιρό τον Τάκη. Ελπίζω στο μέλλον να περάσουμε περισσότερο χρόνο μαζί σε ελληνικές παραλίες και αγγλικές παμπ αντί να ανταλλάσσουμε νύχτες ολόκληρες εμαιλς με απόψεις για τις μεθόδους κατασκευής κυματομορφών.

(Mein besonderer Dank gilt meinem Freund Takis für das unermüdliche monatelange Korrekturlesen meiner Manuskripte und seine zahllosen Verbesserungsvorschläge. Ich bitte Polina und die Kinder um Entschuldigung für die Zeit, die ich ihnen Takis geklaut habe. In Zukunft werden wir hoffentlich mehr unserer Zeit an griechischen Stränden oder in englischen Pubs verbringen anstatt nächtelang per E-Mail über Waveletmethoden zu diskutieren!)

Köln, im Mai 2015

Contents

Zusammenfassung

Diese Arbeit behandelt adaptive Waveletmethoden zur numerischen Lösung semilinearer partieller Differentialgleichungen (PDEs) und Randwertprobleme (BVPs) basierend auf solchen PDEs. Semilineare PDEs sind Spezialfälle aus der Klasse der allgemeinen nichtlinearen PDEs, hier besteht die PDE aus einem linearen Operator und einem nichtlinearen Störungsterm. Im Allgemeinen werden solche BVPs mit iterativen Verfahren gelöst. Es ist daher von höchster Wichtigkeit, effiziente als auch konvergente Lösungsverfahren für die zugrunde liegenden nichtlinearen PDEs zur Verfügung zu haben. Im Gegensatz zu Finite-Elemente-Methoden (FEM) garantieren die adaptive Waveletmethoden aus [33,34], dass Lösungen nichtlinearer PDEs mit konstanten Approximationsraten berechnet werden können. Des Weiteren wurden optimale, d.h. lineare, Komplexitätsabschätzungen für diese adaptiven Lösungsmethoden nachgewiesen. Diese Errungenschaften sind nur möglich, weil Wavelets einen neuartigen Ansatz zur Lösung von BVPs ermöglichen: die PDE wird in ihrer ursprünglichen unendlichdimensionalen Formulierung behandelt. Wavelets sind der ideale Kandidat für diesen Zweck, da sie es erlauben sowohl Funktionen als auch Operatoren in unendlichdimensionalen Banach- und Hilberträumen exakt darzustellen.

Der Zweck adaptiver Verfahren im Lösungsprozess von PDEs ist die Einsetzung von Freiheitsgraden (DOFs) nur dort wo notwendig, d.h. an Stellen wo die exakte Lösung unglatt ist und nur durch eine höhere Anzahl von Funktionskoeffizienten präzise beschrieben werden kann. In diesem Zusammenhang stellen Wavelets eine Basis mit besonderen Eigenschaften für die betrachteten Funktionenräume dar. Die benutzten Wavelets sind stückweise polynomial, haben kompakten Träger und es bestehen Normäquivalenzen zwischen den Funktionenräumen und den ℓ_2 Folgenräumen der Entwicklungskoeffizienten. Dieser neue Ansatz zieht einige Probleme für das Design der numerischen Algorithmen nach sich, welche aber durch eine Struktur der Waveletkoeffizienten, genauer gesagt eine Baumstruktur, komplett ausgeräumt. Diese Baumstruktur bedeutet dabei nur eine geringfügige Einschränkung in der theoretischen Anwendbarkeit in Bezug auf Funktionenräume. Es stellt sich heraus, dass der aufgezeigte Ansatz genau auf den funktionalanalytischen Hintergrund nichtlinearer PDEs passt. Die praktische Umsetzung auf einem Computer erfordert dabei aber eine Einschränkung der unendlichdimensionalen Darstellungen auf endlichdimensionale Quantitäten. Es ist genau dieser Aspekt, der das Leitmotiv dieser Arbeit darstellt. Diese theoretischen Vorgaben wurde im Rahmen dieser Arbeit in einem neuen vollständig dimensionsunabhängigen adaptiven Waveletprogrammpaket umgesetzt. Dieses erlaubt zum ersten Mal die bewiesenen theoretischen Resultate zur numerischen Lösung der oben genannten Randwertprobleme zu nutzen. Für diese Arbeit sind sowohl theoretische als auch numerische Aspekte von höchster Bedeutung.

Großer Wert wurde dabei auf die Optimierung der Geschwindigkeit gelegt, ohne dabei die Möglichkeit zu verlieren, verschiedene numerische Parameter noch zur Laufzeit verändern zu können. Dies bedeutet, dass der Benutzer verschiedenste Aspekte, wie z.B. die verwendete Waveletkonstruktion, austauschen und testen kann, ohne die Software neu kompilieren zu müssen. Der zusätzliche Rechenaufwand für diese Optionen wird während der Laufzeit des Programms durch verschiedene Arten von Zwischenspeichern, z.B. das Abspeichern der Werte des Vorkonditionierers oder die polynomielle Darstellung der mehrdimensionalen Wavelets, so klein wie möglich gehalten. Die Ausnutzung der Struktur in der Konstruktion der Waveleträume verhindert, dass das Zwischenspeichern eine große Menge im Speicher des Computers belegt. Gleichzeitig ermöglicht dies sogar eine Erhöhung der Ausführungsgeschwindigkeit, da die Berechnungen so nur noch ausgeführt werden, wenn sie notwendig sind, und dann auch nur ein einziges Mal. Die essentiellen Randbedingungen in den BVPs werden hier durch Spuroperatoren umgesetzt, was das Problem zu einem Sattelpunktsproblem macht. Diese spezielle Formulierung ist sehr flexibel; insbesondere wechselnde Randbedingungen, wie sie zum Beispiel im Rahmen von Kontrollproblemen mit Dirichlet-Randkontrolle auftreten, können sehr effizient behandelt werden. Ein weiterer Vorteil der Sattelpunktsformulierung ist, dass nicht-Tensorproduktgebiete durch den „Fictitious Domain"-Ansatz behandelt werden können.

Numerische Studien von 2D und 3D BVPs und nichtlinearen PDEs demonstrieren die Möglichkeiten und die Leistungsfähigkeit dieser Algorithmen. Lokale Basistransformationen der Waveletbasen senken dabei die absoluten Konditionszahlen der schon optimal vorkonditionierten Operatoren. Der Effekt dieser Basistransformationen zeigt sich dabei in Verkürzungen der absoluten Laufzeit der eingesetzten Löser; im Idealfall kann dabei eine semilineare PDE innerhalb von Sekundenbruchteilen gelöst werden. Dies kann im einfachsten durch Fall Richardson-Verfahren, wie das Verfahren des steilsten Abstiegs, oder das aufwändigere Newton-Verfahren geschehen. Die BVPs werden mittels eines adaptiven Uzawa-Algorithmus gelöst; dieser erfordert das Lösen einer semilinearen PDE in jedem Schritt. Außerdem wird die Effektivität verschiedener numerischer Lösungsverfahren verglichen und es werden die optimalen Konvergenzraten und Komplexitätsabschätzungen verifiziert. Zusammenfassend präsentiert diese Arbeit zum ersten Mal eine numerisch wettbewerbsfähige Implementierung dieses neuartigen theoretischen Paradigmas zur Lösung von semilinearen elliptischen PDEs in beliebigen Raumdimensionen einschließlich Konvergenz- und Komplexitätsergebnissen.

Abstract

This thesis is concerned with the numerical solution of boundary value problems (BVPs) governed by semilinear elliptic partial differential equations (PDEs). Semilinearity here refers to a special case of nonlinearity, i.e., the case of a linear operator combined with a nonlinear operator acting as a perturbation. In general, such BVPs are solved in an iterative fashion. It is, therefore, of primal importance to develop efficient schemes that guarantee convergence of the numerically approximated PDE solutions towards the exact solution. Unlike the typical finite element method (FEM) theory for the numerical solution of the nonlinear operators, the new adaptive wavelet theory proposed in [33,34] guarantees convergence of adaptive schemes with fixed approximation rates. Furthermore, optimal, i.e., linear, complexity estimates of such adaptive solution methods have been established. These achievements are possible since wavelets allow for a completely new perspective to attack BVPs: namely, to represent PDEs in their original infinite dimensional realm. Wavelets are the ideal candidate for this purpose since they allow to represent functions in infinite-dimensional general Banach or Hilbert spaces and operators on these.

The purpose of adaptivity in the solution process of nonlinear PDEs is to invest extra degrees of freedom (DOFs) only where necessary, i.e., where the exact solution requires a higher number of function coefficients to represent it accurately. Wavelets in this context represent function bases with special analytical properties, e.g., the wavelets considered herein are piecewise polynomials, have compact support and norm equivalences between certain function spaces and the ℓ_2 sequence spaces of expansion coefficients exist. This new paradigm presents nevertheless some problems in the design of practical algorithms. Imposing a certain structure, a tree structure, remedies these problems completely while restricting the applicability of the theoretical scheme only very slightly. It turns out that the considered approach naturally fits the theoretical background of nonlinear PDEs. The practical realization on a computer, however, requires to reduce the relevant ingredients to finite-dimensional quantities. It is this particular aspect that is the guiding principle of this thesis. This theoretical framework is implemented in the course of this thesis in a truly dimensionally unrestricted adaptive wavelet program code, which allows one to harness the proven theoretical results for the first time when numerically solving the above mentioned BVPs. Both theoretical and numerical aspects of the implemented adaptive wavelet schemes are of utmost interest.

In the implementation, great emphasis is put on speed, i.e., overall execution speed and convergence speed, while not sacrificing on the freedom to adapt many important numerical details at runtime and not at the compilation stage. This means that the user can test and choose wavelets perfectly suitable for any specific task without having to rebuild the software. The computational overhead of these freedoms is minimized by caching any interim data, e.g., values for the preconditioners and polynomial representations of wavelets in multiple dimensions. Exploiting the structure in the construction of wavelet spaces prevents this step from becoming a burden on the memory requirements while at the same time providing a huge performance boost because necessary computations are only executed as needed and then only once. The essential BVP boundary conditions are enforced using trace operators, which leads to a saddle point problem formulation. This particular treatment of boundary conditions is very flexible, which especially useful if changing boundary conditions have to be accommodated, e.g., when iteratively solving control problems with Dirichlet boundary control based upon the herein considered PDE operators. Another particular feature is that saddle point problems allow for a variety of different geometrical setups, including fictitious domain approaches.

Numerical studies of 2D and 3D PDEs and BVPs demonstrate the feasibility and performance of the developed schemes. Local transformations of the wavelet basis are employed to lower the absolute condition number of the already optimally preconditioned operators. The effect of these basis transformations can be seen in the absolute runtimes of solution processes, where the semilinear PDEs are solved as fast as in fractions of a second. This task can be accomplished using simple Richardson-style solvers, e.g., the method of steepest descent, or more involved solvers like the Newton's method. The BVPs are solved using an adaptive Uzawa algorithm, which requires repeated solution of semilinear PDE sub-problems. The efficiency of different numerical methods is compared and the theoretical optimal convergence rates and complexity estimates are verified. In summary, this thesis presents for the first time a numerically competitive implementation of a new theoretical paradigm to solve semilinear elliptic PDEs in arbitrary space dimensions with a complete convergence and complexity theory.

Introduction

Problems, Motivations and Applications

At the heart of many systems found in nature and the scientific world lie **Partial Differential Equations (PDEs)**. The fundamental underlying layer of reality, the quantum world, seems to obey only mathematical laws and often defies human common sense here in the macroscopic world. The behavior of elementary particles is best described by the Schrödinger equation [104], which is a partial differential equation, i.e, it contains derivatives of the solution with respect to the space (and time) coordinate(s). Although properties like position and velocity of atomic particles are by nature stochastic, i.e., governed by probabilities, modern quantum theory has made the most accurate predictions and often has the highest explanatory power of any other physical theory ever devised.

Humankind's ability to understand and control matter at the atomic level has produced an unparalleled technological progress during the last century, ranging from the Laser to modern computer chips. The continuation of this scientific progress in the near and far future hinges on our capability to understand the nature of the physical world. An example for the work in this field is the simulation of interaction of the matter within a **semiconductor quantum wire** when excited by a quantum of light [77, 109, 111, 114]. Understanding these interactions could one day lead to secure quantum communication systems and revolutionary new quantum computers.

The **Black-Scholes** partial differential equation, which is used to model the price evolution of pricing options on stocks in mathematical finance [7, 8], has surprisingly many details in common with the Schrödinger equation. The volatility, i.e., the apparent random fluctuation of the price of a stock with time, is modeled and treated as a **Brownian motion** [18], like the random movement of particles in fluids. The influence of this random process is usually represented by a stochastic term called the **Wiener process** [123]. A generalization of this approach leading to the so called **Heston model** can be found in [78]. See [102] for up-to-date numerical approaches to these problems. A good introduction and overview to the numerical treatment of option pricing is the book [135].

As a last example, a number of inverse boundary value problems, originating from various imaging modalities, make extensive use of (repeated) solutions of boundary value problems. A typical example with numerous applications to geo- and biomedical imaging is the **inverse conductivity problem** (a.k.a. Electrical Impedance Tomography (EIT) [136]), i.e., inferring on the interior structure of unobservable material from externally applied electric currents and measured voltages. This is an inverse BVP, the numerical solution of which requires repeated solutions of a partial differential equation. Therein, the Laplacian PDE describes the potential distribution; conditions enforced by the placement of electrodes on the body reduces this to a boundary value problem [85–89]. In medical applications, this technology is now commercially available in intensive care units and in the future its variants may complement other non-invasive scanning techniques like magnetic resonance tomography, without the need for bulky and expensive machinery.

As the above examples show, the need to solve a PDE can come from pure necessity, e.g., when no other way to determine the properties of a physical body is available, to save the costs of running actual experiments or to simulate conditions that are impossible to reproduce in a laboratory. For any of these reasons, physicists, biologists and other scientists seek solutions to PDEs using computers by employing methods and algorithms created by mathematicians and engineers.

Efficient Numerical Solution

The PDE Setting

The most common in applications encountered PDE operators are mathematically categorized into the groups **elliptic**, **hyperbolic** and **parabolic** [75] and PDEs of mixed and changing types are known. Each type of equation requires a different theoretical framework and specialized numerical approaches are necessitated to obtain the numerical solution. The problems considered herein, i.e., operator equations between the Banach or Hilbert space V and its dual V' of the general form

$$\mathcal{F}(u) = f, \quad \mathcal{F} : V \to V',$$

are of **elliptic** type which can be set in a **variational form**. Given a right hand side $f \in V'$, the solution $u \in V$ is sought. The model linear operator in this context is the **Laplacian** $\Delta := \frac{\partial^2}{\partial x_1^2} + \ldots + \frac{\partial^2}{\partial x_n^2}$, which will typically account for the linear part of the considered **semilinear** operators.

The Classical Approach

The classical approach, e.g., using **Finite Element Methods (FEM)** [19], typically approximates the continuous operators and infinite dimensional function spaces by finite dimensional (or discretized) counterparts. This step ensures that these problems accessible by computers, which cannot handle infinite dimensional objects without approximating these objects. A main concern in this approach is the question whether choosing ever larger (but still finite) function spaces then results in a series of solution approximations that converge towards the exact solution of the original PDE in the (infinite dimensional) function space.

In short, a discretization process is said to "work", i.e., is called **convergent**, if, for a grid size $h > 0$, the solution $u_h \in V_h \subset V$ computed in the discretized subspace converges towards the exact solution $u^\star \in V$, i.e, $\|u_h - u^\star\| \to 0$ for $h \to 0$. The norm $\|\cdot\|$ here could be the norm of the space V or a closed subspace $Z \subset V$. The quality of the solution process is here determined by the rate at which $u_h \to u^\star$ depending on the rate at which $h \to 0$. We say it has **convergence rate** $p > 0$, if $\|u_h - u^\star\| \leq C\, h^{-p}$. Such a rate can be expected (by a-priori estimates) if the function $u^\star$ is smooth enough and the space V_h is "close" enough to the space V. In this setting, one cannot expect an error $\|u_h - u^\star\|$ smaller than the **best approximation** $\inf_{v_h \in V_h} \|v_h - u^\star\|$; but it is possible to attain an error in the same order of magnitude. This is clearly the best result one can expect in case the space V_h is chosen fixed. In a uniform (spatially evenly) discretization in dimension n, the number of **degrees of freedom (DOFs)** N is proportional to h^{-n}.

Discretizations of linear operators are often representable as matrices and functions in finite spaces are representable by vectors containing expansion coefficients with respect to a basis for V_h. Therefore, the solution process of a PDE often leads to a **linear system of equations** that can be solved numerically yielding the vector of expansion coefficients $\mathbf{u}_h$.

This coefficient vector, expanded in the basis, produces the sought approximate solution function u_h.

The step from the infinite dimensional problem formulation in V to the discretized finite dimensional representation using the base of the space V_h introduces a new problem: It is not always automatically true that the discretized linear systems are well-posed themselves. For certain kind of problems, **stability** conditions on the discretizations have to be enforced or one could be unable to solve the resulting linear systems of equations. Even if the stability of the discretizations is guaranteed, numerical solution methods could be unable to finish within a reasonable time frame. To accomplish such computations quickly, first one has to employ an **efficient** solution strategy, for example, one that is of optimal linear complexity. In this ideal case, the vector $\mathbf{u}_h$ is obtained with $\mathcal{O}(N)$ operations. Such a solution $\mathbf{u}_h$ then possesses only finite accuracy, no more than the **discretization error accuracy**, which is the highest accuracy that can be accomplished for any given predetermined basis of the space V_h.

The amount of time and computational complexity a numerical solution method needs to compute the solution up to discretization error accuracy can be improved upon by employing **preconditioners**. The purpose of a preconditioner is to lower the (spectral) condition number of discretized operators up to the point where it is bounded uniformly, i.e., the condition number is independent of the discretization grid size h.

Wavelet Discretization

Wavelets have been successfully employed in the realm of signal processing [107], data compression [1], among others. Their main advantage over other discretization techniques is a strong functional analytical background. Of interest are two wavelet bases Ψ and $\widetilde{\Psi}$ which are **biorthogonal**, i.e., the inner product of all functions has the property $\left\langle \Psi, \widetilde{\Psi} \right\rangle = \mathbf{I}$. The **primal base** Ψ is needed explicitly as piecewise polynomials with compact support enabling norm equivalences for a range of values in the positive Sobolev scale. The (often only implicitly given) **dual base** $\widetilde{\Psi}$ can be chosen to maximize **moment conditions** and to assure norm equivalences in the negative Sobolev scale.

The **Riesz-basis property** for this range of Sobolev spaces allows for the ability to **precondition discretized operators uniformly** [46, 96], i.e., to limit the spectral condition over all levels by a common (uniform) limit. In numerical experiments, this enables iterative numerical solution methods in a **nested iteration** scheme to compute the solution up to discretization error accuracy in a uniformly limited number of steps. Combined with the ability to execute each step in a complexity proportional to the number of current unknowns N, this makes possible the computation of solutions in **optimal linear complexity**. The norm equivalences over a range of Sobolev spaces also enable to model Sobolev norms accurately in **control problems** where norms of traces of functions are employed. The natural norms on trace spaces are **fractional Sobolev spaces** and the proper evaluation of these norms is substantial for meticulous modeling of these problems [122].

If one wishes to extend this theory to problem formulations containing **elliptic nonlinear operators** and **(locally) unsmooth functions**, the classical approach no longer fits the problem well. Nonlinear operators are by definition not representable accurately by (linear) matrices. Additionally, one cannot expect high convergence rates for functions of low classical smoothness using linear methods. In this case, uniform discretization grids would supply a lot of DOFs where they might not be needed because of locally smooth patches. All these problems can be addressed by going beyond uniform discretizations and employing adaptive methods.

Adaptive Wavelet Methods – A New Paradigm

To overcome the above described limitations of the classical method, a new approach called **Adaptive Wavelet Method (AWM)** was first proposed in [31] for linear problems and was then extended in [33] to nonlinear problems. The problems considered herein must be set in a **variational form**, i.e., the class of nonlinear problems covered does not encompass hyperbolic conservation laws. A good overview of this complex topic is given in [140], and the current state of the research of fields related to adaptive wavelet methods is presented in some of the articles of [63]. In this setting, one stays within the infinite dimensional problem realm, essentially rendering the issue of finite-dimensional **stability** obsolete. The individual steps of this new paradigm given in [33] read as follows:

(n1) ensure well-posedness of the given variational problem;

(n2) transformation of the infinite dimensional problem into an **equivalent** problem in ℓ_2 which is **well posed** in the Euclidean metric;

(n3) the derivation of an iterative scheme for the infinite dimensional ℓ_2-problem that exhibits a **fixed error reduction per step**;

(n4) numerical realization of the iterative scheme by an **adaptive application** of the involved infinite dimensional operators within some finite dynamically updated accuracy tolerances.

It is important to notice that only in the last stage **(n4)** actual computations are done, which take place in the finite memory setting of a computer and can thus only be based upon a finite number of data points. The first step **(n1)** is not different from the classical discretization method, and the theory of the second step **(n2)** was well established before the development of these methods. The existence of an iterative scheme mentioned in the third step **(n3)** actually has to be proven for each problem, since this is still in the infinite dimensional ℓ_2 space.

The shift in thinking is to view the computations, which cannot be done with arbitrary accuracy, as approximations to the exact (infinite dimensional) computations. The approximation error has to be controlled in such a way that one essentially operates in the infinite dimensional ℓ_2 setting. The obvious obstacle to this approach is the inability to process any numerical operators with infinite accuracy using our present computer systems, which implement floating point arithmetic according to a fixed number of bits [81]. Instead, one emulates here the computations in the infinite dimensional ℓ_2 setting by choosing the DOFs as needed out of the infinite set of DOFs. The determination of this finite set and the computations on said set are then executed as accurately as possible, preferably without further approximations. This principle is called **adaptive inexact operator application**. The main concern in this scheme is that one could break the assumption of **(n3)** because the computations of step **(n4)** are not done sufficiently precise and the convergence properties are thus not retained.

Adaptivity in this context means that one only employs DOFs for the computation and representation of the solution where actually needed. The question how to determine the target index sets was answered by Cohen, Dahmen and DeVore in [31–34]. It is conceptually similar to an a posteriori error estimator used in adaptive finite element methods [19, 119], as the method tries to predict where the local error is not small enough and thus the adaptive set has to be enlarged. Based upon the above mentioned locality and norm equivalence properties of wavelets, the natural data structure in this setup is the **tree**, i.e., a level-wise order is imposed upon the wavelet coefficients that construct a mother-daughter relationship. To ensure optimal convergence rates, successive enlargements of the wavelet indices set during the solution process must be controlled by a nonlinear vector approximation procedure, called **coarsening**, introduced in [16]. For linear operators, an iteration scheme exhibiting optimal rates without coarsening was introduced in [65].

Contrary to the aforementioned classical methods, no discretization space $V_h \subset V$ is chosen fixed at any point. The grid defined by an adaptive wavelet index set is not uniform, i.e., one cannot attribute a single step size h but a hierarchy of step sizes h_j that only apply locally. The **quality of the solution process** is thus not measured with respect to h but with respect to the number of wavelet coefficients N, which is exactly the number of DOFs. Depending on the properties of the PDE, e.g., the domain, right hand side, differential operator and PDE coefficients, one expects the solution $u^\star$ to exhibit a **smoothness order** $s > 0$, i.e., $\inf \|u_N - u^\star\|_V \lesssim N^{-s}$, where the infimum is taken over all functions u_N being describable by N wavelet coefficients arranged in a tree structure. Because of the connection $N^{-s} \equiv \varepsilon$ in the above estimate, it also states that the number of wavelet coefficients N in the adaptive tree of the solution required to attain any target accuracy $\varepsilon > 0$ should be proportional to $\varepsilon^{-1/s}$. The metric to gauge a solution algorithm by is now whether it computes solutions $u(\varepsilon)$ with the same **optimal rate** N^{-s}, i.e., $\|u(\varepsilon) - u^\star\|_V \lesssim N^{-s}$. An attainable accuracy ε in the solution proportional to N^{-s} is therefore the best possible result in this adaptive setting.

The **application of operators** with respect to such adaptive tree structures of wavelet coefficients is an important step in the whole adaptive inexact operator application [35]. Since nonlinear operators, by definition, cannot generally be applied to a linear decomposition exactly, a locally unique representation is sought. Considering the approximation of a function based on wavelets consisting of piecewise polynomials, e.g., B-spline based wavelets, one can represent each wavelet using local polynomials on cells of the underlying domain. Because of the tree structure of the wavelet expansion coefficients, converting such a vector to a polynomial representation constructs polynomials consisting of many overlapping pieces living on different spatial levels. By the local refinement properties of polynomials, this representation can be converted into an unique adaptive one. The application of the operator to these polynomials now has a simple structure due to the locality of the polynomials and many types of operators can be applied exactly to the local polynomials. From these results, the values of the target wavelet index set can be reconstructed. All these steps can be applied in optimal linear complexity [115, 147].

This adaptive strategy has been successfully employed in the context of linear elliptic PDEs [11, 31, 65], saddle point problems of linear elliptic PDEs [40], nonlinear elliptic PDEs on the L-shaped domain [147], parabolic evolution problems [29, 133], eigenvalue problems [128], as well as the Stokes problem [55]. These approaches have an enormous potential for complexity reduction for coupled systems of PDEs as they arise in control problems with PDEs [48, 74].

Additionally, **linear operators**, including trace operators, can be applied adaptively using only the values of bilinear forms. In [10], an evaluation scheme based on [14, 54] for linear and also for certain nonlinear operators was introduced. There, the authors, instead of trying to apply the operator on a fixed index set exactly, approximate the application up to a certain accuracy depending on the accuracy on the prescribed index set. Furthermore, there is a new approach by Schwab and Stevenson concerning the fast evaluation of nonlinear functionals of anisotropic tensor product based wavelet expansions, see [134].

Comparison with Adaptive Finite Element Methods

Adaptive Finite Element Methods (AFEM), see [119], solve by iterating the steps

$$\texttt{SOLVE} \longrightarrow \texttt{ESTIMATE} \longrightarrow \texttt{MARK} \longrightarrow \texttt{REFINE}$$

until the solution is computed up to a preset target tolerance. However, the system of equations numerically solved in the course of a concrete AWM solution method are a subset of the system of equations arising through the reformulation using the full infinite dimensional wavelet basis. Such a "pivot" space does not exist for finite element methods. In the AFEM setting, the **adaptive mesh** of the domain Ω, i.e., a decomposition of the domain Ω into piecewise disjoint patches, and the piecewise polynomials defined on these patches is the main object in the implementation. In contrast, in the AWM setting, the main object of theory and implementation is the **adaptive tree structure** of wavelet coefficients and in this light the adaptive grid is just a tool used in the evaluation process of the (non-)linear PDE operator. In fact, if the operator is linear, the grid generation stage can be skipped and the operator evaluation can be executed using a bilinear form only.

In the `SOLVE` stage, an iterative solution process, e.g., a step in a Richardson scheme, is executed on a finite vector by applying the operator representing the PDE. Next, the purpose of the `ESTIMATE` step is to determine the places where the error is big and thus where new DOFs should be inserted adaptively. In the AFEM scheme this translates to an **error estimator**. This is often done by measuring **smoothness** locally, e.g., by computing the jump of function values across the edges of the elements. The equivalent of the error estimator in AWM is the **tree prediction** algorithm, which combines information on the vector and the operator to determine how many DOFs in the form of wavelet coefficients should be inserted. Thus, `MARK` and `REFINE` steps are executed in the ℓ_2 setting, which does not entail any of the AFEM refinement strategies used for meshes, e.g., longest edge bisection or Dörfler marking.

Lastly, the convergence properties of the above explained iterations of course have to be investigated. For AWM, convergence proofs with guaranteed convergence rates have been established for nonlinear PDEs in [34] and for linear elliptic operators earlier in [31]. Convergence for AFEM was shown for several linear operators, see [30, 110], and for a class of nonlinear operators [80]. However, results concerning (optimal) convergence rates on the other hand have only established for a few cases, usually only for the Poisson problem and similar linear operators [15, 28, 37]. For nonlinear operators, convergence rates of

AFEM have not yet been proven. Furthermore, optimal complexity results, available for linear and nonlinear operators in the wavelet setting, are only available for special cases in AFEM, see [28, 117, 139].

Scope of this Thesis

The goal of this thesis is to work out the theoretical details pertaining to boundary value problems based upon semilinear elliptic PDEs and to present numerical experiments using adaptive wavelet methods. I will investigate the close intertwining of theory and ingredients and the roles and impacts of numerical constants and variables on the theoretical estimates in application in arbitrary space dimension for the first time. To this end, I first summarize some functional analysis details necessary for a good understanding of the following wavelet methods. After giving a short introduction into classical wavelet methods for B-spline wavelets on bounded domains, we elaborate on the adaptive wavelet methods presented in [31–34] and discuss several implementational details of the algorithms presented in those papers and their effect on actual numerical experiments.

I present several algorithms dedicated to applying linear and nonlinear elliptic PDE and trace operators on adaptive wavelet vectors in optimal linear complexity. I discuss several solution methods for semilinear PDEs and boundary value problems based upon said semilinear PDEs and present numerical results for dimensions $n \geq 2$ and varying operators and trace spaces. The goal is to qualitatively confirm and quantitatively investigate the theoretical results in numerical experiments, i.e., show that the produced solutions are attained in optimal complexity $\mathcal{O}(N)$ with $N \sim \varepsilon^{-1/s}$. This work serves as a stepping stone towards adaptive wavelet methods for semilinear elliptic PDE constrained control problems with Dirichlet boundary control.

The LibWavelets Framework

In the course of this thesis, I developed a C++ software package which is tailored to adaptive wavelet methods on general tensor product domains. The earlier mentioned match of the theoretical background and the numerical algorithms essentially forms the perfect basis for the novel approach of dealing with the semilinear PDE operators in its original infinite dimensional realm. Furthermore, I developed and implemented algorithms to apply a range of other operators, e.g., trace operators, Riesz operators and other (linear) PDE operators, within the same theoretical framework. In order to eliminate approximations of integrals when evaluating operators, i.e., a source of inaccuracy during the inexact operator equation in the ℓ_2-space, a method of exactly evaluating the considered semilinear operators was devised for this software (see [115]). For certain linear trace operators, it will be shown how the evaluation of the bilinear form and thus essentially the application of the operator can also carried out by avoiding inexact methods. The adaptive code setup also allows for trace operators that use adaptivity in the determination of the active wavelet coefficients at runtime, thus allowing changes in the trace domain to be easily implemented and tested.

My software is inherently designed completely independent of any fixed dimension while granting the user the freedom to choose wavelet configurations independently for every spatial coordinate. This means the wavelets in each spatial coordinate can therefore differ in smoothness, polynomial order or boundary conditions. This allows the user to tailor the properties of the wavelets specifically to the problem and involved operators, e.g., in case of anisotropic PDE operators which could require higher order wavelets in certain spatial directions. On top of everything, this enables one to model not only simple tensor product domains, e.g., $(0,1)^2 := (0,1) \otimes (0,1)$, but also to model more complicated domains, e.g., the periodic domain $\mathbb{R}/\mathbb{Z} \otimes (0,1)$. This can be used to mimic the effect of an infinitely large physical setup as for the semiconductor setting in [111].

At the same time, a main attention was set onto producing fast and efficient code, i.e., to ensure that execution times and memory requirements stay proportional to the number of unknowns N. I took great care to limit the computational overhead for this versatile setup by extensively caching multidimensional data, e.g., values of the preconditioners or the piecewise polynomial representation of primal basis functions. The potential drawback of this approach, i.e., the cost for setting up the multidimensional data, can be minimized by exploiting the repetitive nature of the translated and dilated basis functions. If optimized correctly, this data can be set up on each program startup without a noticeable (and often not even measurable) delay.

Remark *Proofs of all theorems and statements, which are not given, can be found in the respective references in detail. A proof may be reproduced herein, even though it was originally published in a different publication, to emphasize or explain a certain detail or argument.*

Outline

This thesis is structured as follows:

Section 1
: In Section 1 basic terminology and definitions from functional analysis necessary for a complete understanding of the subject matter are recollected.

Section 2
: This section gives an introduction to wavelets and the theory of multiresolution analysis, the main tool in this work. The established wavelet theory is extended by remarks on improving its qualities in applications by enhanced preconditioning and Riesz operators for norm equivalences with precise constants. This part lays the groundwork for a complete theoretical understanding of the numerical schemes.

Section 3
: Building upon the theory of the previous section, this chapter shows how adaptive methods are introduced in the wavelet realm. This starts by discussing tree structures and then explains how expansion vectors based upon tree structured index sets are at the heart of a series of algorithms implementing the adaptive wavelet methods.

Section 4
: This section explains the theory and implementational details of tree based adaptive wavelet methods. It includes the algorithms to convert wavelet expansion vectors to a piecewise polynomial representation, the application of a (non-)linear operator

and the computation of the values of the dual wavelet expansion coefficients. I conclude by presenting example problems and the numerical results when solving these problems using the previously described algorithms.

Section 5

Again building upon the theory and results of the previous section, the family of problems considered is advanced from a mere (non-)linear PDE to (non-)linear boundary values problems with essential boundary conditions governed by such (non-)linear PDEs. While natural boundary conditions are incorporated into the solution spaces, essential boundary conditions are here enforced using trace operators. Trace operators allow a very flexible treatment of boundary conditions, which is used to accommodate changing boundary conditions when iteratively solving control problems with Dirichlet boundary control. I give an introduction into the general saddle point problem theory and present theoretical and experimental results for several example problems.

Section 6

I close the thesis with a short review of the results of the previous chapters and an outlook on possible future work.

Appendix A

Details on the used wavelet bases and the implementation of the local polynomials can be found in the appendix. This includes refinement and conversion matrices and derivations of the inner product values for the adaptive application of PDE operators.

Appendix B

This section gives a short introduction into the software source code and its design principles. This includes a guideline to the installation of the software and a description of the configuration options available.

Appendix C

Here I list the most important symbols used in this thesis and state some basic definitions.

1 Fundamentals

In this first chapter, we need to recall some fundamental definitions and propositions from functional analysis which are needed in the discussion of the (non-)linear elliptic boundary value problems in the later chapters. This includes function spaces like the Sobolev spaces $H^m(\Omega)$ of Section 1.2 used in the variational weak form of linear elliptic PDEs discussed in Section 1.4. Sobolev spaces $H^s(\Gamma)$ of fractional order are important for the enforcement of boundary conditions using trace operators. In the same way, Besov spaces $B_q^\alpha(L_p(\Omega))$ discussed in Section 1.3 are employed in the solution theory of nonlinear elliptic PDE of Section 1.5.

The contents of the following sections are based on the books [4] and [75]. Other references are stated in place whenever a result is quoted. The content of this section is partially taken from [122].

1.1 Basic Definitions and Vocabulary

Let X, Y be normed linear spaces over the field $\mathbb{R}$.

Definition 1.1 [Linear Operators and Operator Norms]
We denote all **linear operators** *from X to Y by*

$$L(X;\ Y) := \{T : X \to Y;\ T \text{ is continuous and linear}\}. \tag{1.1.1}$$

For any $T \in L(X;\ Y)$, the associated **operator norm** *is defined by*

$$\|T\|_{L(X;\ Y)} := \sup_{x\in X,\ \|x\|_X=1} \|Tx\|_Y, \tag{1.1.2}$$

which is known to be finite for this class of operators (see [4]).

We write $L(X) := L(X;\ X)$ when X and Y coincide.

Definition 1.2 [Banach Spaces and Equivalent Norms]
A **Banach space** *is a complete vector space B with a norm $\|\cdot\|_B$. A Banach space can have several norms, e.g. $\|\cdot\|_{B_1}$ and $\|\cdot\|_{B_2}$, which are called* **equivalent** *if they induce the same topology. This is equivalent to the existence of positive finite constants c and C such that for all $v \in B$*

$$\|v\|_{B_1} \le c\|v\|_{B_2} \text{ and } \|v\|_{B_2} \le C\|v\|_{B_1}, \tag{1.1.3}$$

written shortly as

$$\|v\|_{B_1} \lesssim \|v\|_{B_2} \text{ and } \|v\|_{B_2} \gtrsim \|v\|_{B_1} \qquad \text{or} \qquad \|v\|_{B_1} \sim \|v\|_{B_2}. \tag{1.1.4}$$

Definition 1.3 [Separable Hilbert Space]
A **Hilbert space** *$\mathcal{H}$ is a complete vector space with an inner product $(\cdot,\cdot)_\mathcal{H}$ such that the norm is induced by the inner product as $\|\cdot\|_\mathcal{H} := \sqrt{(\cdot,\cdot)_\mathcal{H}}$. A Hilbert space is called* **separable** *if it contains a countable dense subset, i.e.,*

$$V = \{v_i : i = 1, 2, \ldots\} \subset \mathcal{H}, \quad \text{such that } \operatorname*{clos}_{\mathcal{H}} V = \mathcal{H}. \tag{1.1.5}$$

A Hilbert space is always a Banach space, but the converse does not need to hold. Most spaces relevant for numerical studies are separable since (1.1.5) can equivalently be expressed as

$$\operatorname{dist}(f; V) = 0, \quad \text{for all } f \in \mathcal{H}, \tag{1.1.6}$$

which, in other words, means that every element of $\mathcal{H}$ can be approximated analytically or numerically with arbitrary precision with elements from the space V.

Definition 1.4 [Dual Space]
Let X be a Banach space. The **dual space** *X' of X is the space of all linear continuous functions from X onto the underlying field $\mathbb{R}$. In other words,*

$$X' := L(X;\ \mathbb{R}). \tag{1.1.7}$$

The elements $v' \in X'$ are called **linear functionals**. *The* **dual form** *is defined as $\langle x, x' \rangle_{X \times X'} := x'(x)$.*

In the following, let $\Omega \subset \mathbb{R}^n$ be a bounded domain with piecewise smooth boundary $\partial\Omega$ and Ω being locally on one side. The meaning and degree of smoothness of $\partial\Omega$ will be specified in Section 1.2.2, for now $\partial\Omega$ just shall be considered "sufficiently smooth".
Examples of separable Hilbert spaces are the Lebesgue function spaces $L_p(\Omega)$, $1 \le p < \infty$, and the subspaces $H(\Omega)$ of $L_2(\Omega)$. Note that $L_\infty(\Omega)$ is not separable. Any generic Hilbert space considered in this thesis will be separable.

1.2 Sobolev Spaces

Most of this section is taken from the books of [2, 75, 105]. Some details, especially the Fourier analysis notation, are borrowed from the lecture notes [131].

Definition 1.5 [Lebesgue Space $L_2(\Omega)$]
The space $L_2(\Omega)$ is the space of all real-valued square **Lebesgue integrable** *functions on Ω, i.e.,*

$$v \in L_2(\Omega), \quad \textit{if and only if} \quad \|v\|^2_{L_2(\Omega)} := \int_\Omega v^2(x)\, d\mu < \infty. \tag{1.2.1}$$

where $\mu = \mu(x)$ is the Lebesgue measure. It is equipped with the inner product

$$(u, v)_{L_2(\Omega)} := \int_\Omega u(x) v(x)\, d\mu, \tag{1.2.2}$$

and is a **Hilbert space**.

Functions $u, v \in L_2(\Omega)$ are considered **equal** if $u(x) = v(x)$ holds almost everywhere, i.e., for all $x \in \Omega \setminus A$ and $\mu(A) = 0$.

Remark 1.6 *In the following, $\alpha := (\alpha_1, \ldots, \alpha_n) \in \mathbb{N}_0^n$ is a* **multi-index**. *Its definition, along with that of the classical smoothness spaces C^k, $C_0^\infty(\Omega)$,* **Hölder** *spaces $C^{k,\alpha}$ and* **Lipschitz** *spaces $C^{k,1}$ can be found in Appendix C.*

Definition 1.7 [Weak Derivative]
We say $u \in L_2(\Omega)$ has the **weak derivative** $v =: \partial^\alpha u$, *if* $v \in L_2(\Omega)$ *and*

$$(\phi, v)_{L_2} = (-1)^{|\alpha|} (\partial^\alpha \phi, u)_{L_2}, \quad \text{for all } \phi \in C_0^\infty(\Omega). \tag{1.2.3}$$

Remark 1.8 *For convenience purposes, we will omit the domain from the scalar product like we did in (1.2.3), if this does not create confusion.*

If such a v exists, it is unique (in the L_2-sense). In case $u \in C^m(\Omega)$, the weak derivative corresponds to the classical strong derivative and (1.2.3) follows as an application of **Green's formula**.
We now introduce Sobolev spaces as subspaces of L_2, in which elements possess weak derivatives of specific orders.

Definition 1.9 [Sobolev Space on Ω]
For $m \in \mathbb{N}$ we denote by $H^m(\Omega)$ the **Hilbert space** *of all functions $u \in L_2(\Omega)$ for which the weak derivatives $\partial^\alpha u$ for all $|\alpha| \leq m$ exist. The inner product of this space is given as*

$$(u, v)_{H^m} := \sum_{|\alpha| \leq m} (\partial^\alpha u, \partial^\alpha v)_{L_2} \tag{1.2.4}$$

which is associated to the norm

$$\|u\|_{H^m} := \sqrt{(u, u)_{H^m}} = \sqrt{\sum_{|\alpha| \leq m} \|\partial^\alpha u\|_{L_2}^2}. \tag{1.2.5}$$

A seminorm is given by

$$|u|_{H^m} := \sqrt{\sum_{|\alpha| = m} \|\partial^\alpha u\|_{L_2}^2}. \tag{1.2.6}$$

The Sobolev spaces are obviously nested, i.e., $H^{m+1} \subset H^m$, with the usual definition $H^0 := L_2$. A well known fact is the following

Corollary 1.10 *$C^\infty(\Omega) \cap H^m(\Omega)$ is dense in $H^m(\Omega)$ for $m \in \mathbb{N}_0$.*

The series of spaces $H^m, m \in \mathbb{N}_0$, can be extended to a scale of spaces with continuous smoothness indices, which will be of great importance later. These subspaces of H^m are Sobolev spaces of non-integral order $s \notin \mathbb{N}$ and cannot be characterized by weak derivatives alone as above. Instead, we use the following definition:

Definition 1.11 [Fractional Sobolev Spaces on Ω]
For $s = m + \sigma, m \in \mathbb{N}_0, 0 < \sigma < 1$, we introduce an inner product as

$$(u, v)_{H^s} := (u, v)_{H^m} + \sum_{|\alpha| \leq m} \left(\int_\Omega \int_\Omega \frac{|\partial^\alpha u(x) - \partial^\alpha u(y)||\partial^\alpha v(x) - \partial^\alpha v(y)|}{|x - y|^{n+2\sigma}} \, d\mu(x) \, d\mu(y) \right). \tag{1.2.7}$$

The space $H^s(\Omega)$ is the closure of all functions in $H^m(\Omega)$ for which the norm

$$\|u\|_{H^s} := \sqrt{(u, u)_{H^s}} \tag{1.2.8}$$

is finite. It is thus a Hilbert space.

Remark 1.12 *The Definitions 1.9 and 1.11 also hold in case* $\Omega = \mathbb{R}^n$.

The Sobolev spaces are nested in the following fashion

$$H^{s_1} \subset H^{s_2} \subset L_2, \quad s_1 > s_2 > 0, \tag{1.2.9}$$

for any domain $\Omega \subseteq \mathbb{R}^n$.

An Alternative Characterization

An alternative approach to define fractional Sobolev spaces is given by means of Fourier Analysis. We can define for $f \in L_2(\mathbb{R}^n)$ the **Fourier transform** $\mathcal{F}(f) \in L_2(\mathbb{R}^n)$ as the limit in the L_2 sense of

$$\int_{|\xi|\le M} \exp(\pm 2\pi i x \cdot \xi) f(\xi) d\xi, \quad M \to \infty. \tag{1.2.10}$$

The Fourier transform is an **isomorphism** between $L_2(\mathbb{R}^n)$ and itself, with

$$\|\mathcal{F}(f)\|_{L_2} = \|f\|_{L_2},$$

and the identity

$$\mathcal{F}(\partial^\alpha f) = (2\pi i)^{|\alpha|}\xi^\alpha \mathcal{F}(f), \quad \text{for all } f \in L_2(\mathbb{R}^n)$$

holds. From the above-mentioned remarks we obtain an equivalent characterization of H^m as:

$$H^m(\mathbb{R}^n) = \{v \,|\, \xi^\alpha \mathcal{F}(v) \in L_2(\mathbb{R}^n) \text{ for all } |\alpha| \le m\}. \tag{1.2.11}$$

Obviously, it does not matter in (1.2.11) whether m is an integer or whether it is positive. It may be easily verified that the characterizations $\xi^\alpha \mathcal{F}(v) \in L_2(\mathbb{R}^n)$ and $(1+|\xi|^2)^{s/2}\mathcal{F}(v) \in L_2(\mathbb{R}^n)$ for $|\alpha| \le s$ are equivalent. The latter is predominantly used in the following alternative version of Definition 1.11 for $\Omega = \mathbb{R}^n$.

Definition 1.13 [Fractional Sobolev Space on $\mathbb{R}^n$]
For $s \in \mathbb{R}$ *we define the* **Sobolev space of order s**, $H^s(\mathbb{R}^n)$, *as*

$$H^s(\mathbb{R}^n) := \left\{v \,|\, (1+|\xi|^2)^{s/2}\mathcal{F}(v) \in L_2(\mathbb{R}^n)\right\}, \tag{1.2.12}$$

which is a **Hilbert space** *when endowed with the inner product:*

$$(u,v)_{H^s} := \left((1+|\xi|^2)^{s/2}\mathcal{F}(u), (1+|\xi|^2)^{s/2}\mathcal{F}(v)\right)_{L_2}. \tag{1.2.13}$$

Remark 1.14 *In case* $s \in \mathbb{N}$, *the inner products (1.2.4) and (1.2.13) induce equivalent, but not identical, norms.*

For $u \in H^s(\mathbb{R}^n)$, we can define the **restriction** operator onto $\Omega \subseteq \mathbb{R}^n$,

$$u \to u|_\Omega =: \text{restriction of } u \text{ to } \Omega, \tag{1.2.14}$$

which is continuous and linear. In case $u \in C^k$ holds, the restriction can be defined pointwise.
The space $H^s(\Omega)$ from Definition 1.9 can now be equivalently expressed using the above Definition 1.13 and the following theorem from [105].

Theorem 1.15 *$H^s(\Omega)$ coincides with the space of* **restrictions** *to Ω of the elements of $H^s(\mathbb{R}^n)$.*

Therefore, for any function $u \in H^s(\Omega)$, an element $\widetilde{u} \in H^s(\mathbb{R}^n)$ can be specified which defines u by means of local coordinates on the domain Ω. The approach via Fourier transform is in particular applicable for Sobolev spaces to be defined on periodic domains.

1.2.1 Subspaces $H_0^s \subset H^s$

The spaces $H_0^s(\Omega)$ are normally loosely referred to as elements of the spaces $H^s(\Omega)$ with compact support in Ω. The definition of the spaces $H_0^s(\Omega)$ is an extension of Corollary 1.10.

Definition 1.16 [Sobolev Spaces $H_0^s(\Omega)$]
$H_0^s(\Omega)$ is defined as the closure of $\mathcal{D}(\Omega) := C_0^\infty(\Omega)$ with respect to the norm of $H^s(\Omega)$, i.e.,

$$H_0^s(\Omega) := \left\{ \phi \,|\, \exists \, \{\phi_n\} \in \mathcal{D}(\Omega) \text{ and } \phi_n \to \phi \text{ is a Cauchy sequence in } \| \cdot \|_{H^s(\Omega)} \right\}. \tag{1.2.15}$$

Hence the spaces $H_0^s(\mathbb{R}^n)$ and $H^s(\mathbb{R}^n)$ are equal. In general, the spaces $H_0^s(\Omega)$ are closed subspaces of $H^s(\Omega)$. Specifically, we have

$$H_0^s(\Omega) = H^s(\Omega), \quad 0 \le s \le \frac{1}{2}, \tag{1.2.16}$$

which holds because $\mathcal{D}(\Omega)$ is also dense in $H^s(\Omega)$ for $s \le \frac{1}{2}$, cf. Corollary 1.10. In the other cases we have

$$H_0^s(\Omega) \subsetneq H^s(\Omega), \quad s > \frac{1}{2}, \tag{1.2.17}$$

which means that $H_0^s(\Omega)$ is strictly contained in $H^s(\Omega)$. It is shown in [75] and [73] that one can also characterize the spaces of (1.2.15) as the following family of functions:

$$H_0^s(\Omega) = \left\{ u \,|\, u \in H^s(\Omega), \partial^\alpha u = 0 \text{ on } \partial\Omega, |\alpha| \le s - \frac{1}{2} \right\}. \tag{1.2.18}$$

These spaces also have an important property regarding their dual spaces which will be seen in Section 1.2.3. We now have the following relations between the Sobolev spaces of integral orders:

$$\begin{array}{ccccccccc} L_2(\Omega) & = & H^0(\Omega) & \supset & H^1(\Omega) & \supset & H^2(\Omega) & \supset & \dots \\ & & \| & & \cup & & \cup & & \\ & & H_0^0(\Omega) & \supset & H_0^1(\Omega) & \supset & H_0^2(\Omega) & \supset & \dots \end{array} \tag{1.2.19}$$

All inclusions in the above diagram are dense and the embeddings continuous.

1.2.2 Trace Spaces $H^s(\Gamma)$

Trace spaces and trace operators appear naturally in the treatment of the boundary value problem considered in Section 5. To this end, we need to recall a definition of trace spaces and the extension of classical trace operators onto the space $H^s(\Omega)$.
The **constraint** or **trace** $u|_{\partial\Omega}$ of a function $u \in H^s(\Omega)$ cannot simply be defined pointwise because there is no guarantee that functions in L_2 and H^s can be evaluated at specific points $x \in \Omega$. It also makes no sense to define the trace as the continuous limit when approaching the boundary, because firstly, elements of $H^1(\Omega)$ are generally not continuous, and secondly, $\partial\Omega$ is a manifold in $\mathbb{R}^{n-1}$ and, thus, its measure in $\mathbb{R}^n$ is zero. This means we could have $u = v$ a.e. for $u, v \in L_2(\Omega)$ but $u(x) \neq v(x)$ for all $x \in \partial\Omega$.
The trace of functions in Sobolev spaces is defined through a trace operator and is given in local coordinates on an open cover of the boundary $\partial\Omega$. This definition depends also on regularity conditions of the boundary $\partial\Omega$, which we will now formalize.
Let $\Omega \subset \mathbb{R}^n$ be a domain with Lipschitz boundary $\partial\Omega \in C^{k,1}$ to which Ω lies locally on one side. Also, a fixed section $\Gamma \subseteq \partial\Omega$ should have a positive surface measure. The following local coordinate system is thus well defined :
For any $x \in \partial\Omega$, we can specify a neighborhood $V \subset \mathbb{R}^n$ with new orthogonal coordinates $z = (z', z_n)$ where $z' = (z_1, \dots, z_{n-1})$. Without imposing restrictions, V can be characterized as a cube in these coordinates, i.e.,

$$V = \{(z_1, \dots, z_n) \,|\, |z_j| \leq 1, 1 \leq j \leq n\}, \tag{1.2.20}$$

and the first $n-1$ coordinates z' of z span the space

$$V' := \{(z_1, \dots, z_{n-1}) \,|\, |z_i| \leq 1, 1 \leq j \leq n-1\}. \tag{1.2.21}$$

Let $\Theta = \{\Theta_j \,|\, j = 1, \dots, r\}$ be a family of open bounded sets in $\mathbb{R}^n$, covering $\partial\Omega$, such that, for each j, there exists $\varphi_j \in C^{k,1}(V', \Theta_j)$ with positive Jacobian $J(\varphi_j), 1 \leq j \leq r$ and φ_j is a bijection. Furthermore, we can arrange to have

$$\begin{aligned} \varphi_j &\in C^{k,1}(V_+, \Theta_j \cap \Omega), \quad V_+ := \{(z', z_n) \in V \,|\, z_n < \varphi_j(z')\}, \\ \varphi_j &\in C^{k,1}(V_0, \Theta_j \cap \partial\Omega), \quad V_0 := \{(z', z_n) \in V \,|\, z_n = \varphi_j(z')\}. \end{aligned}$$

because of the preliminary requirements to Ω above. In other words, Ω lies locally below the graph of each φ_j and the graph of φ_j is the boundary of Ω in the patch Θ_j. For each j, the pair (φ_j, Θ_j) is called a **coordinate patch** for the boundary part $\partial\Omega \cap \Theta_j$.

Definition 1.17 [Sobolev Spaces $H^s(\partial\Omega)$]
A distribution u on $\partial\Omega$ is in $H^s(\partial\Omega)$ for any real $|s| \leq k+1$, if and only if

$$u \circ \Phi_j \in H^s(V' \cap \Phi_j^{-1}(\Theta_j \cap \partial\Omega)). \tag{1.2.22}$$

This is a Banach space when equipped with the norm

$$\|u\|^2_{H^s(\partial\Omega)} := \sum_{j=1}^{r} \|y_j \circ \Phi_j\|^2_{H^s(V' \cap \Phi_j^{-1}(\Theta_j \cap \partial\Omega))}. \tag{1.2.23}$$

Remark 1.18 *It was proved in [73] that the above definition is independent of the choice of the system of local maps $\{\varphi_j, \Theta_j\}$.*

The trace space $H^s(\Gamma) \subset H^s(\partial\Omega)$ can be defined analogously by only considering an open cover Θ of Γ, which does not intersect with $\partial\Omega \setminus \Gamma$ (except for parts with zero surface measure), then applying the rest of the definition unchanged.

Remark 1.19 *One can exchange $\partial\Omega \in C^{k,1}$ in the above paragraph by $\partial\Omega \in C^0$ or C^k, the details remain valid if the maps $\{\varphi_j\}$ are adapted appropriately.*

Trace Operators

Trace operators which restrict functions $u \in H^s(\Omega)$ to the boundary can now be constructed as extensions of the classical trace operators of continuous functions,

$$u(x_1, \dots, x_n)|_{x_n=0} := u(x_1, \dots, x_{n-1}, 0). \tag{1.2.24}$$

We summarize here the results of this topic; for details see the books of [75], [105] and [73]. We define for any function $u \in C^{k,1}(\bar{\Omega})$ its traces of normal derivatives by

$$\gamma_j(u) := \left.\frac{\partial^j u}{\partial \nu^j}\right|_\Gamma, \quad 0 \leq j \leq k, \tag{1.2.25}$$

where $\nu = \nu(x)$ is the outward normal on the boundary of Ω which exists a.e. We will be referring to γ_0 when talking about **the** trace operator.

Theorem 1.20 *Assume that $s - 1/2 = m + \sigma$, $0 < \sigma < 1, m \in \mathbb{N}_0$ and $s \leq k+1$. Then the mapping*

$$u \mapsto \{\gamma_0 u, \gamma_1 u, \dots, \gamma_m u\}, \tag{1.2.26}$$

which is defined for $u \in C^{k,1}(\bar{\Omega})$, has a unique continuous extension as an operator from

$$H^s(\Omega) \text{ onto } \prod_{j=0}^{m} H^{s-j-1/2}(\Gamma).$$

We will make no distinction between the classical trace operators and the extensions to Sobolev spaces. In later chapters, we frequently need a classical **Trace Theorem** which holds for domains Ω with Lipschitz continuous boundary $\partial\Omega \in C^{0,1}$.

Theorem 1.21 *For any $u \in H^s(\Omega), 1/2 < s < 3/2$, one can estimate*

$$\|\gamma_0 u\|_{H^{s-1/2}(\Gamma)} \leq c_{T,\Omega}\|u\|_{H^s(\Omega)}. \tag{1.2.27}$$

Conversely, for every $h \in H^{s-1/2}(\Gamma)$, there exists some $u \in H^s(\Omega)$ such that $\gamma_0 u = h$ and

$$\|u\|_{H^s(\Omega)} \leq C_{T,\Omega}\|h\|_{H^{s-1/2}(\Gamma)}. \tag{1.2.28}$$

As before, $c_{T,\Omega}$ and $c_{T,\Omega}$ denote positive finite constants, but, as indicated by their subscript, their value usually depends on properties of the domain Ω.

The range of s extends accordingly if $\partial\Omega$ is more regular.

Extension Operators

We can also give estimates which can be seen as a converse counterpart to the above estimate (1.2.27). These are **Whitney-extension** results which state that any function $u \in H^s(\Omega)$ can be extended to a function $\widetilde{u} \in H^s(\mathbb{R}^n)$ such that $\widetilde{u}|_\Omega = u$ and

$$\|\widetilde{u}\|_{H^s(\mathbb{R}^n)} \leq C_{E,\Omega}\|u\|_{H^s(\Omega)}, \quad s > 0. \tag{1.2.29}$$

This is also true for traces of functions: for any $h \in H^{s-1/2}(\Gamma)$, there exists an extension $\widetilde{h} \in H^{s-1/2}(\partial\Omega)$ such that

$$\|\widetilde{h}\|_{H^{s-1/2}(\partial\Omega)} \leq C_{E,\partial\Omega}\|h\|_{H^{s-1/2}(\Gamma)}, \quad s > 0. \tag{1.2.30}$$

Again, $C_{E,\Omega}$ and $C_{E,\partial\Omega}$ denote (domain dependent) positive finite constants.

1.2.3 Dual of Sobolev Spaces

Recall that the dual space of $H^s(\Omega)$ will generally be denoted by $(H^s(\Omega))'$. The dual space of L_2 is related to L_2 again by the **Riesz Representation Theorem**, i.e., $(L_2)' = L_2$, and the **dual form** is given as

$$\langle u, v\rangle_{L_2\times(L_2)'} := \int_\Omega u(x)\, v(x)\, d\mu, \quad u, v \in L_2. \tag{1.2.31}$$

Remark 1.22 *In the following, we will omit the space specifiers in the dual form and write only $\langle\cdot,\cdot\rangle$ if the exact dual form can be ascertained unambiguously from the arguments.*

Thus, we have trivially

$$(H^0(\Omega))' = (\mathcal{H}_0^0(\Omega))' = (L_2(\Omega))' = L_2(\Omega) \quad \text{for domains } \Omega \subseteq \mathbb{R}^n. \tag{1.2.32}$$

In case $\Omega = \mathbb{R}^n$ we can use Definition 1.13 with arbitrary negative indices to define Sobolev spaces of negative order on all $\mathbb{R}^n$. These spaces $H^{-s}(\mathbb{R}^n)$ are now the dual spaces of $H^s(\mathbb{R}^n)$ as the following result from [105] shows:

Theorem 1.23 *For all $s > 0$ one has*

$$(H^s(\mathbb{R}^n))' = H^{-s}(\mathbb{R}^n). \tag{1.2.33}$$

This is not true when the domain under consideration is bounded. However, we can identify some of these dual spaces with Sobolev spaces of negative order which we define as follows:

Definition 1.24 [Sobolev Spaces $H^{-s}(\Omega)$]
For $\Omega \subset \mathbb{R}^n$ and $s \in \mathbb{R}_+$ we define a norm for $u \in L_2(\Omega)$ by

$$\|u\|_{H^{-s}(\Omega)} := \sup_{v\in\mathcal{H}_0^s(\Omega)} \frac{\langle u, v\rangle_{(L_2)'\times L_2}}{\|v\|_{H_0^s(\Omega)}}, \quad s > 0. \tag{1.2.34}$$

The closure of $L_2(\Omega)$ with respect to this norm is termed $H^{-s}(\Omega) = (H_0^s(\Omega))'$.

The resulting spaces are obviously bigger than $L_2(\Omega)$ and also nested, and we get the following line of inclusions:

$$\ldots \supset H^{-2}(\Omega) \supset H^{-1}(\Omega) \supset L_2(\Omega) \supset H_0^1(\Omega) \supset \mathcal{H}_0^2(\Omega) \supset \ldots . \tag{1.2.35}$$

1.2.4 Regularity Properties

The following theorems from [2] provide information about the relation of the Sobolev spaces $H^m(\Omega)$ to other function spaces, i.e., a function in a different space can be seen as a **representative** (in the Lebesgue sense) of the Sobolev space function.
These embedding theorems must assume some regularity assumptions on the domain $\Omega \subseteq \mathbb{R}^n$. One condition is given for bounded domains Ω by the **(strong) local Lipschitz boundary** $\partial\Omega \in C^{0,1}$. This requires each point $x \in \partial\Omega$ to have a neighborhood $\mathcal{U}(x) \subset \mathbb{R}^n$ whose intersection with $\partial\Omega$ is the graph of a Lipschitz continuous function. It entails that Ω lies locally on one side of its boundary.
The scope of the embeddings for the Sobolev space $H^m(\Omega)$ depend primarily on the value of the **critical Sobolev number**

$$\gamma := m - \frac{n}{2}, \tag{1.2.36}$$

where the value in the denominator is the order of the underlying **Lebesgue** space, here $L_2(\Omega)$. The embedding theorem (Theorem 4.12 in [2]) now distinguish between $\gamma < 0, \gamma = 0$ and $\gamma > 0$.

Theorem 1.25 *If $\Omega \subset \mathbb{R}^n$ has a local Lipschitz boundary, then the following embeddings are continuous:*

- *If $\gamma > 0$, i.e., $2m > n$, then $H^m(\Omega) \hookrightarrow L_q(\Omega)$ for $2 \leq q \leq \infty$.*
- *If $\gamma = 0$, i.e., $2m = n$, then $H^m(\Omega) \hookrightarrow L_q(\Omega)$ for $2 \leq q < \infty$.*
- *If $\gamma < 0$, i.e., $2m < n$, then $H^m(\Omega) \hookrightarrow L_q(\Omega)$ for $2 \leq q \leq p^\star := \frac{2n}{n-2m}$.*

Furthermore, the embedding $H^{m+1}(\Omega) \hookrightarrow H^m(\Omega)$ for $m \in \mathbb{N}_0$, is compact.

These embeddings are still valid if the domain Ω only satisfies the weaker **cone condition**, i.e., there exists a finite cone C such that each $x \in \Omega$ is the vertex of another finite cone C_x contained in Ω and congruent to C.

Remark 1.26 *According to comments in Section 6.2 in [2], the above embeddings are also* **compact** *for* **bounded domains** Ω *"with the exception of certain extreme cases". We will not be dealing with any "extreme" domains in this thesis.*

An even stronger result for the spaces $H_0^s(\Omega)$ is given by

Theorem 1.27 *The embedding $H_0^s(\Omega) \hookrightarrow H_0^t(\Omega)$, for all $s, t \in \mathbb{R}$ with $s > t$, is compact. This holds especially for the case $s = m + 1$, $t = m$ for $m \in \mathbb{N}_0$.*

Of special interest is also the relation of the Sobolev spaces H^s to the classical function spaces C^k. The **Embedding Theorem** by Sobolev establishes this connection.

Theorem 1.28 *If $\Omega \subset \mathbb{R}^n$ has a Lipschitz boundary, then the embedding $H^m(\Omega) \hookrightarrow C^k(\overline{\Omega})$ is continuous for $k \in \mathbb{N}_0$ and $m > k + n/2$.*

Remark 1.29 *The Theorems 1.25 and 1.28 are also valid for the subspaces $H_0^m(\Omega) \subseteq H^m(\Omega)$.*

1.2.5 Tensor Products of Sobolev Spaces

An important way of constructing higher dimensional function spaces is by **tensorizing** lower dimensional spaces, which is exactly the method of construction of higher dimensional wavelets employed in Section 2.4. The notion of tensor product applied to domains is common knowledge, refer to [148] for a strict mathematical definition of the tensor product of Hilbert spaces.

Remark 1.30 *As a convention, we will always read tensor products* **from right to left**, *i.e., in descending order.*

This construction emphasizes certain structure of the generated functions w.r.t. the lower dimensional space axes. It has to be investigated under what conditions a higher dimensional **Sobolev space** $H^r(\Omega)$ of a tensor product domain $\Omega = \bigotimes_{i=1}^n I_i, I_1, \ldots, I_n \subset \mathbb{R}$, can be characterized by tensor products of the Sobolev spaces $H^{r_i}(I_i), i = 1, \ldots, n$. We follow the notation of [71, 72]. Their work is based upon [6, 148]. The book [76] also gives a good introduction and many details on this topic.
For simplicity, we will only depict the following results in two dimensions, they are easily extended to higher space dimensions.

Definition 1.31 [Sobolev Space with Dominating Mixed Derivative]
Let $I_1, I_2 \subset \mathbb{R}$ be two closed intervals and $\Omega := I_2 \otimes I_1$. The **Sobolev space with dominating mixed derivative** $r_1, r_2 \in \mathbb{N}_0$,

$$\mathcal{H}^{(r_1,r_2)}_{mix}(I_2 \otimes I_1) := H^{r_2}(I_2) \otimes H^{r_1}(I_1), \tag{1.2.37}$$

is a Hilbert space with the norm

$$\|f\|_{\mathcal{H}^{(r_1,r_2)}_{mix}} := \sqrt{\|f\|^2_{L_2(\Omega)} + \left\|\frac{\partial^{r_1}}{\partial x_1^{r_1}} f\right\|^2_{L_2(\Omega)} + \left\|\frac{\partial^{r_2}}{\partial x_2^{r_2}} f\right\|^2_{L_2(\Omega)} + \left\|\frac{\partial^{r_1+r_2}}{\partial x_1^{r_1} \partial x_2^{r_2}} f\right\|^2_{L_2(\Omega)}}. \tag{1.2.38}$$

The space $\mathcal{H}^{(r_1,r_2)}_{0,mix}(I_2 \otimes I_1)$ can defined analogously to Definition 1.16.

This space can be highly **anisotropic** in case r_1 and r_2 differ greatly in value. If the smoothness indices are equal, $r = r_1 = r_2$, it is obvious this norm is stronger than (1.2.5) for r alone and the resulting space is thus a (dense) subset of $H^r(\Omega)$:

$$H^r(I_2) \otimes H^r(I_1) \subset H^r(I_2 \otimes I_1).$$

The “original” Sobolev space H^r obviously assures smoothness uniformly, i.e., in all directions to the same extent, but in the sum not beyond the maximum smoothness index, in this sense it is an **isotropic** space. This is emphasized by the following results (see [76]).

Lemma 1.32 *For $r \in \mathbb{N}_0$, the space $H^r(\Omega)$ can be characterized as*

$$H^r(I_2) \otimes L_2(I_1) \cap L_2(I_2) \otimes H^r(I_1). \tag{1.2.39}$$

A version of Definition 1.31 for arbitrary $r_1, r_2 \in \mathbb{R}$ can be found in [79]. Analogous results can also be derived for the spaces $H^r_0(\Omega)$.
It is thus clear that “true” tensor product spaces in the sense of (1.2.37), especially if the individual smoothness indices differ in value, require **anisotropic** methods, but the Sobolev space defined by (1.2.5) can be represented using **isotropic** techniques.

1.3 Besov Spaces

In the course of the adaptive wavelet theory discussed in Section 3, it will become clear that a more finely tunable function space definition than Sobolev spaces $H^m(\Omega)$ is needed. Besov spaces are a generalization of Sobolev spaces, i.e., where, instead of employing weak derivatives, smoothness is measured using finite differences, see [2,64]. Besov spaces appear in the context of **nonlinear approximations**, an introduction to which can be found in [62].

To this end, we define the **difference operator** for functions $u : \Omega \to \mathbb{R}$, $\Omega \subset \mathbb{R}^n$, and the step vector $\mathbf{h} = (h_1, \dots, h_n) \in \mathbb{R}^n$,

$$\Delta_{\mathbf{h}}(u)(\mathbf{x}) := u(\mathbf{x}+\mathbf{h}) - u(\mathbf{x}), \quad \mathbf{x} \in \mathbb{R}^n, \tag{1.3.1}$$

for which inductively holds for $d > 1$

$$\Delta_{\mathbf{h}}^d(u) := \Delta_{\mathbf{h}}^{d-1}(\Delta_{\mathbf{h}}(u)) = \sum_{k=0}^{d} (-1)^{d-k} \binom{d}{k} u(\cdot + k\,\mathbf{h}). \tag{1.3.2}$$

By simple Taylor expansion for a sufficiently smooth function $u \in C^d(\Omega, \mathbb{R})$, it is easily shown that

$$|\Delta_{\mathbf{h}}^d(u)(\mathbf{x})| \to \sum_{|\alpha|=d} \mathbf{h}^\alpha \, \alpha! \, \partial^\alpha u(\mathbf{x}), \qquad \text{with } \mathbf{h}^\alpha := h_1^{\alpha_1} \cdots h_n^{\alpha_d}, \quad \partial^\alpha u(\mathbf{x}) := \partial_1^{\alpha_1} \cdots \partial_n^{\alpha_d} u(\mathbf{x}),$$

holds. For unsmooth functions, subspaces can be obtained by placing conditions on how fast $\|\Delta_{\mathbf{h}}^d(u)\|_{L_p(\Omega)}$ drops to zero as $\mathbf{h} \to 0$. To measure this, we define the **moduli of smoothness**

$$\omega_d(u,t)_p := \sup_{|\mathbf{h}| \le t} \|\Delta_{\mathbf{h}}^d(u)\|_{L_p(\Omega_{d\mathbf{h}})}, \qquad \text{for functions } u \in L_p(\Omega, \mathbb{R}) \text{ and } t > 0, \tag{1.3.3}$$

where $\Omega_{d\mathbf{h}}$ denotes the **line segment** $[\cdot, \cdot + d\mathbf{h}]$, which must be contained in Ω. This definition is sufficient to define **Lipschitz**-like spaces, e.g. by combining all functions for which for a single α holds

$$\sup_{t>0} t^{-\alpha} \omega_d(u,t)_p < \infty, \qquad \text{with } d > \alpha. \tag{1.3.4}$$

Note that this yields exactly the same spaces regardless of the actual value d, as long as $d > \alpha$ holds. The parameter α thus takes the place of the weak derivative order m of Definition 1.9. To define the **Besov spaces** $B_q^\alpha(L_p)$, we require another smoothness parameter $q \in [0, \infty)$, to fine-tune the allowed growth (or rather decay) rate of the function defined by the parameter t.

Definition 1.33 [Besov Space on Ω]
For $0 < q < \infty$, the **Besov seminorm**

$$|u|_{B_q^\alpha(L_p(\Omega))} := \left(\int_{t>0} \left[t^{-\alpha} \omega_d(u,t)_p \right]^q \frac{dt}{t} \right)^{1/q}, \tag{1.3.5}$$

defines the **Besov space** $B_q^\alpha(L_p(\Omega))$ *as all functions of* $L_p(\Omega)$ *for which the norm*

$$\|u\|_{B_q^\alpha(L_p(\Omega))} := \|u\|_{L_p(\Omega)} + |u|_{B_q^\alpha(L_p(\Omega))} \tag{1.3.6}$$

is finite. It can again be shown that the parameter d *yields, as long as* $d > \alpha$ *holds, equivalent norms and thus the same spaces. For the case* $q = \infty$*, (1.3.4) is used as seminorm for* $B_\infty^\alpha(L_p(\Omega))$*.*

Proposition 1.34 *For fixed* p *and* α*, the space* $B_q^\alpha(L_p(\Omega))$ *gets* **larger** *with* **increasing** q *and it holds*

$$|u|_{B_{q_2}^\alpha(L_p(\Omega))} \lesssim |u|_{B_{q_1}^\alpha(L_p(\Omega))}, \qquad \text{for } 0 < q_1 < q_2 < \infty \text{ and } u \in B_{q_1}^\alpha(L_p(\Omega)), \tag{1.3.7}$$

hence it follows $B_{q_1}^\alpha(L_p(\Omega)) \subset B_{q_2}^\alpha(L_p(\Omega))$*.*

The index q is secondary to the indices α, p when it comes to the characterization of the Besov space $B_q^\alpha(L_p(\Omega))$. In fact, many imbedding propositions only depend on α and p and not on q. Besov spaces can also very elegantly be characterized by wavelet expansion sequences, which we will use in Section 3.

1.3.1 Connection to Sobolev Spaces

As the Besov space characterizes smoothness within the Lebesgue spaces $L_p(\Omega)$, there must be some connections to the Sobolev spaces of Section 1.2.
It was shown in [83] that the Besov spaces $B_q^\alpha(L_2(\Omega))$ can be characterized by **K-functional theory** as **intermediate spaces** between the spaces $L_2(\Omega)$ and Sobolev spaces $H^k(\Omega)$: If for two spaces X, Y holds $Y \subset X$, we can define the K-functional for elements $f \in X$ as

$$K(t, f, X, Y) := \inf_{g \in Y} \left(\|f - g\|_X + t\,|g|_Y\right), \quad t > 0, \tag{1.3.8}$$

which is a semi-norm on the space X. We now introduce the **intermediate spaces** as

$$[X, Y]_{\frac{\theta}{\tau}, q} := \left\{ f \in X \mid \|f\|_{\theta/\tau, q; K} < \infty \right\}, \quad 1 \le q \le \infty, 0 < \theta < \tau, \tag{1.3.9}$$

with the norm

$$\|f\|_{\theta/\tau, q; K} := \left(\sum_{j=0}^{\infty} \left[2^{j\theta} K(2^{-j\tau}, f, X, Y) \right]^q \right)^{1/q}. \tag{1.3.10}$$

It is easy to verify that

$$Y \subset [X, Y]_{\theta/\tau, q} \subset X.$$

In this setting, the $L_p(\Omega)$ spaces are interpolation spaces for $X = L_1(\Omega)$, $Y = L_\infty(\Omega)$.
Also, it holds for $k \in \mathbb{N}$ and $\alpha < k$,

$$\left[L_2(\Omega), H^k(\Omega)\right]_{\alpha/k, q} = B_q^\alpha(L_2(\Omega)). \tag{1.3.11}$$

It also follows as a special case for $q = 2$,

$$B_2^\alpha(L_2(\Omega)) = H^\alpha(\Omega), \text{ for all } \alpha \ge 0. \tag{1.3.12}$$

The Besov space parameter q thus has no direct equivalent in the Sobolev scale (1.2.19); but Besov spaces can be used to characterize the fractional Sobolev spaces of Section 1.2.2, see [2]. There are many more relations of Sobolev, Lebesgue and Besov spaces when using the intermediate spaces construction (1.3.9), see [62].

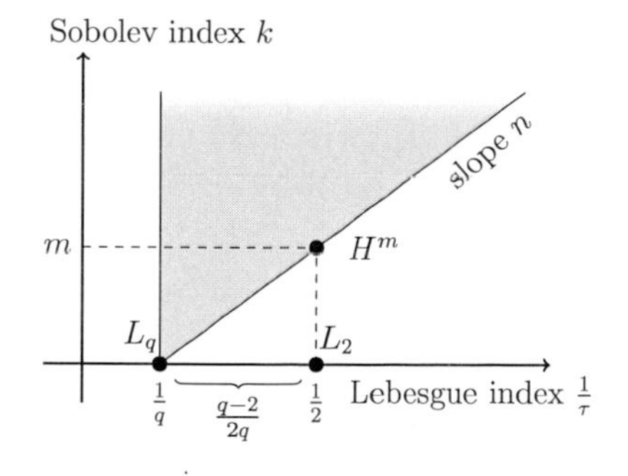

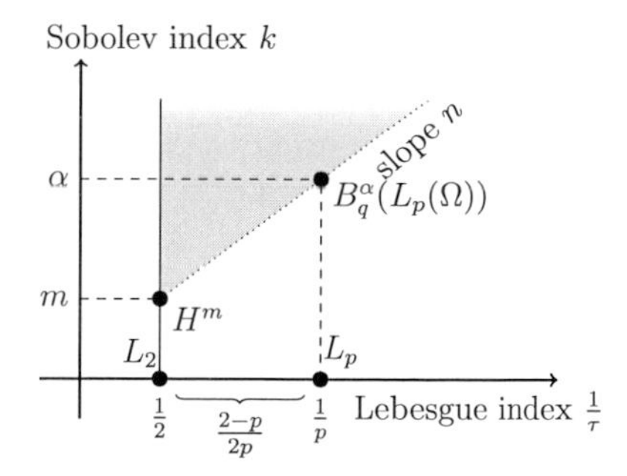

Figure 1.1: Sobolev and Besov embedding diagrams, introduced in [62]. On the abscissa is given the inverse Lebesgue index and on the ordinate axis the order of the weak derivative. The diagram on the left shows the embedding relations given by Theorem 1.25. All Sobolev spaces in the shaded area and on the critical line (of slope n) are continuously (and often compactly) embedded in the space $L_q(\Omega)$. The diagram on the right highlights the Besov spaces embedded in the Sobolev space $H^m(\Omega)$. Each point $(1/p, \alpha)$ in the plane stands for a whole range of Besov spaces $B_q^\alpha(L_p(\Omega))$ because of the undetermined parameter q. The spaces $B_q^\alpha(L_p(\Omega))$ on the critical line of slope n are generally not embedded in $H^m(\Omega)$. Because of the same slope, the right diagram can be seen as an extension of the left diagram with the common point $(1/2, m)$.

Embedding Properties

Common to all cases in Theorem 1.25 is the embedding $H^m(\Omega) \hookrightarrow L_q(\Omega)$ if $m \geq n\frac{q-2}{2q}$, which generalizes to other Sobolev spaces exactly as expected for $p \neq 2$. The main characteristic of the embedding formula is the space dimension n, because it can be interpreted as the **slope** when plotting $1/p$ against m as in Figure 1.1. For a fixed space $L_q(\Omega)$ the family of spaces continuously embedded in it are in the shaded area including the critical cases on the line with slope n.
By (1.3.11), one can interpret the position of the Besov space $B_q^\alpha(L_2(\Omega))$ in the same diagram of the Lebesgue and Sobolev smoothness indices at position $(1/2, \alpha/k)$, and this is easily generalized for other values of p. Since this holds for all values for the parameter q, each point $(1/p, \alpha)$ in the plane therefore stands for a whole family of spaces. Not all theoretical results are valid for all values of the secondary index q, therefore we restrict our further considerations on the special case $q \equiv \tau$, i.e., the spaces $B_\tau^\alpha(L_\tau(\Omega))$. For these spaces hold that for fixed $p \in (0, \infty)$, and τ, α connected by

$$\frac{1}{\tau} = \frac{\alpha}{n} + \frac{1}{p}, \tag{1.3.13}$$

the following interpolation identity holds,

$$[L_p(\Omega), B_\tau^\alpha(L_\tau(\Omega))]_{\theta,q} = B_q^{\theta\alpha}(L_q(\Omega)), \qquad \text{for } \frac{1}{q} = \frac{\theta\,\alpha}{n} + \frac{1}{p}. \tag{1.3.14}$$

Since the Besov spaces characterized by (1.3.13) are all on the line going through $(1/p, 0)$ with slope n, the result (1.3.14) says that interpolating between any two spaces corresponding to points on the line produces another space corresponding to a point on the line. Also, these spaces have the advantage that the embedding property holds for them [62, 124], i.e, the dotted line is solid in Figure 1.1, a property that will be essential when investigating the smoothness properties of solutions to non-linear PDEs in Section 3.1.1.

1.4 Elliptic Partial Differential Equations

At this point, we need to introduce some basic vocabulary from stationary partial differential equations. The weak formulation of an elliptic PDE will be at the heart of the problems considered in later chapters.
In the following $\Omega \subset \mathbb{R}^n$ will always be a bounded domain with a Lipschitzian boundary $\partial\Omega$.

Definition 1.35 [Partial Differential Equation (PDE)]
Let $a_{\alpha,\beta} \in L_\infty(\Omega)$ be bounded (PDE-) coefficient functions satisfying $a_{\alpha,\beta} = a_{\beta,\alpha}$ for all multi-indices α, β with $|\alpha|, |\beta| \leq m$. A **partial differential equation** *of order $2m$,*

$$\mathcal{L}y = f \quad \text{in } \Omega, \tag{1.4.1}$$

is determined by a linear differential operator of order $2m$, i.e.,

$$\mathcal{L} := \sum_{|\alpha|,|\beta| \leq m} (-1)^{|\beta|} \partial^\beta (a_{\alpha,\beta}(x) \partial^\alpha). \tag{1.4.2}$$

We associate to (1.4.2) the polynomial in $\xi = (\xi_1, \dots, \xi_n)$ given by

$$P(\xi, x) := \sum_{|\alpha|=|\beta|=m} a_{\alpha,\beta}(x) \xi^{\alpha+\beta}, \quad \xi^\alpha = \prod_{i=1}^{n} \xi_i^{\alpha_i}. \tag{1.4.3}$$

Definition 1.36 [Elliptic Operator]
The operator $\mathcal{L}$ is said to be **elliptic** *if (1.4.3) satisfies*

$$P(\xi, x) \gtrsim \|\xi\|_2^{2m}, \quad \text{for all } \xi \in \mathbb{R}^n, x \in \Omega. \tag{1.4.4}$$

Example 1.37 *The* **Laplacian operator**

$$\Delta := \sum_{i=1}^{n} \frac{\partial^2}{\partial x_i^2} \tag{1.4.5}$$

of order 2 is elliptic. The operator

$$\mathcal{L} = -\Delta + a_0 I, \quad a_0 \in \mathbb{R}_+, \tag{1.4.6}$$

also satisfies these properties.

It is easily seen that the equation $\mathcal{L}y = f$ for a given right hand side f does not need to have a unique solution. To ensure uniqueness, we have to impose further constraints on the solution space. This is typically done by requiring the solution to attain special boundary values. However, the existence or uniqueness of such boundary values cannot be determined unless we specify which kind of smoothness we require of our solution.

Definition 1.38 [Classical Solution of an Elliptic Boundary Value Problem]
A function $y \in C^{2m}(\Omega) \cap C(\bar{\Omega})$ which solves the **elliptic boundary value problem**

$$\begin{aligned} \mathcal{L}y &= f \qquad in\ \Omega, \\ \frac{\partial^i y}{\partial \nu^i} &= u_i \qquad on\ \partial\Omega, \quad i = 0, \ldots, m-1, \end{aligned} \tag{1.4.7}$$

pointwise for given data f and u_i is called a **classical solution**.

The above boundary conditions are generally classified into two types. Especially for the important case of $m = 1$, i.e., $\mathcal{L}$ is an operator of order 2, for example the **Laplace operator** (1.4.5), one defines

Definition 1.39 [Dirichlet Boundary Conditions]
Constraints of the form

$$y = u \quad on\ \partial\Omega \tag{1.4.8}$$

are called **Dirichlet boundary conditions**.

and

Definition 1.40 [Neumann Boundary Conditions]
Neumann boundary conditions *are of type*

$$\frac{\partial y}{\partial \nu} = u \quad on\ \partial\Omega. \tag{1.4.9}$$

If $u = 0$, the boundary conditions are called **homogeneous**, otherwise **inhomogeneous**.

Example 1.41 *The PDE*

$$\begin{aligned} -\Delta\, y + a_0\, y = f, &\quad in\ \Omega, \\ y = 0, &\quad on\ \partial\Omega, \end{aligned}$$

is called **Helmholtz problem**. *In case $a_0 = 0$ it is called* **Poisson problem**.

The existence of a classical solution cannot be proved for arbitrary right hand sides $f \in C^0(\Omega)$ and $u_i \in C^{2i}(\partial\Omega)$. Therefore, we must extend the solution space of (1.4.7) to spaces which permit solutions and ensure uniqueness.

1.4.1 Variational Problems

In this section, we recall some facts about bilinear forms and operators in the abstract setting of a general Hilbert space $\mathcal{H}$ and its dual $\mathcal{H}'$. In the next section, we recall how a solution approach of the elliptic partial differential equation (1.4.1) can be embedded into this setting, thus benefiting from the results collected here.

Definition 1.42 [Continuous and Elliptic Bilinear Forms]
A symmetric **bilinear form** $a : \mathcal{H} \times \mathcal{H} \to \mathbb{R}$ *is called* **continuous**, *if and only if*

$$a(y, v) \lesssim \|y\|_{\mathcal{H}} \|v\|_{\mathcal{H}} \quad \textit{for all } y, v \in \mathcal{H}. \tag{1.4.10}$$

A continuous bilinear form is called $\mathcal{H}-$**elliptic** *(or* **coercive***), if and only if*

$$a(y, y) \gtrsim \|y\|^2_{\mathcal{H}} \quad \textit{for all } y \in \mathcal{H}. \tag{1.4.11}$$

Let the operator $A : \mathcal{H} \to \mathcal{H}'$ be defined by

$$\langle y, Av \rangle := a(y, v). \tag{1.4.12}$$

Obviously, such a bilinear form is equivalent to the norm of the Hilbert space, i.e.,

$$\sqrt{a(y, y)} \sim \|y\|_{\mathcal{H}} :\Longleftrightarrow c_A \|y\|_{\mathcal{H}} \leq \|A\,y\|_{\mathcal{H}'} \leq C_A \|y\|_{\mathcal{H}'}, \tag{1.4.13}$$

with constants $0 < c_A, C_A < \infty$. By the **Theorem of Lax-Milgram** (see [4]) it is known that A is an isomorphism inducing the norm equivalence

$$\|Ay\|_{\mathcal{H}'} \sim \|y\|_{\mathcal{H}'}, \quad y \in \mathcal{H}. \tag{1.4.14}$$

A **variational problem** can now be phrased like this: Given $f \in \mathcal{H}'$, find $y \in \mathcal{H}$ such that

$$a(y, v) = \langle f, v \rangle\,, \quad \text{for all } v \in \mathcal{H}, \tag{1.4.15}$$

or, equivalently, in operator notation

$$Ay = f. \tag{1.4.16}$$

The Theorem of Lax-Milgram now ascertains a unique solution $y = A^{-1} f$ to problem (1.4.15) which depends continuously on the right hand side f.

1.4.2 Weak Formulation of Second Order Dirichlet and Neumann Problems

Let us return to the elliptic partial differential equation (1.4.7) with (inhomogeneous) Dirichlet and Neumann boundary conditions. The type of problem we will encounter in Section 5 is given in its classical form as

$$\begin{aligned} \mathcal{L}y &= f && \text{in } \Omega, \\ y &= u && \text{on } \Gamma \subset \partial\Omega, \\ \frac{\partial^i y}{\partial \nu^i} &= 0 && \text{on } \partial\Omega \setminus \Gamma, \quad i = 1, \ldots, m-1. \end{aligned} \tag{1.4.17}$$

We now consider the special case $\mathcal{H} = H^m(\Omega)$ with given data $f \in (H^m(\Omega))'$ and $u \in H^{m-1/2}(\Gamma)$. If a classical solution $y \in C^{2m}(\Omega) \cap C(\bar{\Omega})$ exists, multiplication of (1.4.17) with a test function $\phi \in C^\infty(\Omega)$ yields

$$\langle \mathcal{L}y, \phi \rangle = \langle f, \phi \rangle \quad \text{for all } \phi \in C^\infty(\Omega). \tag{1.4.18}$$

The left hand side is used to define the symmetric bilinear form $a(\cdot,\cdot)$ from Section 1.4.1 after using Green's formula,

$$\begin{aligned} a(y, \phi) &:= \langle \mathcal{L}y, \phi \rangle \\ &= \sum_{|\alpha|,|\beta| \le m} (-1)^{|\alpha|} \int_\Omega \phi(x)\, \partial^\alpha (a_{\alpha,\beta}(x)\, \partial^\beta y(x))\, d\mu(x) \qquad (1.4.19) \\ &= \sum_{|\alpha|,|\beta| \le m} \int_\Omega a_{\alpha,\beta}(x)\, (\partial^\alpha y)(x)\, (\partial^\beta \phi)(x)\, d\mu(x). \end{aligned}$$

After applying partial integration $m-$times, the boundary integral terms over $\partial\Omega \setminus \Gamma$ vanish here because of the homogeneous **Neumann boundary conditions** in (1.4.17). Since these boundary conditions are therefore naturally built into the weak formulation, they are called **natural boundary conditions**. The **Dirichlet boundary conditions** must be handled explicitly and are therefore called **essential boundary conditions**. The integral terms over Γ vanish by restricting our test function space to functions with $\phi|_\Gamma = 0$. Since $\phi \in C^\infty(\Omega)$, the trace $\phi|_\Gamma$ is well defined in the classical sense.
The explicit treatment of inhomogeneous Dirichlet boundary conditions will be the objective of Section 5. In short, these will be enforced using the **trace operator** γ_0 of Section 1.2.2 and **Lagrangian multipliers**.
The form $a(\cdot,\cdot)$ defined in (1.4.19) for functions in $C^\infty(\Omega)$ is obviously bilinear and symmetric. We can extend the bilinear form $a(\cdot,\cdot)$ onto $H^m(\Omega) \times H^m(\Omega)$ by Corollary 1.10. The right hand side of (1.4.18) defines a linear functional on the space $H^m(\Omega)$:

$$\langle f, v \rangle = \int_\Omega f(x) v(x)\, d\mu(x). \tag{1.4.20}$$

We will for an instant focus the discussion on the invertibility of the bilinear form (1.4.19) alone, for the special case of homogeneous boundary conditions $u = 0$. In view of Section 1.4.1, the task is then to find the element $y \in H^m(\Omega)$ with $y|_\Gamma = 0$ such that

$$a(y, v) = \langle f, v \rangle \quad \text{for all } v \in H^m. \tag{1.4.21}$$

Definition 1.43 [Weak Solution]
Any function $y \in H^m(\Omega)$ for which (1.4.21) holds will be called **weak solution** *since it is not necessarily in the space $C^{2m}(\Omega) \cap C^0(\bar{\Omega})$.*

Note that the ellipticity of the operator $\mathcal{L}$ does not guarantee the ellipticity of the bilinear form $a(\cdot,\cdot)$. This can only be ensured by requiring additional conditions of the domain Ω and the coefficients $a_{\alpha,\beta}$. One sufficient criterion for ellipticity (1.4.11) in case $m = 1$ is a bounded domain Ω and

$$a_{\alpha,\beta} = 0, \qquad \text{if } |\alpha| + |\beta| \le 1. \tag{1.4.22}$$

In case $m > 1$, additional prerequisites to the coefficient functions $a_{\alpha,\beta}$ are required, see [75].

Remark 1.44 *The Laplacian operator (1.4.5) is of the form (1.4.22) and the induced bilinear form is thus elliptic. This also holds for the operator (1.4.6).*

The other prerequisite of the bilinear form $a(\cdot,\cdot)$ required for the **Theorem of Lax-Milgram** is continuity, which follows directly from the following estimate taken from [75]:

$$|a(y,v)| \leq \sum_{\alpha,\beta} \|a_{\alpha,\beta}\|_{L_\infty(\Omega)} \|y\|_{H^m(\Omega)} \|v\|_{H^m(\Omega)}. \tag{1.4.23}$$

Thus, the homogeneous problem (1.4.17) has a unique solution in $H^m(\Omega)$.

Example 1.45 *The bilinear form for the Helmholtz Problem (Example 1.41) can be written as*

$$a(y,v) = (\nabla y, \nabla v)_{L_2} + a_0\,(y,v)_{L_2}, \tag{1.4.24}$$

where $\nabla := (\frac{\partial}{\partial x_1}, \ldots, \frac{\partial}{\partial x_1})^T$ *denotes the* **gradient**.

1.4.3 Galerkin Method

In the following, let $\mathcal{H}$ be a Hilbert Space, $a(\cdot,\cdot)$ a continuous and elliptic bilinear form on $\mathcal{H} \times \mathcal{H}$ and f a bounded linear functional on $\mathcal{H}$. With the elliptic PDE (1.4.17) now stated as a variational problem in weak formulation (1.4.21), we can go on to solve (1.4.21) numerically. Typically one discretizes (1.4.21) using elements from a finite dimensional closed subspace $S_j \subset \mathcal{H}$ of level j as trial functions. Let Φ_j be a basis for this space $S_j := S(\Phi_j) := \operatorname{span}\{\phi_{j,k} \,|\, k \in \Delta_j\}$ for some ordered finite index set Δ_j. The task is now to find the unique element $y_j \in S_j$ with

$$a(y_j, v_j) = \langle f, v_j \rangle \quad \text{for all } v_j \in S_j. \tag{1.4.25}$$

The coefficients in the expansion $y_j = \sum_{k\in\Delta_j} y_{j,k}\phi_{j,k}$ are compiled in the vector $\mathbf{y}_j := \left((y_{j,k})_{k\in\Delta_j}\right)^T$. Choosing $v_j = \phi_{j,k}$ sequentially for all $k \in \Delta_j$, we obtain the linear system of equations

$$\sum_{k\in\Delta_j} y_{j,k}\, a(\phi_{j,i}, \phi_{j,k}) = \langle f, \phi_{j,i} \rangle, \quad i \in \Delta_j. \tag{1.4.26}$$

Abbreviating $\mathbf{A}_j := \left(a(\phi_{j,i}, \phi_{j,k})\right)_{k,i\in\Delta_j}$ and $\mathbf{f}_j := \langle f, \phi_{j,i} \rangle_{i\in\Delta_j}$ this reads shortly as

$$\mathbf{A}_j\,\mathbf{y}_j = \mathbf{f}_j. \tag{1.4.27}$$

If the bilinear form $a(\cdot,\cdot)$ is symmetric and elliptic, then if follows that $\mathbf{A}_j$ is symmetric positive definite, i.e., $x^T \mathbf{A}_j\, x > 0$ for all $0 \neq x \in \mathbb{R}^{\#\Delta_j}$. In particular, this means that $\mathbf{A}_j$ is non-singular and a unique solution $\mathbf{y}_j = \mathbf{A}_j^{-1}\,\mathbf{f}_j$ to (1.4.27) exists. Note that the vector $\mathbf{y}_j$ determines the unique element $y_j = \sum_{k\in\Delta_j} y_{j,k}\,\phi_{j,k} \in S_j$ which is the solution to (1.4.25).

Definition 1.46 [Discretization Error]
We call the level dependent constant

$$h_j := \operatorname{dist}(y, S_j) := \inf_{v_j \in S_j} \|y - v_j\|_{\mathcal{H}}, \quad y \in \mathcal{H}, \tag{1.4.28}$$

the **discretization error** *of level j (with respect to S_j).*

Of special importance is the following result:

Lemma 1.47 [Céa – Lemma]
Let y resp. y_j denote the solution of the variational problem (1.4.21) resp. (1.4.25) in $\mathcal{H}$ resp. $S_j \subset \mathcal{H}$. Then it follows that

$$\|y - y_j\|_{\mathcal{H}} \lesssim \inf_{v_j \in S_j} \|y - v_j\|_{\mathcal{H}}. \tag{1.4.29}$$

In other words, the Galerkin solution y_j is up to a constant of the same error to the weak solution y as the best approximation of the trial space S_j. The best approximation error is here exactly the discretization error (1.4.28). An immediate consequence is the following

Corollary 1.48 *Let the subspaces S_j be nested and their union dense in $\mathcal{H}$, i.e.,*

$$S_j \subset S_{j+1}, \quad \operatorname{clos}_{\mathcal{H}} \bigcup_j S_j = \mathcal{H}, \tag{1.4.30}$$

then the Galerkin scheme converges, i.e.,

$$\lim_{j \to \infty} \|y - y_j\|_{\mathcal{H}} = 0.$$

1.5 Nonlinear Elliptic Partial Differential Equations

We follow the line of [33, 34], the basic definitions and proposition can also be found in [130,142], among others. This means we concentrate on those nonlinear elliptic operator equations which are locally well-posed. Consider a **nonlinear** stationary operator

$$G : \mathcal{H} \to \mathcal{H}', \tag{1.5.1}$$

mapping from a Hilbert space $\mathcal{H}$ into its dual $\mathcal{H}'$. As before, typical choices of the Hilbert space $\mathcal{H}$ are certain Sobolev spaces, i.e., one can think of $\mathcal{H} = H^m(\Omega)$ on a domain $\Omega \subset \mathbb{R}^n$ with possibly some boundary conditions. This operator will be used as a "perturbation", added to a linear elliptical operator $A : \mathcal{H} \to \mathcal{H}'$ from Section 1.4.1, i.e.,

$$F := A + G, \tag{1.5.2}$$

which together form a **nonlinear partial differential equation**. This type of equation is called **semilinear**. In weak form, the operator (1.5.2) then leads to the following problem formulation: For $f' \in \mathcal{H}'$, find $u \in \mathcal{H}$ with

$$\langle v, F(u) \rangle_{\mathcal{H} \times \mathcal{H}'} := \langle v, Au \rangle_{\mathcal{H} \times \mathcal{H}'} + \langle v, G(u) \rangle_{\mathcal{H} \times \mathcal{H}'} = \langle v, f \rangle_{\mathcal{H} \times \mathcal{H}'} . \tag{1.5.3}$$

We need to know under which conditions such a PDE is still solvable and, ideally, under which conditions the solution is unique.

1.5.1 Nemytskij Operators

It is beneficial to look at nonlinear operators in the context of Nemytskij operators, i.e., a wide class of (possibly non-linear) operators for which a common theory can be derived.

Definition 1.49 [Nemytskij operator]
For a function $\varphi : \Omega \times \mathbb{R} \to \mathbb{R}$, the **Nemytskij operator** $N_\varphi : (\Omega \to \mathbb{R}) \times \Omega \to \mathbb{R}$ *associated to φ is defined by*

$$N_\varphi(u)(x) := \varphi(x, u(x)), \tag{1.5.4}$$

for functions $u : \Omega \to \mathbb{R}$ and points $x \in \Omega$.

Remark 1.50 *This definition can obviously be extended to multiple arguments $u_1(x), \ldots, u_m(x)$ and to function values $u_i(x) \in \mathbb{R}^n$.*

Suitable conditions on the function φ now guarantee practical smoothness properties of the Nemytskij operator N_φ.

Definition 1.51 [Carathéodory conditions]
A function $\varphi : \Omega \times \mathbb{R} \to \mathbb{R}$ is said to satisfy the **Carathéodory conditions** *if*

(i) $\varphi(\cdot, y) : \Omega \to \mathbb{R}$ *is* **measurable** *for every $y \in \mathbb{R}$,*

(ii) $\varphi(x, \cdot) : \mathbb{R} \to \mathbb{R}$ *is* **continuous** *for almost every $x \in \Omega$.*

The following theorem is taken from [127].

Theorem 1.52 *Let $\Omega \subset \mathbb{R}^n$ be a domain, and let $\varphi : \Omega \times \mathbb{R} \to \mathbb{R}$ satisfy the Carathéodory conditions. In addition, let $p \in (1, \infty)$ and $g \in L_q(\Omega)$, with q being the adjoint Lebesgue index to p, be given. If φ satisfies for all x and y*

$$|\varphi(x, y)| \leq C\, |y|^{p-1} + g(x), \tag{1.5.5}$$

then the Nemytskij operator N_φ is a **bounded** *and* **continuous map** *from $L_p(\Omega)$ to $L_q(\Omega)$.*

1.5.2 Well-posed Operator Problems

The generalization of the standard derivative to mappings between Banach spaces $X \to Y$ is called **Fréchet derivative**. For $F : \mathcal{H} \to \mathcal{H}'$, we define the Fréchet derivative $DF : \mathcal{H} \to \mathcal{H}'$ at $z \in \mathcal{H}$ by the duality

$$\langle u, DF(z)\, v \rangle_{\mathcal{H} \times \mathcal{H}'} := \lim_{h \to 0} \frac{1}{h}\, \langle u, F(z + hv) - F(z) \rangle_{\mathcal{H} \times \mathcal{H}'}, \tag{1.5.6}$$

for all $u \in \mathcal{H}$. This leads to the following definition, which will be considered again in Section 3.

Definition 1.53 [Well-posed nonlinear operator equation]
We say an **operator equation** $F(v) = w$ *for $F : \mathcal{H} \to \mathcal{H}'$ is* **well-posed**, *if*

- *F possesses a* **continuous Frechét derivative** $DF \in C(\mathcal{H}, \mathcal{H}')$.
- *There* **exists a solution** $u \in \mathcal{H}$ *to* $F(u) = f$ *for* $f \in \mathcal{H}'$ *and some constants* $0 < c_{z,F} \leq C_{z,F} < \infty$ *such that*

$$c_{z,F}\|v\|_{\mathcal{H}} \leq \|DF(z)\, v\|_{\mathcal{H}'} \leq C_{z,F}\|v\|_{\mathcal{H}} \quad \text{for all } z \in \mathcal{H},\ v \in \mathcal{U}(u), \tag{1.5.7}$$

where $\mathcal{U}(v)$ denotes some neighborhood of u.

For such nonlinear operator equations, well-posed means a solution is **locally unique**.

Remark 1.54 *The Fréchet derivative of a linear operator $A : \mathcal{H} \to \mathcal{H}'$ is again A:*

$$\lim_{h\to 0} \frac{1}{h}\ \langle u, A(z + hv) - A(z)\rangle_{\mathcal{H}\times\mathcal{H}'} = \lim_{h\to 0} \frac{1}{h}\ \langle u, A(hv)\rangle_{\mathcal{H}\times\mathcal{H}'} = \langle u, A(v)\rangle_{\mathcal{H}\times\mathcal{H}'}.$$

Thus, if A is an isomorphism (1.4.14), the equation $Av = w$ for given $w \in \mathcal{H}'$ is well-posed.

We now collect some sufficient prerequisites on the operators F to show well-posedness. First, assume that F is **stable** in the following sense:

$$\|F(u) - F(v)\|_{\mathcal{H}'} \leq C_F(\max\{\|u\|_{\mathcal{H}}, \|v\|_{\mathcal{H}}\})\|u - v\|_{\mathcal{H}}, \quad \text{for all } u, v \in \mathcal{H}, \tag{1.5.8}$$

where $C_F(\cdot)$ is a positive nondecreasing function. This function shall account for the nonlinearity of F, property (1.5.8) is obviously a generalization of Lipschitz continuity. Another important aspect required of the operator to ensure solvability is the following property.

Definition 1.55 [Monotonicity]
We say a mapping $F : \mathcal{H} \to \mathcal{H}'$ is **monotone** *if*

$$\langle u - v, F(u) - F(v)\rangle \geq 0, \quad \text{for all } u, v \in \mathcal{H}. \tag{1.5.9}$$

If the above inequality is strict for $u \neq v$, then F is called **strictly monotone**.

Remark 1.56 *Coercive (1.4.11) linear operators $A : \mathcal{H} \to \mathcal{H}'$ are monotone:*

$$\langle u - v, A(u) - A(v)\rangle = \langle u - v, A(u - v)\rangle = a(u - v, u - v) \gtrsim \|u - v\|_{\mathcal{H}} \geq 0.$$

The next theorem from [33] now takes the above assumptions and deduces that there exists a unique solution to problem (1.5.3).

Theorem 1.57 *If a nonlinear operator $G : \mathcal{H} \to \mathcal{H}'$ is stable (1.5.8) and monotone (1.5.9), then the equation $F := A + G = f$ in weak form (1.5.3) has for every $f \in \mathcal{H}'$ a* **unique** *solution. Moreover, the problem is* **well-posed** *in the sense of (1.5.7) with constants*

$$c_{z,F} := c_A, \quad C_{z,F} := C_A + C_G(\|z\|_{\mathcal{H}}), \tag{1.5.10}$$

where c_A, C_A are the constants from (1.4.13) and $C_G(\cdot)$ is the nondecreasing function defined in (1.5.8).

Proof: Since A, being a linear function, is monotone and assumed to be coercive, so is the operator $F := A + G$. Then the **Browder-Minty Theorem**, see [127, 149], asserts the existence of a solution.
The coercivity of the linear operator A then guarantees uniqueness: If $u, v \in \mathcal{H}$ with $u \neq v$ were two different solutions for $f \in \mathcal{H}'$, i.e., for all $w \in \mathcal{H}$ it holds

$$\langle (A+G)u, w\rangle = \langle f, w\rangle \,,$$
$$\langle (A+G)v, w\rangle = \langle f, w\rangle \,,$$

then it follows for $w := u - v$ by subtracting both equations and employing the estimates of (1.4.11) and (1.5.9),

$$\begin{aligned} 0 &= \langle (A+G)u - (A+G)v, u-v\rangle \\ &= \langle A(u-v), u-v\rangle + \langle G(u) - G(v), u-v\rangle \\ &\gtrsim \|u-v\|_{\mathcal{H}}^2 + 0 > 0 \end{aligned}$$

which is a contradiction.
Similarly, well-posedness of F is confirmed. First, by noting that (1.4.11) and (1.4.10) hold with constants c_A, C_A, respectively, implies

$$c_A \, \|v\|_{\mathcal{H}} \leq \|Av\|_{\mathcal{H}'} \leq C_A \, \|v\|_{\mathcal{H}}$$

and since $DA \equiv A$ by Remark 1.54, it follows that (1.5.7) and (1.5.8) hold with the same constants c_A, C_A. Second, since the operator G is stable (1.5.8), together with definition (1.5.6), we can conclude

$$\begin{aligned} \frac{1}{h}\left(G(z + h\,v) - G(z)\right) &\leq \frac{1}{h} C_G(\max\{\|z + h\,v\|_{\mathcal{H}}, \|z\|_{\mathcal{H}}\}) \, \|h\,v\|_{\mathcal{H}}, \\ &\to C_G(\|z\|_{\mathcal{H}})\|v\|_{\mathcal{H}} \text{ for } h \to 0, \end{aligned}$$

which proves, using $DF(z)\,v = (D(A+G))(z)\,v = A\,v + DG(z)\,v$, the upper bound constant:

$$\begin{aligned} \|DF(z)\,v\|_{\mathcal{H}'} &\leq \|A\,v\|_{\mathcal{H}'} + \|DG(z)\,v\|_{\mathcal{H}'}, \\ &\leq C_A \, \|v\|_{\mathcal{H}} + C_G(\|z\|_{\mathcal{H}}) \, \|v\|_{\mathcal{H}}. \end{aligned}$$

The lower bound follows quickly by noting that monotonicity (1.5.9) implies in (1.5.6) for $u = v$ that $\langle v, DG(z)\,v\rangle \geq 0$, which then proves

$$\langle v, A\,v + DG(z)\,v\rangle = \langle v, A\,v\rangle + \langle v, DG(z)\,v\rangle \geq c_A \|v\|_{\mathcal{H}}^2 + \langle v, DG(z)\,v\rangle \geq c_A \, \|v\|_{\mathcal{H}}^2.$$

This concludes the proof. ∎

Now, all that is left is to identify possible nonlinear operators that are "compatible" in the sense of Definition 1.53 to our elliptic linear PDE operator $A : \mathcal{H} \to \mathcal{H}'$.

1.5.3 Operators of Polynomial Growth

To apply the theory of the previous Section 1.5.2, first a seemingly innocuous assertion has to be considered: For which subspaces $\mathcal{H} \subseteq L_2(\Omega)$ of a bounded domain $\Omega \subset \mathbb{R}^n$ is $G(\cdot)$ a mapping into $\mathcal{H}'$? A first simple observation hints at what the result must look like: The more smoothness the space $\mathcal{H}$ possesses, the more unsmooth the result $G(u), u \in \mathcal{H}$, can become to still act as a bounded linear functional on $\mathcal{H}$. Since the effective smoothness (in the Lebesgue scale) of a Sobolev function depends on the dimension n (it declines in higher dimensions by Theorem 1.25), the maximum smoothness index must be lower for higher dimensions.
The main result of Section 1.5.1 hints that such a result can be proven by considering the class of operators that is bounded as described in equation (1.5.5). Therefore, we now specifically look at operators whose growth is bounded like a polynomial, i.e., operators $G(\cdot)$ which fulfill the following **growth conditions** for some $r \geq 0$ and $s^\star \in \mathbb{N}$,

$$|G^{(s)}(x)| \leq C\,(1+|x|)^{(r-s)_+}, \quad x \in \mathbb{R}, \quad s = 0, 1, \ldots, s^\star, \tag{1.5.11}$$

where G is interpreted as a function from $\mathbb{R}$ to $\mathbb{R}$. Here is set $(a)_+ := \max\{0, a\}$. This condition obviously applies to $G(u) = u^r$ for all $s^\star$ if $r \in \mathbb{N}$ and for $s^\star = \lfloor r \rfloor$ if $r > 0$ but $r \notin \mathbb{N}$.

Remark 1.58 *A detailed analysis of these operators and their properties as mappings of Lebesgue and Sobolev spaces on arbitrary domains can be found in [130].*

The following result from [34] now shows when such a bounded function is indeed a mapping into the dual space of $H^m(\Omega)$ and when the **stability property** (1.5.8) holds true.

Theorem 1.59 *Assume that G satisfies the growth condition (1.5.11) for some $r \geq 0$ and $s^\star \geq 0$. Then G maps from $H^m(\Omega)$ to $(H^m(\Omega))'$ under the restriction*

$$0 \leq r \leq r^\star := \frac{n+2m}{n-2m}, \tag{1.5.12}$$

when $n > 2\,m$, and with no restriction otherwise.
If in addition $s^\star \geq 1$, then we also have under the same restriction

$$\|G(u) - G(v)\|_{(H^m(\Omega))'} \leq C_G(\max\left\{\|u\|_{H^m(\Omega)}, \|v\|_{H^m(\Omega)}\right\})\|u-v\|_{H^m(\Omega)}, \tag{1.5.13}$$

where $C_G(\cdot)$ is a nondecreasing function.

Proof: The proof can be found in [34], but the first part can also be quickly seen by Theorem 1.25. Since $H^m(\Omega) \hookrightarrow L_q(\Omega)$ for $2 \leq q \leq \frac{2n}{n-2m}$ always (the upper bound goes to infinity for $n \to 2m$), it suffices to show $G(u) \in (L_{\frac{2n}{n-2m}}(\Omega))'$ for $u \in H^m(\Omega)$. The adjoint Lebesgue index to $\frac{2n}{n-2m}$ is $\frac{2n}{n+2m}$ and it thus has to hold by (1.5.11) for $s = s^\star = 0$,

$$\int_\Omega |G(u)|^{\frac{2n}{n+2m}}\,d\mu \;\lesssim\; \int_\Omega (1+|u|^r)^{\frac{2n}{n+2m}}\,d\mu \overset{!}{<} \infty.$$

Again, since u is in the space $L_{\frac{2n}{n-2m}}(\Omega)$, this holds for

$$r\frac{2n}{n+2m} \leq \frac{2n}{n-2m} \quad \Longleftrightarrow \quad r \leq \frac{n+2m}{n-2m}.$$

Thus, for $0 \leq r \leq r^\star$ and fixed $w \in H^m(\Omega)$, $G(w) \in L_{\frac{2n}{n+2m}}(\Omega)$ can act as a linear functional

$$\widetilde{w}(\cdot) := \int_\Omega (\cdot)G(w)d\mu \quad : \quad H^m(\Omega) \to \mathbb{R}.$$

Because $u \cdot G(w) \in L_1(\Omega)$, by Hölder's inequality we can bound

$$\begin{aligned}
\sup_{u\in H^m(\Omega)} \frac{\langle \widetilde{w}, u\rangle}{\|u\|_{H^m(\Omega)}} &\leq \sup_{u\in H^m(\Omega)} \frac{\|G(w)\|_{L_{\frac{2n}{n+2m}}(\Omega)}\|u\|_{L_{\frac{2n}{n-2m}}(\Omega)}}{\|u\|_{H^m(\Omega)}} \\
&\leq \sup_{u\in H^m(\Omega)} \frac{\|G(w)\|_{L_{\frac{2n}{n+2m}}(\Omega)}\|u\|_{H^m(\Omega)}}{\|u\|_{H^m(\Omega)}} \\
&= \|G(w)\|_{L_{\frac{2n}{n+2m}}(\Omega)} < \infty.
\end{aligned}$$

Thus $\widetilde{w} \in (H^m(\Omega))'$.
For the proof of the stability property (1.5.13), please turn to [34]. ∎

Remark 1.60 *A generalized version of the above theorem, e.g., applicable to the whole space $\mathbb{R}^n$, can be found in [130].*

Altogether, if a nonlinear operator G fulfills the **monotonicity** assumption (1.5.9) and the **growth condition** (1.5.11) for $s^\star \geq 1$, then, by the assertions of Theorem 1.59 and Theorem 1.57, we can conclude (1.5.3) to have a unique solution. The operator $F = A+G$ is also **strongly monotone**, i.e.,

$$\langle F(u) - F(v), u - v\rangle_{\mathcal{H}'\times\mathcal{H}} \gtrsim \|u - v\|_{\mathcal{H}}^2, \qquad \text{for all } u, v \in \mathcal{H}, \tag{1.5.14}$$

which is a consequence of coercivity (1.4.11) of the linear part A. Therefore, it is sensible to write

$$y = F^{-1}(f), \qquad \text{with } F^{-1} : \mathcal{H}' \to \mathcal{H}.$$

This operator F^{-1} is also nonlinear, but, because of the above properties of F, **strictly monotone**, i.e.,

$$\left\langle \widetilde{u} - \widetilde{v}, F^{-1}(\widetilde{u}) - F^{-1}(\widetilde{v})\right\rangle_{\mathcal{H}'\times\mathcal{H}} > 0, \qquad \text{for all } \widetilde{u}, \widetilde{v} \in \mathcal{H}',$$

and **Lipschitz continuous**, see [149].

Example 1.61 *We now examine the operator $G(u) := u^3$. It satisfies (1.5.11) for all $r, s^\star$ and is thus by Theorem 1.59* **stable** *as a mapping $H^1(\Omega) \to (H^1(\Omega))'$ for dimensions $n \leq 4$ (cf. Remark 1.26).*

The corresponding function $\varphi(x,y) := y^3$ *satisfies the Carathéodory conditions and (1.5.5) for* $4 \leq p < \infty$*. Thus the Nemytskij operator* N_φ *is a bounded and continuous map from* $L_4(\Omega)$ *to* $L_{4/3}(\Omega)$*. Since also* $H^1(\Omega) \hookrightarrow L_4(\Omega)$ *by Theorem 1.25 for* $n \leq 4$*, this means for* $u, v \in H^1(\Omega)$ *follows*

$$(u-v)(G(u)-G(v)) \in L_1(\Omega).$$

This means it holds

$$\langle u-v, G(u)-G(v)\rangle = \int_\Omega (u-v)(G(u)-G(v))d\mu < \infty.$$

Since for $u, v \in L_2(\Omega)$ *holds* u^2, v^2 *and* $(u-v)^2 \geq 0$ *a.e., it follows*

$$\begin{aligned} (u+v)^2 \geq 0 &\Longrightarrow u^2+v^2+u\,v \geq -u\,v, && a.e.,\\ u^2+v^2 \geq 0 &\Longrightarrow u^2+v^2+u\,v \geq u\,v, && a.e.,\\ \Longrightarrow u^2+v^2+u\,v &\geq |u\,v|, && a.e.. \end{aligned}$$

From this we can deduce **monotonicity** *(1.5.9):*

$$\begin{aligned} \int_\Omega (u-v)(G(u)-G(v))d\mu &= \int_\Omega (u-v)(u^3-v^3)d\mu\\ &= \int_\Omega (u-v)^2(u^2+v^2+u\,v)d\mu\\ &\geq \int_\Omega (u-v)^2|u\,v|d\mu\\ &\geq 0,\\ \Longrightarrow (u-v)(G(u)-G(v)) \geq 0\ . \end{aligned}$$

Hence, using an appropriate linear operator A*, the assumptions of Theorem 1.57 are fulfilled and* $F = A + G$ *is* **well-posed** *by Definition 1.53.*
It is easily verified that the operator $G(u) = u^3$ *has the continuous Fréchet derivative* $DG(z)v = 3\,z^2\,v$*, which is for fixed* $z \in H^1(\Omega)$ *a type of* **(square-) weighted identity** *operator. The constants appearing in estimate (1.5.7) are* $C_{z,F} = C_A + 3\|z\|^2_{H^1}$ *and* $c_{z,F} = c_A$*.*

Remark 1.62 *The operator* $G(u) = u^3$ *does arise naturally in real-life applications, e.g., in the modeling of physical superconductors [82, 142]. This* $G(u)$ *will be the nonlinearity of choice for our problems in the later chapters.*

2 Multiresolution Analysis and Wavelets

This section provides an introduction into the theory of wavelets by means of multiresolution analysis.

2.1 Multiscale Decompositions of Function Spaces

The content of this section is based on [45], [99] and [131] and is partially taken from [122]. The function spaces considered will be living on a bounded domain $\Omega \subset \mathbb{R}^n$ with values in $\mathbb{R}$. We first consider the univariate case $n = 1$. In case $\Omega \subset \mathbb{R}^n, n > 1$, a tensor product approach is often sufficient for simple domains. We will discuss tensor products and their application in multiresolution analysis frameworks in Section 2.4.

2.1.1 Basics

Let Δ be a (possibly infinite) index set and $\#\Delta$ its cardinality. Then $\ell_2(\Delta)$ is the **Banach space** of elements $\mathbf{v} \in \ell_2(\Delta)$ for which the norm

$$\|\mathbf{v}\|_{\ell_2(\Delta)} := \left(\sum_{k \in \Delta} |v_k|^2 \right)^{1/2} \tag{2.1.1}$$

is finite. The elements $\mathbf{v} \in \ell_2(\Delta)$ are always regarded as **column vectors** of possibly infinite length. Likewise, we define a (countable) collection of functions Φ in a **Hilbert space** $\mathcal{H}$ as a column vector, whose elements are sorted accordingly and in a fixed order. This enables us to introduce the following shorthand notation for an expansion of Φ with a coefficient vector $\mathbf{c}$,

$$\mathbf{c}^T \Phi := \sum_{\phi \in \Phi} c_\phi \phi. \tag{2.1.2}$$

Recall from Section 1.1 the dual form $\langle v, \widetilde{v} \rangle := \langle v, \widetilde{v} \rangle_{\mathcal{H} \times \mathcal{H}'} := \widetilde{v}(v)$. Consequently, for any $\widetilde{v} \in \mathcal{H}'$, the quantity $\langle \Phi, \widetilde{v} \rangle$ is interpreted as a **column vector** and $\langle \widetilde{v}, \Phi \rangle$ as a **row vector** of expansion coefficients $\langle \phi, \widetilde{v} \rangle$, $\langle \widetilde{v}, \phi \rangle$, $\phi \in \Phi$, respectively. Furthermore, for any two collections $\Phi \subset \mathcal{H}$, $\Theta \subset \mathcal{H}'$ of functions, we frequently work with (possibly infinite) matrices of the form

$$\langle \Phi, \Theta \rangle := (\langle \phi, \theta \rangle)_{\phi \in \Phi, \theta \in \Theta}. \tag{2.1.3}$$

For any finite subset $\Phi \subset \mathcal{H}$ the linear span of Φ is abbreviated as

$$S(\Phi) := \operatorname{span}\{\Phi\}. \tag{2.1.4}$$

In order to make a function $w \in \mathcal{H}$ numerically accessible, its expansion coefficients $\mathbf{w}$ in a basis Φ of $\mathcal{H}$ should be **unique** and **stable**.

Definition 2.1 [Riesz basis of $\mathcal{H}$]
A family $\Phi = \{\phi_k\}_{k \in \mathbb{Z}}$, of elements of a separable Hilbert space $\mathcal{H}$ is called **Riesz basis**, *if and only if the functions in $\{\phi_k\}$ are linearly independent and for every $\mathbf{c} \in \ell_2$ one has*

$$\|\mathbf{c}\|_{\ell_2(\mathbb{Z})} \sim \|\mathbf{c}^T \Phi\|_{\mathcal{H}}, \tag{$\mathcal{S}$)(2.1.5}$$

which is called **Riesz stability** *or just* **stability**.

We will later derive conditions under which multiscale wavelet bases are automatically Riesz bases.

2.1.2 Multiresolution Analysis of $\mathcal{H}$

Recalling Definition 1.3, the inner product of $\mathcal{H}$ is termed $(\cdot,\cdot) := (\cdot,\cdot)_{\mathcal{H}}$, associated with the norm $\|\cdot\|_{\mathcal{H}}$. The elements of $\mathcal{H}$ shall be functions living on a bounded domain $\Omega \subset \mathbb{R}$ with values in $\mathbb{R}$.

Definition 2.2 [Multiresolution Analysis (MRA) of $\mathcal{H}$]
For a fixed parameter $j_0 \in \mathbb{N}_0$, a **multiresolution analysis** $\mathcal{S}$ **of** $\mathcal{H}$ *consists of closed subspaces S_j of $\mathcal{H}$, called multiresolution spaces, which are nested such that their union is dense in $\mathcal{H}$,*

$$S_{j_0} \subset \ldots \subset S_j \subset S_{j+1} \subset \ldots \subset \mathcal{H}, \quad \operatorname*{clos}_{\mathcal{H}}\Big(\bigcup_{j \geq j_0} S_j\Big) = \mathcal{H}. \qquad (\mathcal{R})(2.1.6)$$

Specifically, the multiresolution spaces S_j will be of type

$$S_j := S(\Phi_j), \quad \Phi_j = \{\phi_{j,k} \,|\, k \in \Delta_j\}, \qquad (2.1.7)$$

each defined by a finite dimensional basis Φ_j with Δ_j being a level dependent finite index set. The bases $(\Phi_j)_{j \geq j_0}$ will be assumed to be **uniformly stable** *in the sense of Definition 2.1, i.e., property $(\mathcal{S})$(2.1.5) holds uniformly for every $j \geq j_0$.*

The index j always denotes the **level of resolution**, **refinement level** or **scale** with j_0 being the **coarsest level**. We shall always deal with functions $\phi_{j,k}$ which have the **locality** property, i.e., they are compactly supported with

$$\operatorname{diam} \operatorname{supp} \phi_{j,k} \sim 2^{-j}. \qquad (\mathcal{L})(2.1.8)$$

For this reason the collection Φ_j is termed **single-scale basis**, since all its members live on the same scale j. It follows from $(\mathcal{S})$(2.1.5) with $\mathbf{c} = e_k$ (the k-th unity vector), that the $\phi_{j,k}$ must be scaled such that

$$\|\phi_{j,k}\|_{\mathcal{H}} \sim 1$$

holds. Here, k is called the **positional index** describing the location of the element $\phi_{j,k}$ in the space V. Considering for a moment $\Omega = \mathbb{R}$, the basis functions for Φ_j can be given by translation (by offset k) and dilation (by factor 2^j) of a single function ϕ called the **generator**, i.e.,

$$\phi_{j,k}(x) := (2^j)^{1/2} \phi(2^j x - k), \quad k \in \mathbb{Z}, j \geq j_0. \qquad (2.1.9)$$

In the view of the locality condition $(\mathcal{L})$(2.1.8), (2.1.9) means that $\operatorname{diam}\operatorname{supp} \phi_{j+1,k} \sim \frac{1}{2} \operatorname{diam}\operatorname{supp} \phi_{j,k}$ and therefore Φ_j can model more detail information with increasing level, which was coined by Mallat [106] in the signal processing context as **multiresolution analysis**. Herein ϕ is called the **generator of the MRA** $(\mathcal{R})$(2.1.6).
It is easy to verify that the $\{\phi_{j,k}\}_{j \geq j_0}$ form a Riesz basis for the space S_j with the same

constants as in the case $j = j_0$. Since the MRA spaces are nested, there exists a special sequence of coefficients $\{m_k\}_{k\in\mathbb{Z}} \in \ell_2(\mathbb{Z})$ such that for every $x \in \Omega$

$$\phi(x) = \sum_{k\in\mathbb{Z}} m_k \phi(2x - k).$$

Remark 2.3 *For the* **cardinal B-splines**, *the expansion coefficients can be found in [57], see also Appendix A.*

It follows that such a **refinement relation** can also be expressed for any of the functions $\phi_{j,k}$, $j \geq j_0$, leading to the existence of matrices $\mathbf{M}_{j,0} = (m^{j,0}_{r,k})_{r\in\Delta_{j+1},k\in\Delta_j}$ such that the **two-scale relation**

$$\phi_{j,k} = \sum_{r\in\Delta_{j+1}} m^{j,0}_{r,k} \phi_{j+1,r}, \quad k \in \Delta_j, \tag{2.1.10}$$

is satisfied. The sequence $\mathbf{m}^j_k := (m^{j,0}_{r,k})_{r\in\Delta_{j+1}} \in \ell_2(\Delta_{j+1})$ is called **mask** and each element a **mask coefficient**. Since every function $\phi_{j,k}$ has compact support and only a finite number of functions $\phi_{j+1,k}$ have support intersecting with the support of $\{\phi_{j,k}\}$, non-zero mask coefficients only appear for these functions on level $j + 1$. This means $\mathbf{m}^j_k$ has a uniformly, i.e., level independent, bounded number of non-zero entries. This will be crucial in the application of the **fast wavelet transform** in Section 2.1.3. In the sequel, it will be convenient to write (2.1.10) as a matrix-vector equation

$$\Phi_j = \mathbf{M}^T_{j,0} \Phi_{j+1}. \tag{2.1.11}$$

Thus, the $\mathbf{m}^j_k$ constitute the columns of the matrix $\mathbf{M}_{j,0} \in \mathbb{R}^{(\#\Delta_{j+1})\times(\#\Delta_j)}$. Any family of functions satisfying an equation of this form will be called **refinable**. It is known also for $\Omega \subset \mathbb{R}$ that nestedness $(\mathcal{R})$(2.1.6) and stability $(\mathcal{S})$(2.1.5) alone imply the existence of such matrices (see [45]). Obviously, $\mathbf{M}_{j,0}$ is a linear operator from the space $\ell_2(\Delta_j)$ into the space $\ell_2(\Delta_{j+1})$, i.e., recalling Definition 1.1,

$$\mathbf{M}_{j,0} \in L(\ell_2(\Delta_j);\ \ell_2(\Delta_{j+1})).$$

This matrix is also **uniformly sparse** which means that the number of entries in each row or column are uniformly bounded. Because of the two-scale relation (2.1.10) every $c = \mathbf{c}^T_j \Phi_j$, $\mathbf{c}_j \in \ell_2(\Delta_j)$, has a representation $c = \mathbf{c}^T_{j+1}\Phi_{j+1}$, $\mathbf{c}_{j+1} \in \ell_2(\Delta_{j+1})$, which, in recognition of the norm equivalence $(\mathcal{S})$(2.1.5) applied to both Φ_{j+1} and Φ_j, yields

$$\|\mathbf{c}_j\|_{\ell_2} \sim \|\mathbf{c}^T_j \Phi_j\|_{\mathcal{H}} = \|\mathbf{c}^T_j \mathbf{M}^T_{j,0} \Phi_{j+1}\|_{\mathcal{H}} = \|(\mathbf{M}_{j,0}\mathbf{c}_j)^T \Phi_{j+1}\|_{\mathcal{H}} \sim \|\mathbf{M}_{j,0}\mathbf{c}_j\|_{\ell_2},$$

and consequently, with the definition of operator norm (1.1.2), it follows that

$$\|\mathbf{M}_{j,0}\| = \mathcal{O}(1), \quad j \geq j_0.$$

Because the spaces Φ_j are nested and their infinite union $\mathcal{S}$ is dense in $\mathcal{H}$, a basis for $\mathcal{H}$ can be assembled from the functions which span the complement of two successive spaces Φ_j and Φ_{j+1}, i.e.,

$$S(\Phi_{j+1}) = S(\Phi_j) \oplus S(\Psi_j), \tag{2.1.12}$$

if we define

$$\Psi_j := \{\psi_{j,k} \,|\, k \in \nabla_j\} \subset S(\Phi_{j+1}), \quad \nabla_j := \Delta_{j+1} \setminus \Delta_j. \tag{2.1.13}$$

The complement spaces $\mathrm{W}_j := S(\Psi_j), j \geq j_0$, are called **detail spaces**.

Definition 2.4 [Wavelets]
The basis functions $\psi_{j,k}, j \geq j_0, k \in \mathbb{Z}$, of the detail spaces $\Psi_j, j \geq j_0$, are denoted as **wavelet functions** *or shortly* **wavelets**.

There is more than one way to choose a basis for the space W_j. One option would be to use the orthogonal complement. Of special interests for the case $\Omega = \mathbb{R}$ are those bases of wavelet spaces which can be constructed from a **mother wavelet** ψ by scaling and dilation in the sense of (2.1.9),

$$\psi_{j,k}(x) := (2^j)^{1/2}\psi(2^j x - k), \quad k \in \mathbb{Z}, j \geq j_0. \tag{2.1.14}$$

Thus, if the mother wavelet is compactly supported, the wavelets also satisfy

$$\operatorname{diam}\operatorname{supp}\psi_{j,k} \sim 2^{-j}. \tag{2.1.15}$$

Wavelets of this kind entail a similar band-like structure in $\mathbf{M}_{j,1}$ as seen in $\mathbf{M}_{j,0}$. For this reason, we shall restrict all following discussions to the case of compactly supported generators, which we here also call **scaling functions**, and compactly supported **wavelets**. Moreover, we allow only a finite number of supports of wavelets on one level j to overlap at any fixed point, i.e.,

$$\sup_{x\in\Omega} \# \{(j,k) \,|\, x \in \operatorname{supp}\psi_{j,k}\} \lesssim 1, \text{ for all } j \geq j_0, \tag{2.1.16}$$

where the constant does not depend on the level j.

Remark 2.5 *This property is automatically fulfilled by the conventional wavelet space construction via translation and dilation of a mother wavelet with property ($\mathcal{L}$)(2.1.8).*

Since every $\psi_{j,k} \in \Psi_j$ is also in the space Φ_{j+1} it has a unique representation

$$\psi_{j,k} = \sum_{r\in\Delta_{j+1}} m^{j,1}_{r,k}\phi_{j+1,r}, \quad k \in \nabla_j, \tag{2.1.17}$$

which can again be expressed as a matrix-vector equation of the form

$$\Psi_j = \mathbf{M}^T_{j,1}\Phi_{j+1} \tag{2.1.18}$$

with a matrix $\mathbf{M}_{j,1} \in \mathbb{R}^{(\#\Delta_{j+1})\times(\#\nabla_j)}$. Furthermore, equation (2.1.12) is equivalent to the fact that the linear operator composed of $\mathbf{M}_{j,0}$ and $\mathbf{M}_{j,1}$,

$$\mathbf{M}_j := (\mathbf{M}_{j,0}, \mathbf{M}_{j,1}) : \begin{array}{ccc} \ell_2(\Delta_j) \times \ell_2(\nabla_j) & \longrightarrow & \ell_2(\Delta_{j+1}) \\ (\mathbf{c}, \mathbf{d}) & \longmapsto & \mathbf{M}_{j,0}\mathbf{c} + \mathbf{M}_{j,1}\mathbf{d} \end{array} \tag{2.1.19}$$

is an **invertible** mapping from $\ell_2(\Delta_j \cup \nabla_j)$ onto $\ell_2(\Delta_{j+1})$. The refinement relations (2.1.11) and (2.1.18) combined lead to

$$\begin{pmatrix} \Phi_j \\ \Psi_j \end{pmatrix} = \begin{pmatrix} \mathbf{M}_{j,0}^T \\ \mathbf{M}_{j,1}^T \end{pmatrix} \Phi_{j+1} =: \mathbf{M}_j^T \Phi_{j+1}, \tag{2.1.20}$$

called **decomposition identity**. This means $\mathbf{M}_j$ performs a change of bases in the space Φ_{j+1}. Of course, we want $\mathbf{M}_j$ to have positive traits which can be exploited for numerical purposes, such as sparseness and invertibility.

Definition 2.6 [Stable Decomposition]
If the union $\{\Phi_j \cup \Psi_j\}$ is uniformly stable in the sense of $(\mathcal{S})$(2.1.5), i.e.,

$$\|\mathbf{c}\|_{\ell_2(\Delta_{j+1})} \sim \|(\Phi_j^T, \Psi_j^T)\mathbf{c}\|_{\mathcal{H}},$$

then $\{\Phi_j, \Psi_j\}$ is called a **stable decomposition** *of Φ_{j+1}.*

Note that $\mathbf{M}_j$ as a basis transformation must be invertible. We denote its inverse by $\mathbf{G}_j$, which we conveniently write in block structure as

$$\mathbf{M}_j^{-1} =: \mathbf{G}_j = \begin{pmatrix} \mathbf{G}_{j,0} \\ \mathbf{G}_{j,1} \end{pmatrix}, \tag{2.1.21}$$

with $\mathbf{G}_{j,0} \in \mathbb{R}^{(\#\Delta_j)\times(\#\Delta_{j+1})}$ and $\mathbf{G}_{j,1} \in \mathbb{R}^{(\#\nabla_j)\times(\#\Delta_{j+1})}$. It is known, see [44] for example, that $\{\Phi_j \cup \Psi_j\}$ is uniformly stable if and only if

$$\|\mathbf{M}_j\|, \|\mathbf{G}_j\| = \mathcal{O}(1), \quad j \to \infty. \tag{2.1.22}$$

This condition can be met by any matrix and its inverse with entries whose absolute values are uniformly bounded, e.g., constant, and which are uniformly sparse, i.e., the number of entries in each row and column is independent of j. However, the inverses of sparse matrices are usually densely populated which has made actual construction burdensome in the past. It also draws special attention to the above mentioned choice of the basis Ψ_j, which determines $\mathbf{M}_{j,1}$ through the refinement relation (2.1.17).

Definition 2.7 [Stable Completion]
Any matrix $\mathbf{M}_{j,1}$ which completes $\mathbf{M}_{j,0}$ to a square $(\#\Delta_{j+1}) \times (\#\Delta_{j+1})$ matrix, such that $\mathbf{M}_j$ is invertible and (2.1.22) is satisfied, is called **stable completion**.

In other words, the search for a basis Ψ_j of space W_j, consisting of compactly supported wavelets, can be exchanged for the algebraic search of refinement matrices, which are uniformly sparse with uniformly sparse inverses, too. There is a special type of sparse matrices $\mathbf{M}_j$, whose inverses are automatically sparse, namely, orthogonal matrices.

Definition 2.8 [Orthogonal Wavelets]
The wavelets are called **orthogonal** *if*

$$\langle \Psi_j, \Psi_j \rangle = \mathbf{I}, \tag{2.1.23}$$

which is true, if and only if the special situation occurs that $\mathbf{M}_j$ is **orthogonal**, *that is,*

$$\mathbf{G}_j = \mathbf{M}_j^{-1} = \mathbf{M}_j^T. \tag{2.1.24}$$

Remark 2.9 *Orthogonality will later be extended by the principle of biorthogonality in Section 2.1.4 .*

Considering $\mathbf{G}_j$ again, it is clear that applying $\mathbf{G}_j^T$ on both sides of (2.1.20) results in the so called **reconstruction identity**,

$$\Phi_{j+1} = \mathbf{G}_j^T \begin{pmatrix} \Phi_j \\ \Psi_j \end{pmatrix} = \mathbf{G}_{j,0}^T \Phi_j + \mathbf{G}_{j,1}^T \Psi_j, \tag{2.1.25}$$

which enables us now to freely **change representations** of functions between the **single-scale basis** Φ_{j+1} and the **multiscale basis** $\{\Phi_j \cup \Psi_j\}$.

Remark 2.10 *In case $\Omega \subsetneqq \mathbb{R}$, definitions (2.1.9) and (2.1.14) can be applied only for a limited range of the shifting parameter k. At the boundary $\partial\Omega$ of Ω it might not be applicable at all. Constructions of boundary adapted generators (for various boundary conditions) exist and the assertions of the previous section still hold true in these cases.*

2.1.3 Multiscale Transformation

Repeating (2.1.12), starting with a fixed finest level of resolution J up to the coarsest level j_0, yields a multiscale decomposition for the single-scale space $S_J := S(\Phi_J)$,

$$S(\Phi_J) := S(\Phi_{j_0}) \oplus \bigoplus_{j=j_0}^{J-1} S(\Psi_j). \tag{2.1.26}$$

Thus, every $v \in S(\Phi_J)$ with its **single-scale representation**

$$v = \mathbf{c}_J^T \Phi_J = \sum_{k \in \Delta_J} c_{J,k} \phi_{J,k} \tag{2.1.27}$$

can be written in **multiscale form**

$$v = \mathbf{d}_J^T \Psi_{(J)} := \mathbf{c}_{j_0}^T \Phi_{j_0} + \mathbf{d}_{j_0}^T \Psi_{j_0} + \ldots + \mathbf{d}_{J-1}^T \Psi_{J-1} \tag{2.1.28}$$

with respect to the wavelet basis

$$\Psi_{(J)} := \Phi_{j_0} \cup \bigcup_{j=j_0}^{J-1} \Psi_j = \bigcup_{j=j_0-1}^{J-1} \Psi_j, \quad \Psi_{j_0-1} := \Phi_{j_0}. \tag{2.1.29}$$

We will use the abbreviation

$$\mathbf{d}^T \equiv \mathbf{d}_J^T := (\mathbf{c}_{j_0}^T, \mathbf{d}_{j_0}^T, \ldots, \mathbf{d}_{J-1}^T) \tag{2.1.30}$$

for the multiscale vector and

$$\mathbf{c}^T \equiv \mathbf{c}_{(J)}^T := \mathbf{c}_J^T \tag{2.1.31}$$

for the single-scale coefficients, omitting the index J, if it does not create confusion. The transformation responsible for computing the single-scale coefficients from the multiscale wavelet coefficients is commonly referred to as the **wavelet transform** or **reconstruction algorithm**

$$\mathbf{T}_J : \ell_2(\Delta_J) \longrightarrow \ell_2(\Delta_J), \quad \mathbf{d}_J \mapsto \mathbf{c}_{(J)}, \tag{2.1.32}$$

which, in recognition of the decomposition identity (2.1.20), will involve the application of $\mathbf{M}_j$. In fact, (2.1.19) states

$$\mathbf{c}_j^T \Phi_j + \mathbf{d}_j^T \Psi_j = (\mathbf{M}_{j,0}\mathbf{c}_j + \mathbf{M}_{j,1}\mathbf{d}_j)^T \Phi_{j+1} =: (\mathbf{c}_{j+1})^T \Phi_{j+1},$$

which, if iterated starting from level j_0 to level J, can be visualized as a **pyramid scheme**:

$$\begin{array}{ccccccccc} & \mathbf{M}_{j_0,0} & & \mathbf{M}_{j_0+1,0} & & & & \mathbf{M}_{J-1,0} & \\ \mathbf{c}_{j_0} & \longrightarrow & \mathbf{c}_{j_0+1} & \longrightarrow & \mathbf{c}_{j_0+2} & \longrightarrow \cdots & \mathbf{c}_{J-1} & \longrightarrow & \mathbf{c}_J \\ & \mathbf{M}_{j_0,1} & & \mathbf{M}_{j_0+1,1} & & & & \mathbf{M}_{J-1,1} & \\ & \nearrow & & \nearrow & & \nearrow & & \nearrow & \\ \mathbf{d}_{j_0} & & \mathbf{d}_{j_0+1} & & \mathbf{d}_{j_0+2} & \cdots & \mathbf{d}_{J-1} & & \end{array} \tag{2.1.33}$$

By this scheme, the operator $\mathbf{T}_J$ can be written as a product of level-wise operators

$$\mathbf{T}_J = \mathbf{T}_{J,J-1} \cdots \mathbf{T}_{J,j_0}, \tag{2.1.34}$$

where each factor has the form

$$\mathbf{T}_{J,j} := \begin{pmatrix} \mathbf{M}_j & \mathbf{0} \\ \mathbf{0} & \mathbf{I}_{(\#\Delta_J - \#\Delta_{j+1})} \end{pmatrix} \in \mathbb{R}^{(\#\Delta_J)\times(\#\Delta_J)}. \tag{2.1.35}$$

Conversely, the **inverse wavelet transform**, also known as **decomposition algorithm**,

$$\mathbf{T}_J^{-1} : \ell_2(\Delta_J) \longrightarrow \ell_2(\Delta_J), \quad \mathbf{c}_{(J)} \mapsto \mathbf{d}_J, \tag{2.1.36}$$

can be written in a similar product structure by applying the inverses of the matrices $\mathbf{T}_{J,j}$ in reverse order. The inverses of $\mathbf{T}_{J,j}$ can be constructed as

$$\mathbf{T}_{J,j}^{-1} := \begin{pmatrix} \mathbf{G}_j & \mathbf{0} \\ \mathbf{0} & \mathbf{I}_{(\#\Delta_J - \#\Delta_{j+1})} \end{pmatrix} \in \mathbb{R}^{(\#\Delta_J)\times(\#\Delta_J)}, \tag{2.1.37}$$

and the inverse wavelet transform now takes on the form

$$\mathbf{T}_J^{-1} = \mathbf{T}_{J,j_0}^{-1} \cdots \mathbf{T}_{J,J-1}^{-1}. \tag{2.1.38}$$

The corresponding pyramid scheme is

$$\begin{array}{ccccccccc} & \mathbf{G}_{J-1,0} & & \mathbf{G}_{J-2,0} & & & & \mathbf{G}_{j_0,0} & \\ \mathbf{c}_J & \longrightarrow & \mathbf{c}_{J-1} & \longrightarrow & \mathbf{c}_{J-2} & \longrightarrow \cdots & \mathbf{c}_{j_0+1} & \longrightarrow & \mathbf{c}_{j_0} \\ & \mathbf{G}_{J-1,1} & & \mathbf{G}_{J-2,1} & & & & \mathbf{G}_{j_0,1} & \\ & \searrow & & \searrow & & \searrow & & \searrow & \\ & & \mathbf{d}_{J-1} & & \mathbf{d}_{J-2} & \cdots & & & \mathbf{d}_{j_0} \end{array} \tag{2.1.39}$$

Remark 2.11 *Since $\mathbf{M}_j$ and $\mathbf{G}_j$ have only a uniformly bounded number of non-zero entries in each row and column, each can be applied with a number of arithmetic operations that is of order $\mathcal{O}(\#\Delta_{j+1})$. This obviously also holds for the operators $\mathbf{T}_{J,j}$,$\mathbf{T}_{J,j}^{-1}$. Therefore, the application of operators $\mathbf{T}_J$,$\mathbf{T}_J^{-1}$ will always be computed by successively applying each operator $\mathbf{T}_{J,j}$,$\mathbf{T}_{J,j}^{-1}$. We strongly emphasize that the matrices given by $\mathbf{T}_J$,$\mathbf{T}_J^{-1}$* **are never explicitly computed and stored in computer memory**. *Such an action results in a complexity of $\mathcal{O}(\#\Delta_J \log(\#\Delta_J))$ and thus unnecessary computational overhead.*

Let

$$N_j := \#\Delta_j \tag{2.1.40}$$

be the length of the coefficient vector $\mathbf{c}_j$ on level j. This leads to the following

Proposition 2.12 *The cost of applying $\mathbf{T}_J$ or $\mathbf{T}_J^{-1}$ using the pyramid scheme is of optimal linear complexity, that is, of order $\mathcal{O}(N_J) = \mathcal{O}(\dim S(\Phi_J))$. This justifies the expression* **Fast Wavelet Transform**.

For a proof, see [45].

Remark 2.13 *In contrast, the discrete Fast Fourier Transform needs an overall amount of $\mathcal{O}(N_J \log(N_J))$ arithmetic operations, see [70].*

The fast wavelet transform will play a major part in the representation and fast assembly of operators between Hilbert spaces with wavelet bases, see Section 2.2.4. It will be essential for preconditioning of the systems of linear equations in Section 5.
By $(\mathcal{R})$(2.1.6) and (2.1.26), a basis for the whole space $\mathcal{H}$ can be given by letting $J \to \infty$ in (2.1.29),

$$\Psi := \bigcup_{j=j_0-1}^{\infty} \Psi_j = \{\psi_{j,k} \,|\, (j,k) \in \mathcal{I}\}, \quad \mathcal{I} := \bigcup_{j=j_0-1}^{\infty} \{\{j\} \times \nabla_j\}, \tag{2.1.41}$$

recalling $\Psi_{j_0-1} := \Phi_{j_0}$ and $\nabla_{j_0-1} := \Delta_{j_0}$. For any element $\lambda := (j,k) \in \mathcal{I}$ we define $|\lambda| := j$. The interrelation between $\mathbf{T}_J$ and Ψ is displayed in the next theorem, taken from [45].

Theorem 2.14 *The multiscale Transformations $\mathbf{T}_J$,$\mathbf{T}_J^{-1}$ are well-conditioned,*

$$\|\mathbf{T}_J\|, \|\mathbf{T}_J^{-1}\| = \mathcal{O}(1), \quad J \geq j_0, \tag{2.1.42}$$

if and only if the collection Ψ defined by (2.1.41) is a Riesz-Basis for $\mathcal{H}$, i.e.,

$$\|\mathbf{d}\|_{\ell_2(\mathcal{I})} \sim \|\mathbf{d}^T \Psi\|_{\mathcal{H}}, \quad \text{for all } \mathbf{d} \in \ell_2(\mathcal{I}). \tag{2.1.43}$$

This can be concluded from (2.1.22), see [44].

2.1.4 Dual Multiresolution Analysis of $\mathcal{H}'$

Let Φ_j as before be a Riesz-Basis of a Hilbert space $\mathcal{H}$ decomposed into an MRA as in $(\mathcal{R})$(2.1.6). By the Riesz representation theorem (see [4]), there exists a **dual basis** $\widetilde{\Phi}_j \subset \mathcal{H}'$ in the dual Hilbert space of $\mathcal{H}$. Of course, this basis $\widetilde{\Phi}_j$ is of the same cardinality as Φ_j and is also a **Riesz basis** of $\mathcal{H}'$. Moreover, it is part of a second **multiresolution analysis** $\widetilde{\mathcal{S}}$ of $\mathcal{H}'$, and it holds

$$\left\langle \Phi_j, \widetilde{\Phi}_j \right\rangle = \mathbf{I}. \tag{2.1.44}$$

We define the spaces

$$\widetilde{S}_j := S(\widetilde{\Phi}_j), \quad \widetilde{\Phi}_j := \left\{ \widetilde{\phi}_{j,k} \,|\, k \in \Delta_j \right\}, \tag{2.1.45}$$

where $\widetilde{\Phi}_j$ are designated **dual generator bases**, or just **dual generators**. In this setting, we refer to Φ_j of (2.1.7) as **primal generator/scaling-function bases** or **primal generators/scaling functions**. Furthermore, $P_j : \mathcal{H} \to S_j$ and $\widetilde{P}_j : \mathcal{H}' \to \widetilde{S}_j$ are projectors onto the spaces S_j and $\widetilde{S}_j$ defined by

$$P_j v := \left\langle v, \widetilde{\Phi}_j \right\rangle \Phi_j, \quad v \in \mathcal{H}, \tag{2.1.46}$$

$$\widetilde{P}_j v := \langle v, \Phi_j \rangle \widetilde{\Phi}_j, \quad v \in \mathcal{H}'. \tag{2.1.47}$$

These operators have the **projector property**

$$P_r P_j = P_r, \quad \widetilde{P}_r \widetilde{P}_j = \widetilde{P}_r, \quad r \leq j, \tag{2.1.48}$$

which entails that $P_{j+1} - P_j$ and $\widetilde{P}_{j+1} - \widetilde{P}_j$ are also projectors. We can now define the primal and dual **detail spaces** employing these projectors as

$$\begin{aligned} \mathrm{W}_j &:= \mathrm{Im}(P_{j+1} - P_j), \\ \widetilde{\mathrm{W}}_j &:= \mathrm{Im}(\widetilde{P}_{j+1} - \widetilde{P}_j), \qquad j \geq j_0. \end{aligned} \tag{2.1.49}$$

Setting $\widetilde{P}_{j_0-1} = P_{j_0-1} := 0$, we can write

$$S_{j_0} = \mathrm{W}_{j_0-1} = \mathrm{Im}(P_{j_0} - P_{j_0-1}), \quad \widetilde{S}_{j_0} = \widetilde{\mathrm{W}}_{j_0-1} = \mathrm{Im}(\widetilde{P}_{j_0} - \widetilde{P}_{j_0-1}).$$

The detail spaces W_j can also be expressed by

$$\mathrm{W}_j = S(\Psi_j) = S_{j+1} \cap (\widetilde{S}_j)^{\perp}$$

and concordantly the dual detail spaces as

$$\widetilde{\mathrm{W}}_j = S(\widetilde{\Psi}_j) = \widetilde{S}_{j+1} \cap (S_j)^{\perp}.$$

Nestedness and stability again imply that $\widetilde{\Phi}_j$ is refinable with some matrix $\widetilde{\mathbf{M}}_{j,0}$ similar to (2.1.11),

$$\widetilde{\Phi}_j = \widetilde{\mathbf{M}}_{j,0}^T \widetilde{\Phi}_{j+1}. \tag{2.1.50}$$

The main task is now to not only construct wavelet bases $\left\{\widetilde{\Psi}_j\right\}_{j\geq j_0}$ such that

$$\widetilde{S}_J = S(\widetilde{\Phi}_{j_0}) \cup \bigcup_{j=j_0}^{J} S(\widetilde{\Psi}_j)$$

is an MRA in $\mathcal{H}'$ analogously to (2.1.7), but also to ensure that the following biorthogonality conditions

$$\begin{array}{lll} S(\Phi_j) \perp S(\widetilde{\Psi}_j), & S(\widetilde{\Phi}_j) \perp S(\Psi_j) & j \geq j_0, \\ S(\Psi_j) \perp S(\widetilde{\Psi}_r), & & j \neq r, \end{array} \tag{2.1.51}$$

are satisfied. The connection between the concept of stable completions, the dual generators and wavelets is made by the following theorem taken from [27], see, e.g. [99].

Theorem 2.15 *Suppose that the biorthogonal collections* $\{\Phi_j\}_{j=j_0}^{\infty}, \left\{\widetilde{\Phi}_j\right\}_{j=j_0}^{\infty}$ *are both uniformly stable and refinable with refinement matrices* $\mathbf{M}_{j,0}, \widetilde{\mathbf{M}}_{j,0}$*, e.g.,*

$$\Phi_j = \mathbf{M}_{j,0}^T \Phi_{j+1}, \quad \widetilde{\Phi}_j = \widetilde{\mathbf{M}}_{j,0}^T \widetilde{\Phi}_{j+1},$$

and that they satisfy the duality condition (2.1.44). Assume that $\check{\mathbf{M}}_{j,1}$ *is any stable completion of* $\mathbf{M}_{j,0}$ *such that*

$$\check{\mathbf{M}}_j := (\mathbf{M}_{j,0}, \check{\mathbf{M}}_{j,1}) = \check{\mathbf{G}}_j^{-1}$$

satisfies (2.1.22). Then

$$\mathbf{M}_{j,1} := (\mathbf{I} - \mathbf{M}_{j,0}\widetilde{\mathbf{M}}_{j,0}^T)\check{\mathbf{M}}_{j,1}$$

is also a stable completion of $\mathbf{M}_{j,0}$*, and* $\mathbf{G}_j = \mathbf{M}_j^{-1} = (\mathbf{M}_{j,0}, \mathbf{M}_{j,1})^{-1}$ *has the form*

$$\mathbf{G}_j = \begin{pmatrix} \widetilde{\mathbf{M}}_{j,0}^T \\ \check{\mathbf{G}}_{j,1} \end{pmatrix}.$$

Moreover, the family of functions

$$\Psi_j := \mathbf{M}_j^T \Phi_{j+1}, \quad \widetilde{\Psi}_j := \check{\mathbf{G}}_{j,1} \widetilde{\Phi}_{j+1}$$

form biorthogonal systems

$$\left\langle \Psi_j, \widetilde{\Psi}_j \right\rangle = \mathbf{I}, \quad \left\langle \Psi_j, \widetilde{\Phi}_j \right\rangle = \left\langle \Phi_j, \widetilde{\Psi}_j \right\rangle = \mathbf{0}, \tag{2.1.52}$$

such that

$$S(\Psi_j) \perp S(\widetilde{\Psi}_r), j \neq r, \quad S(\Phi_j) \perp S(\widetilde{\Psi}_j), \quad S(\widetilde{\Phi}_j) \perp S(\Psi_j).$$

Especially (2.1.44) combined with (2.1.52) implies that the wavelet bases

$$\Psi = \bigcup_{j=j_0-1}^{\infty} \Psi_j, \quad \widetilde{\Psi} := \bigcup_{j=j_0-1}^{\infty} \widetilde{\Psi}_j := \widetilde{\Phi}_{j_0} \cup \bigcup_{j=j_0}^{\infty} \widetilde{\Psi}_j, \tag{2.1.53}$$

are biorthogonal,

$$\left\langle \Psi, \widetilde{\Psi} \right\rangle = \mathbf{I}. \qquad (\mathcal{B})(2.1.54)$$

Definition 2.16 [Biorthogonal Wavelets]
Two such Riesz-Bases, i.e, Ψ *of a MRA* $\mathcal{S} \subset \mathcal{H}$ *and* $\widetilde{\Psi}$ *of MRA* $\widetilde{\mathcal{S}} \subset \mathcal{H}'$*, with property* $(\mathcal{B})$*(2.1.54), are called* **biorthogonal wavelets**. Ψ *are called* **primal** *wavelets and* $\widetilde{\Psi}$ **dual** *wavelets.*

Given biorthogonal Ψ, $\widetilde{\Psi}$, every $v \in \mathcal{H}$ has a unique expansion

$$v = \sum_{j=j_0-1}^{\infty} \left\langle v, \widetilde{\Psi}_j \right\rangle \Psi_j =: \sum_{j=j_0-1}^{\infty} \mathbf{v}_j^T \Psi_j =: \mathbf{v}^T \Psi \tag{2.1.55}$$

and every $w \in \mathcal{H}'$ has a corresponding unique expansion

$$w = \sum_{j=j_0-1}^{\infty} \langle w, \Psi_j \rangle \widetilde{\Psi}_j =: \sum_{j=j_0-1}^{\infty} \widetilde{\mathbf{w}}_j^T \widetilde{\Psi}_j =: \widetilde{\mathbf{w}}^T \widetilde{\Psi}, \tag{2.1.56}$$

where (2.1.55) is called **primal expansion** and (2.1.56) **dual expansion**. In case $\mathcal{H} = \mathcal{H}'$ it follows that every $v \in \mathcal{H}$ has two unique expansions

$$v = \sum_{j=j_0-1}^{\infty} \langle v, \Psi_j \rangle \widetilde{\Psi}_j = \sum_{j=j_0-1}^{\infty} \left\langle v, \widetilde{\Psi}_j \right\rangle \Psi_j. \tag{2.1.57}$$

In this case, also the following norm equivalences hold for every $v \in \mathcal{H}$:

$$\|v\|_{\mathcal{H}} \sim \| \left\langle v, \widetilde{\Psi} \right\rangle^T \|_{\ell_2(\mathcal{I})} \sim \| \langle v, \Psi \rangle^T \|_{\ell_2(\mathcal{I})}, \tag{2.1.58}$$

as these expansions satisfy

$$\|v\|_{\mathcal{H}} \sim \| \left\langle v, \widetilde{\Psi} \right\rangle^T \|_{\ell_2(\mathcal{I})}, \quad \|v\|_{\mathcal{H}'} \sim \| \left\langle v, \widetilde{\Psi} \right\rangle^T \|_{\ell_2(\mathcal{I})}. \tag{2.1.59}$$

Two useful expansions can be derived in this case of $\mathcal{H} = \mathcal{H}'$ by applying (2.1.55) for every basis function of $\widetilde{\Psi}$ and (2.1.56) for Ψ,

$$\Psi = \langle \Psi, \Psi \rangle \widetilde{\Psi} = (\Psi, \Psi) \widetilde{\Psi} =: \mathbf{M}_{\mathcal{H}} \widetilde{\Psi}, \tag{2.1.60}$$

$$\widetilde{\Psi} = \left\langle \widetilde{\Psi}, \widetilde{\Psi} \right\rangle \Psi = \left(\widetilde{\Psi}, \widetilde{\Psi} \right) \widetilde{\Psi} =: \widetilde{\mathbf{M}}_{\mathcal{H}'} \Psi, \tag{2.1.61}$$

which, when combined, bring forward an identity for the **Gramian matrices** $\mathbf{M}_{\mathcal{H}}, \widetilde{\mathbf{M}}_{\mathcal{H}'}$,

$$\mathbf{M}_{\mathcal{H}}\widetilde{\mathbf{M}}_{\mathcal{H}'} = (\Psi, \Psi)\left(\widetilde{\Psi}, \widetilde{\Psi}\right) = \mathbf{I}. \tag{2.1.62}$$

Similarly to (2.1.32), the **dual wavelet transformation** is designated as

$$\widetilde{\mathbf{T}}_J : \widetilde{\mathbf{d}}_J \longmapsto \widetilde{\mathbf{c}}_{(J)} \tag{2.1.63}$$

and from the biorthogonality equations (2.1.44) and $(\mathcal{B})$(2.1.54), we can deduct

$$\mathbf{I} = \left\langle \widetilde{\Psi}_{(J)}, \Psi_{(J)} \right\rangle = \left\langle \widetilde{\mathbf{T}}_J^T \widetilde{\Phi}_J, \mathbf{T}_J^T \Phi_J \right\rangle = \widetilde{\mathbf{T}}_J^T \left\langle \widetilde{\Phi}_J, \Phi_J \right\rangle \mathbf{T}_J = \widetilde{\mathbf{T}}_J^T \mathbf{T}_J$$

or $\widetilde{\mathbf{T}}_J = \mathbf{T}_J^{-T}$ and, consequently, $\widetilde{\mathbf{T}}_J^{-1} = \mathbf{T}_J^T$, see Figure 2.1. It follows that $\widetilde{\mathbf{T}}_J$ has the same properties as $\mathbf{T}_J$, e.g., uniform sparseness and uniformly bounded condition numbers.
In Section 2.2.4 we will show how to apply $\mathbf{T}_J$ for preconditioning of linear elliptic operators. To this end, the assembly of operators is first done in terms of the single-scale functions Φ_j, and the fast wavelet transform is used to attain the wavelet representation. Hence, assembly is simple and computation fast.
It should be pointed out that the dual basis functions $\widetilde{\Phi}_j$ and wavelets $\widetilde{\Psi}_j$ are not needed explicitly in this thesis. All which must be known is the dual wavelet transform, which is given by Figure 2.1.

	primal	dual
reconstruction	$\mathbf{T}_J : \mathbf{d} \longmapsto \mathbf{c}$ $\mathbf{T}_J = \widetilde{\mathbf{T}}_J^{-T}$	$\widetilde{\mathbf{T}}_J : \widetilde{\mathbf{d}} \longmapsto \widetilde{\mathbf{c}}$ $\widetilde{\mathbf{T}}_J = \mathbf{T}_J^{-T}$
decomposition	$\mathbf{T}_J^{-1} : \mathbf{c} \longmapsto \mathbf{d}$ $\mathbf{T}_J^{-1} = \widetilde{\mathbf{T}}_J^T$	$\widetilde{\mathbf{T}}_J^{-1} : \widetilde{\mathbf{c}} \longmapsto \widetilde{\mathbf{d}}$ $\widetilde{\mathbf{T}}_J^{-1} = \mathbf{T}_J^T$

Figure 2.1: The primal and dual multiscale transformations.

2.2 Multiresolutions of L_2 and H^s

In view of the application of the MRA framework to PDEs, we now need to consider elements of the Hilbert space $\mathcal{H}$ to be functions $f : \Omega \to \mathbb{R}, \Omega \subset \mathbb{R}$, lying in the function space L_2 or a subspace $H^s \subset L_2$. Let $\mathcal{S}$ be a multiresolution sequence of $\mathcal{H}$ as in Section 2.1.2, possibly incorporating boundary conditions from Ω.
Taking $\mathcal{H} = L_2(\Omega)$ conforms to the case $\mathcal{H} = \mathcal{H}'$ with the dual pairing

$$\langle f, g \rangle_{L_2(\Omega)\times L_2(\Omega)} := \int_\Omega f(x)g(x)d\mu, \quad \text{for all } f, g \in L_2(\Omega).$$

In case $\mathcal{H} = H^s, s > 0$, recall that $\mathcal{H}' = (H^s)', s > 0$, is a significantly larger space than H^s and it holds

$$H^s(\Omega) \subset L_2(\Omega) \subset (H^s(\Omega))', \quad s > 0, \tag{2.2.1}$$

where the embedding is continuous and dense. This identity is an example of a **Gelfand triple**.

2.2.1 Approximation and Regularity Properties

Approximation properties refer to the ability to reproduce certain classes of functions with linear combinations of Φ_j. Of special interest are the spaces of polynomials

$$\Pi_r := \left\{ \sum a_i x^i : 0 \leq i \leq r-1 \right\}. \tag{2.2.2}$$

It will be important in the sequel that there are constants $d, \widetilde{d} \in \mathbb{N}$ such that the space Π_d is contained in $S(\Phi_{j_0})$ and accordingly $\Pi_{\widetilde{d}} \subset S(\widetilde{\Phi}_{j_0})$, and the following identities hold:

$$x^r = \sum_k \left\langle (\cdot)^r, \widetilde{\phi}_{j_0,k}(\cdot) \right\rangle \phi_{j_0,k}(x), \quad r = 0, \ldots, d-1, \qquad (\mathcal{P})(2.2.3)$$

$$x^r = \sum_k \langle (\cdot)^r, \phi_{j_0,k}(\cdot) \rangle \widetilde{\phi}_{j_0,k}(x), \quad r = 0, \ldots, \widetilde{d}-1. \qquad (\widetilde{\mathcal{P}})(2.2.4)$$

Since the spaces $S(\Phi_j), S(\widetilde{\Phi}_j)$ are nested, this also holds true for $S(\Phi_j), S(\widetilde{\Phi}_j), j \geq j_0$. By the biorthogonality conditions (2.1.52), this yields the so called **moment conditions**,

$$\int_\Omega x^r \psi_{j,k}(x) d\mu = 0, \quad r = 0, \ldots, \widetilde{d}-1, \qquad (\mathcal{V})(2.2.5)$$

$$\int_\Omega x^r \widetilde{\psi}_{j,k}(x) d\mu = 0, \quad r = 0, \ldots, d-1, \qquad (\widetilde{\mathcal{V}})(2.2.6)$$

which means that the wavelets $\psi_{j,k}, \widetilde{\psi}_{j,k}$ are orthogonal to all polynomials up to order $\widetilde{d}, d$, respectively. The wavelets $\psi, \widetilde{\psi}_{j,k}$ are said to have $\widetilde{d}$-, d-th order **vanishing moments**.
Now we turn to **regularity** properties, commonly referred to as **smoothness**. We quote from [143] that every generator $\phi \in L_2$ of an MRA $\mathcal{S}$ is also contained in H^s for a certain range $[0, s), s > 0$. We define regularity properties which will play an integral part in the norm equivalence proposition in Section 2.2.2:

Definition 2.17 [Regularity]
The **regularity** *of the MRAs $\mathcal{S}$ and $\widetilde{\mathcal{S}}$ is characterized by*

$$\gamma := \sup \{ s \mid S(\Phi_j) \subset H^s, j \geq j_0 \}, \quad \widetilde{\gamma} := \sup \left\{ s \mid S(\widetilde{\Phi}_j) \subset H^s, j \geq j_0 \right\}. \tag{2.2.7}$$

It is necessary to find the optimal balance between the three properties **regularity** $\gamma, \widetilde{\gamma}$ (2.2.7), **polynomial exactness** d $(\mathcal{P})$(2.2.3) and **vanishing moments** $\widetilde{d}$ $(\mathcal{V})$(2.2.5) of the trial spaces $S(\Phi_j)$ for any problem at hand. The choice of $d, \widetilde{d}$ is not entirely arbitrary. Existence of a compactly supported dual scaling function $\widetilde{\Phi}_j$ was proved in [43] for

$$\widetilde{d} > d, \quad d + \widetilde{d} = \text{even}. \tag{2.2.8}$$

In case of biorthogonal **spline wavelets** it is known that the support of the generators is linked to the **polynomial exactness** (see [36])

$$\operatorname{supp} \phi_{j,k} = \mathcal{O}(d), \quad \operatorname{supp} \widetilde{\phi}_{j,k} = \mathcal{O}\left(\widetilde{d}\right). \tag{2.2.9}$$

This in turn determines the length of the **mask** (2.1.10) and, thus, the constants involving the cost of applying the **fast wavelet transform** (2.1.32).
We will see in the following sections that the wavelet discretization of a differential operator $\mathcal{L} : H^{+t} \to (H^{+t})'$ of order $2t$ requires $\gamma, \widetilde{\gamma} > |t|$.

2.2.2 Norm Equivalences for Sobolev Spaces $H^s \subset L_2$

The inner product $(\cdot,\cdot)_{H^s}$ and norm $\|\cdot\|_{H^s}$ of Sobolev spaces $H^s, s \in \mathbb{R}$, cannot be expressed analytically for arbitrary values of s as in the L_2 case. Therefore, we must resort to norm equivalences which we will introduce and analyze now.
Up to now, Riesz stability is given uniformly for all spaces S_j, $j \geq j_0$, see Definition 2.2. In general, this does not immediately imply stability with respect to several levels, as is needed for infinite sums of elements of these spaces. Results of this kind combined with Theorem 2.14 ensure the Riesz-Basis properties (2.1.43) and (2.1.59) of Ψ. To this end, regularity and approximation properties are required of $\mathcal{S}$ and $\widetilde{\mathcal{S}}$, which are formalized in the following theorem from [46], see also [53] and [96].

Theorem 2.18 *Let $\mathcal{S} := \{S_j\}_{j\geq j_0}$ and $\widetilde{\mathcal{S}} := \left\{\widetilde{S}_j\right\}_{j\geq j_0}$ be multiresolution sequences with bases $\Phi_j,\widetilde{\Phi}_j$ satisfying properties Stability $(\mathcal{S})$(2.1.5), Refinability $(\mathcal{R})$(2.1.6), Locality $(\mathcal{L})$(2.1.8) and Biorthogonality $(\mathcal{B})$(2.1.54) and let $P_j,\widetilde{P}_j$ be defined by (2.1.46), (2.1.47). If $\mathcal{S},\widetilde{\mathcal{S}}$ both satisfy the* **Jackson inequality**

$$\inf_{v_j\in\overline{S}_j} \|v - v_j\|_{L_2(\Omega)} \lesssim 2^{-sj}\|v\|_{H^s(\Omega)}, \quad v \in H^s(\Omega), 0 \leq s \leq \overline{d}, \tag{2.2.10}$$

and the **Bernstein inequality**

$$\|v_j\|_{H^s(\Omega)} \lesssim 2^{sj}\|v_j\|_{L_2(\Omega)}, \quad v_j \in \overline{S}_j, 0 \leq s < \overline{\gamma}, \tag{2.2.11}$$

for the spaces $\overline{S}_j = S_j, \widetilde{S}_j$ with order $\overline{d} := d$, $\widetilde{d}$ and $\overline{\gamma} := \gamma, \widetilde{\gamma}$, respectively, then for

$$0 < \sigma := \min\{d, \gamma\}, \quad 0 < \widetilde{\sigma} := \min\left\{\widetilde{d}, \widetilde{\gamma}\right\}$$

one obtains the norm equivalences

$$\begin{aligned}\|v\|^2_{H^s(\Omega)} &\sim \sum_{j=j_0}^{\infty} \|(P_j - P_{j-1})v\|^2_{H^s(\Omega)} \\ &\sim \sum_{j=j_0}^{\infty} 2^{2sj}\|(P_j - P_{j-1})v\|^2_{L_2(\Omega)} \\ &\sim \sum_{j=j_0-1}^{\infty} 2^{2sj}\| \left\langle v, \widetilde{\Psi}_j\right\rangle^T \|^2_{\ell_2(\nabla_j)}, \quad s \in (-\widetilde{\sigma}, \sigma).\end{aligned} \tag{2.2.12}$$

Note that here we set $H^s = (H^{-s})'$ for $s < 0$.

Remark 2.19 *In particular for $s = 0$, we regain the* **Riesz basis** *property (2.1.58) for $\mathcal{H} = L_2$.*

Remark 2.20 *By interchanging the roles of $\mathcal{S}$ and $\widetilde{\mathcal{S}}$, we obtain*

$$\|v\|^2_{H^s(\Omega)} \sim \sum_{j=j_0}^{\infty} 2^{2sj}\|(\widetilde{P}_j - \widetilde{P}_{j-1})v\|^2_{L_2(\Omega)}, \quad s \in (-\sigma, \widetilde{\sigma}).$$

Remark 2.21 *Usually $\gamma < d$ and $\widetilde{\gamma} < \widetilde{d}$ holds, in which case $\sigma = \gamma$ and $\widetilde{\sigma} = \widetilde{\gamma}$ follows.*

The projectors $\overline{P} = P_j, \widetilde{P}_j$ are uniformly bounded in $H^s(\Omega)$, i.e.,

$$\|\overline{P}v\|_{H^s(\Omega)} \lesssim \|v\|_{H^s(\Omega)}, \quad \text{for all } v \in H^s(\Omega)$$

for s up to $\sigma, \widetilde{\sigma}$, respectively. Theorem 2.18 also implies that the unique wavelet expansions (2.1.55), (2.1.56) in bases $\Psi, \widetilde{\Psi}$ for every $v \in H^{+s}$, $\widetilde{v} \in (H^{+s})'$ satisfy the following **norm equivalences**

$$\|v\|_{H^{+s}} \sim \|\mathbf{D}^{+s}\mathbf{v}\|_{\ell_2(\mathcal{I})}, \quad v = \mathbf{v}^T\Psi = \left\langle v, \widetilde{\Psi} \right\rangle \Psi, \tag{2.2.13}$$

$$\|\widetilde{v}\|_{H^{-s}} \sim \|\mathbf{D}^{-s}\widetilde{\mathbf{v}}\|_{\ell_2(\mathcal{I})}, \quad \widetilde{v} = \widetilde{\mathbf{v}}^T\widetilde{\Psi} = \left\langle \widetilde{v}, \Psi \right\rangle \widetilde{\Psi}, \tag{2.2.14}$$

with the diagonal matrices $\mathbf{D}^{\pm s} = \mathbf{D}_1^{\pm s}$ defined by

$$\left(\mathbf{D}_1^{\pm s}\right)_{\lambda,\lambda'} := 2^{\pm|\lambda| s}\delta_{(\lambda,\lambda')}. \tag{2.2.15}$$

Here we set for any indexes $\lambda = (j,k)$, $\lambda' = (j',k')$ with $|\lambda| = j$,

$$\lambda = \lambda' :\Longleftrightarrow j = j' \wedge k = k'.$$

Remark 2.22 *Other diagonal matrices exist for which the norm equivalences hold. In the following sections, the diagonal matrices $\mathbf{D}^{+s}, \mathbf{D}^{-s}$ should be understood as a placeholder for* **any** *diagonal matrix ensuring (2.2.13) and (2.2.14). The choice of the entries of the diagonal matrix is not important for theoretical considerations, but their impact can be seen in numerical experiments.*

Corollary 2.23 *For any diagonal matrix $\mathbf{D}^{+s}, \mathbf{D}^{-s}$ satisfying (2.2.13),(2.2.14), the wavelet bases $\Psi^s := \mathbf{D}^{-s}\Psi$, $\widetilde{\Psi}^s := \mathbf{D}^{+s}\widetilde{\Psi}$ constitute* **Riesz bases** *for H^{+s}, $(H^{+s})'$, respectively.*

The diagonal scaling can be seen as a smoothing of the wavelet basis for positive Sobolev indices and a roughening for negative indices. In the context of this thesis, an important consequence of norm equivalences is their ability to prove that operators in properly scaled wavelet discretizations are asymptotically optimally preconditioned.

2.2.3 Riesz Stability Properties

As we will see in numerical tests, theoretical equivalence relations have substantial impact on the computed solutions in applications. The computational problem arises from the constants inherent in every norm equivalence which we up to now have gracefully ignored with our convenient short writing symbols like "$\sim$". To get a quantitative measurement of the condition of a **Riesz basis**, we must establish lower and upper bounds for the equivalence relation $(\mathcal{S})$(2.1.5). To this end, we define the **Riesz bounds** for a **Riesz basis** $\Phi \subset \mathcal{H}$,

$$c_\Phi := \sup\left\{c \,|\, c\|\mathbf{v}\|_{\ell_2} \leq \|\mathbf{v}^T\Phi\|_{\mathcal{H}}\right\}, \quad C_\Phi := \inf\left\{C \,|\, C\|\mathbf{v}\|_{\ell_2} \geq \|\mathbf{v}^T\Phi\|_{\mathcal{H}}\right\}, \tag{2.2.16}$$

with which we can rewrite $(\mathcal{S})$(2.1.5) to

$$c_\Phi \|\mathbf{v}\|_{\ell_2} \leq \|\mathbf{v}^T \Phi\|_{\mathcal{H}} \leq C_\Phi \|\mathbf{v}\|_{\ell_2}, \quad \text{for all } \mathbf{v} \in \ell_2. \tag{2.2.17}$$

It follows for a biorthogonal **dual** Riesz basis $\widetilde{\Phi}$, that the stability constants of the dual basis take the values

$$c_{\widetilde{\Phi}} = \frac{1}{C_\Phi}, \quad C_{\widetilde{\Phi}} = \frac{1}{c_\Phi}. \tag{2.2.18}$$

If $\lambda_{\min}(\Phi_j, \Phi_j)$ resp. $\lambda_{\max}(\Phi_j, \Phi_j)$ denote the smallest resp. largest eigenvalue of the **Gramian matrix** $(\Phi_j, \Phi_j) := (\Phi_j, \Phi_j)_{\mathcal{H}}$, it is known (see [91]) that

$$c_{\Phi_j} = \sqrt{\lambda_{\min}(\Phi_j, \Phi_j)} \sim 1, \quad C_{\Phi_j} = \sqrt{\lambda_{\max}(\Phi_j, \Phi_j)} \sim 1. \tag{2.2.19}$$

If the **basis condition**

$$\kappa_{\Phi_j} := \left(\frac{C_{\Phi_j}}{c_{\Phi_j}} \right)^2 \tag{2.2.20}$$

is a large constant, Φ_j is said to be **ill-conditioned**. Unfortunately, the Gramian matrix is not explicitly computable when $\mathcal{H} = H^s$ for arbitrary values of s, and so the quality of the basis cannot be judged accurately. To improve the constants involved in (2.2.17) when evaluating the norm of any element $v \in \mathcal{H}$, we take on a different approach from [24] which builds on the **Riesz Representation Theorem** using **Riesz operators**, which we will discuss in Section 2.2.6.

2.2.4 Operator Representation

We will describe how the wavelet representation of an operator as a successive application of linear operators is constructed. For later purposes, it will be convenient to derive this in a very general setting. Let Ω_A, Ω_B be two open, bounded domains in $\mathbb{R}^n$, not necessarily distinct, and let A be a linear operator

$$A : H^s(\Omega_A) \to (H^t(\Omega_B))', \quad A : v_A \mapsto w_B \tag{2.2.21}$$

from the Sobolev space $H^s(\Omega_A), s \geq 0$, into the dual of the Sobolev space $H^t(\Omega_B), t \geq 0$. Let there be biorthogonal wavelet bases in both spaces at our disposal, both satisfying the **Riesz basis** criterion with diagonal matrices,

$$\begin{aligned} \Psi_A^s &:= \mathbf{D}_A^{-s} \Psi_A \quad \text{(primal) basis of } H^s(\Omega_A), \\ \widetilde{\Psi}_A^s &:= \mathbf{D}_A^{+s} \widetilde{\Psi}_A \quad \text{(dual) basis of } (H^s(\Omega_A))', \end{aligned}$$

and accordingly for $H^t(\Omega_B)$ and $(H^t(\Omega_B))'$. Henceforth, all Sobolev norms are only taken for values which the smoothness of primal and dual bases permit by Theorem 2.18. Now we can express v_A and w_B in terms of these wavelet bases as

$$v_A = \mathbf{v}^T \Psi_A^s := \left\langle v, \widetilde{\Psi}_A^s \right\rangle \Psi_A^s, \quad w_B = \mathbf{w}^T \widetilde{\Psi}_B^t := \left\langle w, \Psi_B^t \right\rangle \widetilde{\Psi}_B^t,$$

and therefore it follows

$$\mathbf{w} = \left\langle \Psi_B^t, w \right\rangle = \left\langle \Psi_B^t, Av \right\rangle = \left\langle \Psi_B^t, A\Psi_A^s \right\rangle \mathbf{v},$$

which can be seen as a discretized infinite-dimensional operator equation

$$\mathbf{A}\mathbf{v} = \mathbf{w}, \tag{2.2.22}$$

upon setting

$$\mathbf{A} := \left\langle \Psi_B^t, A\Psi_A^s \right\rangle. \tag{2.2.23}$$

This is called the **standard representation** of the operator A in wavelet coordinates. Equation (2.2.23) can be reformulated as

$$\mathbf{A} = \left\langle \Psi_B^t, A\Psi_A^s \right\rangle = \mathbf{D}_B^{-t} \left\langle \Psi_B, A\Psi_A \right\rangle \mathbf{D}_A^{-s} = \mathbf{D}_B^{-t}\mathbf{T}_B^T \left\langle \Phi_B, A\Phi_A \right\rangle \mathbf{T}_A\mathbf{D}_A^{-s} \tag{2.2.24}$$

using the **wavelet transform** to express the multiscale bases with respect to the corresponding single-scale bases. It follows that the **adjoint operator** $A' : H^t(\Omega_B) \to (H^s(\Omega_A))'$ defined by

$$\left\langle A'w, v \right\rangle_{(H^s)' \times H^s} := \left\langle w, Av \right\rangle_{H^t \times (H^t)'}, \quad \text{for all } v \in H^s(\Omega_A), w \in H^t(\Omega_B), \tag{2.2.25}$$

then has the representation

$$\mathbf{A}' = \left\langle \Psi_A^s, A'\Psi_B^t \right\rangle = \left\langle A'\Psi_B^t, \Psi_A^s \right\rangle^T = \left\langle \Psi_B^t, A\Psi_A^s \right\rangle^T = \mathbf{A}^T. \tag{2.2.26}$$

Remark 2.24 *In case $A : H^s(\Omega_A) \to H^t(\Omega_B)$ with $s, t \geq 0$, the construction process works accordingly. The role of the primal and dual wavelet bases of $H^t(\Omega_B)$ should be arranged such that the primal side is associated with the positive Sobolev scale index and hence with a smoothing of the wavelet basis.*

2.2.5 Preconditioning

Assuming $A : H^s(\Omega_A) \to (H^t(\Omega_B))'$ is an **isomorphism**, e.g.,

$$\|A\,v\|_{(H^t)'} \sim \|v\|_{H^s}, \quad \text{for all } v \in H^s, \tag{2.2.27}$$

the following theorem can be shown:

Theorem 2.25 *The mapping $\mathbf{A} : \ell_2(\mathcal{I}) \to \ell_2(\mathcal{I})$ from (2.2.24) is an* **isomorphism** *on $\ell_2(\mathcal{I})$,*

$$\|\mathbf{A}\mathbf{v}\|_{\ell_2(\mathcal{I})} \sim \|\mathbf{v}\|_{\ell_2(\mathcal{I})} \sim \|\mathbf{A}^{-1}\mathbf{v}\|_{\ell_2(\mathcal{I})}, \quad \text{for all } \mathbf{v} \in \ell_2(\mathcal{I}), \tag{2.2.28}$$

with bounded spectral condition number

$$\kappa_2(\mathbf{A}) := \|\mathbf{A}\|_2 \|\mathbf{A}^{-1}\|_2 = \mathcal{O}(1). \tag{2.2.29}$$

It is easy to show that with the constants c_A, C_A of (2.2.27) and c_Ψ, C_Ψ of (2.2.17) holds

$$\kappa_2(\mathbf{A}) \leq \frac{C_\Psi^2 \, C_A}{c_\Psi^2 \, c_A}. \tag{2.2.30}$$

Thus, the multiplication by the diagonal matrices $\mathbf{D}_B^{-t}$, $\mathbf{D}_A^{-s}$ has the capability of undoing the effect of A in the sense of the Sobolev scale. Hence, the diagonal matrices can be seen as **preconditioning** of the discretized linear system.
We will later often skip constants (especially of (2.2.17)) within estimates, so we clarify the constants buried within (2.2.28) here:

$$c_{\mathbf{A}} := \frac{C_\Psi}{c_\Psi} C_A, \quad C_{\mathbf{A}} := \frac{c_\Psi}{C_\Psi} c_A. \tag{2.2.31}$$

These can be easily derived using (2.2.17) and (2.2.18).

Diagonal of the Stiffness Matrix

Numerical studies [108] show that the condition number of the finite discretized differential operators $\mathbf{A}$ preconditioned by the application of $\mathbf{D}_1^{-1}$ of (2.2.15) is indeed **uniformly bounded**. Its absolute value can be further reduced by computing the diagonal entries of the unscaled matrix

$$\mathbf{D}_a := \left(a(\psi_{j,k}, \psi_{j,k})\right)_{(j,k)\in\mathcal{I}}, \tag{2.2.32}$$

and using the matrix with entries

$$\mathbf{D}_a^{-s} := \left(((\mathbf{D}_a)_j)^{-s/2} \, \delta_{(j,j')} \delta_{(k,k')} \right)_{(j,k)\in\mathcal{I},(j',k')\in\mathcal{I}} \tag{2.2.33}$$

as the preconditioning operator in Theorem 2.25. This operator $\mathbf{D}_a^{-1}$ can be understood to precondition $\mathbf{A}$ in the **energy norm** $\|\cdot\|_a^2 = a(\cdot,\cdot)$ which explains its effectiveness in applications.

2.2.6 Riesz Operators for H^s

We describe this for the general case $\mathcal{H} = H^{+s}, \mathcal{H}' = (H^{+s})'$ with wavelet bases $\Psi^s, \widetilde{\Psi}^s$. We are now interested in constructing **Riesz maps** $R_\mathcal{H} : \mathcal{H} \to \mathcal{H}'$ defined by

$$\langle R_\mathcal{H} v, w\rangle := (v, w)_\mathcal{H}, \quad \text{for all } v, w \in \mathcal{H}. \tag{2.2.34}$$

The **Riesz Representation Theorem** establishes for Riesz maps $\langle R_\mathcal{H} v, v\rangle_{\mathcal{H}'\times\mathcal{H}} \sim \|R_\mathcal{H} v\|_{\mathcal{H}'} \|v\|_\mathcal{H}$, from which it follows that Riesz operators are, in general, spectrally equivalent to the identity, i.e.,

$$\|R_\mathcal{H}\|_{L(\mathcal{H};\, \mathcal{H}')} \sim 1. \tag{2.2.35}$$

Observe that for any $v = \mathbf{v}^T \Psi^s \in H^s$ one has

$$\|v\|_{H^s}^2 = \left(\mathbf{v}^T\Psi^s, \mathbf{v}^T\Psi^s\right)_{H^s} = \mathbf{v}^T \left(\Psi^s, \Psi^s\right)_{H^s} \mathbf{v} = \mathbf{v}^T \mathbf{M}_{H^s} \mathbf{v}, \tag{2.2.36}$$

where $(\Psi, \Psi)_{H^s} = \mathbf{M}_{H^s}$ is the **Gramian matrix** with respect to the H^s-inner product. Since

$$\|\mathbf{M}_{H^s}^{1/2}\mathbf{v}\|_{\ell_2}^2 = \mathbf{v}^T\mathbf{M}_{H^s}\mathbf{v} \equiv \|\mathbf{R}_{H^s}^{1/2}\mathbf{v}\|_{\ell_2}^2,$$

we conclude that the exact discretization of the Riesz map R_{H^s} would be $\mathbf{R}_{H^s} = \mathbf{M}_{H^s}$.

Remark 2.26 *(i) Recall that the exact* **Gramian matrix** $\mathbf{M}_{H^s}$ *is inaccessible for* $s \notin \mathbb{Z}$.

(ii) Note that $\mathbf{R}_{\mathcal{H}}$ *is symmetric positive definite for any space* $\mathcal{H}$ *and, thus,* $\mathbf{R}_{\mathcal{H}}^{1/2}$ *is always well defined.*

For $s \in \mathbb{Z}$, we can represent $\mathbf{R}_{\mathcal{H}}$ exactly. For example, in the cases $s \in \{0, 1\}$, it follows with the definitions of the L_2-product (1.2.2) and the H^1-product (1.2.4) that we have

$$\mathbf{R}_{L_2} := \mathbf{M}_{L_2} \quad \text{and} \quad \mathbf{R}_{H^1} := \mathbf{D}^{-1}(\mathbf{S}_{H^1} + \mathbf{M}_{L_2})\mathbf{D}^{-1}, \tag{2.2.37}$$

where $\mathbf{S}_{H^1}$ is the **Laplace matrix** and $\mathbf{M}_{L_2}$ the **mass matrix**, for further details see [23]. With these matrices the norm equivalences $\|v\|_{L_2} \sim \|\mathbf{v}\|_{\ell_2}$ and $\|w\|_{H^1} \sim \|\mathbf{w}\|_{\ell_2}$ can be obtained with constants equal to 1 and not of order $\mathcal{O}(1)$.

Riesz Operators based on Scaling

Because we will have to deal with fractional Sobolev spaces in Section 5, we use another construction based on diagonal scaling. Since the Sobolev spaces are nested as in (1.2.9), there is an inclusion operator $\iota : H^s \to (H^s)'$. In wavelet coordinates, this change of bases is merely a diagonal scaling when the wavelet base of the target space is the same as that of the initial space. This is not the case here, but the construction can nevertheless be carried out as follows.
By Theorem 2.18 can we interpret $\widehat{\mathbf{D}}^{+s} := \mathbf{D}_1^{+s}$ as a shifting operator in the Sobolev scale, i.e.,

$$\widehat{\mathbf{D}}^{+s} : H^{t+s} \to H^t, \quad (\widehat{\mathbf{D}}^{+s})_{(j,j')(k,k')} = 2^{+js}\delta_{(j,j')}\delta_{(k,k')}, \tag{2.2.38}$$

so that

$$\widehat{R}_{H^s} := \widehat{\mathbf{D}}^{+2s} : H^s \to (H^s)' \tag{2.2.39}$$

can be used to shift elements of $\mathcal{H}$ into $\mathcal{H}'$. Using the formulation of Section 2.2.4, the standard wavelet representation $\widehat{\mathbf{R}}_{H^s}$ of $\widehat{R}_{H^s}$ is given by

$$\widehat{\mathbf{R}}_{H^s} = \left\langle \Psi^s, \widehat{R}_{H^s}\Psi^s \right\rangle_{H^s \times (H^s)'} \tag{2.2.40}$$

$$= \mathbf{D}^{-s} \left\langle \Psi, \widehat{R}_{H^s}\Psi \right\rangle_{H^s \times (H^s)'} \mathbf{D}^{-s} \tag{2.2.41}$$

$$= \mathbf{D}^{-s} \left\langle \Psi, \widehat{\mathbf{D}}^{+s}\widehat{\mathbf{D}}^{+s}\Psi \right\rangle_{H^s \times (H^s)'} \mathbf{D}^{-s} \tag{2.2.42}$$

$$= \mathbf{D}^{-s}\widehat{\mathbf{D}}^{+s} (\Psi, \Psi)_{L_2} \widehat{\mathbf{D}}^{+s}\mathbf{D}^{-s}. \tag{2.2.43}$$

Remark 2.27 *Note that the Riesz operator scaling* $\widehat{\mathbf{D}}^{+s}$ *and the diagonal scaling* $\mathbf{D}^{-s}$ *could cancel each other out by the use of* $\mathbf{D}^{-s} = \mathbf{D}_1^{-s}$. *However, this does not occur if we choose* $\mathbf{D}^{-s}$ *differently, e.g., see Section 2.2.5.*

Remark 2.28 *The choice* $\widehat{\mathbf{D}}^{+s} = \mathbf{D}_1^{+s}$ *is not without reason. By changing* $\widehat{\mathbf{D}}^{+2s}$, *we change the way we weight functions in the norm equivalences. Although any matrix satisfying the norm equivalence could potentially be used as* $\widehat{\mathbf{D}}^{+2s}$, *the actual choice has an impact on the quality of numerical experiments, see [122].*

The inverse operator $\widehat{R}_{H^s}^{-1} : H^{-s} \to H^{+s}$ is trivially given by $\widehat{\mathbf{D}}^{-2s}$ and can be represented as the inverse of (2.2.43),

$$\begin{aligned} \widehat{\mathbf{R}}_{H^s}^{-1} &= \mathbf{D}^{+s}\widehat{\mathbf{D}}^{-s} \left(\Psi, \Psi\right)_{L_2}^{-1} \widehat{\mathbf{D}}^{-s}\mathbf{D}^{+s} \\ &= \mathbf{D}^{+s}\widehat{\mathbf{D}}^{-s} \left(\widetilde{\Psi}, \widetilde{\Psi}\right)_{L_2} \widehat{\mathbf{D}}^{-s}\mathbf{D}^{+s}, \end{aligned} \tag{2.2.44}$$

where we used identity (2.1.62) in the last step. The case $s = 0$ thus again becomes

$$\begin{aligned} \widehat{\mathbf{R}}_{L_2} &= \mathbf{M}_{L_2} = \left(\Psi, \Psi\right)_{L_2}, \\ \widehat{\mathbf{R}}_{L_2}^{-1} &= \widetilde{\mathbf{M}}_{L_2} = \left(\widetilde{\Psi}, \widetilde{\Psi}\right)_{L_2}. \end{aligned} \tag{2.2.45}$$

Numerical tests show, see [122], that using $\widehat{\mathbf{R}}_{H^s}$ gives better results than using no Riesz operator at all. This observation can be justified by the following deliberation: Fix $0 < c_0, C_0 < \infty$ as the **Riesz bounds** of (2.2.17) for $\mathcal{H} = L_2$. Then it follows

$$\kappa_{L_2} := \left(\frac{C_0}{c_0}\right)^2 = \frac{\lambda_{\max}(\mathbf{M}_{L_2})}{\lambda_{\min}(\mathbf{M}_{L_2})} = \kappa_2(\mathbf{M}_{L_2}) \sim 1, \tag{2.2.46}$$

and for $\mathcal{H} = H^s$ with constants $c_s, C_s < \infty$

$$\kappa_{H^s} := \left(\frac{C_s}{c_s}\right)^2 = \frac{\lambda_{\max}(\mathbf{M}_{H^s})}{\lambda_{\min}(\mathbf{M}_{H^s})} = \kappa_2(\mathbf{M}_{H^s}) \sim 1. \tag{2.2.47}$$

These can easily be combined to give error bounds for the H^s-norm with respect to the ℓ_2-norm of $\|\mathbf{M}_{L_2}^{1/2}\mathbf{v}\|_{\ell_2}$, which is not equal to $\|v\|_{L_2}$, since the coefficient vector $\mathbf{v}$ is scaled by $\mathbf{D}^{+s}$. It holds

$$\frac{c_s}{C_0}\|\mathbf{M}_{L_2}^{1/2}\mathbf{v}\|_{\ell_2} \le \|\mathbf{M}_{H^s}^{1/2}\mathbf{v}\|_{\ell_2} \le \frac{C_s}{c_0}\|\mathbf{M}_{L_2}^{1/2}\mathbf{v}\|_{\ell_2}. \tag{2.2.48}$$

These estimates are not sharp, e.g., for $s = 0$ we should have equality but only obtain equivalence up to the value of κ_{L_2}, and we witness better results in practice than can be predicted here.

Lemma 2.29 *We have for every* $v = \mathbf{v}^T\Psi^s = \left\langle v, \widetilde{\Psi}^s \right\rangle \Psi^s \in H^s$ *the following chain of equivalences:*

$$\|v\|_{H^s} = \|\mathbf{R}_{H^s}^{1/2}\mathbf{v}\|_{\ell_2} \sim \|\widehat{\mathbf{R}}_{H^s}^{1/2}\mathbf{v}\|_{\ell_2} \sim \|\mathbf{R}_{L_2}^{1/2}\mathbf{v}\|_{\ell_2} \sim \|\mathbf{v}\|_{\ell_2}, \quad s \in (-\widetilde{\sigma}, \sigma). \tag{2.2.49}$$

In other words, every one of the operators $\mathbf{R}_{H^s}, \widehat{\mathbf{R}}_{H^s}, \mathbf{R}_{L_2}$ and $\mathbf{I}$ can be used as a Riesz operator for H^s.

Riesz Operators based on Interpolation

The construction of the Riesz operator outlined here was introduced in [23]. We only quote the results and refer to that work for details.
Since the exact Riesz operator for L_2 and H^1 are known, these can be used to construct new Riesz operators by interpolating linearly between them.

Theorem 2.30 *For $s \in [0,1]$, the norm defined by*

$$v = \mathbf{v}^T \Psi \in H^s, \quad \|v\|_s^2 := (1-s)\,\mathbf{v}^T \mathbf{D}^{+s} \mathbf{R}_{L_2} \mathbf{D}^{+s} \mathbf{v} + s\,\mathbf{v}^T \mathbf{D}^{+s} \mathbf{R}_{H^1} \mathbf{D}^{+s} \mathbf{v}, \tag{2.2.50}$$

or alternatively written in the scaled wavelet basis as

$$v = \mathbf{v}^T \Psi^s \in H^s, \quad \|v\|_s^2 = \mathbf{v}^T \left((1-s)\mathbf{M}_{L_2} + s\mathbf{D}^{-1}(\mathbf{S}_{H^1} + \mathbf{M}_{L_2})\mathbf{D}^{-1} \right) \mathbf{v}, \tag{2.2.51}$$

is equal to the standard Sobolev norms for $s \in \{0,1\}$ and equivalent for $s \in (0,1)$. It can be computed in linear time.

In the following, we denote this Riesz operator as

$$\widetilde{\mathbf{R}}_{H^s} := (1-s)\,\mathbf{R}_{L_2} + s\,\mathbf{R}_{H^1}, \quad 0 \le s \le 1. \tag{2.2.52}$$

Just as the summands, the Riesz operator $\widetilde{\mathbf{R}}_{H^s}$ is spectrally equivalent to the identity matrix and thus uniformly well-conditioned.

Remark 2.31 *This construction can be extended for all $s \in \mathbb{R}$ with exact interpolation for all integer $s \in \mathbb{Z}$, see [23].*

These Riesz operators are used in control problems, see [122], to improve the constants in the above norm equivalences. This will somewhat improve the discrepancy between the original analytical problem formulation and the discretized wavelet formulation.

Normalization with Respect to Constant Functions

Now note that the wavelet expansion coefficients of constant functions are exactly zero (the vectors $\mathbf{d}_j$ in (2.1.28)), except for the single-scale expansion coefficients of the minimum level j_0 ($\mathbf{c}_{j_0}$ in (2.1.28)). Thus, the diagonal preconditioner matrix $\widehat{\mathbf{D}}^{+s}$ can in this case effectively be written as a scaling of the identity matrix on the lowest level j_0,

$$\widehat{\mathbf{D}}^{+s}(\mathbf{c}_{j_0}, \mathbf{d}_{j_0}, \ldots, \mathbf{d}_{J-1})^T = 2^{+j_0 s}\, I_{j_0}\, \mathbf{c}_{j_0} = 2^{+j_0 s} \mathbf{c}_{j_0}.$$

To counter this effect, we introduce a simple scaling factor into our Riesz operator,

$$q_s = 2^{2 j_0 s}, \tag{2.2.53}$$

and define the normalized Riesz operator

$$\mathring{\widehat{\mathbf{R}}}_{H^s} := q_{-s} \widehat{\mathbf{R}}_{H^s}. \tag{2.2.54}$$

Thus, the higher the lowest level j_0 in the MRA, the higher the correction factor for the Riesz operator $\widehat{\mathbf{R}}_{H^s}$ to fulfill the norm equivalences of Lemma 2.29 for constant functions. The (for every s) constant factor q_{-s} obviously does not change the spectral elements of $\widehat{\mathbf{R}}_{H^s}$.

2.3 B-Spline Wavelets on the Interval

In the preceding sections we described general properties of wavelet spaces. There exist several constructions of wavelet for the interval $\mathrm{I} = (0, 1)$ which satisfy the list of properties detailed in Section 2.1 and Section 2.2, i.e.,

($\mathcal{S}$)(2.1.5)
The wavelets form a Riesz-Basis for $L_2(\mathrm{I})$, and norm equivalences with respect to Sobolev spaces $H^s(\mathrm{I})$ for a certain range of the smoothness parameter s hold.

($\mathcal{R}$)(2.1.6)
The wavelets are refinable with masks of uniformly bounded length.

($\mathcal{L}$)(2.1.8)
All generators and wavelets on the primal and dual side have compact support.

($\mathcal{B}$)(2.1.54)
The primal and dual wavelets form a biorthogonal pair.

($\mathcal{P}$)(2.2.3)
The primal MRA consists of spline spaces of order up to $d - 1$, and thus has polynomial exactness of order d.

($\widetilde{\mathcal{P}}$)(2.2.4)
The dual MRA has polynomial exactness of order $\widetilde{d} - 1$.

($\mathcal{V}$)(2.2.5)
As a consequence of ($\widetilde{\mathcal{P}}$), the wavelets have $\widetilde{d}$ vanishing moments.

The constants d and $\widetilde{d}$ are preassigned such that they satisfy (2.2.8) before the construction process.

2.3.1 B-Spline Wavelets

The wavelet construction begins by setting up a **Riesz basis** for $L_2(\mathbb{R})$, which is then restricted to the interval $(0, 1)$. Let $\phi^d(x)$ be the **cardinal B-spline** of order $d \in \mathbb{N}$ (see Appendix C for the definition). These B-splines are known to be symmetric and centered around $\mu(d) := (d \mod 2)/2$. We will use these B-splines as **primal generators**, since they are easy to set up, have finite support,

$$\operatorname{supp} \phi^d = [s_1, s_2], \quad s_1 := -\left\lfloor \frac{d}{2} \right\rfloor, s_2 := \left\lceil \frac{d}{2} \right\rceil, \tag{2.3.1}$$

and are known to be **refinable** with mask $\mathbf{a}^d = \{a_k^d\} \in \ell_2$, with

$$a_k^d := 2^{1-d} \binom{d}{k + \lfloor \frac{d}{2} \rfloor}, \quad k = s_1, \ldots, s_2. \tag{2.3.2}$$

These scaling functions also offer the advantage of being scaled correctly in the sense of the Riesz basis property in Definition 2.1. The question whether a refinable **dual basis** exists for any $d \in \mathbb{N}$ has been proved in [36] and we cite this result from [51].

Theorem 2.32 *For each ϕ^d and any $\widetilde{d} \geq d$ with $d + \widetilde{d}$ even, there exists a function $\phi^{d,\widetilde{d}} \in L_2(\mathbb{R})$ such that*

(i) $\phi^{d,\widetilde{d}}$ has compact support, e.g.

$$\operatorname{supp} \phi^{d,\widetilde{d}} = [\widetilde{s}_1, \widetilde{s}_2], \quad \widetilde{s}_1 := s_1 - \widetilde{d} + 1, \widetilde{s}_2 := s_2 + \widetilde{d} - 1$$

(ii) $\phi^{d,\widetilde{d}}$ is also centered around $\mu(d)$.

(iii) $\phi^{d,\widetilde{d}}$ is refinable with finitely supported mask $\widetilde{\mathbf{a}}^d = \{\widetilde{a}_k^d\}_{k=\widetilde{s}_1}^{\widetilde{s}_2}$.

(iv) $\phi^{d,\widetilde{d}}$ is exact of order $\widetilde{d}$.

(v) ϕ^d and $\phi^{d,\widetilde{d}}$ form a **dual pair**, *i.e.,*

$$\left(\phi^d, \phi^{d,\widetilde{d}}(\cdot - k)\right)_{L_2(\mathbb{R})} = \delta_{(0,k)}, \quad k \in \mathbb{Z}.$$

(vi) The regularity of $\phi^{d,\widetilde{d}}$ increases proportionally with $\widetilde{d}$.

Thus, for d, $\widetilde{d}$ fixed, we write $\phi := \phi^d$, $\widetilde{\phi} := \phi^{d,\widetilde{d}}$ and define the generator bases $\Phi_j = \{\phi_{j,k}\}$, $\widetilde{\Phi}_j = \left\{\widetilde{\phi}_{j,k}\right\}$ according to (2.1.9). We also deduce that $\mathcal{S} := \{S(\Phi_j), j \geq j_0\}$ is a **multiresolution analysis** of L_2.
The primal wavelets can now be constructed from the primal basis functions and the dual mask $\widetilde{\mathbf{a}}$ as, see [36],

$$\psi(x) := \sum_{k \in \mathbb{Z}} b_k \phi(2x - k), \quad b_k := (-1)^k \widetilde{a}_{1-k}^d, \tag{2.3.3}$$

and in perfect analogy the dual wavelets can be constructed with help of the mask $\mathbf{a}$ of the primal generators as

$$\widetilde{\psi}(x) := \sum_{k \in \mathbb{Z}} \widetilde{b}_k \widetilde{\phi}(2x - k), \quad \widetilde{b}_k := (-1)^k a_{1-k}^d. \tag{2.3.4}$$

We now define the bases for the complement spaces W_j, $\widetilde{\mathrm{W}}_j$ as

$$\Psi_j = \{\psi_{j,k} \,|\, k \in \mathbb{Z}\}, \quad \widetilde{\Psi}_j = \left\{\widetilde{\psi}_{j,k} \,|\, k \in \mathbb{Z}\right\},$$

with $\psi_{j,k}$, $\widetilde{\psi}_{j,k}$ defined by (2.1.14) using (2.3.3),(2.3.4) as **mother wavelets**. Of course these wavelets could yet be scaled by any factor for numerical purposes.

Corollary 2.33 *The bases $\{\Phi_{j_0}\} \cup \bigcup_{j \geq J} \Psi_j$, $\left\{\widetilde{\Phi}_{j_0}\right\} \cup \bigcup_{j \geq J} \widetilde{\Psi}_j$ are indeed* **biorthogonal wavelet bases** *in the multiresolution framework of Section 2.1.*

Adaptation to the Interval

With proper scaling functions and wavelets on all of $\mathbb{R}$, we can not simply restrict the collections $\Phi_j, \widetilde{\Phi}_j$ to the interval $(0,1)$ in the hope of constructing adapted wavelets, as this would violate **biorthogonality**. Ruling out any $\phi_{j,k}, \widetilde{\phi}_{j,k}$ whose support is not fully contained in $(0,1)$ would lead to primal and dual bases of different cardinality and thus also break biorthogonality. In addition, the approximation property (2.2.10) would not longer hold near the ends of the interval. In view of PDEs, last but not least, we also need to take the boundary conditions themselves into account.
The actual adaptation is done in three steps. First, every function whose support is not fully contained within $(0,1)$ is discarded. Then, new basis functions are inserted at the boundaries to compensate for the reduction. These shall incorporate the boundary conditions and have properties $(\mathcal{V})$ and $(\widetilde{\mathcal{V}})$, thus preserving the **vanishing moments** and **polynomial reproduction** orders d, $\widetilde{d}$ for the new basis. Lastly, the new basis functions are again biorthogonalized by a local basis transformation regarding only the functions near the boundary. Note that this construction yields

$$\#\nabla_j = 2^j, \quad N_j = \#\Delta_j = \mathcal{O}\left(2^j\right), \tag{2.3.5}$$

where the exact value of N_j depends on the boundary adaptations.
This construction process was first proposed in [36]. We will not give a detailed description of the process here, actual implementation results can be found in [51]. A comprehensive overview can also be found in [23, 122] and [126]. We will simply use constructions found in the aforementioned publications.
Details about the later used wavelets can be found in Appendix A.1.

2.3.2 Basis Transformations

Numerical studies show that condition numbers of operators obtained using wavelet discretizations are indeed uniformly bounded, if preconditioned correctly. The involved constants can nevertheless be quite high, and condition numbers of magnitude $10^2 - 10^3$ are seen encountered. We now show some approaches to improve the wavelet bases to achieve lower absolute values of condition numbers and thus faster program executions in applications.

Reducing Boundary Effects

In general, a boundary adapted single scale basis (shown in Section A.1.1), exhibits higher stability constants than the basis without the boundary adapted generators. This means that the absolute values of the condition numbers of differential operators in wavelet discretization of Section A.1.5 are usually higher than those using free boundaries on the interval. A common approach to remedy the situation is the application of a basis transform, i.e.,

$$\Phi'_j := \mathbf{C}_j \Phi_j, \tag{2.3.6}$$

which acts local in the sense that $\mathbf{C}_j$ only affects the boundary blocks, i.e.,

$$\mathbf{C}_j := \begin{pmatrix} C & & \\ & \mathbf{I}_{\#\Delta_j - 2m} & \\ & & C^{\updownarrow} \end{pmatrix} \in \mathbb{R}^{(\#\Delta_j)\times(\#\Delta_j)}, \tag{2.3.7}$$

and the matrix $C \in \mathbb{R}^{m\times m}$ is independent of j. Recall $(C^{\updownarrow})_{m-i,m-j} := (C)_{i,j}$. Obviously, the modified **single-scale basis** implies new refinement matrices $\mathbf{M}'_{j,0}, \mathbf{M}'_{j,1}$:

$$\mathbf{M}'_{j,0} = \mathbf{C}_{j+1}^{-T}\mathbf{M}_{j,0}\mathbf{C}_j^T, \tag{2.3.8}$$

$$\mathbf{M}'_{j,1} = \mathbf{C}_{j+1}^{-T}\mathbf{M}_{j,1}. \tag{2.3.9}$$

The altered multiscale transformation $\mathbf{T}'_J$ can easily be shown to be of the form (remember, $\Psi_{j_0-1} = \Phi_{j_0}$ and $\nabla_{j_0-1} = \Delta_{j_0}$)

$$\mathbf{T}'_J = \mathbf{C}_J^{-T}\mathbf{T}_J \begin{pmatrix} \mathbf{C}_{j_0}^T & \\ & \mathbf{I}_{\#(\Delta_J\setminus\nabla_{j_0-1})} \end{pmatrix}. \tag{2.3.10}$$

Let $\mathcal{L} : \mathcal{H} \to \mathcal{H}'$ be any operator with bases Φ_J and $\Psi_{(J)}$ in $\mathcal{H}$, e.g., $\mathcal{L}$ is the differential operator from Section 1.4.2. The discretized operator $\mathbf{L}$ in standard form (2.2.23) with respect to the wavelet basis $\Psi'_{(J)} = {\mathbf{T}'_J}^T\Phi'_J$ has the representation

$$\mathbf{L}_{\Psi'_{(J)}} = \begin{pmatrix} \mathbf{C}_{j_0-1} & \\ & \mathbf{I}_{\#(\Delta_J\setminus\nabla_{j_0-1})} \end{pmatrix} \mathbf{L}_{\Psi_{(J)}} \begin{pmatrix} \mathbf{C}_{j_0-1}^T & \\ & \mathbf{I}_{\#(\Delta_J\setminus\nabla_{j_0-1})} \end{pmatrix}, \tag{2.3.11}$$

thus, the new operator is obtained from the standard operator matrix by the application of the transformation $\mathbf{C}_{j_0-1}$ on the coarsest level.
A suitable choice for the setup of $\mathbf{C}_{j_0-1}$ can be constructed as follows: We take an upper block

$$\left(\mathbf{L}_{\Psi_{(j_0-1)}}\right)_{i,j=1,\ldots,m} \in \mathbb{R}^{m\times m},$$

with $m \leq \lfloor \#\nabla_{j_0-1}/2 \rfloor$, thereby not changing all of the basis functions of the generator basis. We compute the singular value decomposition of this block, i.e.,

$$\left(\mathbf{L}_{\Psi_{(j_0-1)}}\right)_{i,j=1,\ldots,m} = \mathbf{U}\mathbf{S}\mathbf{U}^T := \mathbf{U} \begin{pmatrix} s_1 & & & \\ & s_2 & & \\ & & \ddots & \\ & & & s_m \end{pmatrix} \mathbf{U}^T, \tag{2.3.12}$$

with an orthogonal matrix $\mathbf{U} \in O(m)$, and we set for $q > 0$,

$$C := \sqrt{q} \begin{pmatrix} 1/\sqrt{s_1} & & & \\ & 1/\sqrt{s_2} & & \\ & & \ddots & \\ & & & 1/\sqrt{s_m} \end{pmatrix} \mathbf{U}^T. \tag{2.3.13}$$

Using this matrix as outlined above yields

$$\mathbf{L}_{\Psi'_{(j_0-1)}} = \mathbf{C}_{j_0-1}\mathbf{L}_{\Psi_{(j_0-1)}}\mathbf{C}_{j_0-1}^T = \left(\begin{array}{c|c|c} q\,\mathbf{I}_c & * & \\ \hline * & \mathbf{L}_\square & * \\ \hline & * & q\,\mathbf{I}_c \end{array}\right). \tag{2.3.14}$$

The middle square block $\mathbf{L}_\square \in \mathbb{R}^{(\#\Delta_{j+1}-2m)\times(\#\Delta_{j+1}-2m)}$ of $\mathbf{L}_{\Psi'_{(j_0-1)}}$ consists of entries that remain unchanged with respect to the unmodified wavelet basis $\Psi_{(J)}$. The blocks marked with asterisks contain new non-zero entries. This basis transformation is particularly cheap in terms of complexity, since it is only used on the coarsest level. The impact of these blocks on the condition number can be influenced by the parameter q.

Remark 2.34 *A Cholesky decomposition can also be used instead of the singular value decomposition. In this case the resulting matrix $\mathbf{L}_{\Psi'_{(j_0-1)}}$ has fewer non-zero values, but no further decrease of the condition number is achieved.*

Operator Adaptation to Preconditioning

We now introduce a basis transformation specifically designed for lowering the absolute values of the condition number of the **stiffness matrix**

$$\mathbf{A}_J := (a(\psi_\lambda, \psi_{\lambda'}))_{\lambda,\lambda'\in\mathcal{I}_J}, \tag{2.3.15}$$

with the **bilinear form** defined in (1.4.24). The condition number of this positive definite symmetric matrix in wavelet discretization depends on the properties of the wavelets as well as on the generator basis. Of course, the condition $\kappa(\mathbf{A}_J)$ can never be smaller than $\kappa(\mathbf{A}_{j_0})$. Therefore, we seek a generator basis adapted to the operator to minimize the absolute value of its condition number.
We make use of an **orthogonal** transformation matrix $\mathrm{O} \in O(\#\Delta_{j_0})$ to create a new, albeit completely equivalent, generator basis for the coarsest level j_0:

$$\Psi'_{j_0-1} := \mathrm{O}^T\,\Phi_{j_0}, \tag{2.3.16}$$

while leaving the higher level generator bases unchanged by this transformation. The resulting MRA

$$\mathrm{O}: \Psi = \{\Psi_{j_0-1}, \Psi_{j_0}, \Psi_{j_0+1}, \ldots\} \quad\longmapsto\quad (\Psi)' := \left\{\Psi'_{j_0-1}, \Psi_{j_0}, \Psi_{j_0+1}, \ldots\right\}. \tag{2.3.17}$$

is also completely equivalent to the original MRA since the orthogonal transformation does not change the stability constants $(\mathcal{S})$(2.1.5). This change obviously requires an adaptation of the **two-scale relation** (2.1.11) for level j_0:

$$\Phi'_{j_0} = (\mathbf{M}'_{j_0,0})^T\Phi_{j_0+1} := (\mathbf{M}_{j_0,0}\mathrm{O})^T\Phi_{j_0+1}. \tag{2.3.18}$$

To still ensure biorthogonality, the dual MRA must also be adapted accordingly:

$$\widetilde{\Phi}'_{j_0} = (\mathbf{G}'_{j_0,0})^T\widetilde{\Phi}_{j_0+1} := (\mathbf{G}_{j_0,0}\mathrm{R})^T\widetilde{\Phi}_{j_0+1}. \tag{2.3.19}$$

To fulfill biorthogonality, it follows $\mathrm{R} = \mathrm{O}^{-T}$ and thus for orthogonal basis transformations $\mathrm{R} \equiv \mathrm{O}$.

From now on, the complete change of bases will always be accomplished by using $\mathbf{M}'_{j_0,0}$ instead of $\mathbf{M}_{j_0,0}$ in the course of the wavelet transform (2.1.32). This is no antagonism, because operators are assembled in the generator base Φ_J and then transform into the wavelet representation by the FWT. This can be implemented by applying O subsequent to the wavelet transform $\mathbf{T}_J$:

$$\mathbf{T}'_J := \mathbf{T}_J \begin{pmatrix} \mathrm{O} & \\ & \mathbf{I}_{\#(\Delta_J \setminus \Delta_{j_0})} \end{pmatrix}. \tag{2.3.20}$$

Theorem 2.14 is still valid as the relations (2.1.42) still hold with the same constants.

Lemma 2.35 *Given the original stiffness matrix* $\mathbf{A}$*, the form of the stiffness matrix* $\mathbf{A}'_J$ *in the wavelet base* Ψ' *is, see [122],*

$$\mathbf{A}'_J = \left(\begin{array}{c|c} \mathrm{O}^T \mathbf{A}_{j_0} \mathrm{O} & (\mathbf{a}\mathrm{O})^T \\ \hline \mathbf{a}\mathrm{O} & \mathbf{A}_{J \setminus j_0} \end{array} \right) \tag{2.3.21}$$

The block $\mathbf{A}_{J\setminus j_0} \in \mathbb{R}^{\#(\Delta_J \setminus \Delta_{j_0}) \times \#(\Delta_J \setminus \Delta_{j_0})}$ *remains unaffected.*

The basis transform induced by an orthogonal matrix does not change the eigenvalues the operator. The trick is to choose the orthogonal matrix $\mathrm{O} \in O(\#\mathcal{I}_{j_0})$ such that $D_{\mathrm{O},j_0} = \mathrm{O}^T \mathbf{A}_{j_0} \mathrm{O}$ is a **diagonal matrix**. This is possible because $\mathbf{A}_{j_0}$ is symmetric positive definite. Fixing this matrix O leads to a diagonal upper block, i.e.,

$$\mathbf{A}'_J = \left(\begin{array}{c|c} \diagdown & (\mathbf{a}\mathrm{O})^T \\ \hline \mathbf{a}\mathrm{O} & \mathbf{A}_{J \setminus j_0} \end{array} \right). \tag{2.3.22}$$

The setup allows for an improved optimal preconditioner for this operator, since any diagonal matrix can easily be preconditioned by its own inverse. The matrix O does not change the spectral elements of $\mathbf{A}'_J$ corresponding to the resolution levels $j > j_0$, hence it has no negative impact. Using any other preconditioner, for example $\mathbf{D}_1^{\pm s}$ of (2.2.15), will usually not result in better preconditioning of the operator $\mathbf{A}'_J$. We define the following diagonal matrix for preconditioning

$$\left(\mathbf{D}^{\pm s}_{\{O,X\}}\right)_{\lambda,\lambda'} := \delta_{(\lambda,\lambda')} \cdot \begin{cases} \left((D_{\mathrm{O},j_0}^{-1/2})_{\lambda,\lambda'}\right)^{\pm s} & |\lambda| = |\lambda'| = j_0 \\ \left(\mathbf{D}_X^{\pm s}\right)_{\lambda,\lambda'} & otherwise \end{cases}, \tag{2.3.23}$$

where $\mathbf{D}_X^{\pm s}$ could be any other preconditioner, for example (2.2.15) or (2.2.33).

Remark 2.36 *The chosen orthogonal matrix* O *will be densely populated. In our wavelet construction of Section A.1.5, the minimum level is* $j_0 = 3$ *and the application of* O *will therefore require 81 floating point multiplications. The wavelet transform* $\mathbf{T}_{J,j_0}$*, on the other hand, requires 95 floating point multiplications. The overhead induced by* O *can thus lead to slightly higher execution times on level* $J = j_0 + 1$*, but is totally negligible on higher levels. Thus, the application of* $\mathbf{A}'_J$ *is still linear in time with respect to the number of unknowns.*

As I have shown in [122], this technique can lower the condition number of $\mathbf{A}_J$ in 1D by several orders of magnitude over all levels and still delivers very good results in several dimensions.

2.4 Multivariate Wavelets

There are several ways to construct wavelets on manifolds in higher dimensions. Of course, a construction of wavelets could be carried out directly, as was done in [146] for L-shaped domains in $\mathbb{R}^2$ and polygonal domains $\Omega \subset \mathbb{R}^n$ in [138] and [95]. For non-tensor-product domains, domain embedding methods are available, see [101] for an example of wavelets on the sphere. Another approach is the embedding of such domains within a hypercube $\Omega \subset (0,1)^n$ and enforcing boundary conditions using **Lagrange multipliers**; this approach will be discussed in Section 5.2.1. Also, many domains can be expressed as a union of tensor product domains after applying domain decomposition strategies, see [26, 53].
Therefore, we focus on the case of Cartesian product domains $\Omega \subset \mathbb{R}^n$. Here, multivariate wavelets can be constructed by tensor products of wavelet bases on the interval. This approach has the advantage of being able to treat the dimension as a variable, i.e., the actual value doesn't have to be specified beforehand. Given a MRA for the interval $\mathrm{I} = (0,1)$, we can use it to form a MRA for the n-dimensional hypercube $\Box := \Box^n := (0,1)^n$ preserving the **regularity** $\gamma, \widetilde{\gamma}$ and **moment conditions** $d, \widetilde{d}$ of Section 2.2.1. Since any problem given on a generally rectangular domain can be scaled to the standard hypercube, this is approach is equivalent to constructing multivariate wavelets on said rectangular domains.

Tensor Products

At the heart of our multidimensional constructions stands the **tensor product**. Tensor products of domains and Hilbert spaces have already been discussed in Section 1.2.5. One of the advantages of tensor products is the ease of the generalization of the involved operators to higher dimensions, see [122]. We distinguish here between three different types of tensor products, depending on the objects being multiplied. The **tensor product of functions** is the construction of a multidimensional function $\phi_{\mathbf{k}} : \mathbb{R}^n \to \mathbb{R}$ with the **multi-index** $\mathbf{k} = (k_1, \dots, k_n) \in \mathbb{N}_0^n$ by a collection of one-dimensional functions $\{\phi_i : \mathbb{R} \to \mathbb{R} \,|\, 1 \leq i \leq n\}$ of the form

$$\phi_{\mathbf{k}}(\mathbf{x}) := \phi_{k_n}(x_n) \otimes \cdots \otimes \phi_{k_1}(x_1) \tag{2.4.1}$$

$$:= \phi_{k_n}(x_n) \cdots \phi_{k_1}(x_1). \tag{2.4.2}$$

This construction can then be extended to sets of functions, e.g. bases, as

$$\Phi \otimes \Psi := \{\phi \otimes \psi \,|\, \phi \in \Phi, \psi \in \Psi\}. \tag{2.4.3}$$

Independently of this tensor product, given $\mathbf{A} = (a_{i,j}) \in \mathbb{R}^{m_2 \times n_2}$ and $\mathbf{B} = (b_{i,j}) \in \mathbb{R}^{m_1 \times n_1}$, the **tensor product of matrices** is

$$\mathbf{A} \otimes \mathbf{B} := \left(\begin{array}{c|c|c} \ddots & \vdots & \iddots \\ \hline \cdots & a_{i,j}\,\mathbf{B} & \cdots \\ \hline \iddots & \vdots & \ddots \end{array}\right) \in R^{(m_1\, m_2) \times (n_1\, n_2)}. \tag{2.4.4}$$

Again, we will always read tensor products **from right to left**, i.e., in descending order. The object associated to the highest dimension is written first, the one associated to the lowest dimension last. Since the product in (2.4.1) is **commutative**, it does not matter if the order is reversed, but this is not true for (2.4.4). Two important rules for the tensor product of matrices are thus

$$\begin{aligned} &(\mathbf{A} \otimes \mathbf{B}) \neq (\mathbf{B} \otimes \mathbf{A}) && \text{(almost always)}, \\ &(\mathbf{A} \otimes \mathbf{B})(\mathbf{C} \otimes \mathbf{D}) = (\mathbf{AC}) \otimes (\mathbf{BD}), && \text{for } \mathbf{A}, \mathbf{C} \in \mathbb{R}^{m_2 \times m_2},\ \mathbf{B}, \mathbf{D} \in \mathbb{R}^{m_1 \times m_1}, \end{aligned} \tag{2.4.5}$$

the second identity meaning that matrix multiplications are **decoupled** in the dimensions. Additional information about tensor products can be found in [76].

2.4.1 Multidimensional Single Scale Basis

Combining n **univariate bases** $\{\Phi_J^l\}_{l=1,\ldots,n}$ into one **multivariate basis**

$$\Phi_{\square,J} := \Phi_{\square^n,J}(\mathbf{x}) := \bigotimes_{l=1}^{n} \Phi_J^l(x_l), \quad \mathbf{x} := (x_1, \ldots, x_n), \tag{2.4.6}$$

it forms a single-scale basis of refinement level J on the domain $\square^n \subset \mathbb{R}^n$. Each function $\phi_{\square,J,\mathbf{k}} \in \Phi_{\square,J}$, with $\mathbf{k} = (k_1, \ldots, k_n)$ now being a **multi-index**, has approximately a support of $2^{-J}(0,1)^n$. If each Φ_J^l is associated with an index set $\Delta_J^l \subset \mathcal{I}^l$, we can associate $\Phi_{\square,J}$ with the index set

$$\Delta_J^{\square} := \Delta_J^1 \times \cdots \times \Delta_J^n. \tag{2.4.7}$$

Remark 2.37 *We will usually focus on the special case where all bases functions coincide, e.g.,*

$$\Phi_J^{l_1} = \Phi_J^{l_2}, \quad \textit{for all } l_1, l_2,$$

and we will write Φ_J instead of Φ_J^l.

2.4.2 Anisotropic Tensor-Product Wavelets

Analogously to the above, we can build tensor products of the wavelet basis $\Psi_{(J)}$ of (2.1.29), i.e.,

$$\Psi_{\square,(J)}^{\text{ani}} := \Psi_{\square,(J)}^{\text{ani}}(\mathbf{x}) := \bigotimes_{i=1}^{n} \Psi_{(J)}(x_i). \tag{2.4.8}$$

Here, functions on different levels in different spatial dimensions are coupled. This construction is thus called **anisotropic**.

Remark 2.38 *It is possible to choose different levels J_i for each dimension $1 \leq i \leq n$ in the construction (2.4.8). We will here use $J_1 = \cdots = J_n \equiv J$, since it would otherwise complicate the notation even further.*

The resulting support of the product function can therefore be quite irregularly distributed in each direction. The wavelet basis functions $\psi^{\text{ani}}_{\mathbf{j},\mathbf{k}}(\mathbf{x}) \in \Psi^{\text{ani}}_{\square,(J)}$ are indexed as

$$\psi^{\text{ani}}_{\lambda}(\mathbf{x}) := \psi^{\text{ani}}_{\mathbf{j},\mathbf{k}}(\mathbf{x}) := \psi^{\text{ani}}_{j_1,\ldots,j_n;k_1,\ldots,k_n}(\mathbf{x}) := \prod_{i=1}^{n} \psi_{j_i,k_i}(x_i). \tag{2.4.9}$$

The wavelet index is thus generalized into the **anisotropic wavelet index** $\lambda = (\mathbf{j},\mathbf{k})$. The tensor product wavelet space analogous to (2.1.53) will be referred to as

$$\Psi^{\text{ani}}_{\square} := \bigcup_{j=j_0-1}^{\infty} \Psi^{\text{ani}}_{\square,j}, \tag{2.4.10}$$

with the infinite index set $\mathcal{I}$, recalling (2.1.53), also given by the external product of the n index sets $\mathcal{I}^i$ of (2.1.41),

$$\mathcal{I} := \mathcal{I}_{\square} := \mathcal{I}^1 \times \cdots \times \mathcal{I}^n. \tag{2.4.11}$$

We can now define the finite linear subspaces $\Psi^{\text{ani}}_{\square,(J)} \subset \Psi^{\text{ani}}_{\square}$ by truncation of the index set $\mathcal{I}_{\square}$ exactly as in (2.1.29). The definition of the finite index set for tensor product wavelets is now

$$\mathcal{I}_{\square,(J)} := \left\{\lambda \in \mathcal{I}_{\square} \mid |||\lambda|||_{\infty} \le J\right\}, \tag{2.4.12}$$

where $|||\lambda|||_{\infty} := \max\{j_1,\ldots,j_n\}$ is the maximum individual level of the anisotropic index. The connection to the wavelet basis $\Phi_{\square,J}$ is established by the wavelet transform $\mathbf{T}_{\square,J}$,

$$\Psi^{\text{ani}}_{\square,(J)} = \left(\mathbf{T}^{\text{ani}}_{\square,J}\right)^T \Phi_{\square,J}. \tag{2.4.13}$$

The tensor product wavelet transform $\mathbf{T}^{\text{ani}}_{\square,J}$ can be constructed as the tensor product of the univariate transformations (2.1.32), i.e.,

$$\mathbf{T}^{\text{ani}}_{\square,J} := \mathbf{T}^{\text{ani}}_{\square,J,J-1} \cdots \mathbf{T}^{\text{ani}}_{\square,J,j_0}, \quad \mathbf{T}^{\text{ani}}_{\square,J,j} := \bigotimes_{i=1}^{n} \mathbf{T}_{J,j}. \tag{2.4.14}$$

The multiplicative cascading structure obviously retains the properties of the multiscale transform. In the same way can this construction process be applied to the dual wavelets and to the inverse wavelet transform. This construction also preserves the biorthogonality of the primal and dual tensor wavelets. The diagonal preconditioning operator (2.2.15) has to be modified, since there is no single level j which is associated to $\lambda = (\mathbf{j},\mathbf{k})$. The norm equivalence was shown in [71, 72] to hold true in this case for

$$\left(\mathbf{D}^{\pm s}_1\right)_{\lambda,\lambda'} := 2^{\pm s|||\lambda|||_{\infty}} \delta_{(\lambda,\lambda')}. \tag{2.4.15}$$

From a computational point of view, the fast wavelet transform (2.4.13) is relatively easy to implement for the full space discretizations of Section 2.5. The tensor product structure allows by (2.4.5) to either apply the transformation level- or dimension-wise. Also, it is possible to choose different maximal levels in different dimensions, for example for highly anisotropic elliptic problems as in [71].

On the downside, the anisotropic FWT is computationally more expensive than the isotropic FWT detailed in Section 2.4.3, because the identity part of the matrices $\mathbf{T}_{J,j}$, given by (2.1.35), is also tensorized in (2.4.14). Another drawback are the complicated **prolongation** and **restriction** operations in a **nested-iteration scheme** [122], because, as seen in Figure 2.3, the new values of higher levels have to be added in between existing values.

2.4.3 Isotropic Tensor-Product Wavelets

The idea of the **isotropic** construction is to only tensorize functions that have the same level. To ease the definition, we introduce another parameter $e \in \{0, 1\}$ and define the one-dimensional isotropic wavelet as

$$\psi^{\text{iso}}_{j,k,e}(x) := \begin{cases} \phi_{j,k}(x), & \text{for } e = 0, k \in \nabla_{j,0}, \\ \psi_{j,k}(x), & \text{for } e = 1, k \in \nabla_{j,1}, \end{cases} \tag{2.4.16}$$

and the type-dependent index sets as

$$\nabla_{j,e} := \begin{cases} \Delta_j, & \text{for } e = 0, \\ \nabla_j, & \text{for } e = 1. \end{cases} \tag{2.4.17}$$

The multi-dimensional isotropic wavelet $\psi^{\text{iso}}_{j,\mathbf{k},\mathbf{e}}$, with $\mathbf{k}, \mathbf{e}$ **multi-indices**, is then constructed as follows:

$$\psi^{\text{iso}}_{j,\mathbf{k},\mathbf{e}}(\mathbf{x}) := \bigotimes_{l=1}^{n} \psi_{j,k_l,e_l}(x_l). \tag{2.4.18}$$

The **generalized wavelet index** $\lambda := (j, \mathbf{k}, \mathbf{e}) \in \mathbb{Z} \times \mathbb{Z}^n \times \mathbb{E}_n$ with $\mathbb{E}_n := \{0, 1\}^n$ thus also carries the knowledge of the different types of functions present in the multi-dimensional wavelet. In case $\mathbf{e} = \mathbf{0}$ this means only single-scale basis functions are coupled. In all the other cases, i.e., $\mathbf{e} \in \mathbb{E}^\star_n := \mathbb{E}_n \backslash \{\mathbf{0}\}$, at least one wavelet function is present in the product. All wavelet indices on the same level $j \geq j_0$ are then combined to form the single wavelet level $\Psi^{\text{iso}}_{\Box,j}$, i.e.,

$$\Psi^{\text{iso}}_{\Box,j} := \bigcup_{\mathbf{e} \in \mathbb{E}^\star_n} \left\{ \psi^{\text{iso}}_{j,\mathbf{k},\mathbf{e}} \,|\, \mathbf{k} \in \nabla_{j,\mathbf{e}} \right\}, \quad \nabla_{j,\mathbf{e}} := \nabla_{j,e_0} \times \ldots \times \nabla_{j,e_n}. \tag{2.4.19}$$

The **isotropic wavelet basis** up to level J is then

$$\Psi^{\text{iso}}_{\Box,(J)} := \left\{ \psi^{\text{iso}}_{j_0,\mathbf{k},\mathbf{0}} \,|\, \mathbf{k} \in \nabla_{j_0,\mathbf{0}} \right\} \cup \bigcup_{j=j_0}^{J-1} \Psi^{\text{iso}}_{\Box,j}. \tag{2.4.20}$$

Note that the first set of functions here have type "$\mathbf{0}$", thus this refers to single-scale functions. As only functions on the same level j are coupled, the structure resembles the single scale wavelet basis more closely than the anisotropic construction. Taking the limit to infinity, all wavelet coefficients are collected in the set

$$\mathcal{I}_\Box := \{(j_0, \mathbf{k}, \mathbf{0}) \,|\, \mathbf{k} \in \nabla_{j_0,\mathbf{0}}\} \cup \bigcup_{j \geq j_0} \bigcup_{\mathbf{e} \in \mathbb{E}^\star_n} \{(j, \mathbf{k}, \mathbf{e}) \,|\, \mathbf{k} \in \nabla_{j,\mathbf{e}}\}. \tag{2.4.21}$$

The finite index set of all wavelet indices of the isotropic wavelet levels are thus

$$\mathcal{I}_{\Box,(J)} := \{(j_0, \mathbf{k}, \mathbf{0}) \,|\, \mathbf{k} \in \nabla_{j_0,\mathbf{0}}\} \cup \bigcup_{j=j_0}^{J-1} \mathcal{I}_j, \text{ with } \mathcal{I}_j := \{\lambda \in \mathcal{I} \,|\, |\lambda| = j\}. \tag{2.4.22}$$

Of course, now a suitable refinement matrices for this basis have to be constructed from their one-dimensional components. To this end, define for $\mathbf{e} \in \mathbb{E}_n$

$$\mathbf{M}^n_{j,\mathbf{e}} := \bigotimes_{l=1}^{n} \mathbf{M}_{j,e_l},$$

with $\mathbf{M}_{j,0}, \mathbf{M}_{j,1}$ from (2.1.11) and (2.1.18). These rectangular matrices make up the building blocks of the isotropic refinement matrix $\mathbf{M}^{\text{iso}}_j$,

$$\mathbf{M}^{\text{iso}}_j := \left(\mathbf{M}^n_{j,(0,0,\ldots,0)}, \mathbf{M}^n_{j,(0,0,\ldots,1)}, \ldots, \mathbf{M}^n_{j,(1,1,\ldots,1)}\right) \in \mathbb{R}^{(\#\Delta_j)^n \times (\#\Delta_j)^n}, \tag{2.4.23}$$

which is again a square matrix. The fast wavelet transform is now defined analogously to (2.1.34) and (2.1.35) as

$$\mathbf{T}^{\text{iso}}_{\square,J} := \mathbf{T}^{\text{iso}}_{\square,J,J-1} \cdots \mathbf{T}^{\text{iso}}_{\square,J,j_0}, \quad \text{with } \mathbf{T}^{\text{iso}}_{\square,J,j} := \begin{pmatrix} \mathbf{M}^{\text{iso}}_j & \\ & \mathbf{I}_{(\#\Delta_J)^n - (\#\Delta_j)^n} \end{pmatrix}, \tag{2.4.24}$$

which leads to the generalized refinement relation

$$\Psi^{\text{iso}}_{\square,(J)} = \left(\mathbf{T}^{\text{iso}}_{\square,J}\right)^T \Phi_{\square,J}.$$

The diagonal preconditioner is unchanged from (2.2.15), i.e.,

$$\left(\mathbf{D}^{\pm s}_{\text{iso}}\right)_{\lambda,\lambda'} := 2^{\pm|\lambda| s}\, \delta_{(\lambda,\lambda')}\,. \tag{2.4.25}$$

Again seen from a computational point of view, the isotropic wavelet construction has a number of advantages compared to the anisotropic construction of Section 2.4.2.
First, just one level variable needed, which means the space required to explicitly save a level variable of a wavelet index $\lambda = (j, \mathbf{k}, \mathbf{e})$ is independent of the dimension. The space required to save the variable $\mathbf{k}$ grows proportionally in the number of **bits** with the dimension, which is always fewer data than required to save several level variables $\mathbf{j}$. Also, the matrix $\mathbf{T}^{\text{iso}}_{\square,J,j}$ consists of tensorized matrices of $\mathbf{M}_{j,0}$, $\mathbf{M}_{j,1}$ only, no identity parts are tensorized. But because the tensorized matrices are not square matrices, the implementation of $\mathbf{T}^{\text{iso}}_{\square,J,j}$ for a full-space discretization is not as straight forward as in the anisotropic case.
Another advantage is the level-wise ordering of the wavelet coefficients as seen in Figure 2.4 which makes **prolongation** and **restriction** operators particularly easy to implement in a **nested-iteration scheme**.

A Comparison

The anisotropic decomposition over four levels $j = j_0, \ldots, j_0 + 3$ in 2D of $\Phi_{j_0+3} \otimes \Phi_{j_0+3}$ leads to the collection

$\Phi_{j_0} \otimes \Phi_{j_0}$	$\Phi_{j_0} \otimes \Psi_{j_0}$	$\Phi_{j_0} \otimes \Psi_{j_0+1}$	$\Phi_{j_0} \otimes \Psi_{j_0+2}$
$\Psi_{j_0} \otimes \Phi_{j_0}$	$\Psi_{j_0} \otimes \Psi_{j_0}$	$\Psi_{j_0} \otimes \Psi_{j_0+1}$	$\Psi_{j_0} \otimes \Psi_{j_0+2}$
$\Psi_{j_0+1} \otimes \Phi_{j_0}$	$\Psi_{j_0+1} \otimes \Psi_{j_0}$	$\Psi_{j_0+1} \otimes \Psi_{j_0+1}$	$\Psi_{j_0+1} \otimes \Psi_{j_0+2}$
$\Psi_{j_0+2} \otimes \Phi_{j_0}$	$\Psi_{j_0+2} \otimes \Psi_{j_0}$	$\Psi_{j_0+2} \otimes \Psi_{j_0+1}$	$\Psi_{j_0+2} \otimes \Psi_{j_0+2}$

of tensor products of wavelets and single scale functions. The decomposition algorithm thereby works on a vector of single scale coefficients (here in 2D written as $(\Phi_J \otimes \Phi_J)$) in the following manner:

$$(\Phi_J \otimes \Phi_J) \xrightarrow{\left(\mathbf{T}^{\text{ani}}_{\square,J,J-1}\right)^{-1}} \left(\begin{array}{c} \Phi_{J-1} \otimes \begin{pmatrix} \Phi_{J-1} \\ \Psi_{J-1} \end{pmatrix} \\ \hline \Psi_{J-1} \otimes \begin{pmatrix} \Phi_{J-1} \\ \Psi_{J-1} \end{pmatrix} \end{array}\right) \xrightarrow{\left(\mathbf{T}^{\text{ani}}_{\square,J,J-2}\right)^{-1}} \left(\begin{array}{c} \Phi_{J-2} \otimes \begin{pmatrix} \Phi_{J-2} \\ \Psi_{J-2} \\ \Psi_{J-1} \end{pmatrix} \\ \hline \Psi_{J-2} \otimes \begin{pmatrix} \Phi_{J-2} \\ \Psi_{J-2} \\ \Psi_{J-1} \end{pmatrix} \\ \hline \Psi_{J-1} \otimes \begin{pmatrix} \Phi_{J-2} \\ \Psi_{J-2} \\ \Psi_{J-1} \end{pmatrix} \end{array}\right) \xrightarrow{\left(\mathbf{T}^{\text{ani}}_{\square,J,J-3}\right)^{-1}} \cdots$$

In each step, all single scale functions (red) are decomposed until only wavelets and single scale functions on the coarsest level are left. As can be seen in the vector, the order of the wavelet coefficients has thus a strong tensor product structure. First, all lower dimensional positional indices are traversed, then, at each overflow, higher dimensional positions are increased. This is also visualized in the following diagram, where the entries in the vector are colored according to their position.

On the other hand, the isotropic decomposition applied to the basis $\Phi_{j_0+3} \otimes \Phi_{j_0+3}$ yields the collection

<table>
<tr><td>$\Phi_{j_0} \otimes \Phi_{j_0}$</td><td>$\Phi_{j_0} \otimes \Psi_{j_0}$</td><td rowspan="2">$\Phi_{j_0+1} \otimes \Psi_{j_0+1}$</td><td rowspan="3">$\Phi_{j_0+2} \otimes \Psi_{j_0+2}$</td></tr>
<tr><td>$\Psi_{j_0} \otimes \Phi_{j_0}$</td><td>$\Psi_{j_0} \otimes \Psi_{j_0}$</td></tr>
<tr><td colspan="2">$\Psi_{j_0+1} \otimes \Phi_{j_0+1}$</td><td>$\Psi_{j_0+1} \otimes \Psi_{j_0+1}$</td></tr>
<tr><td colspan="3">$\Psi_{j_0+2} \otimes \Phi_{j_0+2}$</td><td>$\Psi_{j_0+2} \otimes \Psi_{j_0+2}$</td></tr>
</table>

The isotropic decomposition algorithm similarly produces the structure

$$(\Phi_J \otimes \Phi_J) \xrightarrow{\left(\mathbf{T}^{\text{iso}}_{\square,J,J-1}\right)^{-1}} \left(\begin{array}{c} \Phi_{J-1} \otimes \Phi_{J-1} \\ \hline \Phi_{J-1} \otimes \Psi_{J-1} \\ \hline \Psi_{J-1} \otimes \Phi_{J-1} \\ \hline \Psi_{J-1} \otimes \Psi_{J-1} \end{array}\right) \xrightarrow{\left(\mathbf{T}^{\text{iso}}_{\square,J,J-2}\right)^{-1}} \left(\begin{array}{c} \Phi_{J-2} \otimes \Phi_{J-2} \\ \hline \Phi_{J-2} \otimes \Psi_{J-2} \\ \hline \Psi_{J-2} \otimes \Phi_{J-2} \\ \hline \Psi_{J-2} \otimes \Psi_{J-2} \\ \hline \Phi_{J-1} \otimes \Psi_{J-1} \\ \hline \Psi_{J-1} \otimes \Phi_{J-1} \\ \hline \Psi_{J-1} \otimes \Psi_{J-1} \end{array}\right) \xrightarrow{\left(\mathbf{T}^{\text{iso}}_{\square,J,J-3}\right)^{-1}} \cdots$$

in a vector, since in each step only pure products of single scale functions (red) are refined. In the isotropic case, the dominant pervasive structure is the level j, each level then consists of tensor product bases. This is also visualized in the following Figure 2.4. Here, as in Figure 2.3, each index is colored according to its position.

Incidentally, the anisotropic and isotropic standard preconditioners (2.4.15), (2.4.25) lead to the same values in the matrix $\mathbf{D}^{\pm s}$, only the entries are in different orders. In our numerical experiments for linear elliptic PDEs, the asymptotic limit for this preconditioner is lower for the isotropic construction than for the anisotropic construction. The preconditioner $\mathbf{D}^{\pm s}$ can simply here work better for the isotropic wavelets, even though both constructions promise the same optimal limit $\mathcal{O}(1)$. For example results, see Figure 2.2.

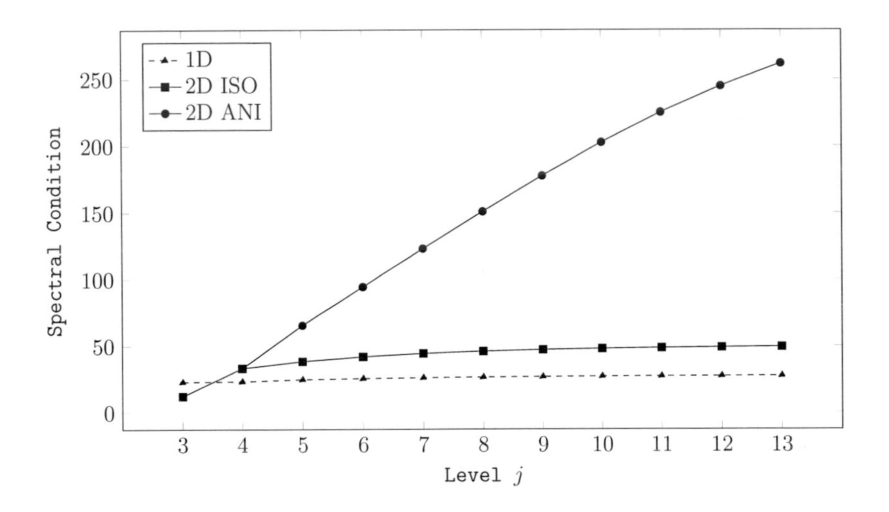

Figure 2.2: Plot of the spectral condition numbers $\kappa_2(\mathbf{A})$ of the H^1-stiffness matrix $\mathbf{A}$ defined by the bilinear form (1.4.24). The matrix $\mathbf{A}$ is discretized in the standard representation (2.2.24) on levels $j = 3, 4, \ldots$ in 1D and 2D using DKU-24 wavelets (see Appendix A.1.5) . The matrices were preconditioned using the standard preconditioners (2.4.15), (2.4.25). Although every series of condition numbers is asymptotically optimally bounded by a constant, the isotropic construction yields lower absolute condition numbers in this case; the boundedness of the 2D anisotropic condition numbers is not yet obvious in this graph.

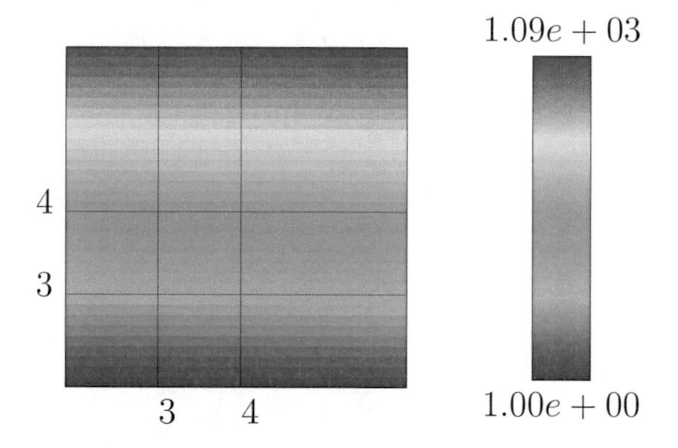

Figure 2.3: Depiction of the order of anisotropic wavelet coefficients in 2D for levels $j = 2, 3, 4$. The first wavelet coefficient has the value 1, the second the value 2, etc, in total there are $33^2 = 1089$ wavelet coefficients. In the anisotropic construction, first all coefficients in the lower dimension (x_1) are traversed, then coefficients of the higher dimension (x_2).

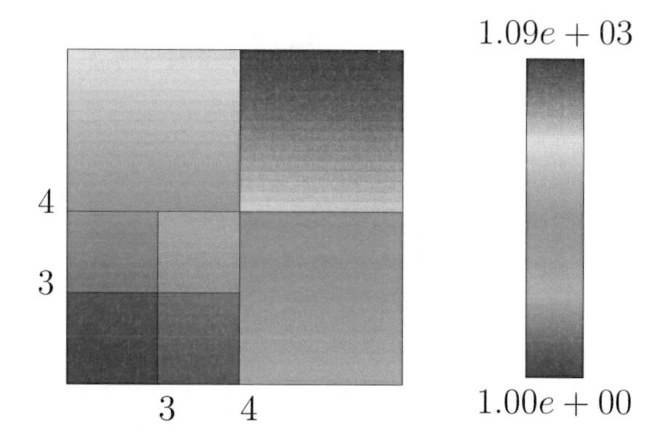

Figure 2.4: Depiction of the order of isotropic wavelet coefficients in 2D for levels $j = 2, 3, 4$. The first wavelet coefficient has the value 1, the second the value 2, etc, in total there are $33^2 = 1089$ wavelet coefficients. In the isotropic construction, the lower levels are traversed first, then higher levels. Within each level block, the indices are traversed by going through the lower dimensions (x_1) first and with each overflow the higher dimension (x_2) is advanced one step.

2.5 Full Space Discretizations

To understand adaptive methods, it is helpful to recapitulate a few details about full space discretizations. Therefore, we now give a brief introduction into full-grid discretizations. The main emphasis of this work lies in **adaptive methods**, see Section 3.
Let $\Psi, \widetilde{\Psi}$ be two biorthogonal wavelet bases as in Section 2.1, Section 2.4.2 or Section 2.4.3. The finite wavelet basis of all levels up to J is denoted by

$$\Psi_{(J)} := \left\{ \psi_\lambda \in \Psi \,|\, \lambda \in \mathcal{I}_{(J)} \right\}. \tag{2.5.1}$$

The, according to Theorem 2.18, scaled wavelet base

$$\Psi^s_{(J)} := \mathbf{D}_J^{-s} \Psi_{(J)} = \left\{ (\mathbf{D}^{-s})_{\lambda,\lambda} \psi_\lambda \,|\, \lambda \in \mathcal{I}_{(J)} \right\} \subset H^s, \tag{2.5.2}$$

constitutes a Riesz basis for $H^s(\Omega)$. The finite diagonal scaling operator $\mathbf{D}_J^{-s}$ is constructed from $\mathbf{D}^s$ by deleting all rows and columns of indexes not in $\mathcal{I}_{(J)}$,

$$\mathbf{D}_J^{-s} := \left(\left(\mathbf{D}^{-s} \right)_{\lambda,\lambda'} \right)_{\lambda \in \mathcal{I}_{(J)}, \lambda' \in \mathcal{I}_{(J)}} \in \mathbb{R}^{\#(\mathcal{I}_{(J)}) \times \#(\mathcal{I}_{(J)})}. \tag{2.5.3}$$

Note that the wavelet space $\Psi_{(J)}$ does not need to be the same as $\Psi_{(J)}$ in (2.1.29), for example in a tensor product setting of Section 2.4. Since definition (2.5.1) extends (2.1.29), it will be used primarily.

Remark 2.39 *This technique creates spaces which are* **linear**, *which makes them easy to handle in applications. The* **linear** *discretization on uniform grids is the optimal case for smooth given data and solutions.*

We can directly conclude from (2.2.10) the value of the **discretization error** (1.4.28) with respect to $\Psi^s_{(J)}$ as $h_J = 2^{-sJ}$.

2.5.1 Best Approximations

An important question in the current setting is to find the **best approximation** $y^\star := y^\star_J$ of an element $y \in H^s(\Omega)$ in the subspace $S(\Psi_{(J)}) \subseteq S(\Psi)$, or to give error bounds for $\|y - y^\star\|_{H^s(\Omega)}$ for any valid value of s. The best approximation $y^\star$ of y is defined as the element for which (1.4.28) is minimized, i.e.,

$$\|y - y^\star\|_{H^s(\Omega)} = \inf_{v \in S(\Psi_{(J)})} \|y - v\|_{H^s(\Omega)}. \tag{2.5.4}$$

The natural candidate for $y^\star$ is obviously the orthogonal projection onto the space $\Psi_{(J)}$ by means of the projectors P_j of (2.1.46). These projectors are used to show error bounds for (2.5.4) given in the following theorem.

Theorem 2.40 *For* $y \in H^s(\Omega), 0 < s \leq d$, *and* $r \leq s$ *one has*

$$\inf_{v \in S(\Psi_{(J)})} \|y - v\|_{H^r} \lesssim 2^{-(s-r)J} \|y\|_{H^s(\Omega)}. \tag{2.5.5}$$

The proof uses only the properties of the projectors P_J and Theorem 2.18 and can be found in [122]. The above result reads for $r = 0$:

$$\inf_{v \in S(\Psi_{(J)})} \|y - v\|_{L_2} \lesssim 2^{-sJ} \|y\|_{H^s(\Omega)}. \tag{2.5.6}$$

This means that the convergence speed of the approximation by wavelet spaces (as $J \to \infty$) is directly linked to the **smoothness** $s > 0$ of the function to be approximated.

2.5.2 Stability of the Discretizations

Since we now have optimally conditioned infinite dimensional wavelet-discretized operators at our disposal, we only need to ensure stability when truncating all multiscale bases Ψ_j above a certain level $j > J$. There are two types of criteria relevant for our problems which ensure stability of the finite discretized systems, both of which we specify now.

Galerkin Stability

We fix a refinement level J at which we wish to find the solution $y_J \in S_J$ of an elliptic differential operator $A : H^{+t} \to (H^{+t})'$ in a **Galerkin scheme** (see Section 1.4.3),

$$\left\langle \Psi^t_{(J)}, A y_J \right\rangle = \left\langle \Psi^t_{(J)}, f \right\rangle, \tag{2.5.7}$$

with respect to the wavelet basis $\Psi^s_{(J)} \subset \Psi \subset \mathcal{H}$ with the finite index set $\mathcal{I}_{(J)} \subset \mathcal{I}$.

Definition 2.41 [Galerkin Stability]
The Galerkin Scheme (2.5.7) is called $(t, -t)$**-stable**, *or* **Galerkin stable**, *if*

$$\|P_J v\|_{H^t} \sim \|\widetilde{P}_J A P_J v\|_{(H^t)'}, \quad v \in S_J, \tag{2.5.8}$$

holds uniformly in J, with the projectors of (2.1.46) and (2.1.47).

Expanding $y_J = \mathbf{y}_J^T \Psi^t_{(J)} = \mathbf{y}_J^T \mathbf{D}_J^{-t} \Psi_{(J)}$ and $f = \left\langle f, \Psi^t_{(J)} \right\rangle \widetilde{\Psi}^t_{(J)} = \mathbf{f}_J^T \widetilde{\Psi}^t_{(J)} = \mathbf{f}_J^T \mathbf{D}_J^{+t} \widetilde{\Psi}_{(J)}$ in (2.5.7), we obtain

$$\mathbf{A}_J := \left\langle \Psi^t_{(J)}, A \Psi^t_{(J)} \right\rangle = \mathbf{D}_J^{-t} \left\langle \Psi_{(J)}, A \Psi_{(J)} \right\rangle \mathbf{D}_J^{-t}, \tag{2.5.9}$$

and analogously to (2.2.22) we can write (2.5.7) as a finite-dimensional discretized operator equation

$$\mathbf{A}_J \mathbf{y}_J = \mathbf{f}_J. \tag{2.5.10}$$

Galerkin stability thus ensures the stability of the finite-dimensional discretized operator.

Proposition 2.42 *If the Galerkin scheme is (t, -t)-stable and it holds*

$$|t| < \gamma, \widetilde{\gamma},$$

then the matrices

$$\mathbf{A}_J = \mathbf{D}_J^{-t} \left\langle \Psi_{(J)}, A \Psi_{(J)} \right\rangle \mathbf{D}_J^{-t} \tag{2.5.11}$$

have uniformly bounded condition numbers.

In other words, Galerkin stability entails that $\mathbf{A}_J$ is still an **isomorphism** on $\ell_2(\mathcal{I}_{(J)})$ **uniformly in** J.

Remark 2.43 *Galerkin stability is trivially satisfied if the operator A is given by $\langle y, Av \rangle = a(y, v)$ as in Definition 1.42 since then $a(v, v) \sim \|v\|^2_{H^t}$ holds. The operator $\mathbf{A}_J$ is then called* **stiffness matrix**.

The LBB-Condition

Galerkin stability is sufficient for the homogeneous version of the elliptic PDE (1.4.7). Additional stability conditions are required during the discretization of an elliptic boundary value problems as **saddle point problems**, because this additionally involves a **trace operator** $B : \mathcal{H} \to \mathcal{K}$, e.g., $\mathcal{H} := H^1(\Omega)$ and $\mathcal{K} := H^{1/2}(\Gamma)$ with $\Gamma \subseteq \Omega$. Choosing finite discretizations $V \subset \mathcal{H}$ and $Q \subset \mathcal{K}$, the Ladysenškaya-Babuška-Brezzi (**LBB**)-condition, or sometimes called **discrete inf-sup condition**, is satisfies if there exists a constant $\widetilde{\alpha} > 0$ for which for the discretized operator $\mathbf{B} : V \to Q$ holds

$$\inf_{\widetilde{q} \in Q'} \sup_{v \in V} \frac{\langle \mathbf{B}v, \widetilde{q} \rangle_{Q \times Q'}}{\|v\|_V \|\widetilde{q}\|_{Q'}} \geq \widetilde{\alpha} > 0. \tag{2.5.12}$$

The LBB-condition can be interpreted as a way of ensuring that no element $\widetilde{q} \in Q'$ is orthogonal to any element $\mathbf{B}v \in Q$ with respect to $\langle \cdot, \cdot \rangle_{Q \times Q'}$ and the constant $\widetilde{\alpha}$ expresses the magnitude of that orthogonality property.

There are several criteria which ensure the validity of the LBB-condition in case of full space discretizations. A prominent result shown in [47] states that the LBB-condition is satisfied whenever the level of discretization is somewhat higher on the domain space V than on the trace space Q.

In the case of **adaptive wavelet methods**, it is no longer necessary to confirm the discrete LBB-condition, see [40]. Since in this case the concept of "discretization level" does not apply, the LBB-condition cannot be properly applied. One can think of applying any operator using adaptive wavelet methods in the infinite ℓ_2 setting, thus inheriting the infinite dimensional stability properties of the operator B. Therefore, we will not go into the details of this topic, details applicable to the current problem formulation can be found in [122].

This concludes our chapter on basic wavelet methods and we now turn to **adaptive wavelet methods**.

3 Adaptive Wavelet Methods based upon Trees

3.1 Introduction

In Section 2 we stated the required properties of the wavelet basis and introduced notations and the well-posed representation of operators in this wavelet setting. The main focus of the classic **full-grid** approach explained in Section 2.5 is laid upon the approximation of **smooth** functions for which linear wavelet spaces are the best choice.
Adaptive wavelet methods offer the means to handle nonlinear elliptic operators and unsmooth right hand sides efficiently. In fact, there are inherent benefits which justify the higher effort of developing adaptive wavelet techniques.

Remark 3.1 *The method described in the following as put forward in [32–34] can generally be applied to* **linear**, **semilinear** *and* **general nonlinear PDE operators**. *In the semilinear case, global convergence towards the unique solution given in Theorem 1.57 is proven. The theoretical considerations in the general nonlinear case are more involved, especially since solutions are only locally unique, but the algorithmic components used to solve these equations are exactly the same used herein for the semilinear PDEs. Hence, I will not generally distinguish between the semilinear and the general nonlinear case; the important theoretical component required is a variational formulation of the considered nonlinear PDE operators.*

3.1.1 The Why of Adaptive Wavelet Methods

The main conceptual difference of **adaptive wavelet methods** compared to finite discretization of Section 2.5 is that one stays formally in the infinite dimensional regime, thus stability concepts, i.e., ensuring the system of equations stay well defined and solvable when restricting the space of test functions to a finite subspace (see Section 2.5.2), are not needed. Instead, the adaptive procedures can be seen as "self-stabilizing", i.e., they inherit the stability properties of the infinite dimensional setting.

Remark 3.2 *Choosing a fixed finite wavelet basis for discretization of an operator equation can be seen as taking a finite subset of rows and columns of the infinite dimensional wavelet discretized system. As no such infinite dimensional discretization exists for* **finite element methods** *(***FEM***), refining a FEM grid does not correspond to adding a subset of rows and columns of an infinite dimensional matrix.*

At the same time, there is also an inherent property stemming from nonlinear approximation theory, which we cite from [62]. An approximation process is called **nonlinear** if the space of approximants is adapted specifically for a given function and not predetermined by some other criteria, e.g., data points placed on a uniform grid.
Examining the properties of functions capable of being approximated using N piecewise polynomials at the error rate N^{-s} for some $s > 0$, these functions can be determined to be contained in classical function spaces [121, 124], e.g., Sobolev and Besov spaces. The main difference is that, using the same number of degrees of freedom, the function space in the nonlinear case is larger than the function space attained using linear methods, see Figure 3.1. In functional analytical terms, the approximation error is bound by a **weaker norm** in the nonlinear case than in the linear case.

This means that the nonlinear adaptive wavelet methods do not approximate smooth functions at a higher rate than classical (linear) methods, but that functions with less smoothness can still be approximated at the same rate. Specifically, assuming the solution $u \in H^m(\Omega)$ of a PDE for some fixed $m \geq 0$, this gives the metric in which the error is measured. In the simplest case, $m = 0$, this means the error is measured using the $L_2(\Omega)$-norm. For the Laplace operator, for which holds $m = 1$ by the remarks from Section 1.4, one measures in the **energy norm** $H^1(\Omega)$.

- Using the **linear spaces** of Section 2.5, let S_N be a space with N DOFs. Then one can expect for $u \in H^r(\Omega)$ with $r > m$ by Lemma 1.47 an approximation rate,

$$\inf_{u_N \in S_N} \|u - u_N\|_{H^m(\Omega)} \lesssim h^{r-m}\|u\|_{H^r(\Omega)} = N^{-(r-m)/n}\|u\|_{H^r(\Omega)},$$

  because of $h \sim N^{-n}$. The value of the smoothness index r is generally limited by the polynomial exactness $(\mathcal{P})$(2.2.3) of the wavelets employed.

- Using **adaptive methods**, the same rate $N^{-(r-m)/n}$ can be achieved whenever $u \in B_p^r(L_p(\Omega))$ for $r > m$ and

$$\frac{1}{p} = \frac{r-m}{n} + \frac{1}{2}, \quad \Longrightarrow \quad r = m + \left(\frac{1}{p} - \frac{1}{2}\right) n =: m + s\,n.$$

  This equation thus again characterizes for fixed dimension n and smoothness index m a line of slope n, cf. Figure 1.1. Denoting by Σ_N the family of sets containing N wavelet coefficients, the approximation error is then bound by

$$\inf_{u_N \in \Sigma_N} \|u - u_N\|_{B_2^m(L_2(\Omega))} \lesssim N^{-(r-m)/n}\|u\|_{B_p^r(L_p(\Omega))}.$$

The spaces $B_p^{m+s\,n}(L_p(\Omega))$ are the largest spaces still continuously embedded in $H^m(\Omega)$, see Section 1.3.1 and Figure 3.1. In summary, using nonlinear methods, we can expect an asymptotic rate N^{-s} if the solution to a PDE is in the Besov space $B_p^{m+s\,n}(L_p(\Omega))$ for some $s > 0$, and we could only achieve the same rate using linear spaces if the solution was in the Sobolev space $H^{m+s\,n}(\Omega)$. Thus, the extra complexity in theory and implementation when using **adaptive wavelet methods** pays off whenever a function has a higher smoothness in the Besov scale than in the Sobolev scale.

3.1.2 The How of Adaptive Wavelet Methods

The infinite dimensional space ℓ_2 poses serious problems for computational schemes because infinite data structures are not manageable in a finite computer system. The challenge is therefore to devise finite approximations without losing the inherent properties of the infinite dimensional realm. The first approach laid out in [32] is based upon **compressible matrices** and iterations of finitely supported (best N-term) approximations. I follow here the later works [33,34], which presented a new paradigm specifically designed for nonlinear variational problems (cf. page 16).
The wavelet discretization of nonlinear operator equations discussed in Section 1.5 in principle works exactly as in Section 2.2.4. The following result from [32] shows under what conditions this discretization is judiciously justified.

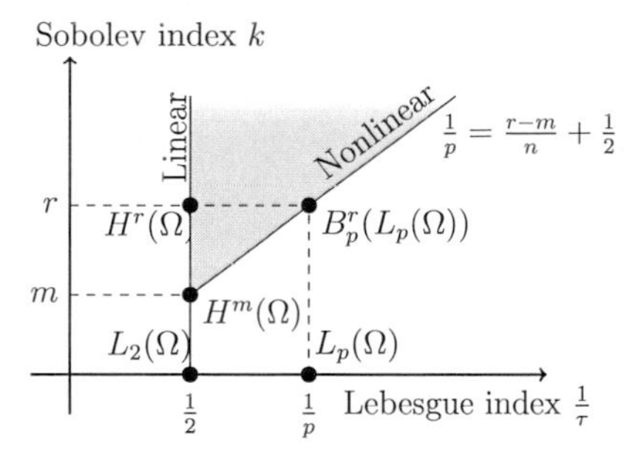

Figure 3.1: Linear vs. Nonlinear Approximation. Spaces corresponding to linear approximation lie on the vertical line, i.e., their smoothness increases uniformly in the Sobolev scale. The line with slope n marks the critical line for continuous embedding in $H^m(\Omega)$ and the spaces corresponding to nonlinear approximation. All spaces in the shaded area are continuously embedded in $H^m(\Omega)$, but not the spaces to the right of the critical line.

Proposition 3.3 *Let $F : \mathcal{H} \to \mathcal{H}'$ be an operator from a Hilbert space $\mathcal{H}$ into its dual $\mathcal{H}'$. Suppose that the problem $F(v) = w$ is well posed, see Definition 1.53, then the reformulated problem $\mathbf{F}(\mathbf{v}) = \mathbf{w}$ is well posed in $\ell_2(\mathcal{I})$.*

Subsequently, the resulting representation of an operator in wavelet coordinates is well conditioned, see [46], if the mapping property (1.4.14) or (1.5.8) holds.
Adaptivity in this context refers to the process of selecting the wavelet coefficients according to their importance which is directly linked to absolute value of the coefficient. The process of choosing the N coefficients with the highest absolute is called the **best N-term** approximation and it is a highly **nonlinear process**. The set of highest valued wavelet indices is generally very unordered and this makes any algorithmic processing of this set difficult. The remedy is to enforce more structure by replacing the general best N-term approximation by the **best tree N-term** approximation, i.e., a **tree structure** is introduced. Unlike the full space discretizations in Section 2.5, the result of an adaptive operator application to an adaptive vector with tree structure is not directly accessible.
The main implementational obstacle in adaptive wavelet methods is the efficient management of these tree based sets and the computation of operator applications in optimal linear complexity w.r.t. the number of wavelet coefficients. The naive idea of inserting all admissible wavelet indices into the input and output vectors and afterwards selecting the relevant non-zero coefficients defies the purpose of the adaptive schemes: The overall complexity of such an approach would make the scheme infeasible and also very quickly impossible to compute in reasonable time. The same argument applies to **pseudo-adaptive methods**, where full-grid data structures are used and **adaptive** procedures are applied to **emulate** adaptive data structure computations using thresholding techniques.
The adaptive application of a (non-)linear operator w.r.t. an adaptive wavelet vector based on a tree is a two step process. First, the output set of the operator application is computed by an algorithm called `PREDICTION`. As the name implies, this algorithm will not produce the perfect, theoretical result but an index set that is just **big enough**, so that the application of the operator can be applied **sufficiently precise**, i.e., up to a

user given error bound.
After this output set is computed, the second step is to compute the values of these wavelet coefficients. In principle, this task could make the whole adaptive scheme infeasible, if the value of any wavelet coefficient of the output vector would depend on **all coefficients** of the input vector **simultaneously**. The remedy to this problem is simple: No coefficient should depend on all coefficients, but only on those that are "nearby".

Definition 3.4 [Local Operator]
An operator $F : \mathcal{H} \to \mathcal{H}'$ is called **local**, *if elements of $\mathcal{H}$ only "interact" through F with other elements of $\mathcal{H}$ if they share* **support**:

$$\operatorname{supp}(v) \cap \operatorname{supp}(u) = \emptyset \Longrightarrow \langle u, F(v) \rangle_{\mathcal{H}\times\mathcal{H}'} = 0, \qquad \textit{for all } u, v \in \mathcal{H}. \tag{3.1.1}$$

Of course the converse does not need to hold, i.e., the value $\langle u, F(v) \rangle$ can be zero even if $\operatorname{supp}(v) \cap \operatorname{supp}(u) \neq \emptyset$. This can easily happen, for example, because of **vanishing moments** $(\mathcal{V})$(2.2.5).

Remark 3.5 *In the following, we assume that all operators $F : \mathcal{H} \to \mathcal{H}'$ are* **local**. *Many operators, e.g., differential operators, considered in practical applications are local so this is not a very constraining assumption.*

There are several types of (local) operators considered in this work, split in linear and nonlinear categories. For **nonlinear operators**, the polynomial representation of the wavelets $(\mathcal{P})$(2.2.3) is used to create a unique local representation living on disjoint cells, cf. (3.4.23). This way, the additive decomposition of the wavelet expansion is not a problem for applying a nonlinear operator. The same technique could also be applied to linear operators and the sum of an nonlinear and linear operator is best dealt with this way. The application of **linear operators** can be implemented differently by avoiding the polynomial representation phase. This is, for example, useful for some inverse operators $F^{-1} : \mathcal{H}' \to \mathcal{H}$, where no polynomial representation is available on $\mathcal{H}'$.

Overview

In Section 3.2.1 we show how to obtain a **tree structured wavelet index set**, a paradigm that is at the heart of the mathematical theory and computational implementation. **Nonlinear** wavelet spaces can be constructed by best (tree) N-**Term approximations**, as explained in Section 3.2.2.
For such tree structured index sets, the result of an application of a local operator with certain decay properties can be computed as recapitulated in Section 3.3. The most important tool is the algorithm `PREDICTION` explained in detail in Section 3.3.3. Another algorithm called `TREE_COARSE` presented in Section 3.3.2 is very important for proving the optimal linear complexity of the numerical solvers given in Section 4.1 by limiting the number of wavelet coefficients over the course of the whole solution process.
In Section 3.4 the transformation of functions in wavelet representation into a representation based on **local polynomials** is explained. These techniques will be combined in Section 3.4.3 to prove the efficient application of **nonlinear operators** in wavelet coordinates up to any given accuracy in optimal linear complexity w.r.t. the number of unknowns.

In Section 3.5, a special treatise is dedicated to the topic of applying **linear operators** based on this adaptive framework, because these can be crucial to the solution process of nonlinear problems, see Section 4.4. A special case of linear operators are **trace operators** which I discuss in Section 3.6. Lastly, I present my own version of an anisotropic adaptive wavelet method in Section 3.7 which is based completely on the isotropic algorithms.

3.2 Nonlinear Wavelet Approximation

We will from now on assume to be in the one-dimensional or **isotropic** multi-dimensional case of Section 2.4.3, so $\mathcal{I}$ is defined by (2.4.21).

Remark 3.6 *The following assertions are independent of the type of wavelet used, as long as the properties given in Section 2.3 are satisfied. Several wavelet constructions applicable to this matter are available and have been implemented, for details see Appendix A.1.*

Let $n \in \mathbb{N}$ be the spatial dimension and $j_0 \in \mathbb{N}$ again be the coarsest level in a multi-level hierarchy.

Remark 3.7 *To ease notation, we drop the $\square$ index representing $(0,1)^n$ from the definitions of Section 2.4.3, the symbol $\square$ will instead later on refer to a cell in a dyadic refinement of the domain $(0,1)^n$ (which could coincide with the whole domain).*

To recapitulate, each index $\lambda := (j, \mathbf{k}, \mathbf{e}) \in \mathcal{I}$ carries information of **refinement level** $j = |\lambda|$, **spatial location** $\mathbf{k} = (k_1, \ldots, k_n)$ and **type** $\mathbf{e} \in \{0,1\}^n$ of the isotropic wavelet. Also to shorten the notation, we abbreviate $S_\lambda := \operatorname{supp}(\psi_\lambda)$.

3.2.1 Tree Structured Index Sets

Although the use of **tree structured index sets** might seem unmotivated, tree structure are naturally found in adaptive methods, e.g., FEM, as **locally refined meshes**.
First of all, we have to state some notations and definitions concerning tree structured index sets. These notions are taken from [16, 33, 34, 111].

Definition 3.8 [Tree]
We call a set $\mathcal{T}$ **tree** *if $\mu \in \mathcal{T}$ implies $\lambda \in \mathcal{T}$ whenever μ is a descendant of λ. That is,*

$$\mu \in \mathcal{T},\ \mu \prec \lambda \Rightarrow \lambda \in \mathcal{T}, \tag{3.2.1}$$

where $\mu \prec \lambda$ expresses that μ is a **descendant** *of λ and $\mu \preceq \lambda$ means that either μ is a descendant of λ or equal to λ. The elements of $\mathcal{T}$ are called* **nodes**.

Remark 3.9 *A tree is usually understood as having only a* **single** *root node, although definitions vary from subject to subject. In our context, we therefore speak of* **tree structured** *sets which means there is a tree like structure but the overall appearance more closely resembles an* **arborescence**, *see [94].*

Definition 3.10 [Children and Parents]
The direct descendants μ of a node λ are called **children** *of λ and λ is referred to as their* **parent**. *The set of all* **children** *of a node $\lambda \in \mathcal{T}$ will be denoted by $\mathcal{C}(\lambda)$, the* **parent** *of λ is $\Pi(\lambda)$.*

Of special interest are of course the extreme cases of the above definition, e.g., parents without children and nodes without parents.

Definition 3.11 [Roots and Leaves]
The nodes of $\mathcal{T}$ which do not have any ancestor, i.e.,

$$\mathcal{N}_0(\mathcal{T}) := \{\lambda \in \mathcal{T} : \nexists\, \mu \in \mathcal{T} \text{ such that } \lambda \prec \mu\}. \tag{3.2.2}$$

are called **root nodes**. *Furthermore, the set of nodes without any children are called* **leaves** *and will be denoted by*

$$\mathcal{L}(\mathcal{T}) := \{\lambda \in \mathcal{T} : \nexists\, \mu \in \mathcal{T} \text{ such that } \mu \prec \lambda\}. \tag{3.2.3}$$

Combining the above definitions brings us to the notion of a proper tree:

Definition 3.12 [Proper Tree, Subtree, Proper Subtree]
A tree $\mathcal{T}$ is called **proper tree**, *if it contains all root nodes. A* **subtree** *$\mathcal{T}' \subset \mathcal{T}$ of a tree $\mathcal{T}$ is called* **proper subtree**, *if all root nodes of $\mathcal{T}$ are also contained in $\mathcal{T}'$. All descendants of a node λ form the* **subtree** *$\mathcal{T}_{[\lambda]} := \{\mu \in \mathcal{T} \,|\, \mu \prec \lambda\}$.*

As we will later see, it is very important to work with proper trees for both theoretical and practical purposes. Therefore, we will assume all trees to be proper trees unless mentioned otherwise.
Another useful theoretical property is the notion of

Definition 3.13 [Expansion Property]
A tree $\mathcal{T}$ is said to have the **expansion property** *if for $\lambda \in \mathcal{T}$ and $\mu \in \mathcal{I}$ holds*

$$\left.\begin{array}{c} |\mu| < |\lambda| \\ S_\mu \cap S_\lambda \neq \emptyset \end{array}\right\} \implies \mu \in \mathcal{T}. \tag{3.2.4}$$

Any tree $\mathcal{T} \subset \mathcal{I}$ can be made into an **expanded tree** $\widetilde{\mathcal{T}} \supseteq \mathcal{T}$ by a simple scheme: Recursively set $\Theta_k(\lambda) := \{\mu \in \mathcal{I} \,|\, S_\mu \cap S_{\mu'} \neq \emptyset$ for any $\mu' \in \Theta_{k-1}(\lambda)\}$ with $\Theta_0(\lambda) := \{\lambda\}$, then

$$\widetilde{\mathcal{T}} := \bigcup_{\lambda \in \mathcal{T}} \bigcup_{k=0}^{|\lambda|} \Theta_k(\lambda), \tag{3.2.5}$$

is an expanded tree.

Tree Structure of Wavelet Indices

By (2.1.29) and (2.4.20), the index set of all scaling functions on the coarsest level j_0 are now set to be

$$\mathcal{I}_{j_0-1} := \{(j_0 - 1, \mathbf{k}, \mathbf{1}) \,|\, \mathbf{k} \in \nabla_{j_0,\mathbf{0}}\}. \tag{3.2.6}$$

Remark 3.14 *This definition deviates from the previous definition (2.4.22) for two reasons:*

1. *The level value $j_0 - 1$ fits better in the tree structure environment.*
2. *When a basis transformation (2.3.16) is being applied to the* **coarsest level** *only, the different representations $(j_0 - 1, \mathbf{k}, \mathbf{1})$ and $(j_0, \mathbf{k}, \mathbf{0})$ of the same indices allow to distinguish between the different bases $\Psi'_{j_0-1} - O^T \Phi_{j_0}$ and Φ_{j_0}, respectively.*

The complete index set given by (2.4.22) can then be written as

$$\mathcal{I} := \mathcal{I}_{j_0-1} \cup \{(j, \mathbf{k}, \mathbf{e}) \,|\, j \geq j_0,\ \mathbf{k} \in \nabla_{j,\mathbf{e}},\ \mathbf{e} \in \mathbb{E}_n^\star\}. \tag{3.2.7}$$

In the following, we show the natural tree structure within these wavelet index sets. This is most easily done by constructing a tree structure for the one-dimensional wavelet base $\Psi_{(J)}$ and then applying these rules coordinate-wise to the isotropic wavelet base $\Psi^{\text{iso}}_{(J)}$. We will only state the list of children, since the determination of the parent is the inverse operation and one just needs to read the definitions from right to left to determine the parent of an index.
Since the number of wavelet functions $\#\nabla_j$ usually is exactly 2^j, the **default wavelet tree structure** is set up using a **dyadic structure**. Each **wavelet node** $\mu = (j, k, 1)$ is given the set of children

$$\mathcal{C}((j, k, 1)) := \{(j + 1, 2k, 1), (j + 1, 2k + 1, 1)\}. \tag{3.2.8}$$

A visualization of this tree structure can be found in Figure A.2. Since isotropic wavelets also contain **single scale** functions on each level, a relation for the bases $\Phi_j \leftrightarrow \Phi_{j+1}$ has to be established, too. Because the relation depends heavily on the boundary conditions of the basis, as this determines the number of functions $\#\Delta_j$, no standard structure can be specified. But it can be noted that the children of a single scale index $(j, k, 0)$ live on level $j + 1$ and have type $e = 0$ again.

Remark 3.15 *Details on this mapping for our boundary adapted hat functions can be found in Appendix A.1.1. For simplicity, one can think of the children of a single scale function as being*

$$\mathcal{C}((j, k, 0)) := \{(j + 1, 2k, 0), (j + 1, 2k + 1, 0)\}. \tag{3.2.9}$$

Definition 3.16 [Isotropic Wavelet Tree Structure]
Each **isotropic wavelet node** $\mu = (j, \mathbf{k}, \mathbf{e}) \in \mathcal{I} \setminus \mathcal{I}_{j_0-1}$ *is associated the set of children*

$$\mathcal{C}(\mu) := \bigotimes_{i=1}^{n} \mathcal{C}(\mu_i), \qquad \text{with } \mu_i := (j, k_i, e_i). \tag{3.2.10}$$

Having established the tree structure for all wavelet levels, it remains to set the root nodes and connect them to the rest of the tree structure. The root nodes are of course the indices of the scaling functions, that is, $\mathcal{I}_{j_0-1}$. These do not fit in Definition 3.16 as they do not contain a single wavelet coefficient, but their children should naturally be all wavelet indices on the next level, i.e., $\{(j_0, \mathbf{k}, \mathbf{e}) \,|\, \mathbf{k} \in \nabla_{j_0,\mathbf{e}}, \mathbf{e} \in \mathbb{E}_n^\star\}$. Hence, an inheritance relation $\Phi_{j_0} \leftrightarrow \Psi_{j_0}$ must be defined. This mapping again depends on the single scale base and can be found in Appendix A.1.1.

Proposition 3.17 *For each $\lambda \in \mathcal{C}(\mu)$ holds that the level is $j+1$ and the type is the type* $\mathbf{e}$ *of the parent.*

We associate now to each root node $\mu = (j_0 - 1, \mathbf{k}, \mathbf{1})$ the children

$$\mathcal{C}(\mu) := \bigcup_{\mathbf{e} \in \mathbb{E}_n^\star} \bigotimes_{i=1}^{n} \mathcal{C}((j_0 - 1, k_i, e_i)), \tag{3.2.11}$$

with

$$\mathcal{C}((j_0 - 1, k_i, e_i)) := \begin{cases} \{(j_0, k_i, 0)\}, & \text{if } e_i = 0, \\ \{(j_0, r_i, 1) \,|\, \text{for some } r_i \in \nabla_{j_0}\}, & \text{if } e_i = 1. \end{cases}$$

Simply put, the children for each wavelet type $\mathbf{e} \in \mathbb{E}_n^\star$ are computed and then the results combined. The children for any type $\mathbf{e} \in \mathbb{E}_n^\star$ in any coordinate $1 \le i \le n$ are here either the result of the mapping $\Phi_{j_0} \leftrightarrow \Psi_{j_0}$ if $e_i = 1$ or the trivial mapping $\Phi_{j_0} \leftrightarrow \Psi_{j_0-1}$ if $e_i = 0$. For theoretical and practical purposes that will become apparent later on, it is important that the wavelet tree exhibits the **Inclusion Property**, i.e.,

$$\mu \prec \lambda \Rightarrow S_\mu \subset S_\lambda \qquad \text{for all } \mu, \lambda \in \mathcal{I} \setminus \mathcal{I}_\mathbf{0}. \tag{3.2.12}$$

Due to the locality of the wavelets ($\mathcal{L}$)(2.1.8) and a bound on the number of intersecting wavelets (2.1.16), one can arrange the tree structure so that the Inclusion property is guaranteed.
Having established the tree structure, we consider $\mathcal{T}$ now to be an ordered subset of the wavelet index set $\mathcal{I}$ of (3.2.7) and a proper tree.

3.2.2 The Best (Tree) N-Term Approximation

In nonlinear wavelet methods, as opposed to the linear setting of Section 2.5, the set of all admissible wavelet coefficients is not fixed given by a highest level $J \ge j_0$. Instead, the set is aspired to be optimal in the sense that only relevant indices are included thus voiding any computational overhead that near-zero coefficient values would entail.
Traditionally, this means that for any given $v = (\mathbf{v}_\lambda)_{\lambda \in \mathcal{I}} \in \ell_2(\mathcal{I})$, the **best N-term approximation**,

$$\sigma_N(\mathbf{v}) := \inf_{\mathbf{w} \in \ell_2(\mathcal{I})} \left\{ \|\mathbf{v} - \mathbf{w}\|_{\ell_2(\mathcal{I})} \,|\, \#S(\mathbf{w}) \le N \right\}, \tag{3.2.13}$$

has to be computed. Here, $S(\mathbf{v}) := \text{supp}(\mathbf{v})$ denotes **support**, i.e., the set of all active wavelet indices. For any given vector $\mathbf{v}$, the minimizing element (which does not have to be unique) can simply be computed by rearranging the coefficients of $\mathbf{v}$ in a **non-decreasing** way,

$$|v_{\lambda_i}| \ge |v_{\lambda_j}| \quad \text{for all } 1 \le i < j \le \#S(\mathbf{v}), \tag{3.2.14}$$

and taking the first N elements $\mathbf{w} := \{v_{\lambda_1}, \ldots, v_{\lambda_N}\}$. Since the elements of this set could theoretically be completely unrelated (although this would be unusual in applications), an approximation preserving **tree structure** is sought. This led to the notion of the **best**

tree N-term approximation, where the **support** of all approximants must have tree structure:

$$\sigma_N^{\text{tree}}(\mathbf{v}) := \inf_{\mathbf{w}\in\ell_2(\mathcal{I})} \left\{ \|\mathbf{v}-\mathbf{w}\|_{\ell_2(\mathcal{I})} \,|\, S(\mathbf{w}) \text{ is a } \mathbf{tree} \text{ and } \#S(\mathbf{w}) \le N \right\}. \tag{3.2.15}$$

Although this tweaked definition helps greatly when considering theoretical problems by introducing a structure, the determination of the minimizing vector $\mathbf{w}$ for given $\mathbf{v}$ is not a simple task, since this would involve searching through all possible subtrees, which would result in exponential complexity w.r.t. $\#S(\mathbf{v})$. Possible strategies to solve this problem will be presented in Section 3.3.2.
The error notion (3.2.15) is now the basis to define approximation spaces:

Definition 3.18 [Tree Approximation Classes $\mathcal{A}^s_{\text{tree}}$]
For $s > 0$, the space in $\ell_2(\mathcal{I})$ of all vectors which can be approximated at a rate of N^{-s} using a tree of N elements is denoted by

$$\mathcal{A}^s_{tree} := \left\{ \mathbf{v} \in \ell_2(\mathcal{I}) \,|\, \sigma_N^{tree}(\mathbf{v}) \lesssim N^{-s} \right\}, \tag{3.2.16}$$

which becomes a **quasi normed space** *with the* **quasi norm**[1],

$$\|\mathbf{v}\|_{\mathcal{A}^s_{tree}} := \sup_{N\in\mathbb{N}} N^s \sigma_N^{tree}(\mathbf{v}). \tag{3.2.17}$$

The space $\mathcal{A}^s_{\text{tree}}$ contains all vectors that can be approximated at rate s using tree-structured index sets. Conversely, since for any $\mathbf{v} \in \mathcal{A}^s_{\text{tree}}$ and $\epsilon > 0$ exists a **best-tree** $\mathcal{T}_N(\mathbf{v}) := \mathcal{T}(\mathbf{v}, \epsilon)$ such that

$$\sigma_N^{\text{tree}}(\mathbf{v}) \equiv \|\mathbf{v} - \mathbf{v}|_{\mathcal{T}_N(\mathbf{v})}\|_{\ell_2(\mathcal{I})} \le \epsilon,$$

the approximation order in (3.2.17) guarantees

$$N \lesssim \|\mathbf{v}\|^{1/s}_{\mathcal{A}^s_{\text{tree}}} \, \epsilon^{-1/s}. \tag{3.2.18}$$

This shows, for any element in $\mathcal{A}^s_{\text{tree}}$, one can generally expect for $N \in \mathbb{N}$ degrees of freedom

$$\textbf{convergence rate} \quad N \sim \epsilon^{-1/s}, \qquad \text{or } \textbf{accuracy} \quad \epsilon \sim N^{-s}, \tag{3.2.19}$$

while approximating using finite trees.
Trivially, any finitely supported vector $\mathbf{v} \in \ell_2(\mathcal{I})$ belongs to $\mathcal{A}^s_{\text{tree}}$ for all $s > 0$.

Remark 3.19 *It was pointed out in [33] that the space $\mathcal{A}^s_{tree}$ is the "right" space to consider for wavelet methods for our class of* **nonlinear** *PDE operators, see Section 1.5. This means theoretical results w.r.t. the norm (3.2.17) are available for complexity and convergence estimates, some of which we will cite in the next sections.*
In case of linear PDEs, the solution is shown to be in the unrestricted space $\mathcal{A}^s$, where the definition is analogous to Definition 3.18, just without the tree structure requirement.

[1] The triangle inequality is replaced by $\|x + y\| \le C(\|x\| + \|y\|)$ for some $C > 1$.

Regularity Properties of the Space $\mathcal{A}^s_{\mathbf{tree}}$

The connection of wavelet expansion to smoothness spaces is given by a **norm equivalence** like Theorem 2.18. For $v = \mathbf{v}^T\Psi \in \mathcal{H}$ the analogue to (2.2.12) is given by (see [62]),

$$\|v\|^q_{B^\alpha_q(L_p(\Omega))} \sim \sum_{j=j_0-1}^{\infty} 2^{jq(\alpha+\frac{n}{2}-\frac{n}{p})} \|\mathbf{v}_j\|^q_{\ell_2(\mathcal{I}_j)} \tag{3.2.20}$$

for $t > 0$. For $p = q = 2$ we retain (2.2.12) in accordance with (1.3.12). The question which smoothness properties the functions of (3.2.16) have to fulfill was answered in [34].

Proposition 3.20 *Let* $v = \mathbf{v}^T\Psi \in H^m(\Omega)$ *for a n-dimensional domain* Ω. *If* $v \in B^{m+sn}_q(L_{\tau'}(\Omega))$ *for* $\frac{1}{\tau'} < s + 1/2$ *and* $0 < q \leq \infty$, *then* $\mathbf{v} \in \mathcal{A}^s_{tree}$. *Specifically, the assertion hold for the case* $q \equiv \tau'$.

This is a somewhat stricter condition than in the general case of the space $\mathcal{A}^s$ where $\frac{1}{\tau'} = s + 1/2$ suffices, see also [34]. But it still allows for the case $q \equiv \tau'$ discussed in Section 1.3.1 and Section 3.1. After these theoretical considerations, we now turn to the actual structures and algorithms used in the setting laid out here.

3.3 Algorithms for Tree Structured Index Sets

In this section, we describe the basic algorithms needed to apply an operator, e.g., a PDE operator as in Section 1.4 or Section 1.5, to an adaptive wavelet expansion described in Section 3.2.1.

3.3.1 The Adaptive Fast Wavelet Transform

Sometimes, computations in wavelet coordinates might be implemented more easily in the single-scale domain. Transforming an adaptive wavelet vector $\mathbf{v}$ on the proper tree $\mathcal{T}$ by the pyramid algorithm (2.1.33) starting from the lowest level $j = j_0$ would always produce a **full** single-scale vector on each next level $j + 1$. Since the sparsity of the adaptive wavelet expansion would thus be lost, we have to re-define what the resulting **single-scale** vector $\mathbf{w} = \mathbf{T}_\mathcal{T}\mathbf{v}$ should actually be.
In our context, a single-scale vector should consist only of **isotropic single-scale** indices, but not necessarily all on the same level j. There should be, however, be no single-scale coefficients present for which all children are also present. This results in a distribution of indices not resembling a tree, but the leaves of a tree.
Such an expansion can of course be easily computed by repeated use of the **reconstruction identities** (2.1.11), (2.1.18) in the isotropic form (2.4.23). Here, even the single-scale function factors of the isotropic wavelets are refined, because otherwise the resulting product of single-scale functions would not be isotropic.
By (2.1.10) and (2.1.17), the isotropic wavelet (2.4.18) can be expressed as

$$\psi_{j,\mathbf{k},\mathbf{e}} = \prod_{i=1}^{n} \sum_{r_i \in \Delta_{j+1}} m^{j,e_i}_{r_i,k_i} \phi_{j+1,r_i} =: \sum_{\mathbf{r} \in \Delta^n_{j+1}} m^{j,\mathbf{e}}_{\mathbf{r},\mathbf{k}} \phi_{j+1,\mathbf{r}} \tag{3.3.1}$$

This **two-scale relation** is now the foundation of the adaptive reconstruction algorithm. First we state a helper algorithm which simply transforms an adaptive wavelet expansion into a expansion of single scale functions:

Algorithm 3.1 Convert a wavelet vector $\mathbf{v}$ on $\mathcal{T}$ into vector of single-scale functions $\mathbf{w}$.

1: **procedure** ADAPTIVE_RECONSTRUCTION($\mathbf{v}$) $\rightarrow \mathbf{w}$
2: $\quad \mathcal{T} \leftarrow S(\mathbf{v})$
3: $\quad j_0 \leftarrow \min\{|\lambda| \mid \lambda \in \mathcal{T}\}$, $J \leftarrow \max\{|\lambda| \mid \lambda \in \mathcal{T}\}$
4: $\quad \mathbf{w} \leftarrow \{\}$ $\quad \triangleright$ Initialize empty (or with zeros on coarsest level j_0)
5: $\quad$ **for** $j = j_0, \ldots, J$ **do**
6: $\quad\quad$ **for all** $\lambda = (j, \mathbf{k}, \mathbf{e}) \in \mathcal{T}_j$ **do**
7: $\quad\quad\quad$ **if** $\mathbf{e} \neq \mathbf{0}$ **then**
8: $\quad\quad\quad\quad \mathbf{w} \leftarrow \mathbf{w} + v_\lambda \sum_{\mathbf{r} \in \Delta_{j+1}^n} m_{\mathbf{r},\mathbf{k}}^{j,\mathbf{e}} \mathbf{b}_{j+1,\mathbf{r}}$ $\quad \triangleright$ Refinement Relation, cf. (3.3.1)
9: $\quad\quad\quad$ **end if**
10: $\quad\quad$ **end for**
11: $\quad$ **end for**
12: $\quad$ **return** $\mathbf{w}$
13: **end procedure**

Here, $\mathbf{b}_{j,\mathbf{r}}$ stands for the **unit vector** in ℓ_2 of the coordinate $\mathbf{r}$. By the nested support property (3.2.12), the support $\mathcal{S} := S(\mathbf{w})$ of the resulting vector $\mathbf{w}$ would again be a tree w.r.t. the single-scale tree structure (3.2.9).
The resulting vector of this algorithm contains many indices for which all refinement indices per (3.3.1) for $\mathbf{e} = \mathbf{0}$ are also present. These elements can be eliminated to arrive at a truly adaptive and **sparse** single-scale representation of a wavelet expansion vector, see Figure 3.2(c), this concept will later be used in Algorithm 3.10. Such an algorithm has been previously presented in [54].

Theorem 3.21 *Algorithm 3.1 finishes in time proportional to the number of input wavelet coefficients $\#S(\mathbf{v})$ and the output satisfies $\#S(\mathbf{w}) \lesssim \#S(\mathbf{v})$.*

Proof: The number of single-scale functions necessary for reconstruction of a wavelet or single-scale function is uniformly bounded, see (2.1.10), (2.1.17). If we designate this constant m, the number of summands in line 8 is bounded by m^n. Since the created addends all have type $\mathbf{e} = \mathbf{0}$, they will not be considered again during execution of the algorithm. ∎

The inverse operation, the **adaptive decomposition**, should reassemble the wavelet vector from the single-scale coefficients generated by Algorithm 3.1. Having knowledge only of the single-scale vector $\mathbf{w}$ and its index set $\mathcal{T}$ would make the decomposition process inefficient, because a single-scale index could be part of wavelets not contained in the tree $\mathcal{T}$. So, to reconstruct only the values of wanted wavelet indices, the target tree $\mathcal{T}$ has to be supplied along with the single-scale data $\mathbf{w}$.
Just as the above relation (3.3.1) can be understood as picking correct columns of the matrix $\mathbf{M}_j$ (2.1.19), the adaptive decomposition works by picking the correct columns

of the matrix $\mathbf{G}_j$ (2.1.21) for every **single-scale** index contained in the input data $\mathbf{w}$ and adds the resulting values to $\mathbf{v}$ for wavelet indices and to $\mathbf{w}$ for single-scale indices, respectively. Mathematically speaking, we use the identity

$$\phi_{j+1,\mathbf{k}} = \prod_{i=1}^{n} \left(\sum_{r_i \in \Delta_j} g^{j,0}_{r,k_i} \phi_{j,r_i} + \sum_{s_i \in \nabla_j} g^{j,1}_{s_i,k_i} \psi_{j,s_i} \right) =: \sum_{\mathbf{r} \in \Delta^n_j} g^{j,\mathbf{0}}_{\mathbf{r},\mathbf{k}} \phi_{j,\mathbf{r}} + \sum_{\substack{\mathbf{e} \in \mathbb{E}^\star_n \\ \mathbf{s} \in \nabla_{j,\mathbf{e}}}} g^{j,\mathbf{e}}_{\mathbf{s},\mathbf{k}} \psi_{j,\mathbf{s},\mathbf{e}}, \quad (3.3.2)$$

which stems from applying (2.1.25) to the isotropic wavelet (2.4.18).

Algorithm 3.2 Adaptive decomposition of single-scale vector $\mathbf{w}$ into vector $\mathbf{v}$ in wavelet coordinates on the proper tree $\mathcal{T}$.

1: **procedure** ADAPTIVE_DECOMPOSITION($\mathbf{w}, \mathcal{T}$) $\rightarrow \mathbf{v}$
2: $\quad \mathcal{S} \leftarrow S(\mathbf{w})$
3: $\quad j_0 \leftarrow \min \{|\lambda| \mid \lambda \in \mathcal{S}\}$, $J \leftarrow \max \{|\lambda| \mid \lambda \in \mathcal{S}\}$
4: $\quad \mathbf{v}_{j_0} \leftarrow \mathbf{0}$ ▷ Initialize with zeros on coarsest level j_0
5: $\quad$ **for** $j = J, \dots, j_0 + 1$ **do**
6: $\quad\quad$ **for all** $\lambda = (j, \mathbf{k}, \mathbf{0}) \in \mathcal{S}$ **do**
7: $\quad\quad\quad D^1_\lambda \leftarrow \{\mathbf{s} \in \nabla_{j-1,\mathbf{e}}, \mathbf{e} \in \mathbb{E}^\star_n \mid g^{j-1,\mathbf{e}}_{\mathbf{s},\mathbf{k}} \neq 0\}$ ▷ Determine per decomposition
8: relation (3.3.2).
9: $\quad\quad\quad$ **for all** $\mathbf{s} \in D^1_\lambda$ with $\mathbf{s} \in \mathcal{T}$ **do**
10: $\quad\quad\quad\quad \mathbf{v} \leftarrow \mathbf{v} + w_\lambda \sum_{\mathbf{s} \in \nabla_{j-1,\mathbf{e}}} g^{j-1,\mathbf{e}}_{\mathbf{s},\mathbf{k}} \mathbf{b}_{j-1,\mathbf{s},\mathbf{e}}$ ▷ Apply decomposition relation
11: $\quad\quad\quad$ **end for**
12: $\quad\quad\quad \mathbf{w} \leftarrow \mathbf{w} \setminus \mathbf{w}_\lambda$ ▷ Delete index λ
13: $\quad\quad$ **end for**
14: $\quad$ **end for**
15: $\quad$ **return** $\mathbf{v}$
16: **end procedure**

Again, $\mathbf{b}_{j,\mathbf{k},\mathbf{e}}$ stands for the **unit vector** in ℓ_2 of the coordinate $(j, \mathbf{k}, \mathbf{e})$. This algorithm is only an exact inverse of Algorithm 3.1, if $\mathcal{T} = S(\mathbf{v})$. Since line 10 will produce all possible types $\mathbf{e} \neq \mathbf{0}$ if $\mathcal{T} = \mathcal{I}$, some might not be already present in $S(\mathbf{v})$ and would be created.

Theorem 3.22 *Algorithm 3.2 finishes in time proportional to the number of input wavelet coefficients $\#S(\mathbf{w})$ and the output satisfies $\#S(\mathbf{v}) \lesssim \#S(\mathbf{w})$.*

Proof: The argument is exactly the same as for Algorithm 3.1, except here all created wavelet indices have type $\mathbf{e} \neq \mathbf{0}$, but the input data has type $\mathbf{e} = \mathbf{0}$. ∎

Remark 3.23 *A few remarks about these algorithms.*

1. *If the single-scale input data is not already present on the coarser levels, for example as in see Figure 3.2 (c), an* **upscaling** *step can be inserted after line 11 by noting the first part of (3.3.2):*

$$\mathbf{w} \leftarrow \mathbf{w} + w_\lambda \sum_{\mathbf{r} \in \Delta^n_{j-1}} g^{j-1,\mathbf{0}}_{\mathbf{r},\mathbf{k}} \mathbf{b}_{j-1,\mathbf{r},\mathbf{e}}$$

This computation will produce the input data for the next coarser level. But this task makes more sense if no target tree $\mathcal{T}$ is given. Because of the exponential decline in cardinality of the involved index set, the complexity of this variant is still linear w.r.t. the input vector, just as in the complexity estimate of the pyramid scheme (2.1.33). If the input vector contains only one non-zero coefficient, the size of the output vector would have a size proportional to the level of the input coefficient, which can be (in theory) arbitrarily high. This problem is circumvented by again assuming a tree structure in the input vector.

2. *Algorithm 3.2 can also be run with single-scale input data not generated by Algorithm 3.1 by simply assuming missing values to be zero.*

3. *To apply the* **dual reconstruction** *or* **decomposition** *algorithms (see Figure 2.1), one simply has to choose the matrices $\widetilde{\mathbf{M}}_j = \mathbf{G}_j^T$ and $\widetilde{\mathbf{G}}_j = \mathbf{M}_j^T$ instead of $\mathbf{M}_j$, $\mathbf{G}_j$, respectively.*

Sample results of these algorithms can be seen in Figure 3.2.

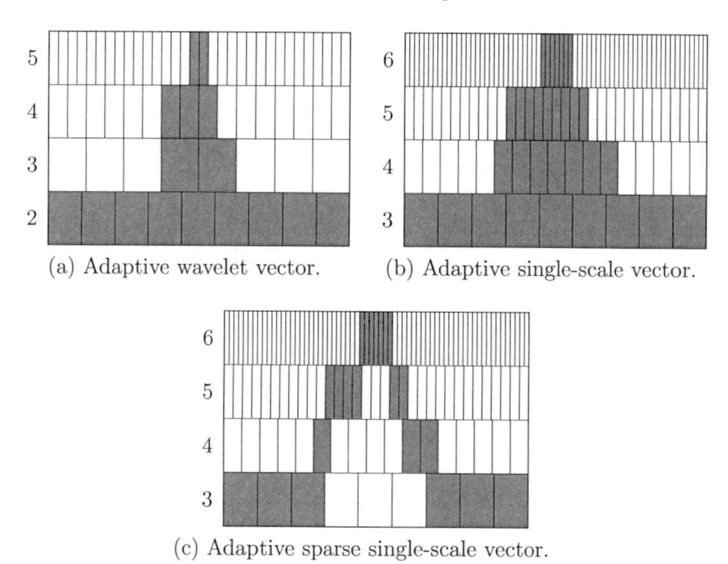

(a) Adaptive wavelet vector.

(b) Adaptive single-scale vector.

(c) Adaptive sparse single-scale vector.

Figure 3.2: In the upper left diagram (a), an arbitrary wavelet expansion for the `DKU22NB` Wavelets is shown. All non-zero coefficients are shown in grey. Subfigure (b) shows the result of Algorithm 3.1 using the wavelet expansion depicted in (a) as input. The bottom diagram (c) presents the result after all indices, for which all refinement indices exist in the vector, were refined. This vector is not based upon a tree, but resembles the outer hull, i.e., the leaves, of some other vector.

3.3.2 Tree Coarsening

The idea of **coarsening** is to determine, for a given proper tree $\mathbf{v} \in \ell_2$, $\mathcal{T} := S(\mathbf{v})$, another, minimally supported, tree-structured $\mathbf{w} \in \ell_2(\mathcal{I})$ such that the error is only increased by a given tolerance $\varepsilon \geq 0$, i.e.

$$\|\mathbf{v} - \mathbf{w}\|_{\ell_2(\mathcal{I})} \leq \varepsilon$$

By the **norm equivalences** (2.2.12) and (3.2.20), this error estimate immediately translates over to an error estimate in the considered function spaces. Conceptually, we purge an input vector of too small values, in particular zero values, to get rid of excess entries. There are several strategies to accomplish this task, the easiest way is to simply remove all expansion coefficients with a value less than the tolerance ε. But this naive approach has the drawback of not preserving the **tree structure** if not only leaves of the tree $\mathcal{T}$ are deleted. Of course any family of wavelet indices can be expanded to become a proper tree, but then the inserted values for these wavelet coefficients are zero, so in the worst case scenario the input tree and the output tree are exactly the same, except for some values set to zero.
To quantify what we accept as appropriate coarsened tree, we need the following two definitions:

Definition 3.24 [Best Tree]
For a given proper tree-structured $\mathbf{v} \in \ell_2(\mathcal{I})$*, a tree* $\mathcal{T}^\star := \mathcal{T}^\star(\eta, \mathbf{v})$ *with* $\eta \geq 0$ *is called* η**-best tree**, *if*

$$\|\mathbf{v} - \mathbf{v}|_{\mathcal{T}^\star}\|_{\ell_2(\mathcal{I})} \leq \eta, \qquad \textit{and} \qquad \#\mathcal{T}^\star = \min\left\{\#\widetilde{\mathcal{T}} \mid \|\mathbf{v} - \mathbf{v}|_{\widetilde{\mathcal{T}}}\|_{\ell_2(\mathcal{I})} \leq \eta \textit{ for any tree } \widetilde{\mathcal{T}}\right\}. \tag{3.3.3}$$

Since the above definition is in practical applications too strict, it is relaxed by introducing a constant $C \geq 1$ with which the tree is allowed to grow:

Definition 3.25 [Near-Best Tree]
A tree $\mathcal{T}' := \mathcal{T}'(\eta, C, \mathbf{v})$ *with* $\eta \geq 0$ *and* $C \geq 1$ *is called* (η, C)**-near-best tree**, *if*

$$\|\mathbf{v} - \mathbf{v}|_{\mathcal{T}'}\|_{\ell_2(\mathcal{I})} \leq \eta, \qquad \textit{and} \qquad \#\mathcal{T}' \leq C\#\mathcal{T}^\star(\eta/C, \mathbf{v}). \tag{3.3.4}$$

For large values $C \gg 1$, the error $\eta/C \to 0$ and thus $\mathcal{T}^\star(\eta/C, \mathbf{v}) \to \mathcal{T}$, while the number of possible nodes in the tree $\mathcal{T}'$ could go to infinity. Values of $C \approx 1$ are thus "better" near-best trees.

The Algorithm

A **coarsening algorithm**, whose output is a **near-best tree**, was proposed in [16]. The idea of the algorithm is not to delete any wavelet coefficients from an input tree, but only to cut whole **branches**, e.g., a node and all its children. This approach entails another problem though: the identification of "unimportant" branches, i.e. branches that can be removed while controlling the error.
To measure the error, we assign the ℓ_2-value of the branch at index $\lambda \in \mathcal{T} \subset \mathcal{I}$, i.e.

$$e(\lambda) := \sum_{\mu \prec \lambda} |v_\mu|^2, \tag{3.3.5}$$

then the error of restricting $\mathbf{v}$ to a **subtree** $\mathcal{T}' \subset \mathcal{T}$, i.e. $\|\mathbf{v} - \mathbf{v}_{\mathcal{T}'}\|_{\ell_2(\mathcal{I})}$, is simply,

$$E(\mathbf{v}, \mathcal{T}') := \sqrt{\sum_{\lambda \in \mathcal{L}(\mathcal{T}')} e(\lambda)}. \tag{3.3.6}$$

Since the coarsened tree is supposed to be minimal in size, it is imperative to find deep branches $\mathcal{T}_{[\lambda]}$ with a small error $e(\lambda)$. But (3.3.5) does not weigh in the depth of the branches, so a modified error functional is sought. In [16], the following error functional was proposed:

$$\widetilde{e}(\lambda) := \begin{cases} e(\lambda), & \text{if } \lambda \in \mathcal{N}_0(\mathcal{T}), \\ \left(\sum_{\mu \in \mathcal{C}(\tau)} e(\mu)\right) \frac{\widetilde{e}(\tau)}{e(\tau) + \widetilde{e}(\tau)}, \text{ with } \tau = \Pi(\lambda), & \text{otherwise.} \end{cases} \tag{3.3.7}$$

The fractional construction of the general term arranges for a monotonous, although not linear, decline in values over the whole tree. An example of a coarsened tree including the values of $e(\lambda)$ and $\widetilde{e}(\lambda)$ can be found in [147]. This functional obviously assigns the same value to each child λ of a node τ, because the general term does not depend on the index λ directly.

The above construction does leave one detail uncertain: How to handle roots without children. Since there is only one value associated with such a leaf, i.e. $|v_\lambda|$, only this value can be used to define an error indicator. This plausible strategy was proposed in [147] and we include it in Algorithm 3.3. But, since we always assume to be handling **proper trees**, it is also a proper strategy to simply always include all roots of the input vector into the output vector. Since we are only talking about a finite number of root nodes of a potentially extremely large tree $\mathcal{T}$, both policies do not affect asymptotic bounds w.r.t. runtime or storage.

Remark 3.26 *A few remarks concerning the implementation are to be given:*

- *The error tree given by (3.3.7) does not need to be assembled. This is because the value $\widetilde{e}(\lambda)$ does only need* **local** *information in the sense that only direct relatives (parents, siblings) are needed to compute it.*

- *On the other hand, the error tree given by (3.3.5) needs* **global** *information, i.e., the value $e(\lambda)$ depends on all the descendants of λ. Assembling each value by each time computing the whole subtree would be inefficient.*

- *Finding the maximum element of the sets $\mathcal{L}, \mathcal{R}$ does not need to be exact. This would require an exact sorting which would lead to an inefficient scheme. An alternative is presented in the next paragraph.*

The following result from [16, 33] shows the result of Algorithm 3.3 is a **near-best** tree and it is computable in **linear** time.

Lemma 3.27 *For $N := S(\mathbf{v})$ the computational cost of* `TREE_COARSE`$(\eta, \mathbf{v}) \to (\mathbf{w}, \mathcal{T}')$ *is proportional to $\mathcal{O}(N)$. The tree $\mathcal{T}'$ is $(\eta, C^\star)$-near-best, where $C^\star > 0$ is a finite constant.*

If for $s > 0$, $\mathbf{u} \in \mathcal{A}^s_{tree}$ and $\|\mathbf{u} - \mathbf{v}\| \leq \eta$ for $\#S(\mathbf{v}) < \infty$, then TREE_COARSE$(2C^\star\eta, \mathbf{v}) \to (\mathbf{w}, \mathcal{T}')$ *satisfies*

$$\#S(\mathbf{w}) \lesssim \|\mathbf{u}\|^{1/s}_{\mathcal{A}^s_{tree}} \eta^{-1/s}, \tag{3.3.8}$$

$$\|\mathbf{u} - \mathbf{w}\|_{\ell_2(\mathcal{I})} \leq (1 + 2C^\star)\eta, \tag{3.3.9}$$

and

$$\|\mathbf{w}\|_{\mathcal{A}^s_{tree}} \lesssim \|\mathbf{u}\|_{\mathcal{A}^s_{tree}}, \tag{3.3.10}$$

where the constants depend only on s when $s \to 0$ and on $C^\star$.

The assertions (3.3.8) – (3.3.10) show that the tree coarsening process does not take away any of the properties of the class $\mathcal{A}^s_{\text{tree}}$: The bound (3.3.10) shows that even a near-best tree of an approximation of a vector in $\mathcal{A}^s_{\text{tree}}$ is still contained in $\mathcal{A}^s_{\text{tree}}$. Estimate (3.3.8) essentially means that the size of the support of $\mathbf{w}$ is proportional to $\eta^{-1/s}$, which is optimal in the sense of (3.2.19). Lastly, (3.3.9) follows obviously after an application of the triangle inequality.

Algorithm 3.3 Adaptive tree coarsening of a vector $\mathbf{v}$ in wavelet coordinates on the proper tree $\mathcal{T}$. Output is a vector $\mathbf{w}$ on tree $\mathcal{T}'$ with $\|\mathbf{v} - \mathbf{w}\| \leq \eta$.

1: **procedure** TREE_COARSE$(\eta, \mathbf{v}) \to (\mathbf{w}, \mathcal{T}')$
2: $\quad \mathcal{T} \leftarrow S(\mathbf{v})$
3: $\quad \mathcal{R} \leftarrow \mathcal{N}_0(\mathcal{T})$ ▷ **List of roots without children**
4: $\quad \mathcal{L} \leftarrow \{(\lambda, \widetilde{e}(\lambda)) \mid \lambda \in \mathcal{N}_0(\mathcal{T}) \setminus \mathcal{R}\}$ ▷ $\mathcal{L}$ is **List of elements to refine**
5: Compute (3.3.7) for all indices
6: $\quad \mathcal{T}' \leftarrow \mathcal{L}$ ▷ Initialize with roots who have children
7: $\quad$ **while** $E(\mathbf{v}, \mathcal{T}') > \eta$ **do** ▷ As long as $\mathcal{L} \neq \emptyset$ and $\mathcal{R} \neq \emptyset$
8: $\quad\quad (\lambda_R, v_R) \leftarrow \arg\max_{\mu \in \mathcal{R}} \widetilde{e}(\mu)$
9: $\quad\quad (\lambda_L, v_L) \leftarrow \arg\max_{\mu \in \mathcal{L}} |v_\mu|$
10: $\quad\quad$ **if** $v_R \geq v_L$ **then**
11: $\quad\quad\quad \mathcal{T}' \leftarrow \mathcal{T}' \cup \{\lambda_R\}$ ▷ Add element of list $\mathcal{R}$
12: $\quad\quad\quad \mathcal{R} \leftarrow \mathcal{R} \setminus \{\lambda_R\}$ ▷ Remove λ_R from list of roots without children
13: $\quad\quad$ **else**
14: $\quad\quad\quad \mathcal{T}' \leftarrow \mathcal{T}' \cup \{\lambda_L\}$ ▷ Add element of list $\mathcal{L}$
15: $\quad\quad\quad \mathcal{L} \leftarrow (\mathcal{L} \setminus \{\lambda_L\}) \cup \{(\mu, \widetilde{e}(\mu)) \mid \mu \in \mathcal{C}(\lambda_L)\}$ ▷ Remove λ_L from list of
16: elements to refine and add all children $\mathcal{C}(\lambda_L)$
17: $\quad\quad$ **end if**
18: $\quad$ **end while**
19: $\quad \mathbf{w} \leftarrow \mathbf{v}|_{\mathcal{T}'}$ ▷ Compute coarsened vector
20: $\quad$ **return** $(\mathbf{w}, \mathcal{T}')$
21: **end procedure**

Quasi-Sorting vs. Exact Sorting

The following well known result, see [92], shows that exact sorting using a **predicate function** cannot be an optimal complexity process. The argument is quite simple, but usually not taught in mathematical courses, so I would like to recapitulate it here:

Proposition 3.28 *Sorting $N \in \mathbb{N}$ numbers one needs generally $\mathcal{O}(N \log N)$ comparisons.*

Proof: Let $T(N)$ be the **average number of comparisons** to sort N numbers. Every decision for sorting by a predicate function gives either **true** or **false**, i.e., two distinct results. Thus, with $T(N)$ operations one can have $2^{T(N)}$ different outcomes, and each one corresponds to a different sorting history. But there are $N!$ different input configurations for N numbers. Since there must be a way to sort any finite set of numbers, the number of possible sorting histories and input configurations must be the same, i.e., by **Stirling's approximation** follows

$$\begin{aligned} 2^{T(N)} \approx N! \quad \Longrightarrow \quad T(N) &\approx \log(N!), \\ &\approx \log\left(\sqrt{2\pi N}\left(\frac{N}{e}\right)^N\right), \\ &\approx \log(N^N) = \log(e^{N \log N}) = N \log(N). \end{aligned}$$

With Landau symbols, this reads $T(N) = \mathcal{O}(N \log N)$, proving the assertion. ∎

This detail increases the complexity of any algorithm including an exact sorting predicate by a logarithmic factor. In our framework of adaptive methods, an exact sorting is not required, a **quasi-sorting** can suffice. This is in line with replacing the concept of the **best tree** (3.3.3) with the **near-best tree** (3.3.4).
Quasi sorting, or **binary binning** as it is called in this context (see [10]), defines exponentially growing **bins** which are used to map to elements in an unordered way:

Definition 3.29 [Binary Bins]
Let $\mathbf{v} = (v_\lambda)_{\lambda \in \mathcal{I}}$, $v_\lambda \in \mathbb{R}$, be a sequence. The j-th **bin** *is defined by*

$$\mathcal{B}_j(\mathbf{v}) := \left\{\lambda \in S(\mathbf{v}) \,|\, 2^{j-1} < |v_\lambda| \leq 2^j\right\}, \quad \text{for } j \in \mathbb{Z}. \tag{3.3.11}$$

Hence, finding the bin to a value $|v_\lambda|$ requires only a evaluation of the logarithm $\log_2(\cdot)$. In return, the bin index j gives an upper and lower bound estimate for the values contained in $\mathcal{B}_j(\mathbf{v})$.

Remark 3.30 *In practice, it does not make much sense to allow bins for all $j \in \mathbb{Z}$.*

- *The lower bound for any practical bin index is given by the* **machine precision** *ε of the used floating point numbers. Even quasi-sorting for values smaller than ε can be omitted since these values can be considered being zero for all intents and purposes. If only values lower than $\eta > 0$ are sought, then $\lfloor \log_2(\eta) \rfloor - 1$ can be used as the lowest bin index.*

- *Since one usually is, as in this coarsening context, interested in the smallest values, restricting the upper bound of the index j can be reasonable as well, since this saves memory required to hold bins and the computation of the bin number can be reduced to checking whether $|v_\lambda| > 2^{J-1}$ holds for the largest bin index J.*
- *In what kind of data structure the indices λ are saved is not very important, as long as insertions, deletions and any kind of access can be done in* **(amortized) constant time.**

A complete coarsening scheme "bin-thresh" only based on this binary binning principle can be found in [10]. In Algorithm 3.3 it only serves as a means to alleviate the complexity of the sorting problem.

3.3.3 Tree Prediction

Here we describe the determination of a suitable tree structured index set $\mathcal{T}'$ which assures that the operator $F : \mathcal{H} \to \mathcal{H}'$ can be applied up to a desired prescribed accuracy ϵ on that index set. That means, one needs to find a tree structured index set $\mathcal{T}' \subset \ell_2(\mathcal{I})$ such that

$$\|\mathbf{F}(\mathbf{v}) - \mathbf{F}(\mathbf{v})|_{\mathcal{T}'}\|_{\ell_2(\mathcal{I})} \lesssim \epsilon. \tag{3.3.12}$$

Moreover, in view of the overall computational complexity, the determination of such an index set shall be arranged in such a way that

a) the set $\mathcal{T}'$ is as small as possible and

b) the method can be applied as efficiently as possible.

Such kind of algorithms, in the wavelet context called `PREDICTION`, are available, see [31,33,34]. The contents of this section are based upon these papers.
Two prerequisites on the wavelet discretized operator $\mathbf{F} : \ell_2(\mathcal{I}) \to \ell_2(\mathcal{I})$ are needed to use the prediction scheme. These are generally stated as follows:

- **Assumption 1**: $\mathbf{F}$ is a **Lipschitz map** from $\ell_2(\mathcal{I})$ into itself, i.e.,

  $$\|\mathbf{F}(\mathbf{u}) - \mathbf{F}(\mathbf{v})\|_{\ell_2(\mathcal{I})} \leq C_1(\sup\{\|\mathbf{u}\|_{\ell_2(\mathcal{I})}, \|\mathbf{v}\|_{\ell_2(\mathcal{I})}\})\|\mathbf{u} - \mathbf{v}\|_{\ell_2(\mathcal{I})}, \tag{3.3.13}$$

  where $C_1(\cdot)$ is a a positive nondecreasing function.

- **Assumption 2**: For every finitely support $\mathbf{v} \in \ell_2(\mathcal{I})$ and $\mathbf{w} := \mathbf{F}(\mathbf{v})$, the operator fulfills the following **decay** estimate with a parameter $\gamma > n/2$:

  $$|w_\lambda| \leq C_2(\|\mathbf{v}\|_{\ell_2(\mathcal{I})}) \sup_{\mu : S_\mu \cap S_\lambda \neq \emptyset} |v_\mu| \, 2^{-\gamma(|\lambda| - |\mu|)}, \tag{3.3.14}$$

  where $\lambda \in \mathcal{I} \setminus \mathcal{N}_0$ and $C_2(\cdot)$ is a positive nondecreasing function.

Remark 3.31 *Assumption 1 follows directly from (1.5.8). It was shown in Theorem 1.59, that this is true for a wide class of operators satisfying the* **nonlinear growth condition** *(1.5.11).*

Assumption 2 is closely related to **operator compression**, see [32], as (3.3.14) predicts how fast elements in the (infinite) matrix $\mathbf{F}$ will tend toward zero, depending on their level difference, or distance from the diagonal.

Proposition 3.32 *If an operator is composed of a sum of discretized operators, i.e.* $\mathbf{F} := \sum_{i=1}^{k} \mathbf{F}_i$, $\mathbf{F}_i : \ell_2(\mathcal{I}) \to \ell_2(\mathcal{I})$, *each fulfilling Assumptions 1 & 2 with individual constants* $\gamma_i > n/2$ *and functions* $C_{1,i}(\cdot)$, $C_{2,i}(\cdot)$, *then both assumptions are still valid for* $\mathbf{F} : \ell_2(\mathcal{I}) \to \ell_2(\mathcal{I})$ *with*

- **Assumption 1**: $C_1(\cdot) := k \max_i C_{i,1}(\cdot)$,
- **Assumption 2**: $\gamma := \min_i \{\gamma_i\}$, $C_2(\cdot) := k \max_i C_{2,i}(\cdot)$.

Proof: The assertion for Assumption 1 simply follows from the triangle inequality:

$$\begin{aligned}
\|\mathbf{F}(\mathbf{u}) - \mathbf{F}(\mathbf{v})\| &= \|\sum_i (\mathbf{F}_i(\mathbf{u}) - \mathbf{F}_i(\mathbf{v}))\| \\
&\leq \sum_i C_{1,i}(\sup\{\|\mathbf{u}\|, \|\mathbf{v}\|\})\|\mathbf{u} - \mathbf{v}\| \\
&\leq \left(\sum_i C_{1,i}(\sup\{\|\mathbf{u}\|, \|\mathbf{v}\|\})\right)\|\mathbf{u} - \mathbf{v}\| \\
&\leq k \max_i C_{1,i}(\sup\{\|\mathbf{u}\|, \|\mathbf{v}\|\})\|\mathbf{u} - \mathbf{v}\|.
\end{aligned}$$

For Assumption 2, it should be noted that in the proof of the prediction theorem in [34], it is always assumed that $|\lambda| \geq |\mu|$, so that the exponent of (3.3.14) is always ≤ 0 since $\gamma > n/2 > 0$. This makes sense as the word "decay" implies a decline of values. Under the assumption $|\lambda| \geq |\mu|$ it follows that

$$\begin{aligned}
(\mathbf{F}(\mathbf{v}))_\lambda = \left(\left(\sum_i \mathbf{F}_i\right)(\mathbf{v})\right)_\lambda &= \sum_i (\mathbf{F}_i(\mathbf{v}))_\lambda \\
&\leq \sum_i C_{2,i}(\|\mathbf{v}\|) \sup_{\mu : S_\mu \cap S_\lambda \neq \emptyset} |v_\mu| \, 2^{-\gamma_i(|\lambda| - |\mu|)} \\
&\leq k \max_i C_{2,i}(\|\mathbf{v}\|) \sup_{\mu : S_\mu \cap S_\lambda \neq \emptyset} |v_\mu| \, 2^{-\min_i \gamma_i(|\lambda| - |\mu|)}
\end{aligned}$$

Even if (3.3.14) must hold for all combinations $\lambda, \mu \in \mathcal{I}$, this can be accomplished by adapting the constants of Assumption 2: Since there are only a finite number of coarser levels for any given wavelet index μ, e.g., $|\mu| - j_0$, and this can be bounded by the maximum level of any index in the finite vector, e.g., $J := \max_{\mu \in S(\mathbf{v})} |\mu|$, it then follows for $|\lambda| < |\mu|$:

$$\max_{\mu \in S(\mathbf{v}), \lambda \in S(\mathbf{w})} \{|\mu| - |\lambda|\} \leq \max_{\mu \in S(\mathbf{v})} |\mu| - \min_{\lambda \in S(\mathbf{w})} |\lambda| \equiv J - j_0.$$

From this follows

$$0 \leq |\mu| - |\lambda| \leq J - j_0 \quad \Longrightarrow \quad 1 \leq 2^{\gamma(|\mu| - |\lambda|)} \leq 2^{\gamma(J - j_0)}.$$

But with this follows for (3.3.14),

$$|w_\lambda| = C_2(\|\mathbf{v}\|)\, 2^{\gamma(J-j_0)} \sup_{\mu:S_\mu \cap S_\lambda \neq \emptyset} |v_\mu|\, 2^{-\gamma(|\lambda|-|\mu|+(J-j_0))}$$
$$=: C_2'(\|\mathbf{v}\|) \sup_{\mu:S_\mu \cap S_\lambda \neq \emptyset} |v_\mu|\, 2^{-\gamma(|\lambda|-|\mu|+(J-j_0))},$$

where the exponent can now only take on values ≤ 0. The equation (3.3.14) is obviously also satisfied setting $C_2'(\|\mathbf{v}\|) := 2^{\gamma(J-j_0)}\, C_2(\|\mathbf{v}\|)$ and the asymptotic behavior for $|\lambda|, |\mu| \to \infty$ is not affected: The constant term is in effect like a perturbed decay parameter γ', with $\gamma' \to \gamma$ in the asymptotics.
Since Assumption 2 is still valid for $C_2'(\cdot)$, the above deliberations can be utilized using this adapted function. ■

The proposition shows what one can understand instinctively: Adding a quickly ($\gamma \gg 1$) and a slowly ($\gamma \approx 1$) decaying operator, the result is still a slowly decaying operator with a little perturbation. The application of this proposition lies in the approach of not handling linear and nonlinear operators separately when a single vector is being applied (as depicted in [33]), but in handling the sum of such operators as a single new operator. This saves resources by reusing the output of Algorithm 3.4 and ensures that all target wavelet coefficients are computed to maximum accuracy.

Remark 3.33 *Assumption 1 is trivially fulfilled for a* **linear bounded operator**, *Assumption 2 must be verified nonetheless, e.g., using Theorem 1.59.*

The question of how the parameter γ of (3.3.14) can be determined was also answered in [34]:

Remark 3.34 *The standard value of the decay parameter from (3.3.14) for many operators $F : H^m(\Omega) \to (H^m(\Omega))'$ and $\Omega \subset \mathbb{R}^n$ using biorthogonal wavelets vanishing moments $d, \tilde{d}$ is given by*

$$\gamma := r + \overline{d} + n/2, \tag{3.3.15}$$

where $\overline{d} := \min\left\{d, \tilde{d}\right\}$ and

- $r := \min\{m, s^\star\}$, *if (1.5.11) holds for an $s^\star \geq 0$ and $m \geq n/2$,*
- $r := \lceil \min\{m, s^\star, \overline{s}\} \rceil$, *if (1.5.11) holds for $0 \leq \overline{s} < r^\star$ of (1.5.12) and $m < n/2$.*

The above can be improved upon if the inequality (1.5.11) can be shown to hold without the function $(\cdot)_+$ in the exponent, which then yields $r := m$.

This concludes the theoretical prerequisites of the scheme, we now describe how the algorithm works.

The Algorithm

Tree Prediction is based upon the idea outlined in Section 3.3.2: Categorize individual wavelet indices according to their "importance" (given by the error functional) and add more indices in the vicinity of the more highly regarded indices. The principle line of thought being, that because of the two locality properties $(\mathcal{L})$(2.1.8) and (3.1.1), the results of the operator application $F(\cdot)$ must only **significantly** influence a limited number of indices which have to be in the vicinity. Additionally, by the norm equivalence Theorem 2.18, indices which are less significant, i.e., if their tree branch was cut, the error made would be small, have lesser influence on their vicinity and thus fewer indices in that vicinity are required in the output tree $\mathcal{T}$.
To this end, the input tree is divided into **layers**, i.e. non-overlapping sets $\widehat{\Delta}_j$ (not to be confused with (2.1.7)), where each layer identifies a group of indices of the same significance, which will correspond to the same depth of refinement in the construction of the predicted tree. So, first the set of $(\varepsilon, C^\star)$-near best trees of the input vector $\mathbf{v} \in \ell_2(\mathcal{I})$, i.e.

$$\mathcal{T}_j := \mathcal{T}\left(\frac{2^j\varepsilon}{j+1}, \mathbf{v}\right), \tag{3.3.16}$$

and the corresponding **expanded trees** $\widetilde{\mathcal{T}}_j$ are computed for $j = 0, \dots, J$. Since $\frac{2^j\varepsilon}{j+1} \to \infty$ for $j \to \infty$, the index $J < \infty$ denotes the smallest index for which $\mathcal{T}_J = \emptyset$. It holds $J \lesssim \log_2(\|\mathbf{v}\|/\varepsilon)$. Then we set

$$\widehat{\Delta}_j := \widetilde{\mathcal{T}}_j \setminus \widetilde{\mathcal{T}}_{j+1}, \tag{3.3.17}$$

where $\widehat{\Delta}_0$ contains very few but significant indexes of $\mathcal{T}$ and $\widehat{\Delta}_{J-1}$ many indexes of lower significance. For each index $\lambda \in \widehat{\Delta}_j$, we now compute the **Influence set** of depth $c\,j \geq 0$, i.e.,

$$\Lambda_\lambda(c\,j) := \{\mu \in \mathcal{I} \mid S_\lambda \cap S_\mu \neq \emptyset \text{ and } |\lambda| \leq |\mu| \leq |\lambda| + c\,j\}, \tag{3.3.18}$$

where the constant c is determined by

$$c := \frac{2}{2\gamma - n} > 0, \tag{3.3.19}$$

where $\gamma > n/2$ is the parameter of (3.3.14). Including indices of lower levels in the Influence set, e.g., $|\mu| < |\lambda|$ is unnecessary, as the **expansion property** (3.2.4) holds. The final predicted tree is then given by

$$\mathcal{T}' := \mathcal{I}_{j_0} \cup \bigcup_{j=0}^{J} \bigcup_{\lambda \in \widehat{\Delta}_j} \Lambda_\lambda(c\,j). \tag{3.3.20}$$

The implementation of the `PREDICTION` can be summarized in these three steps:

1. Calculate the error functional (3.3.7) for whole input tree,

2. Do a quasi-sorting of the indices and associate the refinement constant (3.3.19) for each element,

3. Set up the predicted tree given by the union of all influence sets (3.3.18).

The above algorithm can be executed in optimal complexity, as this result (Theorem 3.4, [34]) shows:

Proposition 3.35 *The number of arithmetic and sorting operations needed to determine the prediction set $\mathcal{T}'$ using the* `PREDICTION` *algorithm can be bounded by a constant multiple of $(\#\mathcal{T}' + \#S(\mathbf{v}))$.*

Although it is not relevant for the operator application on a given index set itself, for the sake of completeness we would like to mention that it can be shown that the size of the generated index set $\mathcal{T}$ is optimal in the sense of being the smallest tree for which (3.3.12) holds.
We cite the main result from [34] :

Theorem 3.36 *Given any $\mathbf{v} \in \ell_2(\mathcal{I})$ and $\mathcal{T}'$ defined by (3.3.20), we have the error estimate*

$$\|\mathbf{F}(\mathbf{v}) - \mathbf{F}(\mathbf{v})|_{\mathcal{T}'}\|_{\ell_2(\mathcal{I})} \lesssim \varepsilon. \tag{3.3.21}$$

Moreover, if $\mathbf{v} \in \mathcal{A}^s_{tree}$ for $0 < s < \frac{2\gamma - n}{2n}$, the estimate

$$\#(\mathcal{T}') \lesssim \|\mathbf{v}\|^{1/s}_{\mathcal{A}^s_{tree}} \varepsilon^{-1/s} + \#(\mathcal{N}_0) \tag{3.3.22}$$

holds. Therefore $\mathbf{F}(\mathbf{v}) \in \mathcal{A}^s_{tree}$ and

$$\|\mathbf{F}(\mathbf{v})\|_{\mathcal{A}^s_{tree}} \lesssim 1 + \|\mathbf{v}\|_{\mathcal{A}^s_{tree}}. \tag{3.3.23}$$

Algorithm 3.4 Tree Prediction for an operator $\mathcal{F}$ (given by γ) and a vector $\mathbf{v}$ in wavelet coordinates on the proper tree $\mathcal{T}$. Output is a tree $\mathcal{T}'$ with $\|\mathbf{v} - \mathbf{w}\| \leq \varepsilon$.

```
1: procedure PREDICTION(ε, v, γ) → T'
2:     T ← S(v)
3:     j ← max{ k ∈ N_0 | 2^k ε/(k+1) < ||v|| }          ▷ Set maximum admissible value
4:     L ← {(λ, ẽ(λ)) | λ ∈ N_0(T)}                      ▷ Error ẽ(λ) defined as in Algorithm 3.3
5:     T' ← L                                            ▷ Initialize with roots
6:     η ← ||v||^2 − Σ_{λ∈L} |v_λ|^2                     ▷ Error up to this point
7:     while √η > ε do
8:         if √η < 2^j ε/(j+1) then
9:             j ← j − 1
10:        end if
11:        (λ, v) ← arg max_{μ∈L} ẽ(μ)                   ▷ Choose maximum element
12:        T' ← T' ∪ Λ_λ(cj)                             ▷ Compute Influence set
13:        L ← (L \ {λ_L}) ∪ {(μ, ẽ(μ)) | μ ∈ C(λ)}      ▷ Remove λ from list and
14:                                                      add all children C(λ)
15:        η ← η − |v_λ|^2                               ▷ Decrease error value
16:    end while
17:    return T'
18: end procedure
```

The constants in these estimates depend only on $\|\mathbf{v}\|_{\mathcal{A}^s_{tree}}$, *the space dimension* n *and the smoothness parameter* s.

The inequality (3.3.21) shows that the resulting tree is big enough such that $\mathbf{F}(\mathbf{v})$ can be computed up to the desired accuracy ε. The estimate (3.3.22) shows that the predicted tree $\mathcal{T}'$ is not unreasonable big, i.e., its size depends on the smoothness parameter s of the input vector $\mathbf{v}$ and the desired accuracy ε. The higher s, the fewer indices are required, the lower the tolerance ε, the bigger the tree. Lastly, (3.3.23) shows that the result is still an element of the smaller class $\mathcal{A}^s_{\text{tree}}$.
At the core of the above Algorithm 3.4 lies the computation of the Influence set (3.3.18). We discuss the details of this set in the next section.

3.3.4 Approximating the Influence Set

A crucial point in the assembly of the predicted tree of Section 3.3.3 is the computation of a wavelet index influence set (3.3.18). As mentioned before, this set is highly dependent on the properties of the wavelet, e.g. polynomial degree, position and level, with boundary adaptations complicating things further. It is not desirable to have to construct an individual influence set depending on all properties of every wavelet. From an implementational point of view, a simple, universally applicable construction of the index set is preferred. A **good approximation** to the influence set should comply with these requirements:

1. Simple implementation, i.e., mostly independent of wavelet position.
2. Computationally cheap, i.e., very low complexity.
3. No excess, i.e., no wavelet index which is not part of the exact influence set should be included.

There are several policies adhering to these conditions. We propose the following alternatives

Children Only Only take into account direct descendants of any wavelet index. This leads to the set

$$\Lambda^1_\lambda(c) := \{\mu \preceq \lambda \in \mathcal{I} \mid |\lambda| \leq |\mu| \leq |\lambda| + c\} \tag{3.3.24}$$

This set is obviously a subset of (3.3.18), but could be very small if the support of the wavelets is large. The biggest advantage of this policy is its simplicity. As it only depends on data of the current index, it is very easy to implement.

Children Of Parent To enlarge the former set, a natural approach is to include all children of the elements' parent. As so, this set engulfs the former set but also includes a sibling and its children.

$$\Lambda^2_\lambda(c) := \{\mu \prec \Pi(\lambda) \in \mathcal{I} \mid |\lambda| \leq |\mu| \leq |\lambda| + c\} \tag{3.3.25}$$

As the former policy, this one is easy to compute, with just one extra computation, i.e., the determination of the parent index. A drawback of this approach is that the set is unsymmetric, i.e., the dyadic structure of the wavelet indices (3.2.8) leads to only one sibling being considered.

Siblings And Their Children To remedy the unbalance of the previous policy, we can directly consider the set of siblings of order $s \in \mathbb{N}_0$ of a wavelet index, i.e.,

$$\mathcal{S}_\lambda(s) := \{(j, \mathbf{k}', \mathbf{e}) \in \mathcal{I} \mid \|\mathbf{k} - \mathbf{k}'\|_\infty \le s\} \quad \text{for } \lambda = (j, \mathbf{k}, \mathbf{e}), \tag{3.3.26}$$

and then adding these and their children, i.e.,

$$\Lambda_\lambda^3(c) := \{\mu \preceq \mathcal{S}_\lambda(s) \mid |\lambda| \le |\mu| \le |\lambda| + c\}. \tag{3.3.27}$$

Here the sibling parameter s has to be chosen in such a way that the set does not grow too large and unnecessary indices are not included. The actual value of the parameter s has to be determined from properties of the wavelets being considered. Otherwise, this policy is still easy to implement, although more involved than the above ones, especially because this set needs to be adapted at the domain boundaries.

Examples for the sets can be found in Figure 3.3. Additional policies could of course be set up to more accurately approximate the proper influence set of any given MRA, but the set (3.3.27) gives already a pretty accurate approximation for our `DKU24NB` wavelets for $s = 1$.

Influence Sets in Multiple Dimensions

Formulation (3.3.18) is completely independent of the underlying problem domain. The definition does not account for the different **isotropic types** $\mathbf{e} \in \{0,1\}^n$ of the wavelet indices (2.4.18). In fact, by (3.3.18) any wavelet indices $\mu \in \mathcal{I}$ with intersecting support to ψ_λ for $\lambda = (j, \mathbf{k}, \mathbf{e}) \in \mathcal{I}$ regardless of the type $\mathbf{e}$ must be taken into account. This entails that any number of wavelets, whose supports cover the whole domain Ω, would lead to the tree $\mathcal{T}$ containing **all** wavelet indices on the next levels (if the level difference c is high enough for all λ). In applications, this would almost always lead to very full trees and is hardly a desired behavior.
Therefore, in higher dimensions, we change the definition (3.3.18) to only include wavelets indices of the **same type**, i.e., direct **descendants** of λ :

$$\Lambda_\lambda(c) := \{\mu \prec \lambda \mid S_\lambda \cap S_\mu \neq \emptyset \text{ and } |\lambda| \le |\mu| \le |\lambda| + c\}. \tag{3.3.28}$$

This setting is also motivated by the idea that the operator $\mathbf{F}$ will preserve a certain structure in the wavelet indices of $\mathbf{u}$ into the resulting wavelet vector $\mathbf{F}(\mathbf{u})$. This should be the case as long as the operator does not rotate or permutate the coordinate axes. For such operators, the original influence set should be considered.
In applications, the tensor product structure of the domain and isotropic wavelet is further used to simplify the above definition to the easily implementable policy

$$\Lambda_\lambda(c) := \left\{ \bigotimes_{i=1}^n \widetilde{\Lambda}_{\lambda_i}(c) \mid \lambda_i := (j, \mathbf{k}_i, \mathbf{e}_i) \right\}, \tag{3.3.29}$$

where $\widetilde{\Lambda}_{\lambda_i}(c)$ is one of the approximations (3.3.24), (3.3.25) or (3.3.27). For this to work, the one-dimensional rules $\widetilde{\Lambda}_{\lambda_i}(c)$ have to be extended from wavelets to single-scale

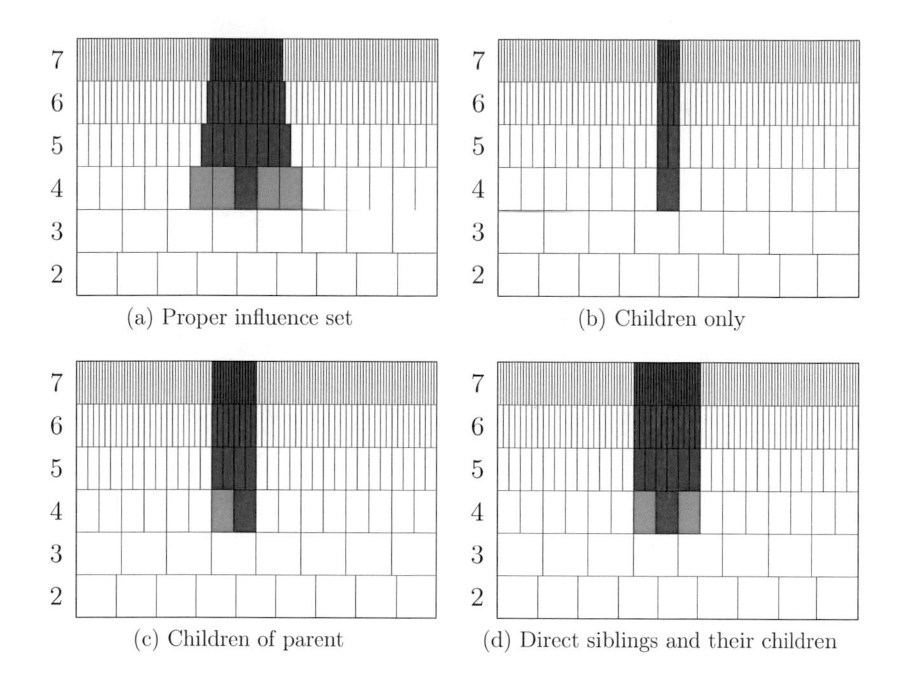

Figure 3.3: Influence set and approximations thereof for the primal wavelet on level $j = 4$, at position $k = 7$ of the DKU24NB MRA. The upper left diagram (a) shows the proper influence set of depth $c = 3$. In each image, the wavelet index $(4, 7, 1)$ is shown in red color, sibling are shown in green color, all children are colored blue. The upper right diagram (b) depicts only the children of this wavelet, the lower left diagram (c) shows all children of the parent $(3, 3, 1)$. In the last diagram (d), the children of the wavelet $(4, 7, 1)$ and its siblings $(4, 6, 1)$ and $(4, 8, 1)$ are selected.

functions, which is easily done. After all, the aforementioned definitions only rely on **level** and **location** or **support**, information which is available for both function types.
The corresponding results of the aforementioned influence sets for the primal boundary adapted hat functions and their tree structure (Appendix A.1.1) can be seen in Figure 3.4. From the study of these images, it is directly evident, that all heuristics except (3.3.24) produce indices outside the exact influence set, i.e. excess indices.

Remark 3.37 *A good balance for the primal isotropic wavelets of type DKU24NB is to choose (3.3.27) for wavelets and (3.3.24) for single scale functions. This policy will be referred to as* **Mixed Sibling/ChildrenOnly**. *In one dimension, it is identical to the* **Siblings and Their Children** *(3.3.27) policy.*

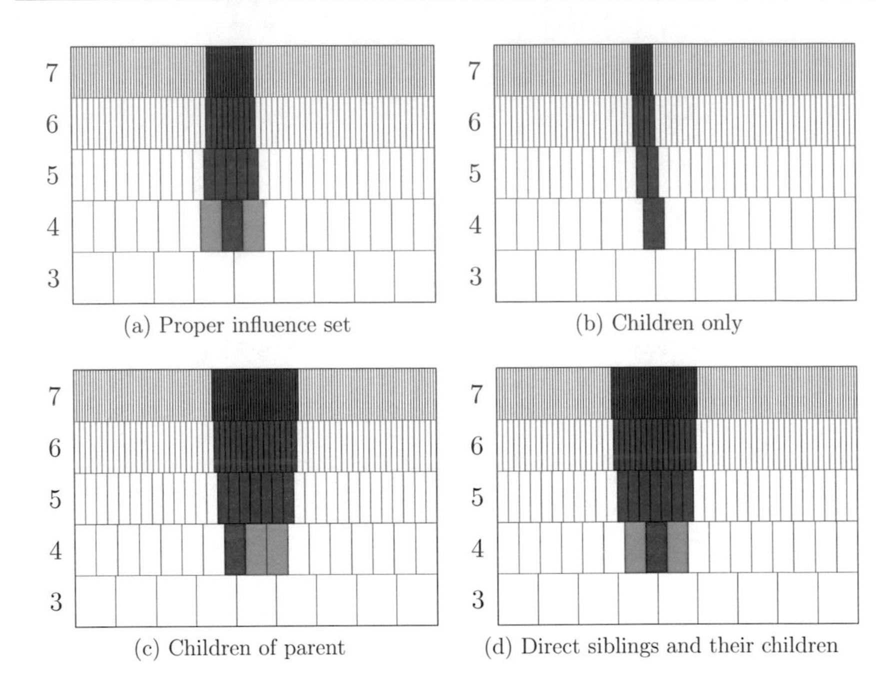

Figure 3.4: Influence set and approximations thereof for the primal single scale function on level $j = 4$, at position $k = 7$ of the DKU24NB MRA. The upper left diagram (a) shows the proper influence set of depth $c = 3$. In each image, the index $(4, 7, 0)$ is shown in red color, sibling are shown in green color, all children are colored blue. The upper right diagram (b) depicts only the children of this single scale function, the lower left diagram (c) shows all children of the parent $(3, 4, 0)$. In the last diagram (d), the children of the function $(4, 7, 0)$ and its siblings $(4, 6, 0)$ and $(4, 8, 0)$ are selected.

Dual Wavelets

The same considerations of the previous paragraph can of course be made for dual wavelets. The tree structure of Section 3.2.1 is not constricted to primal wavelets. While it is possible to define a different tree structure for the dual wavelets, there is little reason to do so. The set of primal and dual indices is exactly the same and if the support inclusion property (3.2.12) is still valid, the tree structure is best simply preserved. The result of the individual approximation policies is then exactly the same as depicted in Figure 3.3 and Figure 3.4, except for the exact influence sets (Diagrams (a)). The support of the dual single scale functions (and therefore the dual wavelets) is by Theorem 2.32 proportional to the dual polynomial exactness $\widetilde{d}$, which can be directly seen in Figure 3.5 and Figure 3.6.

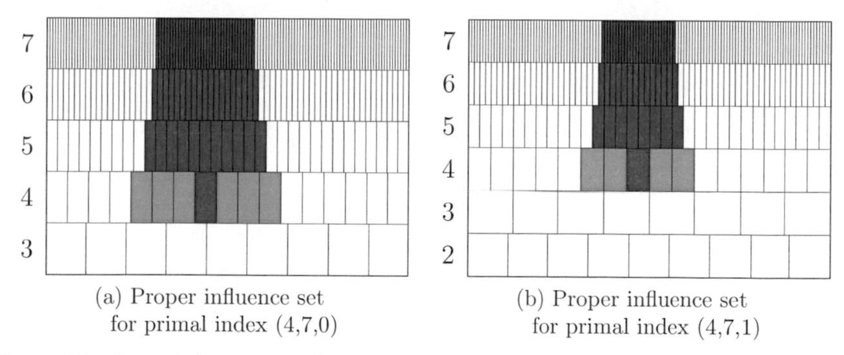

(a) Proper influence set for primal index (4,7,0)

(b) Proper influence set for primal index (4,7,1)

Figure 3.5: Exact influence sets for the dual single scale and wavelet functions of level $j = 4$, position $k = 7$ of the MRA DKU22NB, i.e. $\tilde{d} = 2$.

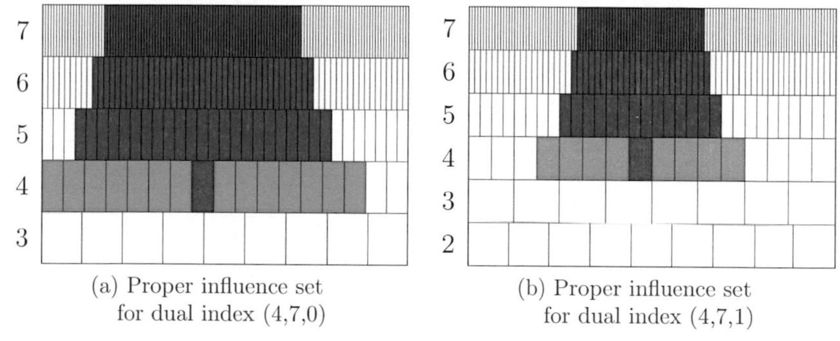

(a) Proper influence set for dual index (4,7,0)

(b) Proper influence set for dual index (4,7,1)

Figure 3.6: Exact influence sets for the dual single scale and wavelet functions of level $j = 4$, position $k = 7$ of the MRA DKU24NB, i.e. $\tilde{d} = 4$.

Runtime Tests

We briefly compare the runtime of each of the refinement policies. This construction of the predicted tree is just a minor step in the calculation of the result vector in (3.3.12). As such, it should also only be a minor computational job and the overall complexity should be linear w.r.t. the resulting vector. Here, we will now only compare the complexity of the individual refinement policies, i.e., we apply all the refinement policies to the same input vectors and compare the execution times. This serves just to show a qualitative comparison, e.g., to estimate the relative complexity of the different policies.
The results depicted in Figure 3.7 and Figure 3.8 were created by filling up a vector on a wavelet level j and then timing the runtime needed to compute the refinements. As the total number of wavelet indices grows exponentially with a factor $\approx 2^d$, the runtime graph on a logarithmic scale is linear. Every index was marked for refinement of depth $c = 3$. As one can observe, the runtime of the tree based policies (3.3.24), (3.3.25), (3.3.27) and Remark 3.37 perform very similarly, with only the exact refinement policy (3.3.18) being more costly in computational terms. This has the simple reason that the tree based polices

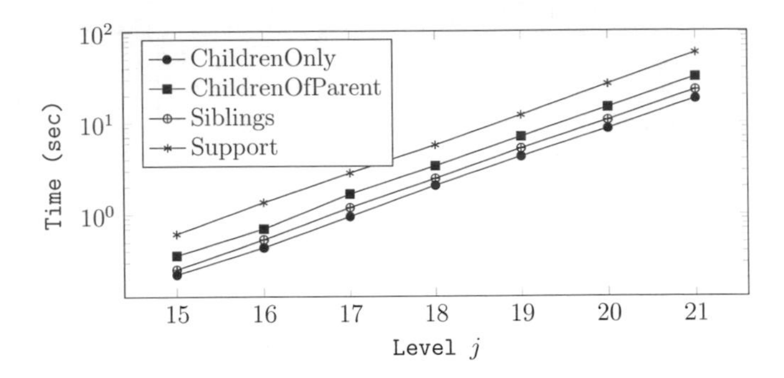

Figure 3.7: Runtime of the different refinement policies for a full vector in 1D of Level j (x-axis) and depth $c = 3$.

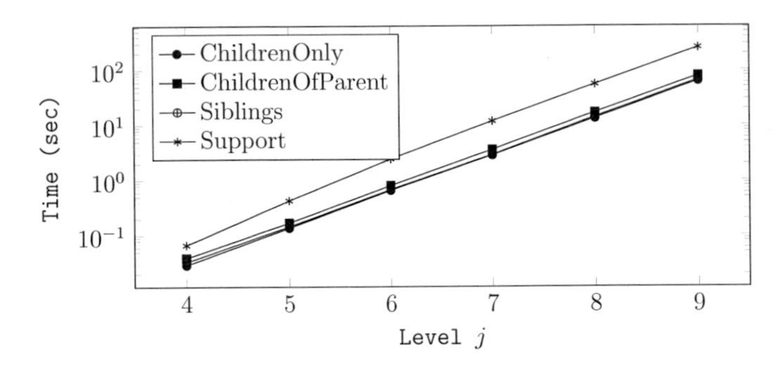

Figure 3.8: Runtime of the different refinement policies for a full vector in 2D of Level j (x-axis) and depth $c = 3$.

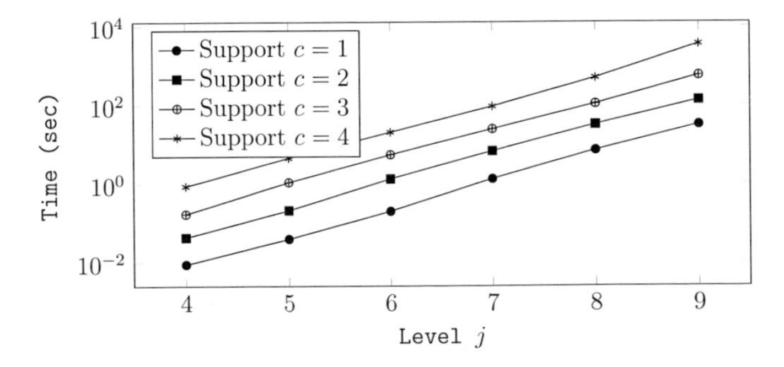

Figure 3.9: Runtime of the exact support refinement policy for a full vector in 2D of Level j (x-axis) and varying depth from $c = 1$ to $c = 4$.

can be applied level-wise, i.e., the tree can simply be traversed from roots to leaves and the elements refined successively, where the maximum is taken when different refinement values for a single element are encountered. The tree structure ensures that no child index can be generated more than once.
On the other hand, with (3.3.18) and (3.3.27), the final set of indices has to be made a proper tree again, since additional indices might be inserted for which the parents are missing in the assembled tree.
The exact refinement policy by support has to compute all indices on all finer levels for a given index and then insert them directly, since there is no other direct connection between these sets. It cannot postpone computation of higher level depths to finer levels with depth c decreased by one. This leads to many indices being inserted several times and explains the increased complexity. The extent of the performance gap thus heavily depends on the refinement depth, which is smaller for lower values $c < 3$, as can be seen in Figure 3.9. Since this is a 2D example, the number of children grows like 2^2 from one refinement depth to the next. Hence, $c = 4$ has to insert $\sim 4^4 = 256$ times as many indices than $c = 1$, which can be clearly recognized in the graph.
In conclusion, we can say that all refinement policies, the approximations and the exact support set, are feasible to use and none is prohibitively expensive.

Now that the first task, the determination of a suitable tree structure for (3.3.12), is completed, we turn to the calculation of the values of the target indices.

3.4 Application of Semilinear Elliptic Operators

In this section we show how to represent a function in wavelet coordinates in terms of local polynomials on disjoint cells. The idea for this reformulation is basically taken from [147] and the contents of this section have already been published in [115].

3.4.1 Adaptive Polynomial Representation

The following discussion is tailored to the specific case of (biorthogonal) spline wavelets, i.e., wavelets that are piecewise polynomials as described in Section 2.3. We begin by introducing the notion of a partition of a domain Ω in the present context.

Definition 3.38 [Partition]
A **partition** $\mathcal{P}(\Omega)$ *of a domain* $\Omega \subset \mathbb{R}^n$ *is a decomposition of* $\overline{\Omega}$ *into closed hypercubes, such that the union of all these subsets is* $\overline{\Omega}$*, i.e.,*

$$\overline{\Omega} = \bigcup_{T \in \mathcal{P}(\Omega)} T.$$

Furthermore, any two pairwise different elements $T, T' \in \mathcal{P}(\Omega)$ *are allowed to either overlap at most in one vertex or in one shared edge of lower dimensionality.*

Of course, a partition can be defined with the aid of other subsets in a similar fashion. For instance, commonly used meshes in finite element methods consists of triangles or tetrahedrons and yield a partition in a natural way. In view of our application and due to the fact that we will consider solely domains consisting of Cartesian products of intervals in $\mathbb{R}$, it suffices to use hypercubes for the partitioning.

Definition 3.39 [Dyadic Partition]
The **dyadic partitioning** *of* $\Omega = [0,1]^n$ *on level* j *is the set*

$$\mathcal{D}_j := \left\{ \prod_{i=1}^{n} [2^{-j}k_i, 2^{-j}(k_i+1)] \mid k_i \in \{0,\dots,2^j-1\},\ 1 \le i \le n \right\} \tag{3.4.1}$$

and its union is denoted by $\mathcal{D} := \bigcup_{j \ge j_0} \mathcal{D}_j$*, where* j_0 *denotes the coarsest level in* $\mathcal{I}$*.*

This means $\mathcal{D}_j$ is the set of all hypercubes with side length 2^{-j} contained in $\overline{\Omega}$. The set $\mathcal{D}$ has a natural tree structure in the following sense: Associate for each element $\square \in \mathcal{D}_j$ the set of children $\mathcal{C}(\square)$ as the set of elements $\square' \in \mathcal{D}_{j+1}$ which are subsets of $\square$, i.e.,

$$\mathcal{C}(\square) := \{\square' \in \mathcal{D}_{j+1} \,|\, \square' \subset \square\}. \tag{3.4.2}$$

Obviously, each element $\square$ has 2^n children and $\square' \subseteq \square$ with $\square, \square' \in \mathcal{D}$ implies that $\square'$ is a descendant of $\square$ or equal to $\square$, i.e., $\square' \preceq \square$.

Piecewise Polynomial Representation

By property $(\mathcal{P})$(2.2.3), each primal wavelet and scaling function can be expressed in terms of piecewise polynomials. Moreover, by $(\mathcal{L})$(2.1.8), the number of polynomials necessary to represent each function is uniformly bounded. We can express any wavelet thus as

$$\psi_\lambda = \sum_i p_i|_{\square_i}, \quad \operatorname{supp}\psi_\lambda = \bigcup_i \square_i, \tag{3.4.3}$$

where each $p_i|_{\square_i}$ is locally defined on the domain $\square_i \subset \mathbb{R}^n$. Due to the special relation (2.1.18) of the wavelets on one level and the scaling functions on the next finer level, these hypercubes have to be on level $|\lambda|+1$, i.e., from the set $\mathcal{D}_{|\lambda|+1}$. For each wavelet, the set of all these **support cubes** is denoted by

$$\mathcal{D}_\lambda := \left\{\square \in \mathcal{D}_{|\lambda|+1} \,|\, \square \subset S_\lambda\right\}, \tag{3.4.4}$$

For a tree structured index set $\mathcal{T}$, we define the set $\mathcal{T}_j$ as the set of all indices on level j, that is,

$$\mathcal{T}_j := \{\lambda \in \mathcal{T} \,|\, |\lambda| = j\}. \tag{3.4.5}$$

Furthermore, the union of supports of all wavelets on level j make up the support of $\mathcal{T}_j$, i.e.,

$$S(\mathcal{T}_j) := \bigcup_{\lambda \in \mathcal{T}_j} S_\lambda, \tag{3.4.6}$$

which can be expressed as the union of the corresponding dyadic hypercubes, i.e.,

$$\mathcal{D}(\mathcal{T}_j) := \{\square \in \mathcal{D}_{j+1} \,|\, \square \subset S(\mathcal{T}_j)\} = \bigcup_{\lambda \in \mathcal{T}_j} \mathcal{D}_\lambda. \tag{3.4.7}$$

These definition can obviously be extended to the whole tree $\mathcal{T}$. The set

$$\mathcal{D}(\mathcal{T}) := \bigcup_{\lambda \in \mathcal{T}} \mathcal{D}_\lambda = \bigcup_{j \geq j_0} \mathcal{D}(\mathcal{T}_j) \tag{3.4.8}$$

contains all dyadic cubes part of the support of any wavelet index contained in tree $\mathcal{T}$. The support of a whole tree $S(\mathcal{T})$ is defined in total analogy by using (3.4.6).
The tree structure (Definition 3.8) combined with the nested support property (3.2.12) of the wavelets entails that any hypercube of $\mathcal{D}(\mathcal{T})$ is contained within a larger hypercube on the next coarser level in $\mathcal{D}(\mathcal{T})$, i.e., it generally holds

$$S(\mathcal{T}_{j+1}) \subseteq S(\mathcal{T}_j). \tag{3.4.9}$$

Thus, the set $\mathcal{D}(\mathcal{T})$ is also a tree in the sense of (3.4.2). This leads to the following statement.

Proposition 3.40 *A tree structured set of wavelet indices $\mathcal{T} \subset \mathcal{I}$ induces a* **tree structure** *of dyadic hypercubes $\mathcal{D}(\mathcal{T})$. Moreover, the leaves $\mathcal{L}(\mathcal{D}(\mathcal{T}))$ form a* **partition** *of the domain Ω if $\mathcal{T}$ is a proper tree.*

To apply assertion (3.4.3) in practice, a polynomial basis to uniquely represent each wavelet must be chosen. In the following considerations, the explicit polynomial representation is not required, only its properties, namely the basis and refinement properties. There exist several possible choices for actual implementation and we refer to Appendix A.3 for details.

Remark 3.41 *For simplicity, we call any basis function of a polynomial space a "monomial", although it does not need to be a monomial in the strict sense.*

Let $p_\square^{\mathbf{t}}$ be a monomial on $\square \subset \mathbb{R}^n$ of order $\mathbf{t} := (t_1, \ldots, t_n) \in \mathbb{N}_0^n$. Considering only polynomials of order less than or equal d, the set of possible exponential indices is given by

$$\mathcal{M}_d^n := \{\mathbf{t} \,|\, 0 \leq t_i \leq d \,, i = 1, \ldots, n\} \,, \tag{3.4.10}$$

which gives a multitude of $(d+1)^n$ different monomials. The set of all these monomials forms a basis of Π_d^n on each $\square$ and we define the vector consisting of all these monomials by

$$\mathbf{P}_\square := (p_\square^{\mathbf{t}})_{\mathbf{t} \in \mathcal{M}_d^n}. \tag{3.4.11}$$

By (3.4.3) therefore exists a vector $\mathbf{q}_{\lambda,\square} \in \mathbb{R}^{(d+1)^n}$ for any wavelet ψ_λ such that holds

$$\psi_\lambda = \sum_{\square \in \mathcal{D}_\lambda} \psi_\lambda|_\square = \sum_{\square \in \mathcal{D}_\lambda} \mathbf{q}_{\lambda,\square}^T \mathbf{P}_\square \chi_\square, \tag{3.4.12}$$

where $\chi_\square$ denotes the characteristic function on $\square$.

Remark 3.42 *If the polynomial basis $\{p_\square^{\mathbf{t}}\}$ is constructed from a choice of monomial basis $\{p^{\mathbf{t}}\}$ of $\mathbb{R}^n$ by scaling and translation onto each domain $\square$, then the coefficients $\mathbf{q}_{\lambda,\square}$ only depend on the type of wavelet and the relative position of $\square$ with respect to S_λ.*

3.4.2 Transformation to Local Polynomial Bases

Now we investigate how to represent a function $v := \mathbf{v}^T \Psi|_{\mathcal{T}} := \sum_{\lambda \in \mathcal{T}} v_\lambda \psi_\lambda$ given its wavelet expansion w.r.t. a tree-structured index set $\mathcal{T}$ by piecewise local polynomials on the cells $\mathcal{D}(\mathcal{T})$.
Considering the function v restricted to the level j, i.e.,

$$v|_{\mathcal{T}_j} := \mathbf{v}_j^T \Psi_j := \sum_{\lambda \in \mathcal{T}_j} v_\lambda \psi_\lambda, \quad \mathbf{v}_j := (v_\lambda)_{\lambda \in \mathcal{T}_j},$$

we can derive its polynomial representation by (3.4.12) as

$$\begin{aligned} \sum_{\lambda \in \mathcal{T}_j} v_\lambda \psi_\lambda &= \sum_{\lambda \in \mathcal{T}_j} v_\lambda \sum_{\Box \in \mathcal{D}_\lambda} \mathbf{q}_{\lambda,\Box}^T \mathbf{P}_\Box \chi_\Box \\ &= \sum_{\Box \in \mathcal{D}(\mathcal{T}_j)} \chi_\Box \sum_{\substack{\lambda \in \mathcal{T}_j, \\ S_\lambda \supset \Box}} (v_\lambda \mathbf{q}_{\lambda,\Box})^T \mathbf{P}_\Box \\ &=: \sum_{\Box \in \mathcal{D}(\mathcal{T}_j)} \chi_\Box (\mathbf{g}_\Box^j)^T \mathbf{P}_\Box. \end{aligned} \tag{3.4.13}$$

In the second step, linearity of the wavelet decomposition enables us to collect the polynomial parts of all wavelets living on $\Box \in \mathcal{D}(\mathcal{T}_j)$ into one vector $\mathbf{g}_\Box^j \in \mathbb{R}^{(d+1)^n}$. Combining this relation for all coefficients of $\mathbf{v}_j$, we arrive at a linear mapping of the wavelet coefficient vector $\mathbf{v}_j$ into all the corresponding polynomial coefficients on $\mathcal{D}(\mathcal{T}_j)$, e.g.,

$$\mathbf{g}_j := (\mathbf{g}_\Box^j)_{\Box \in \mathcal{D}(\mathcal{T}_j)}.$$

This mapping must be representable as a matrix and we denote it as

$$\mathbf{g}_j = \mathbf{G}_j \, \mathbf{v}_j, \quad \mathbf{G}_j \in \mathbb{R}^{(d+1)^n \# \mathcal{D}(\mathcal{T}_j) \times \# \mathcal{T}_j}. \tag{3.4.14}$$

Remark 3.43 *One should consider that, for simplicity, we have chosen to designate all wavelet coefficients on level j of $\mathbf{v}$ as $\mathbf{v}_j$ instead of $\mathbf{v}|_{\mathcal{T}_j}$. Also, the polynomial coefficient vectors $\mathbf{g}_j$ are always to be taken coupled to the dyadic partitioning $\mathcal{D}(\mathcal{T}_j) \subset \mathcal{D}_{j+1}$.*

Furthermore, since polynomials are refinable one can express each polynomial on a cell $\Box \in \mathcal{D}_j$ with respect to polynomials on the next finer cells $\Box' \in \mathcal{D}_{j+1}$ so that (3.4.13) can be expressed on the finer cells as

$$\chi_\Box (\mathbf{g}_\Box^j)^T \mathbf{P}_\Box = \sum_{\Box' \in \mathcal{C}(\Box)} \chi_{\Box'} (\mathbf{M}_{\Box',\Box} \, \mathbf{g}_\Box^j)^T \mathbf{P}_{\Box'}, \tag{3.4.15}$$

with a $(d+1)^n \times (d+1)^n-$ **refinement matrix** $\mathbf{M}_{\Box',\Box}$. Again, the matrices $\mathbf{M}_{\Box',\Box}$ do not directly depend on the cells $\Box$ and $\Box'$ but only on the position of the child $\Box' \subset \Box$ with respect to $\Box$. That is, there are only $\#\mathcal{C}(\Box) = 2^n$ different refinement matrices $\mathbf{M}_{\Box',\Box}$.

Due to this refinement relation, we can expand (3.4.13) into a two-level representation, e.g.,

$$\sum_{\lambda\in\mathcal{T}_j} v_\lambda\psi_\lambda = \sum_{\square\in\mathcal{D}(\mathcal{T}_j)\cap\mathcal{L}(\mathcal{D}(\mathcal{T}))} \chi_\square(\mathbf{g}_\square^j)^T\mathbf{P}_\square + \sum_{\square'\in\mathcal{D}(\mathcal{T}_{j+1})} \chi_{\square'}(\mathbf{M}_{\square',\Pi(\square')}\mathbf{g}_{\Pi(\square')}^j)^T\mathbf{P}_{\square'}, \tag{3.4.16}$$

where $\Pi(\square')$ denotes the parent of $\square'$. That means, each polynomial on a cell $\square\in\mathcal{D}(\mathcal{T}_j)$ with a child contained in $\mathcal{D}(\mathcal{T}_{j+1})$ will be refined and expressed in terms of polynomials on these finer subcells. Considering now two levels, $v|_{\mathcal{T}_j\cup\mathcal{T}_{j+1}} = v|_{\mathcal{T}_j} + v|_{\mathcal{T}_{j+1}}$,

$$\sum_{\lambda\in\mathcal{T}_j\cup\mathcal{T}_{j+1}} v_\lambda\psi_\lambda = \sum_{\square\in\mathcal{D}(\mathcal{T}_j)\cap\mathcal{L}(\mathcal{D}(\mathcal{T}))} \chi_\square(\mathbf{g}_\square^j)^T\mathbf{P}_\square + \sum_{\square'\in\mathcal{D}(\mathcal{T}_{j+1})} \chi_{\square'}\left(\mathbf{M}_{\square',\Pi(\square')}\mathbf{g}_{\Pi(\square')}^j + \mathbf{g}_{\square'}^{j+1}\right)^T\mathbf{P}_{\square'} \tag{3.4.17}$$

shows the motivation for the refinement in (3.4.16), namely, the possibility to combine coefficients on different levels. The representation (3.4.17) is now comprised only of non-overlapping cells. Repeating this procedure yields an adaptive polynomial representation of the entire function,

$$v = \sum_{\lambda\in\mathcal{T}} v_\lambda\psi_\lambda = \sum_{\square\in\mathcal{L}(\mathcal{D}(\mathcal{T}))} \chi_\square\mathbf{g}_\square^T\mathbf{P}_\square, \tag{3.4.18}$$

where the coefficients $\mathbf{g}_\square$ can be computed from the $\mathbf{g}_\square^j$ via the recursive application of the refinement relations (3.4.16) and (3.4.17).
In summary, the multilevel wavelet expansion $v = \sum_{\lambda\in\mathcal{T}} v_\lambda\psi_\lambda$ can be transformed into a **polynomial representation** w.r.t. polynomials living on the leaves of the tree $\mathcal{D}(\mathcal{T})$ only, whose supports are thus pairwise **disjoint**.
The overall transformation, which transforms the wavelet coefficients $\mathbf{v}$ into the polynomial coefficients $\mathbf{g} := (\mathbf{g}_\square)_{\square\in\mathcal{L}(\mathcal{D}(\mathcal{T}))}$ will be denoted analogously to (3.4.14) by

$$\mathbf{g} = \mathbf{G}_\mathcal{T}\mathbf{v}. \tag{3.4.19}$$

With $\mathbf{P}_\mathcal{T} := (\chi_\square\mathbf{P}_\square)_{\square\in\mathcal{L}(\mathcal{D}(\mathcal{T}))}$ denoting the local polynomial bases on the leaves, it holds

$$\mathbf{g}^T\mathbf{P}_\mathcal{T} = (\mathbf{G}_\mathcal{T}\mathbf{v})^T\mathbf{P}_\mathcal{T} = \mathbf{v}^T\mathbf{G}_\mathcal{T}^T\mathbf{P}_\mathcal{T}, \tag{3.4.20}$$

which means $(\psi_\lambda)_{\lambda\in\mathcal{T}} = \mathbf{G}_\mathcal{T}^T\mathbf{P}_\mathcal{T}$. Finally, we state the algorithm which performs the transformation from wavelet coordinates into the local polynomial coordinates in Algorithm 3.5 An important aspect of course is the efficiency of the algorithm, that is, the computational effort needed to perform the algorithm. This question is answered by the following proposition cited from [147].

Proposition 3.44 *Suppose that the input vector $\mathbf{v}\in\ell_2(\mathcal{I})$ is finite and that the underlying index set $\mathcal{T} := S(\mathbf{v})$ is a proper tree. Then the number of arithmetic operations needed to compute the output of Algorithm 3.5 is proportional to the size of the input vector $\mathbf{v}$, i.e., the transformation from wavelet coordinates into polynomial coordinates can be applied in $\mathcal{O}(\#\mathcal{T})$ arithmetic operations.*

This result shows that the transformation is of optimal computational complexity, since the computational effort is linear in the size of data.

Remark 3.45 *It should be noted that the size of the coefficient vector of the polynomial representation can be larger than the coefficient vector of the wavelet representation. Nevertheless, its size can be estimated to be*

$$\#\mathbf{g} \lesssim \#\mathbf{v}, \qquad \mathbf{g}^T \mathbf{P}_{\mathcal{T}} = \mathbf{v}^T \Psi|_{\mathcal{T}},$$

with a constant only depending on property $(\mathcal{P})$(2.2.3) of the wavelet basis.

Algorithm 3.5 Transformation from wavelet representation $\mathbf{v}$ on $\mathcal{T}$ to local unique polynomial representation $\mathbf{g}$ on $\mathcal{D}(\mathcal{T})$.

1: **procedure** `WTREE_2_PTREE`($\mathbf{v}$) $\to \mathbf{g}$
2: $\quad \mathcal{T} := S(\mathbf{v})$, $j_0 \leftarrow \min\{|\lambda| : \lambda \in \mathcal{T}\}$, $J \leftarrow \max\{|\lambda| : \lambda \in \mathcal{T}\}$
3: $\quad$ Calculate $\mathbf{g}_{j_0} = \mathbf{G}_{j_0} \mathbf{v}_{j_0}$ $\qquad \triangleright$ According to (3.4.13)
4: $\quad \mathbf{g} \leftarrow \mathbf{g}_{j_0}$
5: $\quad$ **for** $j = j_0 + 1, \ldots, J$ **do**
6: $\qquad$ Calculate $\mathbf{g}_j = \mathbf{G}_j \mathbf{v}_j$ $\qquad \triangleright$ According to (3.4.13)
7: $\qquad$ **for all** $\Box \in \mathcal{D}(\mathcal{T}_{j-1})$ with $\Box \notin \mathcal{L}(\mathcal{D}(\mathcal{T}))$ **do**
8: $\qquad\quad \mathbf{g} \leftarrow \mathbf{g} \setminus (\mathbf{g}_{j-1})_\Box$ $\qquad \triangleright$ Considering $\mathbf{g}$ as a set
9: $\qquad\quad$ **for all** $\Box' \in \mathcal{C}(\Box)$ **do**
10: $\qquad\qquad (\mathbf{g}_j)_{\Box'} \leftarrow (\mathbf{g}_j)_{\Box'} + \mathbf{M}_{\Box',\Box}(\mathbf{g}_{j-1})_\Box$ $\qquad \triangleright$ According to (3.4.15)
11: $\qquad\quad$ **end for**
12: $\qquad$ **end for**
13: $\qquad \mathbf{g} \leftarrow \mathbf{g} \cup \mathbf{g}_j$
14: $\quad$ **end for**
15: $\quad$ **return g**
16: **end procedure**

3.4.3 Adaptive Nonlinear Operator Application

In the previous section we have seen how to represent a function in wavelet coordinates in terms of local polynomials. We now show how to apply a **nonlinear local operator** $F : \mathcal{H} \to \mathcal{H}'$ on the basis of this polynomial representation.
In the following, let the wavelet coefficients of $\mathbf{v}$ form a proper tree $\mathcal{T}$ and let $\mathcal{T}'$ be another proper tree, generated by `PREDICTION` of Section 3.3.3. The task is now the efficient application of the operator $\mathbf{F}$ in wavelet coordinates to $\mathbf{v}$ on the given index set $\mathcal{T}'$. Recalling the definition of the operator $\mathbf{F}$ in Section 2.2.4, we recognize that we need to evaluate dual forms or inner products, depending on the operator, for all $\lambda \in \mathcal{T}'$ with the given function $v = \mathbf{v}^T \Psi$.

Remark 3.46 *The first idea that comes to mind to calculate each coefficient $\langle \psi_\lambda, F(v) \rangle$ is probably* **numerical quadrature**. *This approach might not be optimal since the integrands, i.e., the piecewise polynomial wavelets, are in general not globally smooth and due to the multilevel structure of the wavelet basis, point evaluations are quite expensive. Moreover, this obviously could lead to an inexact calculation of the desired vector*

$(\langle \psi_\lambda, F(v)\rangle)_{\lambda\in\mathcal{T}'}$, *which in addition gives rise to theoretical difficulties in the overall context of adaptive solution methods.*

Given Remark 3.46, we pursue a different direction. The idea is to exploit the smoothness of the polynomials in the polynomial representation and the fact that the construction in Section 3.4.2 yields a representation w.r.t. polynomials on disjoint cells which do not overlap on different levels.
Using the representation (3.4.13), we obtain for each $\lambda \in \mathcal{T}'$,

$$\begin{aligned}(\mathbf{F}(\mathbf{v}))_\lambda = \langle \psi_\lambda, F(v)\rangle &= \langle \sum_{\square\in\mathcal{D}_\lambda} \chi_\square \mathbf{q}^T_{\lambda,\square}\mathbf{P}_\square, F(v)\rangle \\ &= \sum_{\square\in\mathcal{D}_\lambda} \mathbf{q}^T_{\lambda,\square}\langle \chi_\square \mathbf{P}_\square, F(v)\rangle,\end{aligned} \tag{3.4.21}$$

meaning we can evaluate the operator w.r.t. our monomial basis $\mathbf{P}_\square = (p^{\mathbf{t}}_\square)_{\mathbf{t}}$ and then reassemble the value of the wavelet expansion coefficient using the knowledge of $\mathbf{q}_{\lambda,\square}$. Therefore, recalling the overall transformations (3.4.19) and (3.4.20), the target vector in wavelet coordinates can be written as

$$\begin{aligned}(\langle \psi_\lambda, F(v)\rangle)_{\lambda\in\mathcal{T}'} &= \langle (\psi_\lambda)_{\lambda\in\mathcal{T}'}, F(v)\rangle \\ &= \langle \mathbf{G}^T_{\mathcal{T}'}\mathbf{P}_{\mathcal{T}'}, F(v)\rangle \\ &= \mathbf{G}^T_{\mathcal{T}'}\langle \mathbf{P}_{\mathcal{T}'}, F(v)\rangle.\end{aligned} \tag{3.4.22}$$

Thus, we have split up the evaluations $(\langle \psi_\lambda, F(v)\rangle)_{\lambda\in\mathcal{T}'}$ into the simpler evaluations $(\langle \chi_\square p^{\mathbf{t}}_\square, F(v)\rangle)_{\square\in\mathcal{L}(\mathcal{D}(\mathcal{T}')),\mathbf{t}\in\mathcal{M}^n_d}$ on each cell $\square$ combined with an application of the linear operator $\mathbf{G}^T_{\mathcal{T}'}$.
Postponing discussion of the operator $\mathbf{G}^T_{\mathcal{T}'}$ till Section 3.4.5, we first handle the computation of the inner products $\langle \mathbf{P}_{\mathcal{T}'}, F(v)\rangle$ for each $\square \in \mathcal{D}(\mathcal{T}')$. To simplify these calculations, the result of the operator, $F(v)$, should be expressed in terms of polynomials on the same cells as the polynomials in the left hand side of the expression, i.e., $\mathcal{L}(\mathcal{D}(\mathcal{T}'))$. In other words, although only the **input** data v is given, $F(v)$ should be computed on the partition $\mathcal{L}(\mathcal{D}(\mathcal{T}'))$ of the **output** vector.

Remark 3.47 *To this end, we possibly have to enlarge the tree $\mathcal{T}'$ to include $\mathcal{T}$ but this is usually unnecessary because $\mathcal{T} \subset \mathcal{T}'$ holds in order to fulfill (3.3.12). From this directly follows $\mathcal{D}(\mathcal{T}) \subset \mathcal{D}(\mathcal{T}')$ and we thus can refine the partition determined by $\mathcal{T}$ to match the partition of $\mathcal{T}'$ where needed. The complexity of the operation obviously depends on the number of children of each cell and the number of refinement steps. The complexity to refine a single polynomial can be bounded as a constant only depending on the dimension and polynomial degree, but does not depend on the data itself. Overall, this step is necessary but can be neglected when estimating the overall complexity of the nonlinear operator application.*

Using the representation (3.4.18) for v on $\mathcal{T}'$, it follows for any $\square \in \mathcal{L}(\mathcal{D}(\mathcal{T}'))$,

$$\begin{aligned}\langle \chi_\square \mathbf{P}_\square, F(v)\rangle &= \langle \chi_\square \mathbf{P}_\square, F(\sum_{\square'\in\mathcal{L}(\mathcal{D}(\mathcal{T}'))} \chi_{\square'}\mathbf{g}^T_{\square'}\mathbf{P}_{\square'})\rangle \\ &= \left\langle \chi_\square \mathbf{P}_\square, F\left(\chi_\square \mathbf{g}^T_\square \mathbf{P}_\square\right)\right\rangle,\end{aligned} \tag{3.4.23}$$

since the cells are disjoint and the operator F is assumed to be local. The advantage of this representation is that we only have to evaluate the dual forms on disjoint cells $\square \in \mathcal{L}(\mathcal{D}(\mathcal{T}'))$ and only with respect to our smooth polynomial basis $\mathbf{P}_\square$. Moreover, we only need to determine the operator application $F(\chi_\square \mathbf{g}_\square^T \mathbf{P}_\square)$ with respect to **one local polynomial** $\chi_\square \mathbf{g}_\square^T \mathbf{P}_\square$ instead of a superposition of several overlapping wavelets on different levels.

Remark 3.48 *For many operators, especially polynomial, e.g., $F(u) := u^3$, and linear PDE operators, the result $F(p_\square^{\mathbf{t}})$ is again representable in the basis $\mathbf{P}_\square$. This has the advantage, that the internal products of monomials can be calculated analytically and therefore quadrature errors can be avoided. This is preferred to quadrature rules as their errors can become an impediment for convergence in the overall context of adaptive solution methods.*

It is now apparent that the **unique**, **non-overlapping** polynomial representation of Section 3.4.2 easily enables the **exact** and **fast** application of local (non-)linear operators by simply traversing $\mathcal{L}(\mathcal{D}(\mathcal{T}'))$ and using (3.4.23) on each cell.
To conclude, we present the following algorithm implementing the operator application $F(v)$ on the partition $\mathcal{L}(\mathcal{D}(\mathcal{T}'))$ of the domain Ω of the (predicted) proper tree $\mathcal{T}'$.

Algorithm 3.6 Application of operator F on $\mathcal{L}(\mathcal{D}(\mathcal{T}'))$ based upon polynomial representation $\mathbf{g}$ of v computed by Algorithm 3.5 refined to match $\mathcal{D}(\mathcal{T}')$.

```
1: procedure APPLY_POLY_OP(F,g) → I
2:     for all □ ∈ L((D(T')) do
3:         for all t ∈ M_d^n do
4:             I_{□,t} ← ⟨p_□^t, F(g_□^T P_□)⟩        ▷ Evaluate exactly or by quadrature
5:         end for
6:     end for
7:     return I
8: end procedure
```

We now elaborate on the computation of the individual operator applications on the hypercubes $\square$.

3.4.4 Reference Element Operator Applications

To actually be able to produce numerical results, we first have to choose a basis for the polynomial space $\mathbf{P}_\square$ from (3.4.3) for $\square \in \mathcal{D}_j$ defined in (3.4.1). A common approach is to use tensor product constructions, i.e., for $\mathbf{x} \in \square$,

$$p_\square^{\mathbf{t}}(\mathbf{x}) := \prod_{1 \le i \le n} p_{j,k_i}^{t_i}(x_i), \quad \mathbf{t} := (t_1, \ldots, t_n) \in \mathbb{N}_0^n. \tag{3.4.24}$$

The actual choice of the 1D monomial bases $\{p_{j,k_i}^{t_i}(x_i)\}$ is not important in the following discussion. All that has to be known about the basis is how to represent given piecewise polynomial functions in it and the value of integrals of the form

$$\int_\square F(p_\square^{\mathbf{t}}(\mathbf{x})) p_\square^{\mathbf{r}}(\mathbf{x}) d\mathbf{x}, \quad p_\square^{\mathbf{t}}, p_\square^{\mathbf{r}} \in \mathbf{P}_\square. \tag{3.4.25}$$

Suitable constructions can be found in Appendix A.3.

The Nonlinear Operators $F(v) := v^s$

As an example of a nonlinear local operator, we discuss the set up of the operator $F(v) := v^s$, $s \in \mathbb{N}$. We will show how these nonlinear operators can be applied **exactly**, i.e. how the inner products $\langle \chi_\Box \mathbf{P}_\Box, F(\chi_\Box \mathbf{g}_\Box^T \mathbf{P}_\Box) \rangle$ can be calculated analytically. To evaluate the expression $\langle \chi_\Box \mathbf{P}_\Box, F(\chi_\Box \mathbf{g}_\Box^T \mathbf{P}_\Box) \rangle$, the multinomial theorem can be used to expand the right hand side expression as

$$
\begin{aligned}
(\chi_\Box \mathbf{g}_\Box^T \mathbf{P}_\Box)^s &= \chi_\Box \left(\sum_{i=1}^m (\mathbf{g}_\Box)_i \, (\mathbf{P}_\Box)_i \right)^s \\
&= \chi_\Box \left(\sum_{s_1+\cdots+s_m=s} \frac{s!}{s_1! \cdots s_m!} \prod_{1 \le i \le m} ((\mathbf{g}_\Box)_i \, (\mathbf{P}_\Box)_i)^{s_i} \right) \\
&=: \chi_\Box \left(\sum_{s_1+\cdots+s_m=s} \mathbf{Y}(\mathbf{g}_\Box; s; s_1, \ldots, s_m) \prod_{1 \le i \le m} ((\mathbf{P}_\Box)_i)^{s_i} \right),
\end{aligned} \tag{3.4.26}
$$

with $m := \#\mathcal{M}_d^n = (d+1)^n$ and $\mathbf{Y}(\mathbf{g}_\Box; s; s_1, \ldots, s_m) := \frac{s!}{s_1! \cdots s_m!} \prod_{1 \le i \le m} (\mathbf{g}_\Box)_i^{s_i}$, where $(\mathbf{P}_\Box)_i$ and $(\mathbf{g}_\Box)_i$ specify the i-th element of $\mathbf{P}_\Box$ and $\mathbf{g}_\Box$, respectively. Inserting this expression into the dual forms yields

$$
\begin{aligned}
&\langle \chi_\Box \mathbf{P}_\Box, (\chi_\Box \mathbf{g}_\Box^T \mathbf{P}_\Box)^s \rangle \\
&= \langle \chi_\Box \mathbf{P}_\Box, \chi_\Box \left(\sum_{s_1+\cdots+s_m=s} \mathbf{Y}(\mathbf{g}_\Box; s; s_1, \ldots, s_m) \prod_{1 \le i \le m} ((\mathbf{P}_\Box)_i)^{s_i} \right) \rangle \\
&= \sum_{s_1+\cdots+s_m=s} \mathbf{Y}(\mathbf{g}_\Box; s; s_1, \ldots, s_m) \left\langle \chi_\Box \mathbf{P}_\Box, \chi_\Box \prod_{1 \le i \le m} ((\mathbf{P}_\Box)_i)^{s_i} \right\rangle .
\end{aligned} \tag{3.4.27}
$$

That is, one solely needs to calculate the inner products of local monomials of certain degrees depending on the exponent s and the degree of the polynomials determined by the underlying wavelet basis. Obviously, these inner products can be calculated independently of the application of the operator itself.

The Laplace Operator

To exemplify that the polynomials representation method is not only applicable to operators that itself can be expressed as polynomials, we look at the stereotypical PDE operator, the operator $-\Delta$, see formula (1.4.5). In weak form, this leads to the bilinear form $a(v, w) := \int \nabla v \cdot \nabla w \, d\mu$, which can simply be transferred to any $\Box \subset \Omega$. There, the gradient acts as a mapping from the polynomial basis $\mathbf{P}_\Box$ of $\mathcal{M}_d^n$ to the polynomial basis $\mathbf{P}_\Box$ of $\mathcal{M}_{d-1}^n$. For a local representation $\mathbf{g}_\Box^T \mathbf{P}_\Box$, probing with another basis function

$(\mathbf{P}_\square)_i$, the linearity of the gradient then leads to

$$\begin{aligned} a((\mathbf{P}_\square)_t, \mathbf{g}_\square^T \mathbf{P}_\square) &= \int_\square \nabla (\mathbf{P}_\square)_t \cdot \nabla \left(\mathbf{g}_\square^T \mathbf{P}_\square\right) d\mu \\ &= \int_\square \nabla (\mathbf{P}_\square)_t \cdot \nabla \left(\sum_m (\mathbf{g}_\square)_m (\mathbf{P}_\square)_m \right) d\mu \\ &= \sum_m (\mathbf{g}_\square)_m \int_\square \nabla (\mathbf{P}_\square)_t \cdot \nabla (\mathbf{P}_\square)_m \, d\mu \end{aligned} \tag{3.4.28}$$

So, as before, the results of the operator w.r.t. the polynomial basis can be precomputed analytically and only a linear term depending on the input vector $\mathbf{g}_\square$ has to be evaluated. In higher dimensions, the tensor product structure entails that the operator can not simply be expressed as a tensor product, but as a sum of tensor products:

$$\int_\square \nabla p_\square^{\mathbf{t}}(\mathbf{x}) \cdot \nabla p_\square^{\mathbf{m}}(\mathbf{x}) d\mu(\mathbf{x}) \tag{3.4.29}$$

$$= \int_\square \sum_{i=1}^n \left(\frac{\partial}{\partial x_i} p_{j,\mathbf{k}_i}^{\mathbf{t}_i}(x_i) \prod_{r \neq i} p_{j,\mathbf{k}_r}^{\mathbf{t}_r}(x_r) \frac{\partial}{\partial x_i} p_{j,\mathbf{k}_i}^{\mathbf{m}_i}(x_i) \prod_{s \neq i} p_{j,\mathbf{k}_s}^{\mathbf{m}_s}(x_s) \right) d\mu(\mathbf{x}) \tag{3.4.30}$$

$$= \sum_{i=1}^n \left(\int_{2^{-j}\mathbf{k}_i}^{2^{-j}(\mathbf{k}_i+1)} \frac{\partial}{\partial x_i} p_{j,\mathbf{k}_i}^{\mathbf{t}_i}(x_i) \frac{\partial}{\partial x_i} p_{j,\mathbf{k}_i}^{\mathbf{m}_i}(x_i) d\mathbf{x}_i \prod_{r \neq i} \int_{2^{-j}\mathbf{k}_r}^{2^{-j}(\mathbf{k}_r+1)} p_{j,\mathbf{k}_r}^{\mathbf{t}_r}(x_r) p_{j,\mathbf{k}_r}^{\mathbf{m}_r}(x_r) d\mathbf{x}_r \right). \tag{3.4.31}$$

Each individual term can then be evaluated according to one-dimensional formulas.

Having computed the result of operator F w.r.t. the polynomial basis on $\mathcal{L}(\mathcal{D}(\mathcal{T}'))$, we now need to compute the expressions $\langle \psi_\lambda, F(v) \rangle$ w.r.t. all wavelets of the tree $\mathcal{T}'$.

3.4.5 Reconstruction of Target Wavelet Indices

Assuming a proper tree $\mathcal{T}'$ and a vector $\mathbf{I} = \left(p_\square^{\mathbf{t}}, F(\mathbf{g}_\square^T \mathbf{P}_\square)\right)_\square$ for all $\square \in \mathcal{L}(\mathcal{D}(\mathcal{T}'))$, we now want to to reconstruct the values $\langle \psi_\lambda, F(v) \rangle$ of the vector $\mathbf{F}(\mathbf{v})$.
Combining the refinement relation (3.4.15) with (3.4.21), we obtain the relation

$$\begin{aligned} \langle \psi_\lambda, F(v) \rangle = & \sum_{\square \in \mathcal{D}_\lambda \cap \mathcal{L}(\mathcal{D}(\mathcal{T}'))} \mathbf{q}_{\square,\lambda}^T \langle \chi_\square \mathbf{P}_\square, F(v) \rangle \\ & + \sum_{\square \in \mathcal{D}_\lambda \setminus \mathcal{L}(\mathcal{D}(\mathcal{T}'))} \mathbf{q}_{\square,\lambda}^T \sum_{\square' \in \mathcal{C}(\square)} \mathbf{M}_{\square',\square}^T \langle \chi_{\square'} \mathbf{P}_{\square'}, F(v) \rangle, \end{aligned} \tag{3.4.32}$$

which shows how to calculate the target values from data given not only on cells $\square \subset S_\lambda$ but also from children $\square' \subset \square$, as long as both sets form a partition of S_λ. Repeating the argument gives rise to the following Algorithm 3.7 performing the transformation from the output of the operator application in polynomial representation $\mathbf{I}$ to the representation in terms of wavelets.

Algorithm 3.7 Decomposition of output data $\mathbf{I}$ of Algorithm 3.6 in order to retain wavelet coefficients for all $\lambda \in \mathcal{T}'$.

1: **procedure** DECOMPOSE($\mathbf{I}$,$\mathcal{T}'$) $\rightarrow \mathbf{w}$
2: $\quad j_0 \leftarrow \min\{|\lambda| \,|\, \lambda \in \mathcal{T}'\}$, $J \leftarrow \max\{|\lambda| \,|\, \lambda \in \mathcal{T}'\}$
3: $\quad$ **for** $j = J, \dots, j_0$ **do**
4: $\quad\quad$ **for all** $\square \in \mathcal{D}(\mathcal{T}_j') \setminus \mathcal{L}(\mathcal{D}(\mathcal{T}'))$ **do**
5: $\quad\quad\quad \mathbf{I}_\square \leftarrow \sum_{\square' \in \mathcal{C}(\square)} \mathbf{M}_{\square',\square}^T \mathbf{I}_{\square'}$ $\quad$ ▷ Upscaling of values, cf. (3.4.32)
6: $\quad\quad$ **end for**
7: $\quad\quad$ **for all** $\lambda \in \mathcal{T}_j'$ **do**
8: $\quad\quad\quad w_\lambda \leftarrow \sum_{\square \in \mathcal{D}_\lambda} \mathbf{q}_{\lambda,\square}^T \mathbf{I}_\square$ $\quad$ ▷ Compute coefficient value, cf. (3.4.32)
9: $\quad\quad$ **end for**
10: $\quad$ **end for**
11: $\quad$ **return w**
12: **end procedure**

In words, the scheme assembles from the dual forms w.r.t. the polynomial basis, the vector of dual forms w.r.t. wavelets on one level, then constructs the results w.r.t the polynomial basis on the next coarser level and repeats the process. By (3.4.22), the application of Algorithm 3.7 on the tree $\mathcal{T}'$ is denoted by $\mathbf{G}_{\mathcal{T}'}^T$.

Remark 3.49 *As shown, the transformation* $\mathbf{G}_{\mathcal{T}'}^T$ *can be interpreted as a reverse application of Algorithm 3.5 in some sense. In wavelet terms, Algorithm 3.5 implements the* **primal direct fast wavelet transform** *and Algorithm 3.7 the dual inverse, i.e.* **transposed**, *version.*

This completes the assemblage of algorithms necessary for the adaptive application of nonlinear operators.

3.4.6 The Nonlinear Apply Scheme

We now state the overall APPLY_NONLINEAR scheme performing the determination of a suitable wavelet index set combined with the application w.r.t. local polynomials and state a computational complexity result.
The implementation of the APPLY_NONLINEAR scheme can be summarized in these steps:

1. Depending on an input vector $\mathbf{v}$ and target accuracy $\varepsilon > 0$, predict a target set $\mathcal{T}'$,
2. Assemble the adaptive local piecewise polynomial representation $\mathbf{g}$ on $\mathcal{D}(\mathcal{T})$ of $\mathbf{v}$ w.r.t. $\mathcal{T}'$,
3. Apply the operator F on each $\square \in \mathcal{D}(\mathcal{T}')$,
4. Reconstruct the output tree wavelet coefficients by decomposing each one.

The complete APPLY_NONLINEAR scheme thus reads as follows:

Algorithm 3.8 Adaptive approximate operator application F on a finitely supported $\mathbf{v} \in \ell_2(\mathcal{I})$ up to accuracy ϵ.

1: **procedure** APPLY_NONLINEAR(ϵ, F, γ, $\mathbf{v}$) $\rightarrow \mathbf{w}$
2: $\quad \mathcal{T}' \leftarrow$ PREDICTION(ϵ, $\mathbf{v}$, γ) ▷ Ensure (3.3.12)
3: $\quad \mathcal{T}' \leftarrow \mathcal{T}' \cup S(\mathbf{v})$ ▷ Ensure $S(\mathbf{v}) \subset \mathcal{T}'$
4: $\quad$ **for all** $\mu \in \mathcal{T}' \setminus S(\mathbf{v})$ **do** ▷ Set up $\mathbf{v}$ on $\mathcal{T}'$
5: $\quad\quad v_\mu \leftarrow 0$ ▷ Insert wavelet coefficients
6: $\quad$ **end for**
7: $\quad \mathbf{g} \leftarrow$ WTREE_2_PTREE($\mathbf{v}$)
8: $\quad \mathbf{I} \leftarrow$ APPLY_POLY_OP(F, $\mathbf{g}$)
9: $\quad \mathbf{w} \leftarrow$ DECOMPOSE($\mathbf{I}$, $\mathcal{T}'$)
10: $\quad$ **return** $\mathbf{w}$
11: **end procedure**

We now consider the efficiency of the APPLY_NONLINEAR scheme, that is, the computational effort needed to perform the algorithm.

Theorem 3.50 *Suppose that we have a routine PREDICTION according to Proposition 3.35 satisfying (3.3.12) at hand and that the dual forms $\langle p_\square^{\mathbf{t}}, F(\mathbf{g}_\square^T \mathbf{P}_\square)\rangle$ can be calculated exactly in a constant number of arithmetic operations which do not depend on $\square$, then the number of arithmetic and sorting operations needed to calculate the approximation $\mathbf{w}$ by the APPLY_NONLINEAR scheme Algorithm 3.8 applied to $\mathbf{v} \in \ell_2(\mathcal{I})$ can be bounded by a constant multiple of its output size $\#S(\mathbf{w})$. Moreover, it holds that*

$$\|\mathbf{F}(\mathbf{v}) - \mathbf{F}(\mathbf{v})|_{\mathcal{T}}\|_{\ell_2(\mathcal{I})} \lesssim \epsilon. \tag{3.4.33}$$

Proof: The PREDICTION step needs $\mathcal{O}(\#(\mathcal{T} \setminus S(\mathbf{v})) + \#S(\mathbf{v})) = \mathcal{O}(\#\mathcal{T})$ arithmetic and sorting operations, according to Proposition 3.35, where we have used that the output set $\mathcal{T}$ is not yet enlarged in this step.
Applying the transformation to local polynomials in WTREE_2_PTREE as well needs $\mathcal{O}(\#\mathcal{T})$ operations according to Proposition 3.44, since the input vector $\mathbf{v}$ is expressed on $\mathcal{T}$, see step 3.
In APPLY_POLY_OP one has to calculate $\#\mathcal{M}_p \times \#\mathcal{L}(\mathcal{D}(\mathcal{T}))$ of the expressions $\langle p_\square^{\mathbf{t}}, F(\mathbf{g}_\square^T \mathbf{P}_\square)\rangle$, where each can be calculated in constant time by assumption. Here $\#\mathcal{M}_p$ is a fixed constant only depending on the underlying wavelet and its polynomial representation. Moreover, it was shown in [147] that $\#\mathcal{L}(\mathcal{D}(\mathcal{T})) \lesssim \#\mathcal{T}$ so that APPLY_POLY_OP can be applied in $\mathcal{O}(\#\mathcal{T})$ arithmetic operation.
In the DECOMPOSE step, we have to apply the matrices $\mathbf{M}_{\square',\square}$ a number of $\#C(\square)$ times on each cell $\square \in \mathcal{D}(\mathcal{T})\setminus\mathcal{L}(\mathcal{D}(\mathcal{T}))$, where the number of rows and columns are fixed, depending again only on the underlying wavelet and its polynomial representation. Additionally, we have to calculate $\mathbf{q}_{\lambda,\square}^T \mathbf{I}_\square$ for each index $\lambda \in \mathcal{T}$ and $\square \in \mathcal{P}_\lambda$ a total of $\#\mathcal{P}_\lambda \lesssim 1$ times. That is, one needs $\mathcal{O}(\#(\mathcal{D}(\mathcal{T}) \setminus \mathcal{L}(\mathcal{D}(\mathcal{T}))) + \#\mathcal{T}) \lesssim \mathcal{O}(\#\mathcal{T})$ operations to perform the DECOMPOSE scheme, where we made use of the fact that $\#\mathcal{D}(\mathcal{T}) \lesssim \#\mathcal{L}(\mathcal{D}(\mathcal{T})) \lesssim \#\mathcal{T}$, see [147].
Since each consecutive step of the algorithm can thus be executed in $\mathcal{O}(\#\mathcal{T})$ operations, it also follows for the overall computational complexity, i.e., $\mathcal{O}(\#\mathcal{T}) = \mathcal{O}(\#S(\mathbf{w}))$. The estimation (3.4.33) is a direct consequence of (3.3.12), since the dual forms $\langle p_\square^{\mathbf{t}}, F(\mathbf{g}_\square^T \mathbf{P}_\square)\rangle$,

and therefore the entire operator application on the fixed index set $\mathcal{T}$, can be applied exactly. ∎

This result shows that Algorithm 3.8 can be applied with asymptotic optimal computational complexity and yields an approximation up to a desired accuracy ϵ. The costs of applying the operators depicted in Section 3.4.4 and Algorithms 3.5 and 3.7 can be found in Figures 3.10 through 3.12. A more detailed view of the actual execution can be found in Figure 3.13.
It is clearly noticeable that each algorithm and the PDE polynomial operators are asymptotically linear in complexity. Of the Algorithms 3.4, 3.5 and 3.7, the prediction algorithm executes the fastest. The prediction algorithm runs very fast here because the input vector contains only a few hundred, up to a thousand, wavelet coefficients in any case. The complexity thus stems from the construction of the Influence set (3.3.18), especially since here the exact sets are constructed. In comparison, the complexity of the decomposition procedure is here slightly higher than the complexity of the construction of the adaptive polynomial. This is due to the increased complexity of constructing the adaptive polynomial of the input vector within the polynomial described by the output vector. To save memory, the combined vector of Algorithm 3 is never constructed, but it is checked for every wavelet index of the output if it is contained in the input vector. If the wavelet index exists, its value is used, otherwise the value is assumed to be zero. This implementation saves the insertion of a lot of zero valued wavelet coefficients at the expense of a higher execution cost. But if the values were inserted as in Algorithm 3, this step should also be counted when measuring the complexity of Algorithm 3.5, which would entail the same computational overhead coupled with the memory overhead, so this approach seems favorable.
In contrast, the application of the PDE operators on the piecewise polynomials is approximately an order of magnitude faster than the algorithms involving wavelet coefficients. This is due to the design decision for our code to being able to choose the wavelet type dynamically at runtime, but the type of polynomials used is chosen at compile time and stays fixed. Hence, computations in the realm of polynomial data structures are very quickly executable compared to the wavelet realm. But this is also only possible because of some caching strategies of the multi-dimensional polynomials, the repeated computation of these tensor products would otherwise seriously impact efficiency.
Studying the PDE operator diagrams, it is clear that the application of linear operators is quicker than nonlinear operators, which was to be expected. But also, all of the operators show a linear asymptotic rate.
This concludes the application of nonlinear operators, we now turn to specialized implementations for the application of linear operator in the present context.

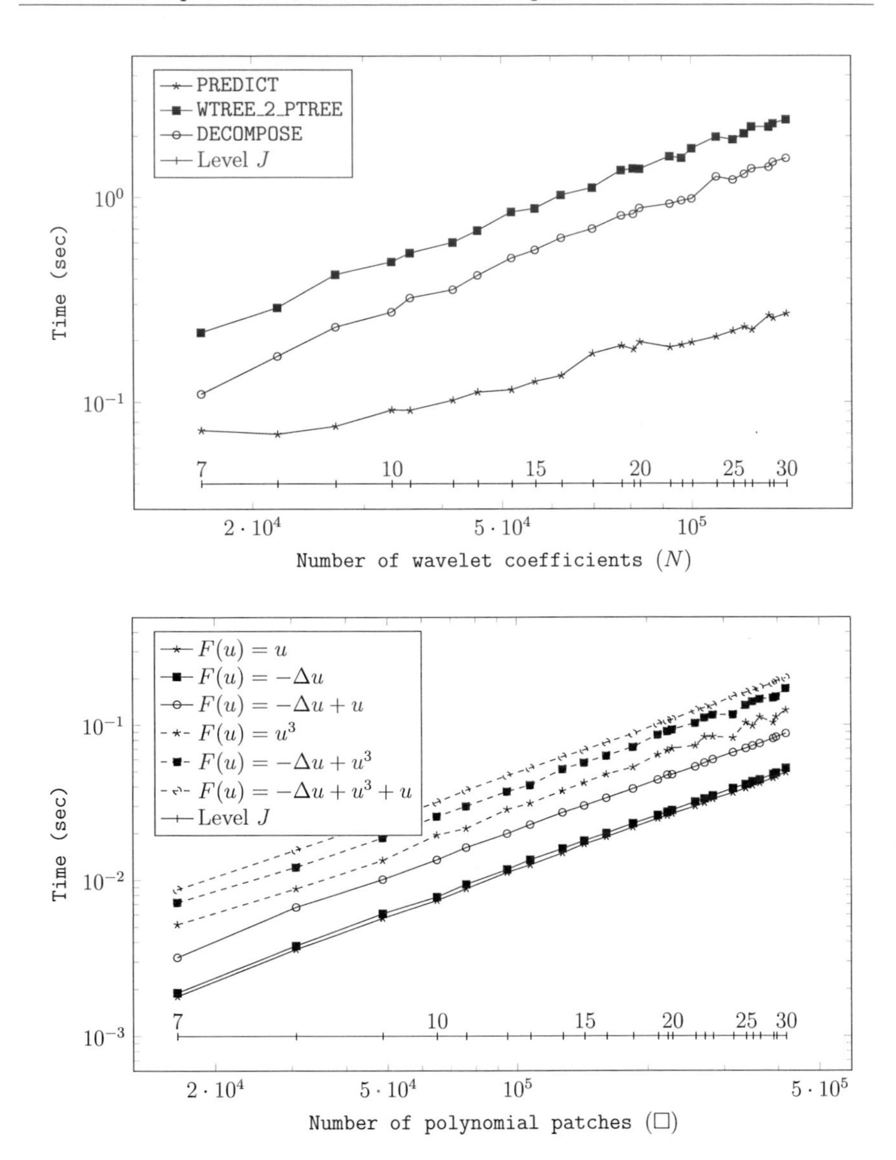

Figure 3.10: Times of the algorithms and PDE operators for the DKU22NB wavelet in 2D. For the application, a vector on a single level J with three random wavelet coefficients on the highest wavelet level is made into a proper tree with random coefficient values and then used as the input vector for the algorithms. In each case, ten such vectors are constructed and the execution time measured. The tree prediction is run with $\gamma := 4$ and $\varepsilon := 0.001$.

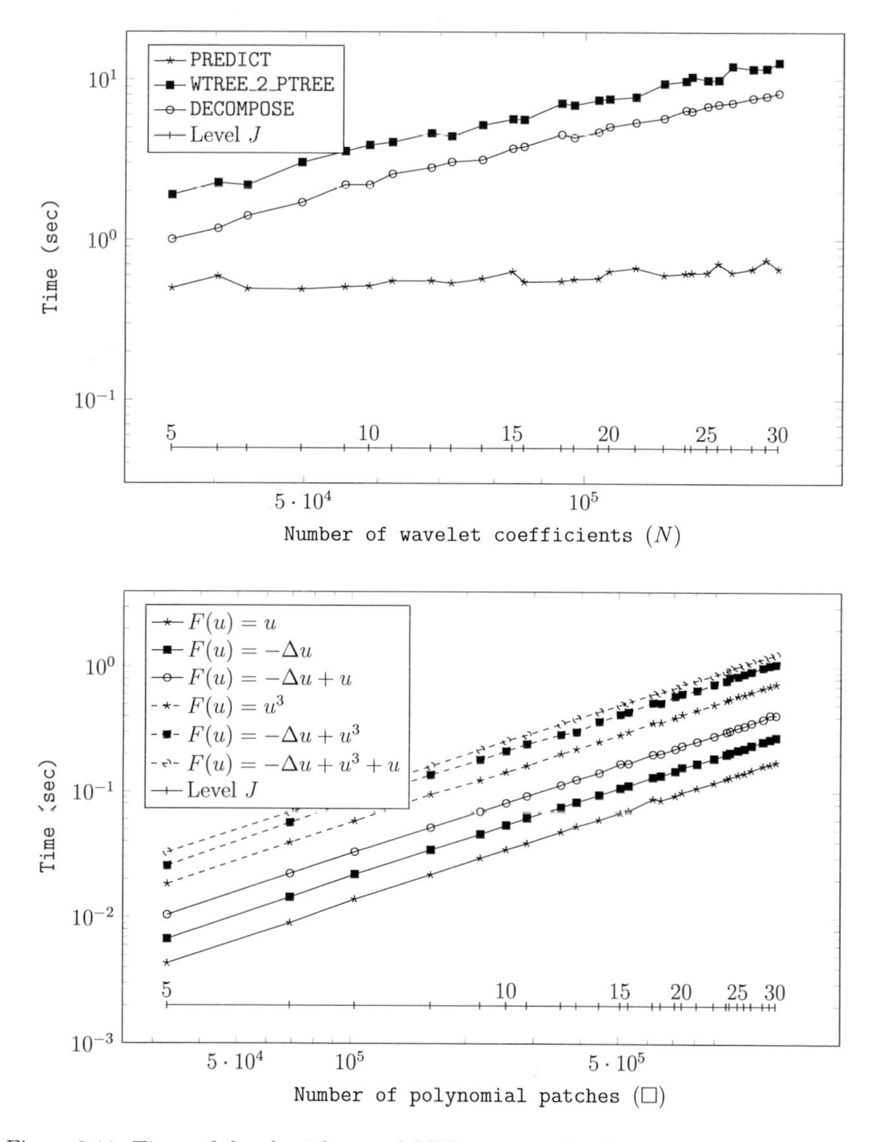

Figure 3.11: Times of the algorithms and PDE operators for the DKU22NB wavelet in 3D. For the application, a vector on a single level J with three random wavelet coefficients on the highest wavelet level is made a proper tree with random coefficient values and then used as the input vector for the algorithms. In each case, ten such vectors are constructed and the execution time measured. The tree prediction is run with $\gamma := 6$ and $\varepsilon := 0.001$.

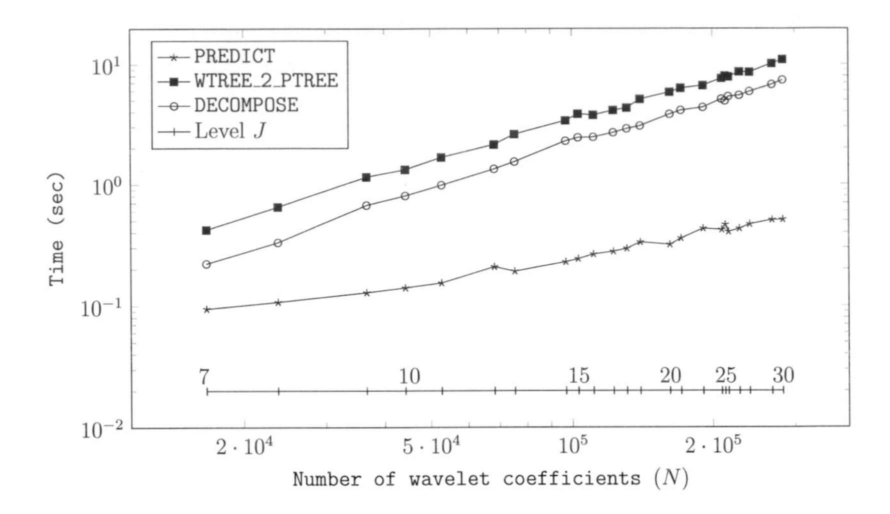

Figure 3.12: Execution times of the algorithms for the DKU24NB wavelet in 2D. The data points were computed in the same manner as before, the only difference here is the higher order of the dual wavelets. As this also enlarges the support of the primal wavelets, it entails a higher complexity in the Algorithms 3.5 and 3.7. For comparison, see Figure 3.10. The tree prediction is run with $\gamma := 4$ and $\varepsilon := 0.001$. As the execution time of the polynomial operators $F(\cdot)$ is independent of the wavelets employed, these values would be en par with the ones displayed in Figure 3.10 and are omitted.

Remark 3.51 *Diagram 3.13 shows that most work must be invested in constructing the adaptive polynomial representation from the adaptive wavelet vector (Algorithm 3.5), closely followed by the decomposition of the target wavelet indices from the integral values on the polynomial grid (Algorithm 3.7). The tree prediction algorithm (Algorithm 3.4) is comparatively quickly computable, but is still beaten by the computation of the bilinear forms (3.4.27) and (3.4.29) during Algorithm 3.6.*
The cost of some mathematically relevant sub-procedures of these algorithms are also highlighted. It is clearly noticeable that computations done upon existing data structures are cheap compared to creating or destroying said (adaptive) data structures. Some techniques to lower the overhead imposed by these data structures can be found in Appendix B.3. The basis computations referred to in the reconstruction/decomposition algorithm have already been streamlined to minimize computational and memory overhead, but it is simply a fact that the number of polynomial pieces of a wavelet (3.4.12) grows exponentially in the dimension, due to the tensor product structure (2.4.18). The representation of these polynomial pieces can be precomputed, but the impact of having to traverse all pieces cannot be avoided. The unmarked parts of the pie charts above represents all the other commands in the algorithms, e.g., auxiliary calculations and operations executed for memory management.

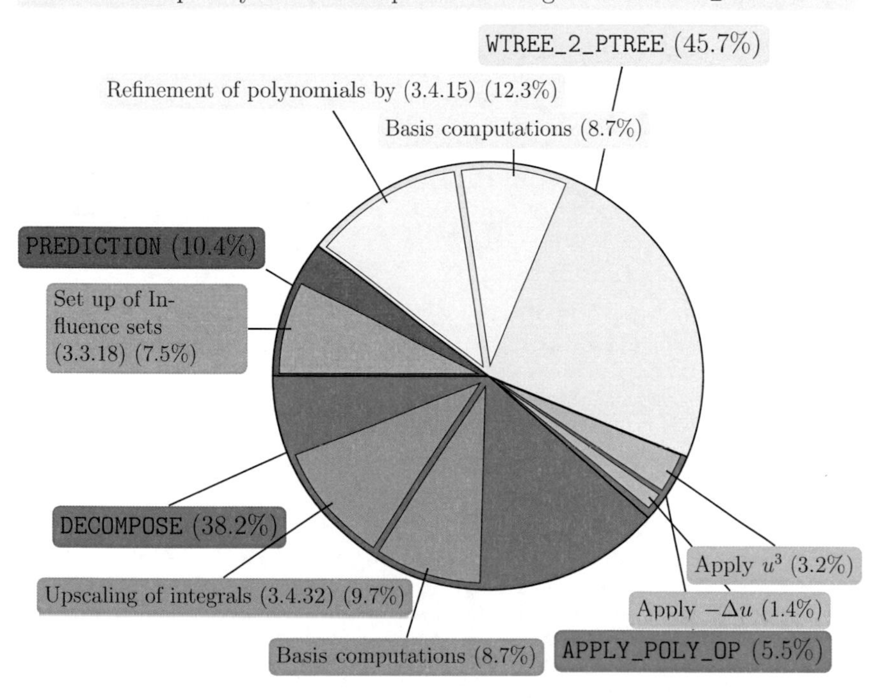

Figure 3.13: Runtime percentages of the individual algorithms during the execution of APPLY_NONLINEAR (Algorithm 3.8). The data was gathered during the computation of the solution of Section 4.6 using `valgrind`. This tool gathers profiling data on a procedural level, i.e., on functions that the code execution "jumps into". The chart does not represent a single execution of APPLY_NONLINEAR, but the accumulated data of the overall program run. As the presentation only shows relative percentages, the data can also be interpreted as the average of the individual calls. 100% here refers to the overall execution time and all percentages are relative to this number.

3.5 Application of Linear Operators

Although the current framework of adaptive wavelet methods was primarily developed for the solution of **nonlinear** problems, **linear** sub- or side-problems could be accommodated as well. Since linear operators are just a small subset of the larger class of nonlinear operators, the machinery of Section 3.4 is of course applicable. Apart from these tools, there is a prior generation of adaptive wavelet methods available specifically designed for linear operators, see [31].

Remark 3.52 *Converting an adaptive vector, which only stores non-zero wavelet indices, into a vector used in full-grid methods, which stores all elements, and using the standard wavelet operator representation of Section 2.2.4 is usually not feasible. This is because the required memory depends exponentially on the highest level with a non-zero coefficient. However, this method can be used if the effective level is not too high. In this case, this approach usually has the advantage to lead quickly to very accurate results, which is especially tempting for applications of inverse linear operators.*

Our main focus here is to not only use as much of the data structures and algorithms developed for nonlinear operators to minimize overhead, but also to exploit the advantageous properties of a linear operator during its application.
For example, in case of the **control problems** discussed in [122], linear operators are encountered in the following situations:

- First, incorporating **Riesz Operators** (Section 2.2.6) to control the norm with which to minimize leads to direct and inverse applications of said operator during the solution process. The exact Riesz Operators for L_2 and H^1 given in (2.2.37) are obviously linear. Moreover, they are **local** operators (3.1.1) and so is the **dual** Riesz Operator for L_2 given by (2.2.45). This is not true for the inverse stiffness matrix and thus the inverse of $\mathbf{R}_{H^1}$, therefore, in this context, the application of $\mathbf{R}_{H^1}^{-1}$ is best done by using a numerical solution method. We will discuss these operators in detail in Section 3.5.1 and Section 3.5.3.

- Second, in the adjoint problem and the application of Newton's method to the operator $F(u) = u^3$, an **adaptively weighted** mass matrix will emerge. This is an operator based upon the bilinear form

$$a_z(u, v) := \int_\Omega z^2 \, u \, v \, d\mu, \qquad \text{for } u, v \in \mathcal{H},$$

  for a fixed given function $z \in \mathcal{H}$. We discuss this operator subsequently after the Riesz operator section in Section 3.5.4.

- Lastly, the **trace operator** $\gamma_0 : H^s(\Omega) \to H^{s-1/2}(\Gamma)$ and its adjoint have to be applied during the solution process. These operators (distinguishing between the parts $\Gamma \subset \partial\Omega$) do not fall in the same category as the above, because they do not map from a space into its dual. Instead, these operators can here be applied by a simple exclusion criteria and a subsequent adaptive wavelet transform. This will be discussed in Section 3.6.

Common to all approaches mentioned here is the fact, that, for a linear operator, each wavelet coefficient of the input vector can be considered **independently** of all other wavelet coefficients and each result of the operator application can (e.g., as a last step) be combined into a single resulting output vector. This does not mean that these applications are trivial though, because the result of the operator w.r.t. to a single wavelet index might depend on any number of other wavelet indices in the input tree, thus potentially resulting in an overall nonlinear complexity.

3.5.1 Evaluation Algorithms for Linear Operators

We discuss here the application of the **Mass Matrix** and **Stiffness Matrix** (2.2.37) and how the other Riesz Operators of Section 2.2.6 can be applied.
The main deviation from the application of nonlinear operator in Section 3.4.3 is to skip the polynomial representation altogether and stay in the wavelet domain all the time. This has the advantage, that also operators can still be applied if employed wavelets are not piecewise polynomial, e.g., if the wavelet expansion is w.r.t. the **dual** wavelets of Section 2.1.4. This is for example the case for the **inverse Mass Matrix** discussed in Section 3.5.3.

The General Task of Applying a Linear Operator

Let $A : V \to V'$ be a **linear** operator given by the **bilinear form** $a(\cdot,\cdot)$ and V be a space on the domain $\Omega \subset \mathbb{R}^n$ equipped with biorthogonal wavelet bases Ψ, $\widetilde{\Psi}$. To compute $\mathbf{A}\,\mathbf{v} = \mathbf{w}$, with $\mathbf{v} \in \ell_2$ based on the **proper tree** $\mathcal{T}$, we still assume that $\mathbf{w}$ was generated by predicting the output tree set $\mathcal{T}'$ as before. Herein, we are mainly concerned about calculating the values of the entries of the output vector $\mathbf{w}$. The main idea is to facilitate property (3.1.1) to calculate the entries by the following identity:

$$(\mathbf{w})_\lambda = a(\psi_\lambda, \mathbf{v}^T\Psi) = \sum_{\mu\in\mathcal{I}\,:\mathrm{supp}\,\psi_\lambda\cap\mathrm{supp}\,\psi_\mu\neq\emptyset} a(\psi_\lambda, \psi_\mu). \tag{3.5.1}$$

In order to iterate over all elements $w_\lambda \in \mathbf{w}$, the following steps are executed:

1. Generate the **Dependency set**, i.e., the set of all coefficients

$$\Upsilon_\lambda(\mathcal{T}) := \{\mu \in \mathcal{T} \mid \mathrm{supp}\,\psi_\mu \cap \mathrm{supp}\,\psi_\lambda \neq \emptyset\}. \tag{3.5.2}$$

2. For all $\mu \in \Upsilon_\lambda(S(\mathbf{v}))$, calculate $v_\mu \cdot a(\psi_\lambda, \psi_\mu)$ and add to w_λ.

Because of the linearity of the operator A, this amounts to applying $\mathbf{A}$ to $\mathbf{v}$. In essence, should the discretized operator $\mathbf{A}$ in wavelet coordinates be directly available (and not by the scheme (2.2.24)), this process mimics the matrix-vector multiplication using all unit vectors in the full space discretization.

Remark 3.53 *The Dependency set looks very similar to the* **Influence set** *(3.3.18), but the Influence set $\Lambda_\lambda(c)$ contains indices on* **higher** *levels, whereas the* **Dependency set** *will later contain only indices on* **lower** *levels. Moreover, it can be understood as an* **inverse** *of the Influence set, because the Dependency set will be made up of all indices μ, whose (infinite) Influence set contains the index λ.*

It should be noted that for $\mathcal{T} \subset \mathcal{I}$, it holds

$$\Upsilon_\lambda(\mathcal{T}) = \Upsilon_\lambda(\mathcal{I}) \cap \mathcal{T}, \tag{3.5.3}$$

which is the actual way of determining $\Upsilon_\lambda(\mathcal{T})$ in applications. This is simply because the set $\mathcal{T}$ is usually not fixed and determined at runtime, whereas $\mathcal{I}$ is known beforehand. As equation (3.5.1) demands, we digress to cover the creation of arbitrary values $a(\psi_\lambda, \psi_\mu)$.

Calculating Arbitrary Values for Bilinear Forms

Very often, the manual calculation of bilinear form values $a(\cdot,\cdot)$ for the **linear** operator $A : \mathcal{H} \to \mathcal{H}'$ is only feasible for **same-level single-scale** functions due to the high number of possible combinations of spatial locations, e.g., see Section 3.5.2. But even if the operator is only available on a single level J, then all values of all **coarser levels** $j \leq J$ can be computed by applying the fast wavelet transform. By studying (2.1.33) and (2.1.39), it holds

$$\mathbf{M}_{j+1,1}^T \, \mathbf{M}_{j+2,0}^T \cdots \mathbf{M}_{J,0}^T \left\langle \Phi_J, A\Phi_J \right\rangle = \left\langle \Psi_j, A\Phi_J \right\rangle. \tag{3.5.4}$$

Thus, by only applying the fast wavelet transform from the left, all possible combinations $\left\langle \Phi_j, A\Phi_J \right\rangle, \left\langle \Psi_j, A\Phi_J \right\rangle$ of the levels $j_0 \leq j \leq J$ can be computed successively. This can obviously be extended by applying appropriate terms $\mathbf{M}_{j,0}/\mathbf{M}_{j,1}$ from the right, but it would usually be unnecessary for symmetric bilinear forms, except when $\langle \cdot, A\Psi_J \rangle$ is sought.
Assuming the bilinear form to be **translation invariant**, i.e., the result does not depend on the absolute position of the involved functions but only on their **relative positions**, the resulting matrix would contain many repeated values in different columns, and thus only very few columns of these matrices actually have to be assembled. These columns can be found automatically by categorizing the involved functions $\phi_{j,k}$ and then noting their possible positions.

Example 3.54 *The boundary adapted piecewise linear hat functions (Appendix A.1.1) are divided into three types given in (A.1.3).*

Saving the indices and the values of the column bands for all possible combinations of functions is all the information needed to compute the value of the bilinear form $a(\cdot, \phi_{J,k})$ for any fixed function $\phi_{J,k}$: For any level difference $J - j \geq 0$, the appropriate matrix band has to be loaded (depending on the category of the function) and the value of the offset k has to be taken into account. Usually, this simply means that the matrix band is **shifted** two positions per one offset to the original (assembly) location index.

Remark 3.55 *The above deductions have more applications than just the evaluation of single-scale functions:*

- *If the bilinear form is* **local** *and the converse of (3.1.1) also holds, a non-zero value in a certain row of a column of the matrix $\langle \Phi_j, A\Phi_J \rangle$ then gives the information that the supports of these functions must be overlapping. This gives rise to an easily computable* **Support Operator**, *which can answer whether two arbitrary functions share support in $\mathcal{O}(1)$ time.*

- *If the converse of (3.1.1) does not hold and the functions involved are piecewise polynomials, e.g., primal functions, and the function on the finer level is a wavelet, then gaps in the band due to* **vanishing moments** *$(\widetilde{\mathcal{V}})$(2.2.6) will occur. This can only happen if the level difference is high enough so a wavelet is evaluated w.r.t. to a single polynomial piece of the other function. Since, by the dyadic construction (2.1.14), the number of functions living a fixed interval on level j grows like 2^{J-j}, this means the gaps of zeros double from each level J to the next $(J+1)$.*

The **Support Operator** *is useful for Algorithm 3.4 of Section 3.3.3 and Algorithms 3.9, 3.10 below.*

Although a fully working version can be derived using the techniques presented in Section 2.5, one should seek for an alternative implementation plan regarding the details of this paragraph.

A Simple Single-Scale Evaluation Scheme

We assume here to be in the same setting as in Section 3.4.3, that is the **input** and **target** trees are given (the latter one by Algorithm 3.4), and a linear operator $A : \mathcal{H} \to \mathcal{H}'$ gives a **bilinear form** $a(\cdot, \cdot)$. By (3.5.1) and (3.5.2), the main tasks for implementing this scheme are now:

1. Quick identification of the Dependency set Υ_λ for any ψ_λ.
2. Efficient evaluation of the bilinear form $a(\cdot, \cdot)$ for all necessary functions.

The first task can be done independent of the operator, see the previous paragraph. The second task depends heavily of the operator under consideration and we will talk about possible strategies for the different operators in the respective sections later on.

The exact **Dependency set** $\Upsilon_\lambda(\mathcal{I})$, as the Influence set, depends on the wavelet type (polynomial degree, regularity properties, etc) and the level and position of the individual wavelet. Of course this means we could **approximate** the Dependency set as we did with the Influence set in Section 3.3.4, but, as we will later see, this is unnecessary for our operators. We will now rather concentrate on the intrinsic properties of the Dependency set:
First of all, the set can be quite large and its setup could be infeasible because the number of children, which, by the **Inclusion Property** (3.2.12) are always part of the Dependency set, grows exponentially with the level difference. To remedy the situation, we focus here on **symmetric operators** $A = A'$, i.e., $a(u, v) = a(v, u)$, which enables us to only consider wavelet indices with equal or lower level than the wavelet indices in the output data. Since the output tree given by $S(\mathbf{w})$ is a superset of $S(\mathbf{v})$, $S(\mathbf{v})$ can be expanded to match $S(\mathbf{w})$, just as in Algorithm 3.8. Then, if all possible combinations of input and output coefficients was evaluated, $a(\phi_\lambda, \phi_\mu)$ and $a(\phi_\mu, \phi_\lambda)$ would be evaluated independently for $\lambda \neq \mu$. Since their values are equal by symmetry, we choose to only calculate the value once and apply it to the coefficients given by λ and μ simultaneously when the other, e.g., λ, turns up while traversing the Dependency set of one of them, e.g.,

μ. This means only indices on lower levels have to be considered, i.e.,

$$\overline{\Upsilon}_\lambda(\mathcal{T}) := \{\mu \in \mathcal{T} \text{ with } |\mu| \leq |\lambda| \,|\, \operatorname{supp}\psi_\mu \cap \operatorname{supp}\psi_\lambda \neq \emptyset\}, \tag{3.5.5}$$

and because of the finite intersection property (2.1.16), the number of elements in Υ_λ would then be **uniformly** bounded per level.
By the refinement relation (2.1.18), a wavelet can be expressed on higher levels only by single-scale functions, not by wavelets. Given **input** $v \in \mathcal{H}$ and **target** $w \in \mathcal{H}'$, this then yields by (3.5.1) and $v = \mathbf{v}^T\Psi \equiv \sum_j \mathbf{v}_{\Phi,j}^T\Phi_j$ with (3.3.1),

$$(\mathbf{w})_\lambda = a(\psi_\lambda, \sum_j \mathbf{v}_{\Phi,j}^T\Phi_j) = \sum_{\mathbf{r}\in\Delta_{j+1}^n} m_{\mathbf{r},\mathbf{k}}^{j,\mathbf{e}} \sum_{\mu:\operatorname{supp}\phi_{j+1,\mathbf{r}}\cap\operatorname{supp}\phi_\mu\neq\emptyset} a(\phi_{j+1,\mathbf{r}}, \phi_\mu).$$

Remark 3.56 *The important detail here is, that, although the output vector* $\mathbf{w}$ *corresponds to a* **dual** *wavelet expansion, the adaptive reconstruction has to be performed w.r.t. the primal wavelet basis, same as the* **primal** *input vector. Of course, if* $A : \mathcal{H}' \to \mathcal{H}$*, then all roles are reversed.*

The application of the **adaptive wavelet transform** (Algorithm 3.1) to both the input and the output vector has several computational advantages:

1. Primal single scale functions have smaller support, see Section 2.3, and thus the Dependency set (with respect to the single-scale tree structure, e.g., (3.2.9)) would contain fewer elements.
2. During the second step of evaluating the bilinear form, in wavelet coordinates the bilinear form has to be evaluated for wavelets and single-scale functions in all combinations. For single-scale representations only the combinations of single-scale functions is required.
3. Only a single computation of the evaluation data is required to apply an operator w.r.t. wavelets of different **dual** polynomial exactness $\widetilde{d}$ (of which also the exact form of the primal wavelets depend).

Examples for the Dependency sets in wavelet and single-scale representation can be seen in Figure 3.14(a).

Remark 3.57 *Of course, the calculations are feasible for both the single-scale and the wavelet representation of the vectors. In essence, it is about choosing higher memory requirements (wavelet representation) or longer computation time (single-scale representation).*

After calculating all values of the bilinear form $a(\cdot,\cdot)$ for all elements of the target vector in single-scale coordinates, the values of the output vector $\mathbf{w}$ in wavelet coordinates have to be reconstructed by Algorithm 3.2.

Theorem 3.58 *Algorithm 3.9 finishes for a vector of size* $N := \#S(\mathbf{w})$*, where* $\mathbf{w}$ *is the output vector, in time* $\mathcal{O}(N\log(N))$.

Proof: Since the adaptive reconstruction of line 11 creates a vector with $\mathcal{O}(N)$ single scale coefficients and every coefficient is accessed exactly once in 12, it suffices to show the size of $\Upsilon_\lambda(S(\mathbf{t}))$ is $\mathcal{O}(\log(N))$.
The result of Algorithm 3.1 can have many functions on different levels intersecting over any given point, but the number of wavelets or single-scale functions on a single level j intersecting any given point is uniformly bounded by (2.1.16). Calling this number $m \in \mathbb{N}$, the number of wavelet coefficients in $\Upsilon_\lambda(S(\mathbf{t}))$ is bounded by $m^n(J - j_0)$, where $j_0 := \min\{|\lambda| \mid \lambda \in S(\mathbf{t})\}$ and $J := \max\{|\lambda| \mid \lambda \in S(\mathbf{t})\}$. Since $\#\Delta_J^n \sim 2^{nJ}$, follows $(J - j_0) \sim \log(\Delta_J)$. ∎

This is a worst case scenario estimate, but because of the tree structure, no levels of $\mathbf{r}, \mathbf{t}$ will be completely empty. But since the logarithmic factor will only show up for $J \gg j_0$ (it could be considered a constant factor for fixed J) and it is more easily recognizable for vectors with many elements, it should not be notable in applications with really sparse vectors.

Remark 3.59 *A few remarks about Algorithm 3.9.*

- *The "search phase" of determining the Dependency set $\Upsilon_\lambda(S(\mathbf{t}))$ can be shortened by also using the sparse representation for the input vector $\mathbf{v}$ and then further refining it to match the target vector $\mathbf{w}$ more closely. Then, only one or two level differences*

Algorithm 3.9 Simple application of a symmetric linear operator $A : \mathcal{H} \to \mathcal{H}'$ to a wavelet vector $\mathbf{v}$ up to accuracy ϵ.

```
1: procedure APPLY_LINEAR_SINGLE_SCALE_SIMPLE(ε, A, γ, v) → w
2:     T ← S(v)
3:     T' ← PREDICTION( ε, v, γ )                       ▷ Ensure (3.3.12)
4:     T' ← T' ∪ T                                     ▷ Ensure S(v) ⊂ T
5:     for all μ ∈ T' do                                ▷ Set up w on T'
6:         w_μ ← 0                                      ▷ Insert wavelet coefficients
7:     end for
8:     t ← ADAPTIVE_RECONSTRUCTION( v )         ▷ Create single-scale representation
9:                                                      of input vector
10:    r ← ADAPTIVE_RECONSTRUCTION( w )         ▷ Create single-scale representation
11:                                                     of output vector
12:    for all λ ∈ S(r) do
13:        for all μ ∈ Υ_λ(S(t)) do
14:            r_λ ← r_λ + t_μ a(φ_μ, φ_λ)                 ▷ Apply linear operator
15:            if |μ| ≠ |λ| then
16:                r_μ ← r_μ + t_λ a(φ_μ, φ_λ)    ▷ Apply symmetric part of linear operator
17:            end if
18:        end for
19:    end for
20:    w ← ADAPTIVE_DECOMPOSITION( r, T' ) ▷ Assemble vector in wavelet coordinates
21:    return w
22: end procedure
```

would have to be checked, which would make it possible to bound the cost for assembling the Dependency set by a constant (which depends on the basis functions). The disadvantage would be higher memory cost (for storing the reconstructed values) and the runtime of the refinement computations. Asymptotically, this would give an optimal linear complexity estimate. This idea is the basis for the next adaptive algorithm.

- *Refining the input vector* $\mathbf{v}$ *can also alleviate another use-oriented concern: Computing the value of the bilinear form for all possible combinations of functions might prove burdensome. Fewer single-scale indices means fewer cases to compute. The use of this technique depends heavily upon the actual bilinear form and cannot be discussed judiciously in general.*

All that is left now, is to determine the Dependency set for any single-scale index and the computations of the value of the linear operator. These two tasks can actually be processed in a single step: By determining the value of the bilinear form $a(\cdot, \phi_\lambda)$ for all functions ϕ_μ on coarser levels, the Dependency set Υ_λ is simply given by the set of indices for which the evaluation of the bilinear form turns out to be non-zero. This step can be executed once for any bilinear form and the results loaded on program startup. The cost of accessing a value of the bilinear form can thus be managed by a simple lookup in a table or similar data structure.
In application, the Dependency set (3.5.5) can thus be replaced by an **operator adapted version**, i.e.,

$$\overline{\Upsilon}^a_\lambda(\mathcal{T}) := \{\mu \in \mathcal{T} \text{ with } |\mu| \leq |\lambda| \,|\, a(\psi_\mu, \psi_\lambda) \neq 0\} . \tag{3.5.6}$$

This change could lower the constants involved in Theorem 3.58 if $\overline{\Upsilon}^a_\lambda(\mathcal{T})$ is much smaller than $\overline{\Upsilon}_\lambda(\mathcal{T})$ (for example due to **vanishing moments** $(\mathcal{V})$(2.2.5)). But the asymptotics and thus the complexity of Algorithm 3.9 remain unchanged unless the value of the bilinear form turns out to be zero for all coefficients on most (except for a uniformly bounded number) levels.

An Adaptive Single-Scale Evaluation Scheme

A disadvantage of using (3.5.4) to calculate bilinear form values for large level differences $J \gg j_0$ is the exponential memory requirement needed for the full vectors. The necessity to use an adaptive scheme to evaluate the bilinear form is thus obvious. We present here an easily usable algorithm which traverses the vectors level-wise from root to leaves. This algorithm was independently formulated and presented in [90], yet in a more general setting.
We assume to have an input $\mathbf{v}$ and target isotropic wavelet vector $\mathbf{w}$ based on trees $\mathcal{T}$, $\mathcal{T}'$, converted into single-scale vectors $\mathcal{R}$, $\mathcal{R}'$, respectively. The whole idea is to ascendingly traverse the target vector $\mathcal{R}'$ level-wise and categorize all its coefficients into these two categories:

- All target functions $\lambda = (j, \mathbf{k}) \in \mathcal{R}'$ not intersecting any functions on the next level $\mathcal{R}_{j+1}$:

$$\Theta(\mathcal{R}'_j, \mathcal{R}_{j+1}) := \left\{\lambda \in \mathcal{R}'_j \,|\, S(\phi_\lambda) \cap S(\phi_\mu) = \emptyset \text{ for any } \mu \in \mathcal{R}_{j+1}\right\} . \tag{3.5.7}$$

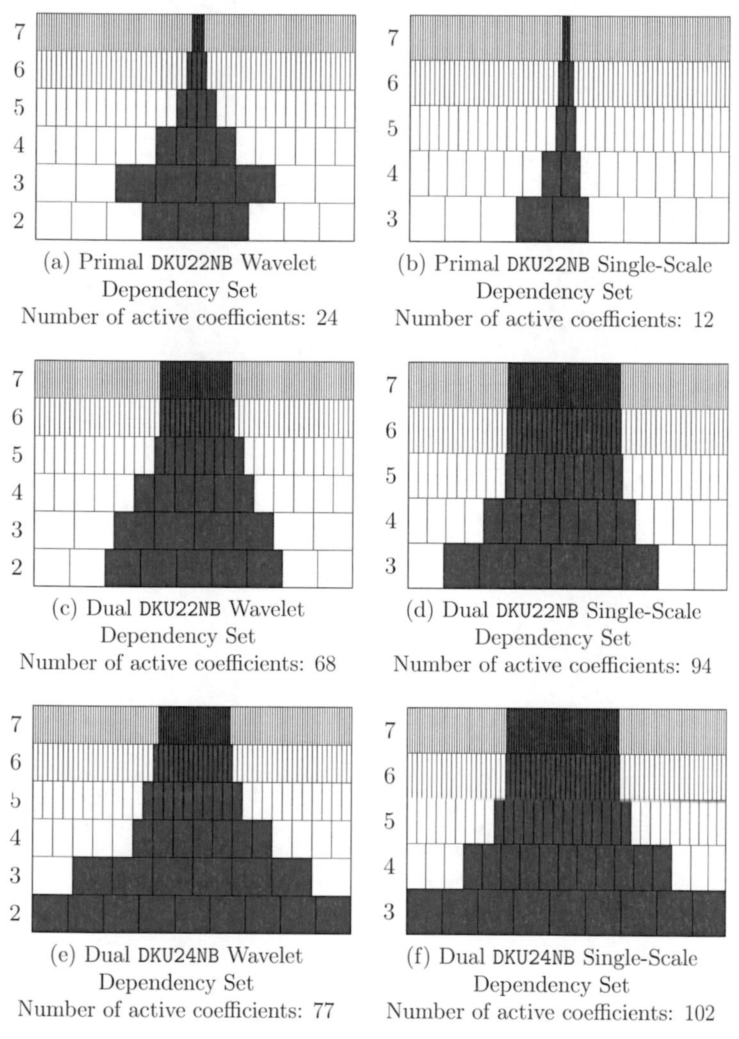

(a) Primal DKU22NB Wavelet Dependency Set
Number of active coefficients: 24

(b) Primal DKU22NB Single-Scale Dependency Set
Number of active coefficients: 12

(c) Dual DKU22NB Wavelet Dependency Set
Number of active coefficients: 68

(d) Dual DKU22NB Single-Scale Dependency Set
Number of active coefficients: 94

(e) Dual DKU24NB Wavelet Dependency Set
Number of active coefficients: 77

(f) Dual DKU24NB Single-Scale Dependency Set
Number of active coefficients: 102

Figure 3.14: The diagrams in the left column, (a), (c) and (e), show the Dependency set for the wavelet coefficient $\lambda = (7, 64, 1)$, for the primal and dual wavelets of type DKU22NB and DKU24NB. The right column, diagrams (b), (d) and (f), depicts the Dependency sets of single scale functions for the single-scale coefficient $(7, 62, 0)$, for the primal and dual single-scale functions of order $d = 2$ and $\widetilde{d} = 2, 4$ from top to bottom. The coefficients are each is drawn in red color, all indices of the Dependency set on the same or lower level are marked in blue color. As expected, the sets grow when the regularity is increased as this increases the support of the involved functions.

- For the rest of the functions the bilinear form cannot be calculated on the current level. The set of these functions is designated

$$\overline{\Theta}(\mathcal{R}'_j, \mathcal{R}_{j+1}) := \mathcal{R}'_j \setminus \Theta(\mathcal{R}'_j, \mathcal{R}_{j+1}). \tag{3.5.8}$$

The complexity of having to evaluate the bilinear form w.r.t. to functions on different levels in Algorithm 3.9 is here replaced with the determination whether functions on adjacent levels are sharing support. The algorithm then consists of the following steps:

(1) All output coefficients of $\overline{\Theta}(\mathcal{R}'_j, \mathcal{R}_{j+1})$ are transferred to the next level using the reconstruction identity (3.3.1), together with any input data needed to accurately evaluate the bilinear form for these functions on the next level:

$$\text{The set } \Theta(\mathcal{R}_j, \overline{\Theta}(\mathcal{R}'_j, \mathcal{R}_{j+1})) \text{ has to be refined.}$$

(2) For all coefficients of $\Theta(\mathcal{R}'_j, \mathcal{R}_{j+1})$ holds that the bilinear form can be evaluated **exactly** w.r.t. ϕ_λ. Here, the bilinear form must only be evaluated w.r.t. functions on the same level $j = |\lambda|$, because any data of coarser levels was refined and thus retained until it was represented on the current level.

(3) The evaluation process is then restarted on the next finer level $j+1$ and any reconstructed coefficients, which are omitted in step (1), are later reconstructed using the decomposition identities (3.3.2).

The algorithm hinges on the following property of the involved function representations:

Proposition 3.60 *During the execution of the above steps, the support of all functions on any higher level is a subset of the support of all the functions on the current level, see (3.4.9).*

Proof: Proof is done by induction. Since (3.4.9) is valid for a proper isotropic wavelet tree, the nested support property is still valid after the application of Algorithm 3.1, because the adaptive reconstruction of an isotropic wavelet $\psi^{\text{iso}}_{j,\mathbf{k},\mathbf{e}}$ on level j consists only of single-scale functions $\phi_{j+1,\mathbf{k}'}$ on level $j+1$.
The induction step is simply true because the current levels $\mathcal{R}_j$, $\mathcal{R}'_j$, are enlarged in the steps (1) and (2) and never made smaller. ∎

This assertion makes it possible to not need to check all finer level for intersecting functions but only the next level. One of the main tasks during the execution is thus for $\lambda = (j, \mathbf{k})$ the determination of all indices $\mu = \{(j+1, \mathbf{k}') \,|\, \mathbf{k}'\}$ for which $S(\phi_\lambda) \cap S(\phi_\mu) \neq \emptyset$ for the refinement steps, and the indices $\mu = \{(j, \mathbf{k}') \,|\, \mathbf{k}'\}$ for the evaluation of the bilinear form $a(\phi_\lambda, \phi_\mu)$. This is done in $\mathcal{O}(1)$ by the aforementioned **Support Operator** of Remark 3.55.

Algorithm 3.10 Adaptive application of a linear operator $A : \mathcal{H} \to \mathcal{H}'$ to a wavelet vector $\mathbf{v}$ up to accuracy ϵ.

1: **procedure** APPLY_LINEAR_SINGLE_SCALE_ADAPTIVE(ϵ, A, γ, $\mathbf{v}$) $\to \mathbf{w}$
2: $\mathcal{T} \leftarrow S(\mathbf{v})$
3: $\mathcal{T}' \leftarrow$ PREDICTION(ϵ, $\mathbf{v}$, γ) ▷ Ensure (3.3.12)
4: $\mathcal{T}' \leftarrow \mathcal{T}' \cup \mathcal{T}$ ▷ Ensure $S(\mathbf{v}) \subset \mathcal{T}$
5: **for all** $\mu \in \mathcal{T}' \setminus \mathcal{T}$ **do** ▷ Set up $\mathbf{v}$ on $\mathcal{T}'$
6: $v_\mu \leftarrow 0$ ▷ Insert wavelet coefficients
7: **end for**
8: **for all** $\mu \in \mathcal{T}'$ **do** ▷ Set up $\mathbf{w}$ on $\mathcal{T}'$
9: $w_\mu \leftarrow 0$ ▷ Insert wavelet coefficients
10: **end for**
11: $\mathbf{r} \leftarrow$ ADAPTIVE_RECONSTRUCTION($\mathbf{w}$) ▷ Create single-scale representation
12: of output vector
13: $\mathbf{t} \leftarrow$ ADAPTIVE_RECONSTRUCTION($\mathbf{v}$) ▷ Create single-scale representation
14: of input vector
15: $\mathcal{R} \leftarrow S(\mathbf{t})$, $\mathcal{R}' \leftarrow S(\mathbf{r})$
16: $j_0 \leftarrow \min\{|\lambda| \mid \lambda \in \mathcal{R}'\}$, $J \leftarrow \max\{|\lambda| \mid \lambda \in \mathcal{R}'\}$ ▷ Minimum and Maximum Levels
17: **for all** $j = j_0, \ldots, J$ **do**
18: **for all** $\lambda \in \mathcal{R}'_j$ **do**
19: **if** $\lambda \in \Theta(\mathcal{R}'_j, \mathcal{R}_{j+1})$ **then** ▷ Coefficient λ can be computed
20: **for all** $\mu \in \overline{\Theta}(\mathcal{R}_j, \{\lambda\})$ **do**
21: $r_\lambda \leftarrow r_\lambda + t_\mu a(\phi_\mu, \phi_\lambda)$ ▷ Apply linear operator
22: **end for**
23: **else** ▷ Coefficient λ must be transferred to the next level
24: $\mathbf{t} \leftarrow \mathbf{t} +$ ADAPTIVE_RECONSTRUCTION($\overline{\Theta}(\mathcal{R}_j, \{\lambda\})$)
25: ▷ These new coefficients must
26: not be considered in line 19 for the set $\mathcal{R}_{j+1}$
27: $\mathbf{r} \leftarrow \mathbf{r} +$ ADAPTIVE_RECONSTRUCTION($\{r_\lambda\}$) ▷ Insert with zero value
28: into vector
29: **end if**
30: **end for**
31: **end for**
32: $\mathbf{w} \leftarrow$ ADAPTIVE_DECOMPOSITION($\mathbf{r}$, $\mathcal{T}'$) ▷ Assemble vector in wavelet coordinates
33: **return** $\mathbf{w}$
34: **end procedure**

The advantage of this more complex evaluation algorithm is clearly shown in the next theorem:

Theorem 3.61 *Algorithm 3.10 finishes for a vector of size $N := \#S(\mathbf{w})$, where $\mathbf{w}$ is the output vector, in time $\mathcal{O}(N)$.*

Proof: Because of the results of Theorem 3.21 and Theorem 3.22, it is only necessary to estimate the number of calls to evaluate the bilinear form in line 21. Without any refinements in line 27, the number of evaluations during the algorithm is trivially $\mathcal{O}(N)$. Calling the constant of (2.1.16) again $m \in \mathbb{N}$, applying the reconstruction identity (3.3.1) creates at most m^n coefficients on the next level. This is still an independent constant and thus at most (if all $\lambda \in \mathcal{R}'_j$ have to be refined) still $\mathcal{O}(\#\mathcal{R}'_j)$ new coefficients could be inserted. Repeating this thought, over all levels, this leads to

$$\sum_{j=j_0}^{J} \mathcal{O}(\#\mathcal{R}'_j) = \mathcal{O}(\#\mathcal{R}'_J) = \mathcal{O}(N),$$

number of reconstructed coefficients. As all other involved procedures, e.g., checking whether $\lambda \in \Theta(\mathcal{R}'_j, \mathcal{R}_{j+1})$ does hold or not, are applicable in $\mathcal{O}(1)$, the overall complexity is $\mathcal{O}(N)$. ■

We compare the results of Theorem 3.58 and Theorem 3.61 in the Figures 3.15 and 3.16. In all cases,the tree prediction was executed using the exact support, i.e., (3.3.18). Since the support of the dual wavelets is larger for $\tilde{d} = 4$ than for $\tilde{d} = 2$, this policy creates larger vectors for $\tilde{d} = 4$ if the other parameters to the algorithm are otherwise chosen equally. This explains the increase in complexity for these wavelets visible in Figure 3.16. As is illustrated in these figures, both algorithms behave as proclaimed, but the asymptotics of the logarithmic factor shown in Theorem 3.58 is only visible for very high numbers of coefficients, i.e., for many levels $J \gg j_0$. This is probably due to the fact that, for vectors that are full at lower levels but sparse in higher levels, Algorithm 3.9 will have no "misses" in (3.5.3) on lower levels but Algorithm 3.10 must refine all coefficients. This explains the first parallel part of the graphs, as Algorithm 3.9 is at its best efficiency and Algorithm 3.10 is at its worst. In other words, the constants in the Landau notation of Theorem 3.58 and Theorem 3.61 were changing throughout the tests.
Although Algorithm 3.9 has worse asymptotics, it runs faster for many dimensions $n \gg 1$ than Algorithm 3.10, especially for sparse vector. This is understandable as the constant number of single-scale functions needed to reconstruct a single function in n dimensions rises exponentially (m^n) and this must be reflected in the absolute complexity of the algorithm. On the other hand, as stated for Algorithm 3.10, for only a fixed number of levels $J \approx j_0$, the logarithmic factor can be regarded as a constant and this constant grows linearly w.r.t. the number of dimensions n. Another problem in comparing these algorithms is the computation of the Dependency set $\Upsilon_\lambda(\mathcal{I})$ for level differences > 20. The computations take too long to be made "on the fly" and even precomputing this data becomes unfeasible at some point. But without this data, the algorithm cannot function and therefore no comparison is then possible.
In summary, Algorithm 3.10 is superior to Algorithm 3.9 in almost any way, except when

applying sparse vectors in high dimensions. Since vectors can be easily checked for these conditions, the application algorithm can be chosen depending on the vector, in the hope of minimizing execution time.

Remark 3.62 *The results of Figures 3.15 and 3.16 cannot be compared to the prior results shown in Figures 3.10 to 3.12. The timing data were created using different options for the compiler and the programs ran on different computers. In general, no apply algorithm beats any other algorithm in any situation. All that should be judged is the performance in typical use cases. For these, Algorithm 3.8 is often superior w.r.t. to execution times. This stems from the fact, that Algorithms 3.9 and 3.10 need to access many more elements directly, which always entails evaluations of hash functions (see Appendix B.3). In contrast, during Algorithm 3.8, the set of all indices is traversed using iterators, which works without the hash function. But to again make the point, basing the evaluation of the operator on the bilinear form, operators can be applied if the employed wavelets are not piecewise polynomials.*

We now discuss the evaluation of bilinear forms for the linear operators relevant to the control problem discussed in [122].

3.5.2 Bilinear Forms

We now show how to calculate the exact values of common bilinear forms usable in Algorithms Algorithm 3.9 and Algorithm 3.10.

The Mass Matrix

To calculate the value of the bilinear form of the mass matrix,

$$a_0(v, w) = \int_\Omega v(\mathbf{x})\, w(\mathbf{x})\, d\mu(\mathbf{x}), \tag{3.5.9}$$

for all combinations of single-scale functions $\phi_\lambda \in \Phi_{j_1}$, $\phi_\mu \in \Phi_{j_2}$, we first note a few facts which will make this task much easier:

1. The tensor product structure of the multi-dimensional single-scale basis (2.4.6) leads to a decoupling of the bilinear form w.r.t. the dimension. Therefore, we restrict ourselves here to one-dimensional considerations.

2. Since the single-scale functions are created by translation and dilation from a mother function (2.1.9) with a normalization w.r.t. the L_2-norm, the value of the bilinear form (3.5.9) only depends on the **difference** in **location** and **scale**, i.e.,

$$a_0(\phi_{j_1,k_1}, \phi_{j_2,k_2}) = \widetilde{a}_0(|j_1 - j_2|, |2^{-j_1}k_1 - 2^{-j_2}k_2|). \tag{3.5.10}$$

Without loss of generality, we assume $j_1 \geq j_2$ and define the level difference

$$p := j_1 - j_2 \geq 0. \tag{3.5.11}$$

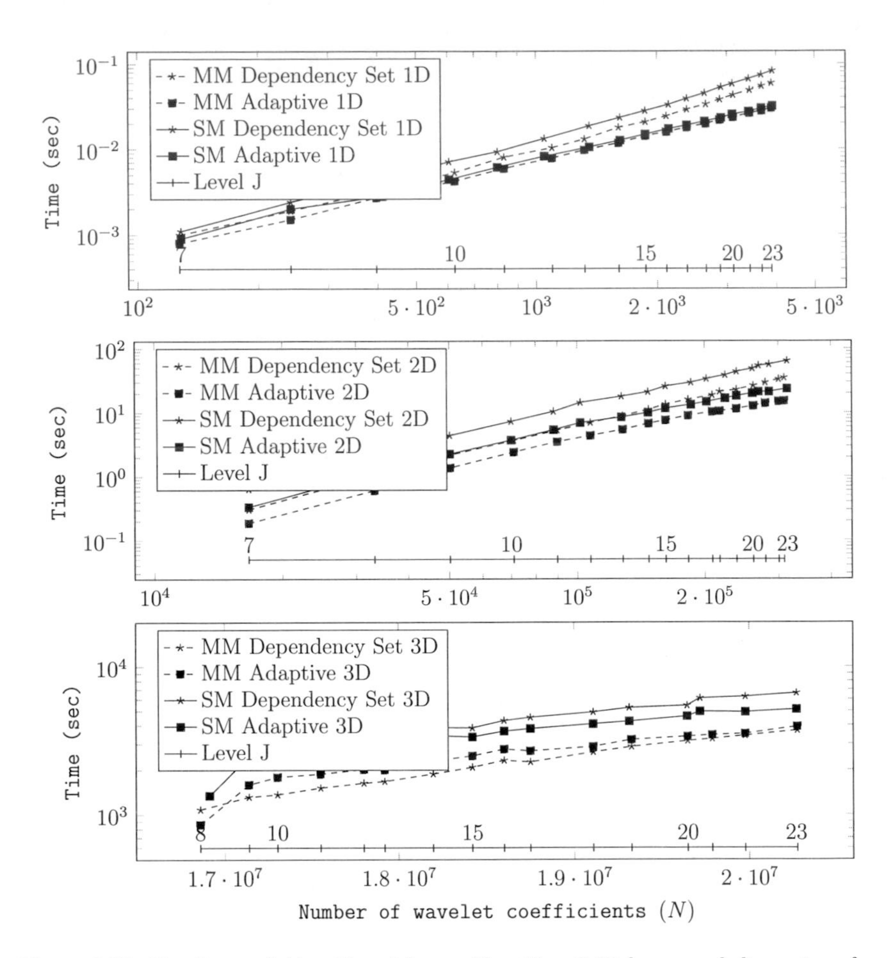

Figure 3.15: Runtimes of Algorithm 3.9 vs. Algorithm 3.10 for several dimensions for the Mass Matrix bilinear form (3.5.13) and Stiffness Matrix bilinear form (3.5.16), implemented using DKU22NB wavelets. For the application, a vector on a single level J with ten random wavelet coefficients on the highest wavelet level is made a proper tree with random coefficient values and then used as the input vector for the algorithms. In each case, ten such vectors are constructed and the execution time measured. Each data point thus corresponds to a single level, but since the vectors are randomized, these vectors are not full but increasingly sparse w.r.t. dimension and level. Because tree prediction with $\gamma := 4$ and $\varepsilon := 0.001$ usually constructs Influence sets with depth 3, the adaptive vectors will be full for levels $J \leq j_0 + 3$, and still very much filled up for several levels. The data points for the different operators were not always calculated using the same vectors, therefore the data points often do not line up perfectly.

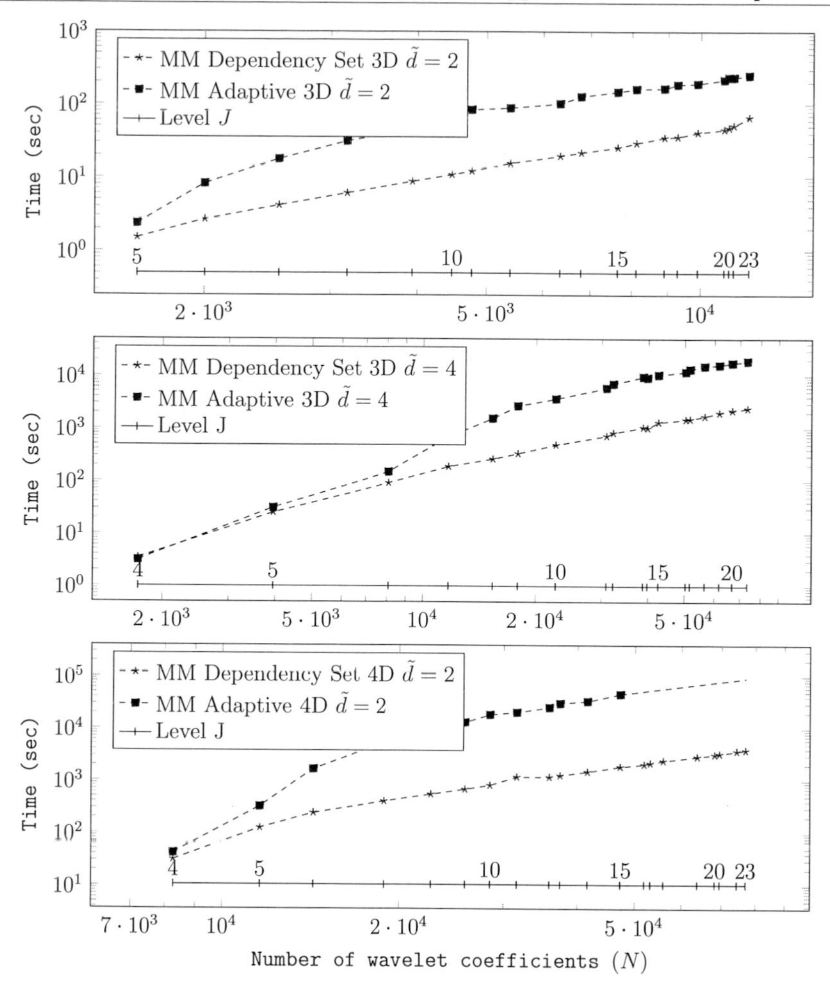

Figure 3.16: Time of Algorithm 3.9 vs. Algorithm 3.10 for several dimensions for the Mass Matrix bilinear form (3.5.13), implemented using dual DKU22NB and DKU24NB wavelets. For the application, a vector on a single level J with three random wavelet coefficients on the highest wavelet level is made a proper tree with random coefficient values and then used as the input vector for the algorithms. In each case, ten such vectors are constructed and the execution time measured. Each data point thus corresponds to a single level, but since the vectors are randomized, these vectors are not full but increasingly sparse w.r.t. dimension and level. The parameters for tree prediction were $\gamma := 10$ and $\varepsilon := 0.1$, which leads to very sparse vectors. In this case, the advantage clearly goes to Algorithm 3.9, it is more efficient for highly sparse vectors with coefficients, as long as the data of the bilinear form for the respective level difference is available. The dashed line shows extrapolated values calculated from a linear fit of the five previous data points.

Since these spaces Φ_j are set up on a dyadic grid according to (2.1.9), each grid point is also contained on all finer grids. The spatial distance can thus be measured on the finer grid exactly by counting the intermediate grid points, which follows from $|2^{-j_1}k_1 - 2^{-j_2}k_2| = 2^{-j_1}|k_1 - 2^{j_1-j_2}k_2| = h|k_1 - 2^{j_1-j_2}k_2|$ to be

$$q := |k_1 - 2^p k_2|. \tag{3.5.12}$$

This also assumes the functions $\phi_{j,k}$ are **symmetric** around their center $2^{-j}k$, which is true for the constructions of Section 2.3.

Example 3.63 *We consider the case of piecewise linear hat functions (A.1.2) without boundary adaptations. It can then easily be shown to hold*

$$a_0(\phi_{j_1,k_1}, \phi_{j_2,k_2}) = \widetilde{a}_0(p,q) := 2^{-\frac{3}{2}p} \begin{cases} 2^p - \frac{1}{3}, & q = 0, \\ 2^p - q, & 0 < q < 2^p \\ \frac{1}{6}, & q = 2^p, \\ 0, & \text{otherwise.} \end{cases}$$

Looking from the perspective of the function on the higher level j_1, the number of hat functions on coarser levels with overlapping support is bounded by 3*, and usually is only* 2*. The boundary adaptations complicate the above formula only slightly, e.g., only in the cases where the supports of any function would partially pass out of the domain.*

For higher dimensions, the matrix bands are determined independently for each spatial position and them combined via tensor product. The values of the bilinear forms are simply multiplied to give the value of the multi-dimensional bilinear form:

$$a_0(\phi_{\mathbf{j},\mathbf{k}}, \phi_{\mathbf{j}',\mathbf{k}'}) := \prod_{i=1}^{n} a_0(\phi_{j_i,k_i}, \phi_{j_i',k_i'}). \tag{3.5.13}$$

In particular, basis functions of different kind can easily be combined here in different spatial directions.

Remark 3.64 *The actual implementation of this procedure is a bit more technical because functions with large supports and boundary adaptations might need special treatment when both levels are very small, i.e., $j \approx J \approx j_0$, and the supports of functions on different sides of the interval intersect. Nevertheless, the idea is the same.*

The Stiffness Matrix

The bilinear form of the stiffness matrix can be described as a sum of bilinear forms of the kind seen in the previous paragraph. Precisely, it holds

$$a(u,v) = \int_\Omega \nabla v(\mathbf{x}) \cdot \nabla w(\mathbf{x}) + v(\mathbf{x})\, w(\mathbf{x})\, d\mu(\mathbf{x}), \tag{3.5.14}$$

Since the mass matrix part is already dealt with, we now focus on the **Laplace** part

$$a_1(u,v) = \int_\Omega \nabla v(\mathbf{x}) \cdot \nabla w(\mathbf{x})\, d\mu(\mathbf{x}). \tag{3.5.15}$$

By (2.1.9), it follows for two one-dimensional single scale functions

$$\begin{aligned} a_1(\phi_{j_1,k_1}, \phi_{j_2,k_2}) &= \int_\Omega \nabla\phi_{j_1,k_1}(x) \cdot \nabla\phi_{j_2,k_2}(x)\, d\mu(x) \\ &= \int_\Omega 2^{j_1/2} 2^{j_1} \phi'(2^{j_1}x - k_1) \cdot 2^{j_2/2} 2^{j_2} \phi'(2^{j_2}x - k_2)\, dx \\ &= 2^{j_1+j_2} \int_\Omega 2^{j_1/2} \phi'(2^{j_1}x - k_1) \cdot 2^{j_2/2} \phi'(2^{j_2}x - k_2)\, dx \end{aligned}$$

Except for the extra factor of $2^{j_1+j_2}$, this expression has the same structure as (3.5.10) and can be dealt in the same way. The factor $2^{j_1+j_2}$ can simply be applied after the value of the integral has been determined.

Example 3.65 *For the same functions as in Example 3.63 holds here*

$$a_1(\phi_{j_1,k_1}, \phi_{j_2,k_2}) = 2^{j_1+j_2}\widetilde{a}_1(p,q) := 2^{j_1+j_2}\, 2^{-\frac{1}{2}p} \begin{cases} 2, & q = 0, \\ 0, & 0 < q < 2^p \\ -1, & q = 2^p, \\ 0, & \textit{otherwise} \end{cases}$$

Any value of the multi-dimensional Laplace operator bilinear form for tensor products of single-scale functions can simply be calculated by

$$a_1\left(\phi_{\mathbf{j},\mathbf{k}}, \phi_{\mathbf{j}',\mathbf{k}'}\right) := \sum_{i=1}^{n} a_1(\phi_{j_i,k_i}, \phi_{j_i',k_i'}) \prod_{\substack{r=1 \\ r\neq i}}^{n} a_0(\phi_{j_r,k_r}, \phi_{j_r',k_r'}). \tag{3.5.16}$$

Just as before, the comments of Remark 3.64 apply.

Riesz Operators

The Riesz Operator $\widetilde{\mathbf{R}}_{H^s}$ of (2.2.52) can, for $s \in (0,1)$, be implemented easily as a combination of the bilinear forms (3.5.13) and (3.5.16):

$$\begin{aligned} a_s((\phi_{\mathbf{j},\mathbf{k}}, \phi_{\mathbf{j}',\mathbf{k}'})) &:= (1-s)\, a_0((\phi_{\mathbf{j},\mathbf{k}}, \phi_{\mathbf{j}',\mathbf{k}'})) + s\, a_1((\phi_{\mathbf{j},\mathbf{k}}, \phi_{\mathbf{j}',\mathbf{k}'})) \\ &= s \sum_{i=1}^{n} a_1(\phi_{j_i,k_i}, \phi_{j_i',k_i'}) \prod_{\substack{r=1 \\ r\neq i}}^{n} a_0(\phi_{j_r,k_r}, \phi_{j_r',k_r'}) + a_0\left(\phi_{\mathbf{j},\mathbf{k}}, \phi_{\mathbf{j}',\mathbf{k}'}\right). \end{aligned} \tag{3.5.17}$$

The normalization w.r.t. constant function of (2.2.53) can also be easily applied, since this is only a single constant factor.

3.5.3 Inverses of Linear Operators

The application of the inverse of a linear operator is often also sought in applications. Linear operators share some common properties with their inverse counterparts. For example, the inverses must themselves be linear operators again and if an operator written in matrix form is symmetric, its inverse must also be symmetric. There are, however, also properties that do not carry over,e.g., the inverses of sparse matrices are generally not sparse. Ideally, the inverse of an operator would exhibit all the same properties as the original operator. In this setting, this situation can arise when the inverse operator of an operator w.r.t. some wavelet basis Ψ can be expressed w.r.t. the dual basis $\widetilde{\Psi}$ of (2.1.44).

The Dual Mass Matrix

The dual mass matrix is, theoretically, directly available by calculating the values of the bilinear form (3.5.9). This can be done using a technique called **refineable integrals**, which was put forth in [52, 97]. That theory was designed to compute integrals on whole domain $\mathbb{R}$ (even though the involved functions have compact support) and thus does not deal with specific boundary adaptations. Some boundary adaptations can be emulated by truncations of the domain, which can be implemented by artificially inserting a **characteristic function** into the integral. The refinable integrals theory is thus firstly applicable to general wavelet construction of Section 2.3.1, which is usually the construction used in case of periodic boundary conditions.

Remark 3.66 *A computer program evaluating these refinable integrals was first implemented in [98]; a `MATLAB` implementation was developed recently and is available online [129]. Additionally, a `C++` version is included in the software developed for this thesis.*

We here use another, more direct way of calculating the dual mass matrix entries: The idea is that the **single-scale** mass matrix $\widetilde{\mathbf{M}}_{\Phi_J}$ is a **band matrix** and the values are exactly reproduced when the **FWT** is applied to decompose the matrix once, i.e.,

$$\widetilde{\mathbf{M}}_{\Phi_J} = \left(\widetilde{\mathbf{T}}_{J+1}^T \widetilde{\mathbf{M}}_{\Phi_{J+1}} \widetilde{\mathbf{T}}_{J+1}\right)\Big|_{\Delta_J \times \Delta_J} = \left(\begin{array}{c|c} \widetilde{\mathbf{M}}_{\Phi_J} & \cdots \\ \hline \vdots & \ddots \end{array}\right)\Bigg|_{\Delta_J \times \Delta_J}. \tag{3.5.18}$$

The individual entries in the matrix bands can thus be seen as the solution of an "eigenvalue" problem. Each non-zero matrix entry of $\widetilde{\mathbf{M}}_J$ is here considered an unknown and therefore there are exactly as many unknowns as equations. In the case of periodic boundary conditions, though, the system might not be uniquely solvable, but another equation can be derived because the sum of any row or column must be exactly equal to 1. The necessary calculations can easily be executed by a **Computer Algebra System**, e.g., Mathematica, and if the entries of $\widetilde{\mathbf{T}}_J$ are available exactly as fractions, then the entries of the mass matrix can be computed as fractions as well. The complexity of these computations can be significantly reduced when repeating values, e.g., due to symmetry and translation, are recognized and excluded.
The computed values for our used dual single-scale functions can be found in Appendix A.1.2 and Appendix A.1.3. Then, with the approach in (3.5.4), any possible value of the bilinear form is computable. The **dual mass matrix** is therefore available exactly.

The Dual vs. the Inverse Mass Matrix

The inverse of the mass matrix $\mathbf{M}_{\mathcal{H}}^{-1}$ is by (2.1.62) the mass matrix of the dual wavelets $\mathbf{M}_{\mathcal{H}'}$. But this is no longer true if the **finite dimensional** mass matrices $\widetilde{\mathbf{M}}_J$, $\mathbf{M}_J^{-1}$ are considered. Simply put, if $Q_J : \ell_2(\mathcal{I}) \to \ell_2(\Delta_J)$ is the projector that deletes all elements with level $> J$, then holds

$$\mathbf{M}_J = \left(\Psi_{(J)}, \Psi_{(J)}\right) = \left(Q_J \Psi, Q_J \Psi\right) = Q_J \left(\Psi, \Psi\right) Q_J',$$

and

$$\widetilde{\mathbf{M}}_J = \left(\widetilde{\Psi}_{(J)}, \widetilde{\Psi}_{(J)}\right) = \left(Q_J\widetilde{\Psi}, Q_J\widetilde{\Psi}\right) = Q_J\left(\widetilde{\Psi}, \widetilde{\Psi}\right) Q'_J,$$

so that, because the extension operator $Q'_J : \ell_2(\Delta_J) \to \ell_2(\mathcal{I})$ can not be the inverse of Q_J as information is lost, follows

$$\mathbf{M}_J\widetilde{\mathbf{M}}_J = Q_J\left(\widetilde{\Psi}, \widetilde{\Psi}\right) Q'_J\, Q_J\left(\Psi, \Psi\right) Q'_J \neq \mathbf{I}_J.$$

The infinite dual mass matrix is thus the infinite **inverse mass matrix**, but this is not true in the finite case. Even worse, the actual values of the inverse mass matrices $\mathbf{M}_J^{-1}$ are different for all values of the level J, i.e., $\mathbf{M}_{J+1}^{-1}$ is not an exact extension of the matrix $\mathbf{M}_J^{-1}$. Hence, in an algorithmic implementation, the values of each matrix $\mathbf{M}_J^{-1}$, $J = j_0, j_0+1, \ldots$, would need to be computed and saved separately.
Since the application of the involved projectors in computing $\mathbf{M}_J^{-1}$ can be seen as approximating the uppermost block $\left(\widetilde{\Psi}_{j_0}, \widetilde{\Psi}_{j_0}\right)$ (as in (3.5.18)) using functions of up to level J, i.e., $\Psi_{(J)}$, then one can directly deduce that

$$\left(\mathbf{M}_J^{-1}\right)\big|_{\Delta_{j_0}\times\Delta_{j_0}} \to \widetilde{\mathbf{M}}_{j_0}, \qquad \text{for } J \to \infty.$$

In the application of a theoretically infinite vector, this motivates to use $\widetilde{\mathbf{M}}_J$ instead of $\mathbf{M}_J^{-1}$.

Inverses of Stiffness Matrix and Riesz Operators

The inverse A_J^{-1} of the stiffness matrix is **not sparse**, i.e., $a_1^{-1}(\cdot,\cdot)$ is **not a local operator**, therefore the storage of the values of this operator is not feasible. Also, since no analytic form for the inverse bilinear form is available, these operators can only be applied through application of direct or iterative solvers, see Section 4.1.

3.5.4 A Square Weighted Mass Matrix

Here we refer to the operator based upon the bilinear form

$$a_z(u,v) := \int_\Omega z^2\, u\, v\, d\mu, \qquad \text{for } u, v \in \mathcal{H}, \tag{3.5.19}$$

for a fixed given function $z \in \mathcal{H}$, which corresponds to the operator application

$$\mathbf{A_z} : \ell_2(\mathcal{I}) \to \ell_2(\mathcal{I}), \quad \mathbf{A_z}(\mathbf{u}) = \mathbf{v}. \tag{3.5.20}$$

This operator will emerge from the **derivative** of the nonlinear operator $G(v) = v^3$, needed in the course of **Newton's method**.
We impose here a single constraint on the parameters u, v and z: They shall all be **primal**, and therefore piecewise polynomial,**functions**. Evaluating the bilinear form (3.5.19) for all combinations of functions u, v, z is theoretically possible but precomputing all these values - even only for single-scale functions as in Example 3.63 or Example 3.65 - quickly becomes problematic. The task becomes much more feasible by assuming all functions to live on the same level j, which is all that is necessary for Algorithm 3.10.

Example 3.67 *For the same functions as in Example 3.63, setting $u = \phi_{j,k_1}, v = \phi_{j,k_2}$ and $z = \phi_{j,k_3}$, the supports of all functions must overlap over an area of non-zero measure. Also, symmetry arguments further diminish the number of possible combinations. By this process of elimination, three distinct non-trivial combinations of the positions k_1, k_2, k_3 remain:*

$$k_1 = k_2 = k_3, \quad \text{and} \quad k_1 = k_2 \wedge |k_3 - k_1| = 1, \quad \text{and} \quad |k_1 - k_2| = 1 \wedge (k_3 = k_1 \vee k_3 = k_2).$$

For these sets of functions, one can easily verify that holds for $z = \phi_{j,k_3}$,

$$a_z(\phi_{j,k_1}, \phi_{j,k_2}) = 2^j \begin{cases} \frac{2}{5}, & k_1 = k_2 = k_3, \\ \frac{1}{30}, & k_1 = k_2 \wedge k_3 = k_1 \pm 1, \\ \frac{1}{20}, & k_1 = k_2 \pm 1 \wedge (k_3 = k_1 \vee k_3 = k_2), \\ 0, & \text{otherwise}. \end{cases}$$

Since the value of the inner product of two single scale functions is ~ 1, the product of four of these functions must produce a value $\sim 2^j$.

The main problem in this course is now how to ensure that for any proper trees $\mathbf{u}, \mathbf{v}, \mathbf{z} \in \ell_2(\mathcal{I})$, the evaluation algorithm will have the finest data of the input data, e.g., $\mathbf{u}$ and $\mathbf{z}$, while traversing the output tree, e.g, $\mathbf{v}$. Algorithm 3.10 does solve this problem for two vectors, but not for a third one, by assuming $S(\mathbf{v}) \supset S(\mathbf{u})$. A possible strategy is therefore either also assuming $S(\mathbf{z}) \subset S(\mathbf{v})$ or, if this is not true, extending $S(\mathbf{v})$ to include $S(\mathbf{z})$ and then, after the application, to drop the additional coefficients.
As it will be shown in Section 4.4.2, we can here assume $\mathbf{z}$ is not unrelated to the vectors $\mathbf{u}$,$\mathbf{v}$, but that, in view of (3.5.20), holds

$$S(\mathbf{z}) \subseteq S(\mathbf{u}) \subseteq S(\mathbf{v}). \tag{3.5.21}$$

Thus, the support of all vectors can be expanded to cover $S(\mathbf{v})$ and Algorithm 3.10 can run with minor changes, i.e., tracking the information stated by $\mathbf{z}$. Another approach is to use the techniques of Section 3.4.3, generate the adaptive polynomial representation (3.4.8) $S(\mathcal{T})$ of the largest vector $\mathbf{v}$ and then construct the information of $\mathbf{z}$ and $\mathbf{u}$ on the same set of support cubes. Again, this is perfectly possible as long as (3.5.21) is true. Thus, Algorithm 3.6 can be executed to apply the bilinear form (3.5.19). Again, as in Section 3.4.4, the reference element application w.r.t. the local polynomial basis $\mathbf{P}_\square$ (3.4.11) can be expressed as a data dependent and a base dependent part. By denoting $\mathbf{g} = \mathbf{G}_\mathcal{T}\mathbf{u}$ and $\mathbf{h} = \mathbf{G}_\mathcal{T}\mathbf{z}$ for $\mathcal{T} := S(\mathbf{v})$ the polynomial coefficients as in (3.4.19), it follows

$$\begin{aligned}
&(\chi_\square \mathbf{h}_\square^T \mathbf{P}_\square)^2 (\chi_\square \mathbf{g}_\square^T \mathbf{P}_\square) \\
&= \chi_\square \left(\sum_{i=1}^m (\mathbf{h}_\square)_i \, (\mathbf{P}_\square)_i \right)^2 \left(\sum_{i=1}^m (\mathbf{g}_\square)_i \, (\mathbf{P}_\square)_i \right) \\
&= \chi_\square \left(\sum_{s_1+\cdots+s_m=2} \mathbf{Y}(\mathbf{h}_\square; 2; s_1, \ldots, s_m) \prod_{1\le i\le m} ((\mathbf{P}_\square)_i)^{s_i} \right) \left(\sum_{i=1}^m (\mathbf{g}_\square)_i \, (\mathbf{P}_\square)_i \right), \\
&=: \chi_\square \left(\sum_{s_1+\cdots+s_m=3} \mathbf{Z}(\mathbf{g}_\square, \mathbf{h}_\square; 3; s_1, \ldots, s_m) \prod_{1\le i\le m} ((\mathbf{P}_\square)_i)^{s_i} \right),
\end{aligned} \tag{3.5.22}$$

and thus

$$
\begin{aligned}
&\langle \chi_\Box \mathbf{P}_\Box, (\chi_\Box \mathbf{h}_\Box^T \mathbf{P}_\Box)^2 (\chi_\Box \mathbf{g}_\Box^T \mathbf{P}_\Box) \rangle \\
&= \langle \chi_\Box \mathbf{P}_\Box, \chi_\Box \left(\sum_{s_1+\cdots+s_m=3} \mathbf{Z}(\mathbf{g}_\Box, \mathbf{h}_\Box; 3; s_1, \ldots, s_m) \prod_{1 \le i \le m} ((\mathbf{P}_\Box)_i)^{s_i} \right) \rangle \\
&= \sum_{s_1+\cdots+s_m=3} \mathbf{Z}(\mathbf{g}_\Box, \mathbf{h}_\Box; 3; s_1, \ldots, s_m) \left\langle \chi_\Box \mathbf{P}_\Box, \chi_\Box \prod_{1 \le i \le m} ((\mathbf{P}_\Box)_i)^{s_i} \right\rangle . \qquad (3.5.23)
\end{aligned}
$$

Hence, the inner products of the polynomial basis can be precomputed and only the terms $\mathbf{Z}(\mathbf{g}_\Box, \mathbf{h}_\Box; 3; s_1, \ldots, s_m)$ have to be computed with the data of $\mathbf{u}$ and $\mathbf{z}$.

We now turn to the application of Trace operators, which are local, linear operators, but they require special treatment nonetheless.

3.6 Trace Operators

The application of trace operators, although linear operators by Section 1.2.2, does not directly fit into the framework of the previous section. The trace operator γ_0 differs evidently from the operators considered in Section 3.5.1 since it does not map from a space $\mathcal{H} = H^1(\Omega)$ into its dual $\mathcal{H}'$, but into another primal space $\mathcal{V} = H^{1/2}(\Gamma)$, which is not even based on the same domain. Here, we assume $\Omega \subset \mathbb{R}^n$ to be a **domain** and $\Gamma \subseteq \partial\Omega \subset \mathbb{R}^{n-1}$ to be a part of its **boundary**. The linear trace operator $\gamma_0(\cdot) := (\cdot)|_\Gamma$ of (1.2.25) leads to the continuous bilinear form

$$
b(v, \widetilde{q}) := \langle \gamma_0 \, v, \widetilde{q} \rangle_{\mathcal{V} \times \mathcal{V}'} = \int_\Gamma (v)|_\Gamma \cdot \widetilde{q} \, ds, \qquad (3.6.1)
$$

which is well-defined on $\mathcal{H} \times \mathcal{V} := H^1(\Omega) \times (H^{1/2}(\Gamma))'$. The operator $B : \mathcal{H} \to \mathcal{V}$ and its adjoint $B' : \mathcal{V}' \to \mathcal{H}'$ are defined by the bilinear form (3.6.1) according to (1.4.12), i.e.,

$$
\langle v, B' \widetilde{q} \rangle_{\mathcal{H} \times \mathcal{H}'} = \langle B v, \widetilde{q} \rangle_{\mathcal{V} \times \mathcal{V}'} := b(v, \widetilde{q}). \qquad (3.6.2)
$$

Therefore, γ_0 is not an operator in the sense of Section 3.3.3 and Algorithm 3.4 is not used here. Rather, the output of the operator application is calculated directly from the input tree. If this can be done without the employment of approximation, e.g., quadrature rules, then the the result is **not approximate**, i.e., it does not depend on any **tolerance** ε.

Remark 3.68 *There are conceptually at least two ways to compute the trace operator application:*

1. *The direct approach is to compute for each wavelet index λ the trace $\psi_\lambda^\Omega|_\Gamma$, i.e, determine all wavelets $\widetilde{\psi}^\Gamma$ for which $b(\psi_\lambda^\Omega, \widetilde{\psi}^\Gamma) \neq 0$ and save the whole set in a data structure. The application of the trace operator is then a simple lookup in all these data structures and summing up all values of the bilinear form. The application of the adjoint operator here means that the wavelet $\widetilde{\psi}^\Gamma$ is given and thus the search direction is reversed, but the course of action is the same. This technique is very flexible w.r.t. complex domains and is used in the example of Section 5.5.*

2. *The disadvantage of the above direct method is that a lot of data has to be generated and held available. In special cases, the generation of the trace data, i.e., wavelet indices and bilinear form values, can be avoided. If the trace $\psi_\lambda^\Omega|_\Gamma$ is expressible in the wavelet base on the trace space (maybe after an adaptive decomposition), then the bilinear form can be evaluated exactly. The disadvantage of this approach is the non-trivial application of the adjoint trace operator because several wavelets ψ_λ^Ω might have the same trace. Thus, the set of all applicable wavelets on Ω for each $\widetilde{\psi}^\Gamma$ might have to be produced algorithmically.*

Of course, a hybrid of these two approaches might also be applicable in certain cases.

3.6.1 Trace Operators Parallel to the Coordinate Axes

We now assume that $\Omega = \square^n$ and γ_0 is merely a restriction onto a lower dimension along the Cartesian product axes, for some coordinate point $c \in [0,1]$, for $1 \leq i \leq n$. The case $\Gamma \subset \partial\Omega$ is then simply the special case $c \in \{0,1\}$.

Adaptive Application

The operator will be applied to a vector $\mathbf{v}$, which shall be supported on the tree $\mathcal{T} \subset \ell_2(\mathcal{I})$, and since the operator is linear, we can simply apply it to each individual wavelet index. Then follows for an n-dimensional isotropic wavelet (2.4.18) of the index $\lambda = (j, \mathbf{k}, \mathbf{e})$,

$$\begin{aligned}
\gamma_0(\psi_{(j,\mathbf{k},\mathbf{e})}(\mathbf{x})) &= \gamma_0\,(\psi_{\lambda_1}(x_1)\cdots\psi_{\lambda_n}(x_n)) \\
&= \psi_{\lambda_1}(x_1)\cdots\psi_{\lambda_{i-1}}(x_{i-1})\,(\gamma_0\psi_{\lambda_i}(x_i))\,\psi_{\lambda_{i+1}}(x_{i+1})\cdots\psi_{\lambda_n}(x_n) \\
&= \psi_{\lambda_1}(x_1)\cdots\psi_{\lambda_{i-1}}(x_{i-1})\,\psi_{\lambda_i}(x_i)|_\Gamma\,\psi_{\lambda_{i+1}}(x_{i+1})\cdots\psi_{\lambda_n}(x_n) \\
&= \psi_{\lambda_i}(c)\bigotimes_{\substack{\ell=1\\ \ell\neq i}}^{n}\psi_{(j,k_\ell,e_\ell)}(x_\ell) \\
&=: \psi_{\lambda_i}(c)\,\psi_{\lambda'}(\mathbf{x}'),
\end{aligned} \tag{3.6.3}$$

with $\lambda_i := (j, k_i, e_i)$ and $\lambda' := \lambda \setminus \lambda_i$.

Remark 3.69 *The above calculation is done in exactly the same fashion for anisotropic wavelets.*

The resulting $n-1$-dimensional isotropic wavelet index λ' will inherit valid level and location values from the index λ, but the index $\mathbf{e}' := \mathbf{e} \setminus e_i$ could possibly be all zeros, if e_i was the only type coordinate with a non-zero value. This means $\mathbf{e}' \notin \mathbb{E}^\star_{n-1}$ is a possible outcome of this calculation. If $j \neq j_0 - 1$, then ψ'_λ is not a valid isotropic wavelet, but simply a $n-1$ dimensional **single-scale** function. Applying this to each individual index $\lambda \in \mathcal{T}$, the resulting set of wavelet indices must be sorted w.r.t. the type and all single-scale functions must be decomposed into proper isotropic wavelet indices, see Algorithm 3.2.
In operator form, **biorthogonality** (2.1.51) gives

$$\left\langle B\psi_\lambda^\Omega, \widetilde{\psi}_\mu^\Gamma\right\rangle = \left\langle \psi_{\lambda_i}(c)\,\psi_{\lambda'}^\Gamma, \widetilde{\psi}_\mu^\Gamma\right\rangle = \psi_{\lambda_i}(c)\,\delta_{\lambda',\mu}, \quad \text{with } \delta_{\lambda',\mu} := \begin{cases} 1, & \text{if } \lambda' = \mu, \\ 0, & \text{otherwise.}\end{cases} \tag{3.6.4}$$

The operator B thus simply encodes whether a wavelet ψ_λ^Ω shares support with a trace space wavelet ψ_μ^Γ. The above thoughts gives rise to the following Algorithm 3.11.

Algorithm 3.11 Adaptive application of a trace operator $B : \mathcal{H} \to \mathcal{V}$ to a wavelet vector $\mathbf{v}$.

1: **procedure** APPLY_TRACE_OPERATOR($(^*\varepsilon^*)$, B, $\mathbf{v}$) $\to \mathbf{w}$
2: $\quad \mathcal{T} \leftarrow S(\mathbf{v})$
3: $\quad j_0 \leftarrow \min\{|\lambda| \,|\, \lambda \in \mathcal{T}\}$
4: $\quad J \leftarrow \max\{|\lambda| \,|\, \lambda \in \mathcal{T}\}$
5: $\quad \mathbf{w} \leftarrow \{\}$ ▷ Initialize empty (or with zeros on coarsest level j_0)
6: $\quad \mathbf{t} \leftarrow \{\}$ ▷ Temporary data; Initialize empty
7: $\quad$ **for** $j = j_0, \ldots, J$ **do**
8: $\quad\quad$ **for all** $\lambda = (j, \mathbf{k}, \mathbf{e}) \in \mathcal{T}$ **do**
9: $\quad\quad\quad \psi_{\lambda_i}(c)\, \psi_{(j,\mathbf{k}',\mathbf{e}')} \leftarrow B\psi_\lambda$ ▷ Apply Trace Operator
10: $\quad\quad\quad$ **if** $\mathbf{e}' \neq \mathbf{0}$ **then**
11: $\quad\quad\quad\quad w_{\lambda'} \leftarrow w_{\lambda'} + \psi_{\lambda_i}(c)$ ▷ Add wavelets to $\mathbf{w}$
12: $\quad\quad\quad$ **else**
13: $\quad\quad\quad\quad t_{\lambda'} \leftarrow t_{\lambda'} + \psi_{\lambda_i}(c)$ ▷ Add single scale functions to $\mathbf{t}$
14: $\quad\quad\quad$ **end if**
15: $\quad\quad$ **end for**
16: $\quad$ **end for**
17: $\quad \mathbf{w} \leftarrow \mathbf{w} +$ ADAPTIVE_DECOMPOSITION($\mathbf{t}$, $\mathcal{I}$) ▷ Decompose into wavelets
18: $\quad$ w.r.t. **PRIMAL** Wavelets
19: $\quad$ **Ensure** $\mathbf{w}$ is a **proper Tree**
20: $\quad$ **return** $\mathbf{w}$
21: **end procedure**

Theorem 3.70 *Algorithm 3.11 is applicable in linear time w.r.t. the number of elements in the input vector* $\mathbf{v}$.

Proof: Because of the tree structure of the input vector $\mathbf{v}$, any element with support intersecting the trace space will have a parent with the same property. Also, the type $\mathbf{e}$ of the parent is the same function type, because this is a prerequisite of the tree structure, see Proposition 3.17. As such, both the vector $\mathbf{t}$ and $\mathbf{w}$ will have tree structure after the main loop. The first part of the algorithm will thus finish in $\mathcal{O}(\#S(\mathbf{v}))$ steps.
In the worst case, all wavelet coefficients would be put into the single-scale tree $\mathbf{t}$, which therefore must contain at most $\#S(\mathbf{v})$ elements. In any case, its tree structure is determined by the tree structure of the input vector. The decomposition in line 17 will finish in time $\mathcal{O}(\#S(\mathbf{v}))$ by Theorem 3.22 and Remark 3.23. The last step of ensuring the output tree to be a proper tree should only insert few missing elements, i.e., the root elements. It is in any case a process of linear complexity.
The whole algorithm will thus run in linear time $\mathcal{O}(\#S(\mathbf{v}))$. ∎

Remark 3.71 *Because any point $c \in [0,1]$ will only intersect a finite amount of single-scale and wavelet functions on each level j, see (2.1.16), all possible location indices and their trace values can be precomputed in constant time for each level $j \geq j_0$. By looking up into this precomputed list, only the relevant wavelet indices from the input vector* $\mathbf{v}$ *can be selected, thus voiding the computational complexity of all non-relevant wavelet indices. By only selecting these indices, the complexity could be* **super-linear** *with respect to the input vector, i.e., linear with a constant factor ≤ 1.*

Adaptive Application of Adjoint Operator

The adjoint operator $B' : \mathcal{V}' \to \mathcal{H}'$, with $\mathcal{H} = H^1(\Omega)$ and $\mathcal{V} = H^{1/2}(\Gamma)$, extends a vector from the trace space onto the domain space, both w.r.t. dual wavelet expansions. By (3.6.2), **all** possible (proper) wavelet indices on the domain space, which are "connected" in the sense of (3.6.4), have to be produced. This means a dual wavelet on the trace space is expanded with single-scale and wavelet functions, but a dual single-scale function is only expanded using wavelets, because the result must be a proper dual wavelet on $\mathcal{H}'$:

$$B'\widetilde{\psi}^{\Gamma}_{\lambda'} = \psi_{\lambda_i}(c)\,\widetilde{\psi}^{\Omega}_{\lambda}, \qquad \text{for } \lambda' = (j, \mathbf{k}', \mathbf{e}'), \quad \mathbf{k}' \in \mathbb{Z}^{n-1}, \mathbf{e}' \in \mathbb{E}_{n-1}, \tag{3.6.5}$$

where the target index λ is any index in the set

$$\mathcal{M}_{\lambda'} := \begin{cases} \{(j, \mathbf{k}' \otimes \{k_i\}, \lambda' \otimes \{0\}), (j, \mathbf{k}' \otimes \{k_i\}, \lambda' \otimes \{1\})\}, & \text{if } \mathbf{e}' \neq 0 \vee (\mathbf{e}' = 0 \wedge j = j_0), \\ \{(j, \mathbf{k}' \otimes \{k_i\}, \lambda' \otimes \{1\})\}, & \text{if } \mathbf{e}' = 0 \wedge j > j_0. \end{cases} \tag{3.6.6}$$

As the location indices k_i have to be taken those, that, by (3.6.3), have non-zero values in the bilinear form (3.6.4). This family of indices is simply computed as explained in Remark 3.71.

But this is not the complete application, as then no dual wavelets in $\mathcal{H}'$ could be produced which consist of single-scale functions on levels $j > j_0$ on the trace space. To create these wavelets, the input vector has to be **reconstructed**, see Algorithm 3.1, after the application on each level w.r.t. the **dual MRA** to the next higher level. In a sense, the adaptive decomposition of Algorithm 3.11 has to be "undone" (B and B' are not inverse to each other).

Therefore, the number of output coefficients produced from a single input wavelet coefficient is theoretically unlimited. The reconstruction process thus must stop at some point and it is prudent to make the criteria depend on the input vector data, same as with Algorithm 3.11. In computations, we denote the maximum level containing a non-zero wavelet coefficient the **effective maximum level**, and refining functions beyond this level does not bring new information, as there is none already present on these levels. This is akin to the situation of the PDE application algorithms, where refinement of the input vector beforehand is also theoretically possible, but would just blow up the vector and thus increase the complexity without adding any new information.

But since the refinement step is necessary, it is important to look at the implications: Since we are assuming to be working with proper trees, and thus all coefficients of the coarsest single-scale level j_0 are present, the refinement of this level data of the vector will always fill up the next level completely with single-scale coefficients. This will make computations impossible for very high levels, therefore it is imperative to delete all elements

of the input vector $\mathbf{v}$ during execution that are too small, e.g., below machine precision, to contribute effectively to the output vector in the forthcoming computations.
This explains all necessary details of the application of B' and the final scheme is shown in Algorithm 3.12.

Algorithm 3.12 Adaptive application of the adjoint of a trace operator $B' : \mathcal{V}' \to \mathcal{H}'$ to a wavelet vector $\mathbf{v}$.

```
 1: procedure APPLY_TRACE_OPERATOR_ADJOINT((*ε*), B', v) → w
 2:     T ← S(v)
 3:     j0 ← min {|λ| | λ ∈ T}                                  ▷ Minimum level
 4:     J ← max {|λ| | λ ∈ T}                         ▷ Effective maximum level
 5:     w ← {}          ▷ Initialize empty (or with zeros on coarsest level j0)
 6:     for j = j0, ..., J do
 7:         T_j ← S(v_j)
 8:         for all λ' ∈ T_j do
 9:             if |v_λ'| ≤ ε then           ▷ ε stands for the machine precision
10:                 v ← v \ v_λ'            ▷ Delete coefficients of small absolute value
11:             else
12:                 M_λ' ← B' ψ̃_λ'                          ▷ Construct set (3.6.6)
13:                 for all λ ∈ M_λ' do
14:                     w_λ ← w_λ + ψ_λi(c)            ▷ Compute bilinear form (3.6.4)
15:                 end for
16:             end if
17:         end for
18:         v ← v + ADAPTIVE_RECONSTRUCTION( v_j, I_{j+1} )  ▷ Reconstruct Level j of v
19:                                                       w.r.t. DUAL Wavelets
20:     end for
21:     Ensure w is a proper Tree
22:     return w
23: end procedure
```

Remark 3.72 *The fully discretized operator* $\mathbf{B}$ *is not a square matrix, but a highly rectangular matrix with* $N : N^2$ *side lengths. Although* $\mathbf{B}$ *is uniformly sparse, see [122], the complexity to apply* $\mathbf{B}$ *or* $\mathbf{B}^T$ *must always be measured w.r.t. to the longer side, which here corresponds to the output vector* $\mathbf{w}$.

Theorem 3.73 *Algorithm 3.12 is applicable in linear time w.r.t. the number of elements in the output vector* $\mathbf{w}$. *If the input vector contains* $N := \#S(\mathbf{v})$ *elements, then the output vector contains at most* $\mathcal{O}(N^2)$ *elements.*

Proof: Assuming the input vector to be sparse, e.g., containing $N := \mathcal{O}(J - j0)$ elements, where j_0 is the coarsest level and J the finest level, the application of B' in line 14 will produce at most 2 wavelet coefficients by (3.6.5).
In the next step, line 18, a uniformly bounded number of new coefficients will be created through reconstruction on level $j + 1$, we designate this number m. Thus, in the next iteration of the loop, the number of coefficients on the next level $j+1$ is at most increased

by m for any element on level j. The actual number in application is much lower as some coefficients will already be present and many may be inserted several times. In the next reconstruction step, the repeated reconstruction will only produce another finite number of new elements, since the refinement functions of a wavelet share support, they will also have common refinement functions. Overall, the number of coefficients created by a single coefficient λ is thus of order $\mathcal{O}(J - |\lambda|)$ and for N input coefficients the output vector will contain $\mathcal{O}(N(J - j0)) = \mathcal{O}(N^2)$ elements. ∎

Before going further into the details of solving PDEs using the tools of this section, we discuss how the isotropic framework can serve as a means of dealing with anisotropic wavelet schemes.

3.7 Anisotropic Adaptive Wavelet Methods

Although the previous sections only dealt with **isotropic** wavelet constructions, it is possible to use the procedures for adaptive **anisotropic** wavelet constructions. As we shall see, this approach does not yield algorithms of **optimal linear complexity**, but it is an easy way to produce results for the anisotropic case, because it mainly applies already existing proven isotropic algorithms.

3.7.1 A Tree Structure for Anisotropic Wavelet Indices

As explained in Section 2.4.2, we consider the **effective level** J of an anisotropic level vector $\mathbf{j} := \{j_1, \ldots, j_n\}$ to be $J := \max\{j_1, \ldots, j_n\}$.

Remark 3.74 *In theory, the* **function type** $\mathbf{e} \in \mathbb{E}^\star$ *can be omitted in the anisotropic case, but in practice it has several computational advantages to explicitly save it. For example, see the second point in Remark 3.14. The expended memory is just one bit per dimension and thus carries not much weight.*

Then, a tree structure for **anisotropic wavelet indices** $\lambda = (\mathbf{j}, \mathbf{k}, \mathbf{e})$ seems to follow naturally:
The general idea can be easily understood by studying Figure 3.17, where the tree structure (in the direction of increasing levels) is marked by the blue arrows. The point is, that only components on the highest level J (which must represent a wavelet) have children on the next level, the other components remain unchanged. This directly infers the following properties, which coincide with the isotropic tree structure (compare Proposition 3.17):

- The **effective level** of a child index is always increased by one.
- The **function type** ($\mathbf{e} \in \mathbb{E}_n^\star$) of a children index is the same as the parent index.
- The **support inclusion property** (3.2.12) is inferred directly from the one-dimensional counterparts.

If several components are on the highest level J, the children have to be in all combinations of the children sets and the wavelet itself. This corresponds to the corner blocks of the wavelet diagrams and is the same concept as the root node children (3.2.11). In general, if

$1 \leq m \leq n$ coordinates are on the highest level J, then the children live in $2^m - 1$ different blocks (in the wavelet diagram). Because a strict general definition is very technical, we will here assume that the first m components of the anisotropic wavelet index $\mu = (\mathbf{j}, \mathbf{k}, \mathbf{e})$ are on level J.
With the one-dimensional children relation (3.2.8) and

$$\mathcal{C}_i((j,k,e)) := \begin{cases} \{(j,k,e)\}, & \text{for } i = 0, \\ \mathcal{C}((j,k,e)), & \text{for } i = 1, \end{cases}$$

then we can define the children of μ to be

$$\mathcal{C}(\mu) := \left\{ \bigcup_{t \in \mathbb{E}_m^\star} \bigotimes_{i=1}^{m} \mathcal{C}_{t_i}(\mu_i) \right\} \otimes \bigotimes_{i=m+1}^{n} \mu_i, \quad \text{with } \mu_i := (j_i, k_i, e_i). \tag{3.7.1}$$

Lastly, the root nodes are constructed just as in the isotropic construction, see (3.2.6), and the same connection to the first wavelet level (3.2.11) is used.

3.7.2 Conversion Algorithms

To apply a nonlinear operator as shown in Section 3.4.3, it is necessary to construct a locally unique polynomial of a wavelet vector as described in Section 3.4.
The problem is simply that, by definition, an **anisotropic wavelet index** $\lambda = (\mathbf{j}, \mathbf{k}, \mathbf{e})$ can exist on different levels $\mathbf{j} = (j_1, \ldots, j_n)$ in each spatial dimension and thus on very rectangular domains, an example can be seen in Figure 3.18.
It is not possible to construct a single dyadic partitioning $\mathcal{D}_\mathbf{j}$ as in (3.4.1) with the same properties for all combinations of levels. Even if one specific combination $(j_1, \ldots, j_n)$ is chosen, the problem is that all possible combinations of levels must exist simultaneously on the same domain. Altogether, the computation of overlappings especially would be really cumbersome.
The construction of a locally unique **isotropic polynomial** from an anisotropic wavelet tree could either be done directly, or by an intermediate step where the anisotropic vector is converted into an isotropic one without loss of information. Both ways depend on the fact that information on a coarser level can be completely transferred on the next finer level, be it in wavelet or polynomial representation.
The second approach, the construction of an isotropic representation of an anisotropic adaptive wavelet vector seems favorable, as this does not require the development of new techniques to represent anisotropic piecewise polynomials. In contrast, the representation of isotropic wavelet indices in the realm of anisotropic wavelet indices is intrinsically possible, the level multi-index $\mathbf{j}$ would then simply be uniformly filled: $\mathbf{j} = (j, \ldots, j)$.

Remark 3.75 *The conversion algorithms depicted in this section are meant to serve as a "plug-in" just before the application of a nonlinear operator $F : \mathcal{H} \to \mathcal{H}'$ in Algorithm 3.8. Therefore, we assume here that only a single vector (created by prediction) holds the input data exactly and its elements are the indices for which the applied operator must be computed.*

The idea of the conversion algorithm of an anisotropic vector to an isotropic one is thus very simple: If an element v_λ is not actually isotropic, i.e. $\mathbf{j} \neq |\lambda|$ (component-wise), then

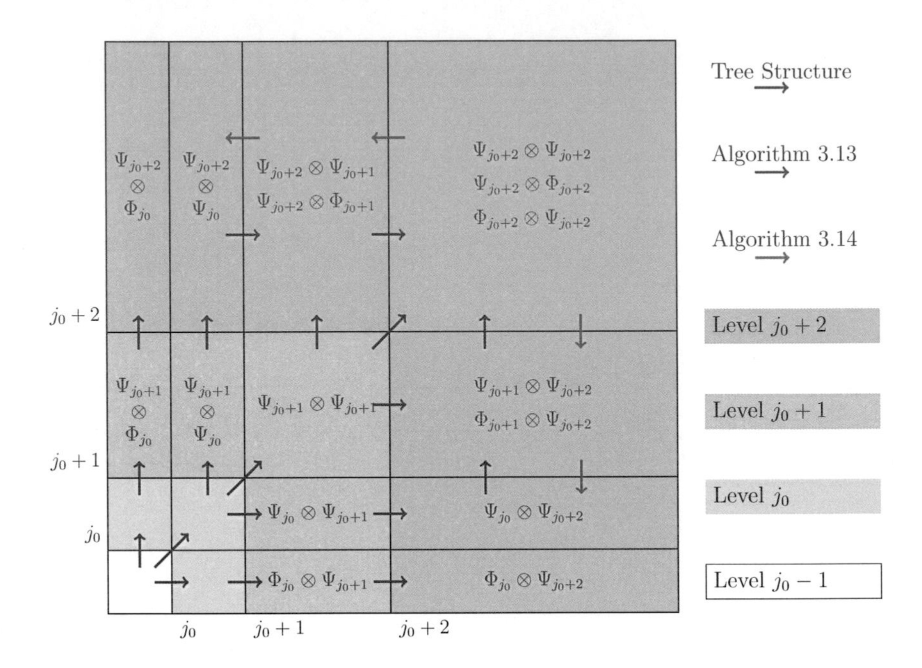

Figure 3.17: Anisotropic 2D wavelet diagram with tree structure and algorithmic connections of levels marked. The blue bases correspond to the tree structure, the red to the two algorithms. Since there is only limited space, only three blocks are labeled. But these show, that the reconstruction algorithm produces functions of different type $((0,1),(1,0))$ which are in this diagram combined into a single level.

those components with $j < |\lambda|$ are reconstructed (2.1.17) **independently** until their level matches $|\lambda|$:

$$\mathcal{M}(\psi_{j,k,e}, J) := \begin{cases} \{\psi_{j,k,e}\}, & \text{if } j = J, \\ \{m_{r,k}^{j,0}\psi_{j+1,r,0} \,|\, r \in \Delta_{j+1}\}, & \text{if } j < J. \end{cases} \tag{3.7.2}$$

This means the complexity of this scheme depends on the maximum level difference

$$\Delta J := \max\{j_1, \ldots, j_n\} - \min\{j_1, \ldots, j_n\}, \tag{3.7.3}$$

which, since $\max\{j_1, \ldots, j_n\} = \mathcal{O}(J)$ and $\min\{j_1, \ldots, j_n\} = \mathcal{O}(1)$ and $J = \mathcal{O}(\log N_J)$, because of (2.3.5), generally gives a **logarithmic** work complexity for treating each wavelet index.

Remark 3.76 *In the algorithm, in line 8, the tensor product of sets of functions must be expanded and all possible combinations have to be traversed.*

The process of operation of Algorithm 3.13 can simply be understood as **towards** the diagonal blocks in the wavelet diagrams, see Figure 3.17. The output of the algorithm

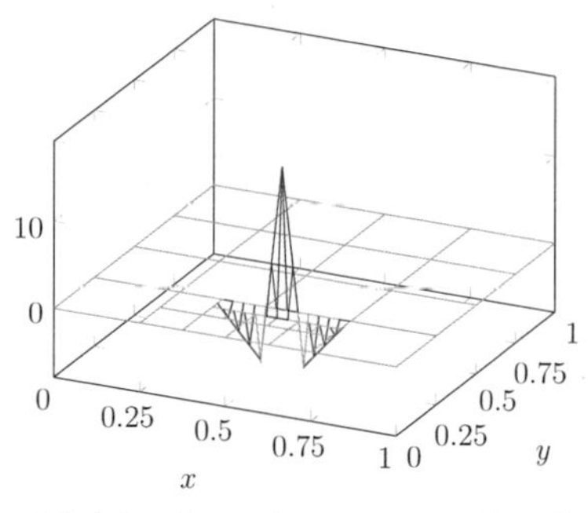

(a) Adaptive anisotropic wavelet of index $((3,5),(4,7),(1,0))$.

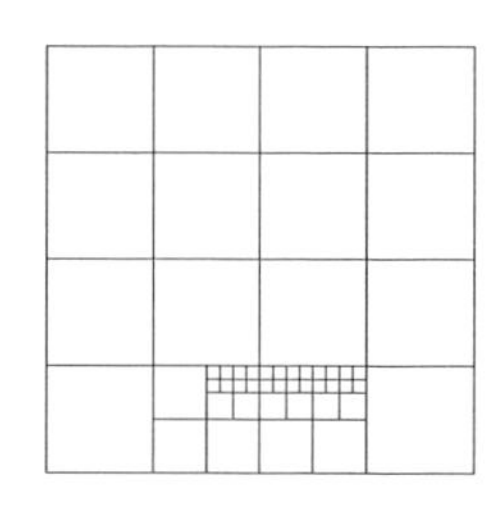

(b) Underlying isotropic grid.

Figure 3.18: An anisotropic DKU-22 Wavelet and the corresponding grid, composed of isotropic square patches.

is **not** necessarily a **proper isotropic wavelet vector**, because the parents of elements created by the reconstruction operator (3.7.2) might not also be created by the algorithm. But this is not a problem for Algorithm 3.5, as long as the anisotropic vector has a tree structure and thus (3.4.9) holds also for the output of Algorithm 3.13.
After the application of the nonlinear operator, the isotropic dual expansion vector $\widetilde{\mathbf{u}}$ must be converted back into the anisotropic dual vector. To this end, any single scale function must be decomposed back into wavelets and single-scale functions.

$$\widetilde{\mathcal{M}}(\widetilde{\psi}_{j,k,e}) := \begin{cases} \left\{\widetilde{\psi}_{j,k,e}\right\}, & \text{if } e = 1, \\ \left\{\widetilde{m}^{j,0}_{r,k}\widetilde{\psi}_{j-1,r,0} \,|\, r \in \Delta_{j-1}\right\} \cup \left\{\widetilde{m}^{j,1}_{s,k}\widetilde{\psi}_{j-1,s,1} \,|\, s \in \nabla_{j-1}\right\}, & \text{if } e = 0 \wedge j > j_0, \\ \left\{\widetilde{\psi}_{j,k,e}\right\}, & \text{if } e = 0 \wedge j = j_0. \end{cases} \tag{3.7.4}$$

This step is repeated until no component any more contains a single-scale function and as long as the level is greater than the coarsest level j_0. During the process, functions, which are not needed to compute the desired output, will be computed. This effect can be minimized if the intermediate reconstructed results of Algorithm 3.13 were saved and the required elements could be identified. On the other hand, since the decomposition is in the direction of lower levels, the number of possible elements drops exponentially, and the arguments of Proposition 2.12 could be applied. Overall, the effect is the same as with the previous algorithm: The overall complexity depends on the maximum level difference (3.7.3) and this introduces a logarithmic factor w.r.t. the number of indices.
The direction of operation of Algorithm 3.14 can be seen in Figure 3.17 as **away** from the diagonal blocks, toward the opposing sides.

This concludes the description of the application of operators and we now turn to so far omitted computational topics about the numerical treatment of adaptive wavelet methods.

Algorithm 3.13 Convert an anisotropic wavelet vector $\mathbf{v}$ on $\mathcal{T}$ into a vector of only isotropic functions $\mathbf{u}$.

```
 1: procedure ANISOTROPIC_TO_ISOTROPIC(v) → u
 2:     T ← S(v)
 3:     j_0 ← min{|λ| | λ ∈ T}, J ← max{|λ| | λ ∈ T}
 4:     u ← {}                                        ▷ Initialize empty
 5:     for j = j_0, ..., J do
 6:         for all λ = (j, k, e) ∈ T_j do            ▷ For all components
 7:             if j ≠ j then                         ▷ If λ is not isotropic
 8:                 v ← v + v_λ ⊗_{i=1}^n M(ψ_{λ_i}, |λ|)   ▷ Refinement with
 9:                                                   maximum Level, cf. (3.7.2)
10:             else
11:                 μ ← (j, k, e)                     ▷ Convert to isotropic index
12:                 u ← u + (u_μ ← v_λ)
13:             end if
14:         end for
15:     end for
16:     return u
17: end procedure
```

Algorithm 3.14 Reconstruct values from an isotropic wavelet vector $\widetilde{\mathbf{u}}$ into the anisotropic vector $\widetilde{\mathbf{v}}$ on $\mathcal{T}$.

```
 1: procedure RECONSTRUCT_ANISOTROPIC_FROM_ISOTROPIC(ũ, ṽ)
 2:     T ← S(ũ)
 3:     j_0 ← min{|λ| | λ ∈ T}, J ← max{|λ| | λ ∈ T}
 4:     ṽ ← 0                                         ▷ Set all elements to zero
 5:     for j = J, ..., j_0 do
 6:         for all λ = (j, k, e) ∈ (S(ũ))_j do
 7:             if e ≠ 1 then                         ▷ Check if anisotropic wavelet
 8:                 ũ ← ũ + ũ_λ ⊗_{i=1}^n M̃(ψ̃_{λ_i})   ▷ Decomposition Identity, cf. (3.7.4)
 9:             else
10:                 if λ ∈ S(ṽ) then                  ▷ Update existing elements only
11:                     ṽ ← ṽ + ũ_λ
12:                 end if
13:             end if
14:             ũ ← ũ \ ũ_λ                           ▷ Delete index λ
15:         end for
16:     end for
17: end procedure
```

4 Numerics of Adaptive Wavelet Methods

On the next few pages, we discuss some of the **numerical details** in solving an operator equation $\mathbf{F}(\mathbf{u}) = \mathbf{f}$ (1.5.3) originating from a nonlinear operator $F : \mathcal{H} \rightarrow \mathcal{H}'$ in the setting of Section 1.5. For some **implementational details**, please turn to Appendix B.

4.1 Iterative Solvers

One implementation detail we have not addressed so far is what kind of **iterative solvers** will be used in order to solve our (nonlinear) problems in the ℓ_2 realm. Direct solvers disqualify because firstly nonlinear problems cannot be accurately represented using matrices and since the possible number of wavelet coefficients is only limited by accessible memory, even setting up sparse matrices is prohibitive by the memory requirements.
The most basic iterative solvers operate on the **residual**, i.e.,

$$\mathbf{R}(\mathbf{u}) := \mathbf{F}(\mathbf{u}) - \mathbf{f}, \tag{4.1.1}$$

which needs to be made smaller than a given tolerance $\varepsilon > 0$, i.e., $\|\mathbf{R}(\mathbf{u})\|_{\ell_2(\mathcal{I})} \leq \varepsilon$, to have sufficiently solved the equation $\mathbf{F}(\mathbf{u}) = \mathbf{f}$. In its simplest form, a general explicit iterative solver can be written in the form

$$\mathbf{u}^{i+1} = \mathbf{u}^i + \mathbf{C}_i(\mathbf{R}(\mathbf{u}^i)), \quad i \in \mathbb{N}_0, \tag{4.1.2}$$

with a **step operator** $\mathbf{C}_i$ and given initial vector $\mathbf{u}^0 \in \ell_2(\mathcal{I})$. The choice of the **step operator** $\mathbf{C}_i$ is critical in determining the properties of the resulting numerical scheme and different types of equation only allow for specific solver types:

- Richardson Iteration $\mathbf{C}_i := \alpha\mathbf{I}$, for a constant $\alpha \in \mathbb{R}$ and all $i \in \mathbb{N}_0$,
- Gradient Iteration $\mathbf{C}_i := \alpha_i\mathbf{I}$, for a step dependent $\alpha_i \in \mathbb{R}$,
- Newton Iteration $\mathbf{C}_i := D\mathbf{R}(\mathbf{u}^i)^{-1}$, where $D\mathbf{R}(\mathbf{u}^i)$ denotes the Fréchet derivative (1.5.6).

One is obviously interested in a **convergent** iteration, that is, to **reduce** the **exact error**, i.e.,

$$\mathbf{e}(\mathbf{u}^i) := \|\mathbf{u}^\star - \mathbf{u}^i\|_{\ell_2(\mathcal{I})}, \quad \text{with } \mathbf{u}^\star \in \ell_2(\mathcal{I}) \text{ being the exact solution}, \tag{4.1.3}$$

in each step of the iteration, i.e.,

$$\mathbf{e}(\mathbf{u}^{i+1}) \leq \rho\, \mathbf{e}(\mathbf{u}^i)^p, \quad \text{with } \rho < 1 \text{ and } p \geq 1. \tag{4.1.4}$$

The case $p = 1$ is called **linear convergence**, $p > 1$ **super-linear convergence** and $p = 2$ **quadratic** convergence. If (4.1.2) converges for any $\mathbf{u}^0 \in \ell_2(\mathcal{I})$, the scheme is called **globally convergent**, if $\mathbf{u}^0$ has to be sufficiently close to converge to the exact solution $\mathbf{u}^\star$ already, i.e., $\mathbf{u}^0 \in \mathcal{U}(\mathbf{u}^\star)$, then it is called **locally convergent**. Numerically, the **reduction factor**

$$\rho_{i+1,i} := \frac{\mathbf{e}(\mathbf{u}^{i+1})}{\mathbf{e}(\mathbf{u}^i)} \tag{4.1.5}$$

can not be determined directly unless $\mathbf{u}^\star$ is available. Instead, one usually computes the **residual reduction factor**,

$$\widetilde{\rho}_{i+1,i} := \frac{\|\mathbf{R}(\mathbf{u}^{i+1})\|_{\ell_2}}{\|\mathbf{R}(\mathbf{u}^{i})\|_{\ell_2}}, \tag{4.1.6}$$

for which holds for a **nonlinear stable operator** (1.5.8) after wavelet discretization,

$$\|\mathbf{R}(\mathbf{u}^i)\|_{\ell_2} = \|\mathbf{F}(\mathbf{u}^i) - \mathbf{F}(\mathbf{u}^\star)\|_{\ell_2} \lesssim \|\mathbf{u}^i - \mathbf{u}^\star\|_{\ell_2} = \mathbf{e}(\mathbf{u}^i),$$

and thus $\widetilde{\rho}_{i+1,i}$ is not comparable to $\rho_{i+1,i}$. For a **linear stable operator** it holds $\|\mathbf{R}(\mathbf{u}^i)\|_{\ell_2} \sim \mathbf{e}(\mathbf{u}^i)$ by (2.2.28) and thus $\widetilde{\rho}_{i+1,i} \sim \rho_{i+1,i}$.
The theoretical considerations of this family of solvers for finite linear problems can be found in any numerics textbook, e.g., [58, 70]; but for the adaptive, infinite ℓ_2-setting, an applicable theory to this setting was introduced in [33] and further developed in [34].
We briefly review a few of the solvers for our types of (non-)linear equations, the details can be found in [33, 147].

4.1.1 The General Scheme

In a nested-iteration scheme employed in **full-grid** discretizations, the first step is usually to gain an approximate, but very precise, solution on a very coarse discretization level. This can, in case of a linear problem, usually be very effectively accomplished by assembling the matrix representation of the linear problem and then solving it using a direct solver, e.g., the QR-/LU-decomposition. After such an initial solution $\mathbf{u}_{j_0}$ has been acquired, the nested-iteration works its way upwards for all levels $j = j_0 + 1, \ldots, J$, in each step decreasing the level-wise error ε_i by a constant factor, usually $\frac{1}{2}$. Then, when a certain error threshold is undershot, i.e., $\varepsilon_J < \varepsilon$, the nested-iteration stops.
In the **adaptive** wavelet realm, the iterative solvers in itself can be seen as sets of nested iterations: The accuracy goal $\varepsilon > 0$ is achieved starting from an initial accuracy $\varepsilon_0 \geq \varepsilon$ and reducing the value in each step by a constant factor until ε is achieved. Additionally, a **coarsening** of the iterands is required after several steps to prevent an accumulation of too many indices and a bloating of the vectors. In analytical terms, the vector must be pulled towards the **best tree N-term approximation** for the scheme to remain of asymptotically optimal complexity. For linear operators, a solver without coarsening was devised in [65].
All considered solvers therefore follow a common structure:

(i) **Init:** Choose a tolerance $\varepsilon > 0$, a finite initial vector $\mathbf{u}^0 \in \ell_2(\mathcal{I})$ and set $\varepsilon_0 := \|\mathbf{u}^\star - \mathbf{u}^0\|_{\ell_2}$ and $i := 0$.

(ii) **Loop:** While $\varepsilon_i > \varepsilon$, do

(ii.1) **Error Reduction:** Choose an $\hat{\varepsilon}_{i+1} \leq \varepsilon_i$ and compute a new solution $\hat{\mathbf{u}}^{i+1}$ satisfying

$$\|\mathbf{u}^\star - \hat{\mathbf{u}}^{i+1}\|_{\ell_2} \leq \hat{\varepsilon}_{i+1}.$$

(ii.2) **Coarse:** If necessary, delete as many coefficients as possible from $\hat{\mathbf{u}}^{i+1}$, but do not significantly increase the error, i.e., compute $\mathbf{u}^{i+1}$ with

$$\|\hat{\mathbf{u}}^{i+1} - \mathbf{u}^{i+1}\|_{\ell_2} \leq C\hat{\varepsilon}_{i+1} \text{ with a } C \sim 1.$$

Otherwise, set $\mathbf{u}^{i+1} := \hat{\mathbf{u}}^{i+1}$ and continue.

(ii.3) **Error Estimation:** Compute the error $\varepsilon_{i+1} := \|\mathbf{u}^\star - \mathbf{u}^{i+1}\|_{\ell_2}$. Set i to $i+1$ and continue from (ii).

(iii) **Finish:** Return $\mathbf{u}^i$.

Since the exact solution $\mathbf{u}^\star \in \ell_2(\mathcal{I})$ is usually not available (except for benchmarks), the error $\|\mathbf{u}^\star - \mathbf{u}^i\|_{\ell_2}$ is usually estimated using $\|\mathbf{R}(\mathbf{u}^i)\|_{\ell_2}$ or other means. For the actual algorithms we therefore need algorithms to compute the residual $\mathbf{R}(\mathbf{u}^i)$ and this also requires computations of the right hand side, hence we discuss these two matter in the next sections.

4.1.2 The Right Hand Side

The right hand side of the (nonlinear) equation $\mathbf{F}(\mathbf{u}) = \mathbf{f}$ must also be constructed in the adaptive wavelet setting. For a given function $f = \mathbf{f}^T\widetilde{\Psi} \in \mathcal{H}'$ and bound $\varepsilon > 0$, we are interested in a way to compute a **best tree N-term approximation** of this function, i.e.,

$$\|\mathbf{f} - \mathbf{f}_\varepsilon\|_{\ell_2} \leq \varepsilon. \tag{4.1.7}$$

The problem here is two-fold:

1. The first step is determining for **any** function $f \in \mathcal{H}'$ a set, preferably a tree structured set $\mathcal{T}$, which contains the relevant wavelet indices.

2. The second step is then calculating for any $\lambda \in \mathcal{T}$ the wavelet coefficients $\mathbf{f}_\lambda = \langle f, \psi_\lambda \rangle$.

Executing the second step is not very difficult once a suitable set $\mathcal{T}$ is available. The approach to determining $\mathcal{T}$ depends on the properties of the function f. There are usually two types of right hand sides: Analytical functions and measured data. The latter is usually given as data points and is thus already discretized. From this discretization a wavelet representation can be computed and this would result in a vector $\mathbf{f}$.
For explicitly expressible functions, all individual wavelet coefficients $\mathbf{f}_\lambda$ could be computed analytically and thus exactly. Using quadrature rules especially designed for wavelets, see [12, 13], one can compute these wavelet expansion coefficients quickly. However, as we will later see, we will need to generate these wavelet vectors often and for varying values ε_k. No matter how fast the computation of an individual wavelet vector is, with increasing number of invocations it becomes more economical to precompute the vector $\mathbf{f}$ once and save it, so that the values do not need to be computed over and over again, but can be read once from disk into memory.

Remark 4.1 *The vector* $\mathbf{f} \in \ell_2(\mathcal{I})$ *can potentially contain only a finite number of non-zero wavelet coefficients. But non-smooth functions, i.e., functions with cusps or discontinuities, will generally have an infinite number of non-zero wavelet coefficients, since their features cannot be represented exactly using piecewise polynomial scaling functions* $\phi_{j,k}$*. Since the accuracy of a numerical solution to a problem is always bound by machine precision, it does not make sense to demand infinitely accurate right hand side data and thus cutting off at some level is permissive. This cut-off error must simply be so small, that the sought tolerance of the solution can be achieved. By basic error analysis arguments, it does not make sense to ask for more accuracy in the solution than is given in the right hand side data.*

Construction of Right Hand Side Vectors

A first strategy to set up a wavelet vector is to simply define the values of the wavelet expansion coefficients directly, i.e.,

$$\mathbf{f}_\lambda := \widehat{f}(\lambda) \equiv \widehat{f}(j, \mathbf{k}, \mathbf{e}), \qquad \text{for some function } \widehat{f} : \mathcal{I} \to \mathbb{R} \text{ and } \lambda := (j, \mathbf{k}, \mathbf{e}),$$

which would correspond to a function, but not necessarily to one that is expressible in an explicit way. Using norm equivalences (2.2.12) and (3.2.20), this is, however, a good approach to construct functions contained in a specific **Sobolev** $H^s(\Omega)$ or **Besov** $B_q^\alpha(L_p)$ space.
Another way of computing the wavelet expansion coefficients is to use **full-grid** techniques to compute the values and then transform them into **adaptive data structures**. If the wavelet norm equivalences of Theorem 2.18 are applicable for the function f, we can conclude from (2.2.12) that a full vector approximation $\mathbf{f}_J$ for a level $J \geq j_0$ of a function $f \in L_2 \subset H^{-1}$ results in an error of $\sim 2^{-J}$. The major drawback of this approach is, of course, the excessive memory requirements and the exponentially growing complexity because of the exponentially growing number of single scale functions (2.3.5). Nevertheless, it is a good way to compute right hand data quickly and reliably, especially for globally smooth functions.
In our context, piecewise polynomial functions, exactly represented using the dyadic grids $\mathcal{D}_j$ (3.4.1), are an interesting special case. Such piecewise functions are hard to accurately describe using the tools of full-grid code, since there one usually only uses single-scale functions $\phi_{j,k}$ to set up functions. Piecewise polynomial right hand side vectors can easily be set up using the techniques developed in Section 3.4.2 and we present a complete adaptive strategy to compute such wavelet vectors next.

Adaptive Assembly for Piecewise Polynomial Functions

A class of functions that commends itself for consideration as right hand sides by the developed techniques of Section 3.4.1 is the class of **piecewise polynomials**. This group contains some interesting candidates for right hand sides, e.g., piecewise constants with discontinuities.

Since the computation of dual values involves evaluating integrals of the form

$$\int_\Omega f\,\psi_\lambda\,d\mu = \int_{\operatorname{supp}\psi_\lambda} f\,\psi_\lambda\,d\mu = \sum_{\square\in\mathcal{D}_\lambda}\int_\square f|_\square\,\mathbf{q}_{\lambda,\square}^T\mathbf{P}_\square\chi_\square\,d\mu, \tag{4.1.8}$$

which again comes back to computing the integrals (3.4.25), just with an "operator" representing the function f. For piecewise polynomial f, the evaluations of these inner products can again be executed **exactly** by analytically computing the integrals. Using Algorithm 3.7 of Section 3.4.5, it is possible to construct the values of the dual expansion coefficients of a tree $\mathcal{T}\subset\mathcal{I}$ as long as the adaptive grid constructed by Algorithm 3.5 w.r.t. the tree $\mathcal{T}$ is **finer** then the grid on which the piecewise polynomial f is constructed. This has the conceptual advantage of being able to compute the wavelet coefficients of $f=\mathbf{f}^T\widetilde{\Psi}$ exactly (up to machine precision).
The adaptive determination of the active wavelet coefficients is governed by a simple observation: For structured data, the distribution of the active wavelet coefficients is not arbitrary but follows the structure of the data. For piecewise polynomials, the important details are areas with discontinuities and high gradients. Smooth, e.g., constant or linear, patches are easily exactly representable using single scale functions on the coarsest level. The set of all relevant wavelet coefficients thus naturally forms a **tree structure**.

Remark 4.2 *By (2.2.34), the* **dual wavelet coefficients** *can also be calculated using a* **primal** *expansion of the function and applying the* **Riesz operator** $R_{\mathcal{H}}:\mathcal{H}\to\mathcal{H}'$. *Since the Riesz operator by its definition is a* **local** *operator, this also entails that the dual coefficients will mimic the structure of the primal coefficients.*

Thus it is reasonable to directly construct the tree structure of the adaptive wavelet coefficients by only considering children of already present indices and prioritizing these depending on the absolute value of the already computed indices. This approach clearly imitates the construction of Algorithm 3.4, the tree prediction.

Remark 4.3 *Since wavelet coefficients are scaled to conform to a specific Sobolev space* H^s *by norm equivalences (2.2.12), it suffices to compute the values w.r.t.* L_2 *and apply the proper scaling afterwards. This way, the values of any function only need to be calculated once for different Sobolev scalings, as long as the same wavelets are used.*

Our approach here can be understood to solve the simple equation $u=f$, where $f\in L_2(\Omega)$ is known and $u\in L_2(\Omega)$ is unknown. Since both the primal and dual wavelets constitute Riesz bases for $L_2(\Omega)$, the proper application for Algorithm 3.4 would be to assume f to be given as a dual wavelet expansion and u as a primal wavelet expansion vector. The operator involved in this equation would thus be the Riesz operator of $L_2(\Omega)$, the mass matrix. In the next step, then u would be projected onto the dual side again by means of the dual mass matrix. To save the application of two operators that cancel each other out, we directly use the predicted tree to compute the dual wavelet coefficients.
Due to the nature of the prediction algorithm, there will be some coefficients very close to zero absolute value, thus the result should be purged of these values again, i.e., a coarsening step is applied to the computed output tree.
Another advantage of the piecewise polynomial class of functions is the possibility to

exactly calculate the L_2 and H^1-norm of the polynomial and use this data to estimate the error w.r.t. a finite approximation $\mathbf{f}_\eta \in \ell_2(\mathcal{I})$, e.g.,

$$\left| \|f\|_{\mathcal{H}} - \|\mathbf{f}_\eta^T \widetilde{\Psi}\|_{\mathcal{H}} \right| \leq \eta. \tag{4.1.9}$$

The terms on the left hand side can be easily and accurately calculated from the data, but one is usually rather interested in the norm of the difference, not in the difference of the norms stated above. That is, one is interested in estimates of the form

$$\|f - \mathbf{f}_\eta^T \widetilde{\Psi}\|_{\mathcal{H}} \leq \eta, \tag{4.1.10}$$

but this would require knowledge of the **uncomputed** wavelet coefficients $\mathbf{f} - \mathbf{f}_\eta$. It is easily deducible from the triangle inequality that holds

$$\left| \|f\|_{\mathcal{H}} - \|\mathbf{f}_\eta^T \widetilde{\Psi}\|_{\mathcal{H}} \right| \leq \|f - \mathbf{f}_\eta^T \widetilde{\Psi}\|_{\mathcal{H}}, \tag{4.1.11}$$

but the converse, which would be significant in this context, cannot be true universally. The following Lemma shows that a different equivalence estimate can be accomplished in the wavelet context.

Lemma 4.4 *For a finite approximation $\mathbf{f}_\eta^T \widetilde{\Psi} \in \mathcal{H}$ to a function $f \in \mathcal{H}$ holds*

$$\sqrt{\left| \|f\|_{\mathcal{H}}^2 - \|\mathbf{f}_\eta^T \widetilde{\Psi}\|_{\mathcal{H}}^2 \right|} \quad \sim \quad \|f - \mathbf{f}_\eta^T \widetilde{\Psi}\|_{\mathcal{H}}. \tag{4.1.12}$$

Proof: By the norm equivalences (2.2.12) holds with $\mathcal{T} := S(\mathbf{f}_\eta)$, $f = \mathbf{f}^T \widetilde{\Psi}$ and $f_\lambda := \langle f, \psi_\lambda \rangle$

$$\|f\|_{\mathcal{H}}^2 - \|\mathbf{f}_\eta^T \widetilde{\Psi}\|_{\mathcal{H}}^2 \sim \sum_{\lambda \in \mathcal{I}} f_\lambda^2 - \sum_{\lambda \in \mathcal{T}} f_\lambda^2 = \sum_{\lambda \in \mathcal{I} \setminus \mathcal{T}} f_\lambda^2 \sim \| \sum_{\lambda \in \mathcal{I} \setminus \mathcal{T}} f_\lambda \widetilde{\psi}_\lambda \|_{\mathcal{H}}^2 = \|f - \mathbf{f}_\eta^T \widetilde{\Psi}\|_{\mathcal{H}}^2,$$

where the wavelets and coefficients are assumed to be scaled according to the norm of the space $\mathcal{H}$. ∎

Thus, for a finite wavelet vector $\mathbf{f}_\eta$ satisfying $\sqrt{\left| \|f\|_{\mathcal{H}}^2 - \|\mathbf{f}_\eta^T \widetilde{\Psi}\|_{\mathcal{H}}^2 \right|} \lesssim \eta$ also holds

$$\|f - \mathbf{f}_\eta^T \widetilde{\Psi}\|_{\mathcal{H}} \lesssim \eta. \tag{4.1.13}$$

The knowledge of the error constant $\eta > 0$ will be needed for the final algorithm to accurately construct tree approximations with accuracy $\varepsilon \geq \eta$ of f from the vector $\mathbf{f}_\eta$.

Remark 4.5 *The norm equivalences in the above proof can be made exact using the Riesz operator $R_{\mathcal{H}}$ of the space $\mathcal{H}$. But, in the proof, this introduces mixed terms of the form $\mathbf{f}_\eta^T \mathbf{R}_{\mathcal{H}}(\mathbf{f} - \mathbf{f}_\eta)$ which do not necessarily evaluate to zero like $\mathbf{f}_\eta^T(\mathbf{f} - \mathbf{f}_\eta)$. Since the norm $\|f\|_{\mathcal{H}}^2$ will be evaluated from the piecewise polynomial and is thus exact, it makes sense to calculate the norm $\|\mathbf{f}_\eta^T \widetilde{\Psi}\|_{\mathcal{H}}^2$ exactly, too, using the Riesz operator for the Hilbert space $\mathcal{H}$.*

Remark 4.6 *The above assertions hold in the wavelet domain without change for* **dual spaces** $\mathcal{H}'$*, but an evaluation of the dual norm for a polynomial* $\|f\|_{\mathcal{H}'}$ *is not as easily computable. From the definition of the norm, see (1.2.34), it follows for* $f \in \mathcal{H}$ *that* $\|f\|_{\mathcal{H}'} \leq \|f\|_{\mathcal{H}}$ *holds. Thus, the estimate (4.1.13) then is also valid for* $\mathcal{H}'$ *and the computable value* $\|f - \mathbf{f}_\eta^T \widetilde{\Psi}\|_{\mathcal{H}}$ *can be used as an upper bound in the final algorithm below. This strategy is also justified in the wavelet domain, as the (uncomputed) wavelet coefficients of the error* $\mathbf{f} - \mathbf{f}_\eta$ *will be scaled with values of values* ≤ 1*, see (2.2.15), when the Sobolev index is shifted towards higher negative values. Thus,* $\|\mathbf{D}^{-s}(\mathbf{f} - \mathbf{f}_\eta)\|_{\ell_2} \leq \|\mathbf{f} - \mathbf{f}_\eta\|_{\ell_2}$ *for* $s \geq 0$ *.*

The accurate computation of the norm $\|\mathbf{f}_\eta\|_{\mathcal{H}}$ and the precursor wavelet vectors leading to become $\mathbf{f}_\eta$ are an important detail of Algorithm 4.2, as this determines when the algorithm finishes. As stated in Section 2.2.6, the Riesz operators are known and often directly accessible for Sobolev spaces $\mathcal{H} = H^m(\Omega)$ of integer orders $m \in \mathbb{Z}$. Unfortunately, the work required to accurately determine the norm is higher than the work required to set up the values of the vector itself. To speed this algorithm up, one can choose to calculate the norm only after $\delta \leq \eta$ holds, beforehand it is unlikely that $|\xi - \tau| \leq \eta$ holds (except for very smooth functions).
Nevertheless, Algorithm 4.2 can only be executed in "real time", i.e., within a few seconds, if $\eta \gtrsim 10^{-3}$, more precise right hand sides should be precomputed and saved for later use. We evaluate the accurateness and efficiency of Algorithm 4.2 in Section 4.6.

The Final Algorithm

We now assume the computation of the right hand side is completed and the result is given by a finite vector $\mathbf{f}_\eta$, where η describes the discretization error of the data in the sense of (4.1.7). As shown in the previous paragraph, such a vector $\mathbf{f}_\eta$ can be precomputed for fixed η and loaded on demand in applications. Then an approximation up to a tolerance $\beta > \eta$ of this vector can be computed by Algorithm 3.3,

$$\mathbf{f}_{\eta,\beta} := \texttt{TREE\_COARSE}(\beta - \eta, \mathbf{f}_\eta),$$

which satisfies $\|\mathbf{f}_\eta - \mathbf{f}_{\eta,\beta}\| \leq \beta - \eta$ and already has a tree structure. For a given sought tolerance $\varepsilon > \eta$ then follows

$$\begin{aligned} \| \langle f, \Psi \rangle - \mathbf{f}_{\eta,\varepsilon}\|_{\ell_2} &\leq \underbrace{\| \langle f, \Psi \rangle - \mathbf{f}_\eta\|_{\ell_2}}_{\leq \eta} + \underbrace{\|\mathbf{f}_\eta - \mathbf{f}_{\eta,\varepsilon}\|_{\ell_2}}_{\leq \varepsilon - \eta} \\ &\leq \eta + \varepsilon - \eta = \varepsilon \end{aligned} \tag{4.1.14}$$

Algorithm 4.1 Adaptive set up of a wavelet vector $\mathbf{f}_\varepsilon$ of the function f and accuracy $\varepsilon > 0$.

```
1: procedure RHS( ε, f ) → f_ε
2:     Load f_η                              ▷ With any η ≤ ε
3:     f_{η,ε} ← TREE_COARSE(ε − η, f_η)
4:     return f_{η,ε}
5: end procedure
```

Algorithm 4.2 Adaptive construction of a wavelet vector $\mathbf{f}_\xi$ of the piecewise polynomial function f for an accuracy $\eta > 0$. The returned expansion wavelet vector $\mathbf{f}_\xi$ will obey $\xi = \|f - \mathbf{f}_\xi^T \widetilde{\Psi}\|_{\mathcal{H}} \lesssim \eta$.

1: **procedure** ADAPTIVE_CONSTRUCTION_RHS(η, f) $\rightarrow$ ($\mathbf{f}_\xi, \xi$)
2: $\quad\gamma \leftarrow 4$ ▷ Choose a decay parameter value for Algorithm 3.4
3: $\quad\tau \leftarrow \|f\|_{\mathcal{H}}$ ▷ Compute norm of polynomial
4: $\quad$// Construct an initial wavelet vector
5: $\quad\mathbf{v} \leftarrow \{(J, \mathbf{k}, \mathbf{0}) \,|\, \mathbf{k} \in \Delta_J^n\}$ ▷ Choose a level J high enough so that the grid
6: $\quad$ given by $\{\Phi_J\}$ is finer than the grid on which f is given
7: $\quad\mathbf{g} \leftarrow$ WTREE_2_PTREE($\mathbf{v}$) ▷ Construct target grid by Algorithm 3.5
8: $\quad$CALCULATE_INTEGRALS($\mathbf{g}, S(\mathbf{v}), f$)
9: $\quad\mathbf{v} \leftarrow$ DECOMPOSE($\mathbf{g}, S(\mathbf{v})$) ▷ Assemble values of wavelet coefficients
10: $\quad\delta \leftarrow 2^{-J}$ ▷ Initial tolerance for tree prediction
11: $\quad\sigma \leftarrow$ CALCULATE_NORM($\delta/2, \mathbf{v}$) ▷ Calculate norm exactly using Riesz operators
12: $\quad$**while** $\sqrt{|\sigma^2 - \tau^2|} > \eta$ **do**
13: $\quad\quad\delta \leftarrow \delta/2$
14: $\quad\quad\mathcal{T} \leftarrow$ PREDICTION($\delta, \mathbf{v}, \gamma$) ▷ Apply Algorithm 3.4, $\mathcal{T}$ is a tree
15: $\quad\quad\mathbf{w} \leftarrow \{\}$ ▷ Initialize $\mathbf{w}$ empty
16: $\quad\quad$**for all** $\mu \in \mathcal{T}$ **do** ▷ Set up $\mathbf{w}$ on $\mathcal{T}$
17: $\quad\quad\quad w_\mu \leftarrow 0$ ▷ Insert wavelet coefficients
18: $\quad\quad$**end for**
19: $\quad\quad\mathbf{g} \leftarrow$ WTREE_2_PTREE($\mathbf{w}$) ▷ Construct target grid by Algorithm 3.5
20: $\quad\quad$CALCULATE_INTEGRALS($\mathbf{g}, \mathcal{T}, f$)
21: $\quad\quad\mathbf{w} \leftarrow$ DECOMPOSE($\mathbf{g}, \mathcal{T}$) ▷ Assemble values of wavelet coefficients
22: $\quad\quad\mathbf{v} \leftarrow$ TREE_COARSE($\delta/4, \mathbf{w}$) ▷ Rid $\mathbf{w}$ of very small values; overwrite $\mathbf{v}$
23: $\quad\quad\sigma \leftarrow$ CALCULATE_NORM($\delta/2, \mathbf{v}$) ▷ Calculate norm using Riesz operators
24: $\quad$**end while**
25: $\quad$**return** $(\mathbf{v}, \sqrt{|\sigma^2 - \tau^2|})$
26: **end procedure**
27:
28: // Calculate Integral values in $\mathbf{g}$, see (4.1.8)
29: **procedure** CALCULATE_INTEGRALS($\mathbf{g}$, $\mathcal{T}$, f)
30: $\quad$**for all** $\square \in \mathcal{L}(\mathcal{D}(\mathcal{T}))$ **do** ▷ Traverse all individual patches
31: $\quad\quad$**for all** $\mathbf{t} \in \mathcal{M}_d^n$ **do** ▷ Traverse all basis functions
32: $\quad\quad\quad\mathbf{g}_{\square,\mathbf{t}} \leftarrow \langle p_\square^{\mathbf{t}}, f|_\square \rangle$ ▷ Refine polynomial f to match target grid and
33: $\quad\quad\quad$ evaluate exactly or by quadrature
34: $\quad\quad$**end for**
35: $\quad$**end for**
36: **end procedure**
37:
38: // Calculate norm of wavelet vector $\mathbf{v}$ with an accuracy ε
39: **procedure** CALCULATE_NORM(ε, $\mathbf{v}$) $\rightarrow \nu$
40: $\quad$**Set up** R, γ ▷ Riesz operator for $\mathcal{H}$
41: $\quad\mathbf{w} \leftarrow$ APPLY($\varepsilon, R, \gamma, \mathbf{v}$) ▷ Apply Riesz operator using an appropriate algorithm
42: $\quad\nu \leftarrow \sqrt{(\mathbf{v}, \mathbf{w})}$ ▷ Calculate scalar product and norm
43: $\quad$**return** ν
44: **end procedure**

Of course this estimate also works for many different variations of the involved constants, i.e., the error could also be equilibrated to $\varepsilon/2$ between the two addends. It should just be noted that one cannot acquire an accuracy w.r.t. an original function f smaller than η once $\mathbf{f}_\eta$ is selected. In case there exists no $\mathbf{f}_\eta$ with $\eta \leq \varepsilon$, then simply the whole vector without coarsening is returned.
In summary, our algorithm to set up a right hand side vector is presented in Algorithm 4.1.

4.1.3 The Residual

Of central importance for any solution method is the numerical evaluation of the **residual** (4.1.1), because this is the minimization target for many schemes. The evaluation procedure here is based upon the simple calculation,

$$\begin{aligned}\| \langle R(v), \Psi\rangle - \mathbf{R}(\mathbf{v})\|_{\ell_2} &= \| \langle F(v) - f, \Psi\rangle - (\mathbf{F}(\mathbf{v}) - \mathbf{f})\|_{\ell_2}, \\ &\leq \|(\langle F(v), \Psi\rangle - \mathbf{F}(\mathbf{v})) - (\langle f, \Psi\rangle - \mathbf{f})\|_{\ell_2}, \\ &\leq \underbrace{\| \langle F(v), \Psi\rangle - \mathbf{F}(\mathbf{v})\|_{\ell_2}}_{\leq \varepsilon/2} + \underbrace{\| \langle f, \Psi\rangle - \mathbf{f}\|_{\ell_2}}_{\leq \varepsilon/2} \leq \varepsilon,\end{aligned}$$

where the first term corresponds to the function `APPLY` and the second term to `RHS`. This evaluation algorithm thus works for any kind of operator F and right hand side f.

Algorithm 4.3 Adaptive computation of the residual $\mathbf{R}(\mathbf{v}) = \mathbf{F}(\mathbf{v}) - \mathbf{f}$ for a proper tree $\mathbf{v}$ up to accuracy $\varepsilon > 0$.

1: **procedure** RESIDUAL(ε, $\mathbf{F}$, $\mathbf{f}$, $\mathbf{v}$) $\to \mathbf{r}_\varepsilon$
2: $\quad \mathbf{w} \leftarrow$ APPLY($\varepsilon/2, \mathbf{F}, \mathbf{v}$) ▷ APPLY stands for one of the algorithms presented,
3: whichever is appropriate for this $\mathbf{F}$
4: $\quad \mathbf{g} \leftarrow$ RHS($\varepsilon/2, \mathbf{f}$) ▷ $\mathbf{g}$ and $\mathbf{w}$ are both based on trees, but not
5: necessarily with the same support
6: $\quad \mathbf{r}_\varepsilon \leftarrow \mathbf{w} - \mathbf{g}$ ▷ Do not delete canceled out values,
7: then $\mathbf{r}_\varepsilon$ is a proper tree
8: $\quad$ **return** $\mathbf{r}_\varepsilon$
9: **end procedure**

Since it is the goal of most solvers to solve an equation by minimizing the residual, it is important to be able to control the output of Algorithm 4.3 in applications.

Theorem 4.7 *Algorithm 4.3 is $s^\star$-**sparse**, that is:*
Whenever the exact solution $\mathbf{u}^\star$ belongs to $\mathcal{A}^s_{tree}$ for some $s < s^\star$, one has for any finitely supported input $\mathbf{v}$ and any tolerance $\varepsilon > 0$ that the output of $\mathbf{r}_\varepsilon :=$ RESIDUAL$(\varepsilon, \mathbf{F}, \mathbf{f}, \mathbf{v})$ satisfies

$$\begin{aligned}\#\operatorname{supp}\mathbf{r}_\varepsilon &\lesssim \varepsilon^{-1/s}\left(1 + \|\mathbf{v}\|^{1/s}_{\mathcal{A}^s_{tree}} + \|\mathbf{u}^\star\|^{1/s}_{\mathcal{A}^s_{tree}}\right), \\ \|\mathbf{r}_\varepsilon\|_{\mathcal{A}^s_{tree}} &\lesssim 1 + \|\mathbf{v}\|_{\mathcal{A}^s_{tree}} + \|\mathbf{u}^\star\|_{\mathcal{A}^s_{tree}},\end{aligned}$$

where the constant depends only on s when $s \to s^\star$. Moreover, the number of operations needed to compute $\mathbf{r}_\varepsilon$ stays proportional to

$$\varepsilon^{-1/s}\left(1 + \|\mathbf{v}\|_{\mathcal{A}^{1/s}_{tree}} + \|\mathbf{u}^\star\|_{\mathcal{A}^{1/s}_{tree}}\right) + \#S(\mathbf{v}).$$

The proof can be found in [33]. It basically just combines the result from Lemma 3.27 and Theorem 3.36. An important detail is that the limit $s^\star$ is again given by $s < \frac{2\gamma-n}{2n}$ from Theorem 3.36 where γ is the decay constant in (3.3.14) and $n \in \mathbb{N}$ represents the dimension. The $s^\star$-sparsity property of the solution $\mathbf{u}^\star$ is transferred through the operator $\mathbf{F}$ onto the right hand side by the simple connection of the equation to solve: $\mathbf{F}(\mathbf{u}^\star) = \mathbf{f}$.

4.1.4 Convergence Rates

The importance of all these theorems culminates in the following result from [33]:

Theorem 4.8 *Let $F = A + G : H^m(\Omega) \to (H^m(\Omega))'$ be a semilinear operator with a linear part A and a non-linear part G. Suppose that for the operator G holds (3.3.14) for $\gamma > n/2$ and the same for A with $\sigma \geq \gamma$. If the solution $u = \sum_{\lambda \in \mathcal{T}} u_\lambda \psi_\lambda$ satisfies $\mathbf{u} \in \mathcal{A}^s_{tree}$ for some $s < s^\star := \frac{2\gamma-n}{2n}$, then the approximate solution u_ε computed by the algorithms in Section 4.2 and Section 4.4 after finitely many steps satisfies*

$$\|u_\varepsilon - u\| \lesssim \varepsilon,$$

and it holds

$$\#\mathit{flops}, \#\operatorname{supp}\mathbf{u}_\varepsilon \lesssim \varepsilon^{-1/s}(1 + \|\mathbf{u}\|^{1/s}_{\mathcal{A}^s_{tree}}),$$

and

$$\|\mathbf{u}_\varepsilon\|_{\mathcal{A}^s_{tree}} \lesssim \|\mathbf{u}\|_{\mathcal{A}^s_{tree}},$$

where the constants depend only on $\|\mathbf{u}\|_{\ell_2}$, the constants of (2.2.17), (1.5.8), (2.2.28) and on s when $s \to s^\star$.

The best **approximation rate** to expect for a semilinear operator is therefore

$$s^\star = \frac{2\gamma - n}{2n} = \frac{\gamma}{n} - \frac{1}{2}. \tag{4.1.15}$$

By Proposition 3.32, in a sum different operators, the smallest exponent γ_i dominates the overall approximation rate.

Besov Regularity

The prerequisite of the wavelet expansion vector $\mathbf{u} \in \mathcal{A}^s_{\text{tree}}$ is by Proposition 3.20 equivalent to the solution u being in the Besov space $B^{m+sn}_\tau(L_\tau(\Omega))$ for $\tau > (s + 1/2)^{-1}$. Of course the regularity of the solution can not be known directly, it has to be inferred from the regularity of the right hand side. For the Laplace's equation on polygonal domains in $\mathbb{R}^n$, this topic was discussed in [38, 41]. In general, it can be said that one gains smoothness of order 2, just as in the Sobolev scale. For example, the following inference was shown

$$f \in H^s(\Omega), s > -1/2, \quad \Longrightarrow \quad u \in B^{\alpha+3/2}_\tau(L_\tau(\Omega)), \quad 0 < \alpha < s + \frac{1}{2}, \quad \frac{1}{\tau} = \frac{\alpha}{2} + \frac{1}{2}.$$

For sufficiently smooth right hand sides, the adaptive scheme will converge automatically at the highest possible rate, which is here limited by (3.3.15).

Determination of the Decay Parameter γ

The theoretical values of the parameter γ of (3.3.14) were predicted in Remark 3.34. The actual values can be determined in two ways: A theoretical proof for a specific bilinear form $a(\cdot,\cdot)$ is of course the desired source for these values. If a proof is not available, it is relatively easy to construct the formula for $\gamma(n)$ if the operator is given by a bilinear form composed of one-dimensional terms. For the linear operators $a_0(u,v) = \int u\, v\, d\mu$ and $a_1(u,v) = \int \nabla u \cdot \nabla v\, d\mu$ discussed in Example 3.63 and Example 3.65, the determination of the values of γ seems straightforward: By **bilinearity** follows from (3.3.1),

$$a(\psi_{j,\mathbf{k},\mathbf{e}}, \psi_{j',\mathbf{k}',\mathbf{e}'}) = \sum_{\mathbf{r}\in\Delta^n_{j+1}} \sum_{\mathbf{r}'\in\Delta^n_{j'+1}} m^{j,\mathbf{e}}_{\mathbf{r},\mathbf{k}}\, m^{j',\mathbf{e}'}_{\mathbf{r}',\mathbf{k}'}\, a(\phi_{j+1,\mathbf{r}}, \phi_{j'+1,\mathbf{r}'}).$$

Thus, the decay parameter should be ascertainable from noting the decay of the bilinear form w.r.t. single scale functions, i.e., $a(\phi_{j+1,\mathbf{r}}, \phi_{j'+1,\mathbf{r}'})$. In Example 3.63, the maximum value of the exponent is $\gamma = \frac{1}{2}$ and this translates to a value of $\gamma = n/2$ by (3.5.13) for dimensions $n \geq 1$, see Remark 4.9. But this determination is not accurate as it does not take **vanishing moments** $(\mathcal{V})$(2.2.5) of the wavelets into account, a property single-scale functions lack.
Since the exact value of γ for a bilinear form is not easily proven, we determine the values experimentally: Fixing one wavelet ψ_λ, we can calculate the value of the bilinear form for all wavelets $\psi_\nu \in \overline{\Upsilon}^a_\lambda(\mathcal{I})$ with $|\nu| = |\lambda| + p$ (see (3.5.6)). One has to use the correct scaling $\mathbf{D}_1^{-1}$ (2.2.15) though, and this scaling affects the asymptotics greatly because each factor of 2^{-j} effectively increases γ by 1, see the asymptotics in Figure 4.3. By relating the maximum value $|a(\psi_\lambda, \psi_\nu)|$ in these sets for different values of p, we can determine an **upper limit** for the decay parameter γ: Unless all possible wavelets ψ_λ, $\lambda \in \mathcal{I}$, and all possible sets $\overline{\Upsilon}^a_\lambda(\mathcal{I})$ are checked, the value of γ could be higher or lower in value, but only the **smallest value** can be used as a decay constant for all cases.
From the diagrams in Figure 4.1, we deduce the values of Table 4.1.

Remark 4.9 *It is easily possible to compute the decay parameter γ for multi-dimensional bilinear forms that are given in terms of one-dimensional bilinear forms. For this, one needs the individual decay parameters $\gamma_{i,o}$, $i,o \in \{0,1\}$, where each index corresponds to a function type, i.e., using (2.4.16),*

$$a(\psi_{j,k,i}, \psi_{j',k',o}) \lesssim 2^{-\gamma_{i,o}|j'-j|}.$$

Since the combinations for the types i and o are finite, these values are easily computable from experiments without ambiguity. To calculate the multi-dimensional decay parameters, one has to note two different rules:

1. *The decay parameters of products of bilinear forms have to be added, i.e.,*

$$a(\psi_\lambda, \psi_\nu) := a_1(\psi_{\lambda_1}, \psi_{\nu_1})\, a_2(\psi_{\lambda_2}, \psi_{\nu_2}) \lesssim 2^{-\gamma_1||\lambda_1|-|\nu_1||}\, 2^{-\gamma_2||\lambda_2|-|\nu_2||} = 2^{-(\gamma_1+\gamma_2)||\lambda|-|\nu||},$$

where λ_i and ν_i are the components as in (3.3.29). This identity comes easily by noting that $|\lambda_i| = |\lambda|$ and $|\nu_i| = |\nu|$ for $i = 1, 2$.

2. In a sum of bilinear forms, the smallest decay parameter has to be chosen, i.e.,

$$a(\psi_\lambda, \psi_\nu) := a_1(\psi_\lambda, \psi_\nu) + a_2(\psi_\lambda, \psi_\nu) \lesssim 2^{-\min(\gamma_1,\gamma_2)||\lambda|-|\nu||}.$$

This is a direct consequence of Proposition 3.32.

The only complication in several dimensions $n \geq 1$ is that the type $\mathbf{e} \in \mathbb{E}_n = \{0,1\}^n$ of the wavelets can attain up to 2^n different states. But the type $\mathbf{e} = \mathbf{0}$ can only be used on the coarsest level j_0 and can thus be ignored.

Example 4.10 *For the mass matrix bilinear form (3.5.9), the values are easily ascertainable from Figure 4.2:*

$\gamma_{i,o}$	$e_o = 0$	$e_o = 1$
$e_i = 0$	$1/2$	$3/2$
$e_i = 1$	$1/2$	$3/2$

Since the multi-dimensional bilinear form of the mass matrix (3.5.13) is a direct product of one-dimensional bilinear forms, we just have to determine the lowest possible sum of values $\gamma_{i,o}$ for types $\mathbf{e}_o \in \mathbb{E}_n^*$. This is obviously the case if $\mathbf{e}_{o,t} = 0$ for all but one component. Then holds for $n \geq 1$,

$$\gamma = \gamma(n) = \frac{3}{2} + (n-1)\frac{1}{2} = 1 + \frac{n}{2}. \tag{4.1.16}$$

These values are exactly the ones depicted in Figure 4.3 and the preconditioner for $H^1(\Omega)$ increases the order by 1 as shown in Figure 4.1.

Example 4.11 *For the Laplace operator bilinear form (3.5.15), the values from Figure 4.2 are:*

$\gamma_{i,o}$	$e_o = 0$	$e_o = 1$
$e_i = 0$	$-1/2$	$-1/2$
$e_i = 1$	$-1/2$	$-1/2$

To compute $\gamma(n)$ for the bilinear form (3.5.16), we first note that the sum goes over all possible combinations and, thus, by the first rule of Remark 4.9, only one addend has to be considered. The same line of thought as in Example 4.10 can thus be applied. For $n \geq 1$ hence holds

$$\gamma = \gamma(n) = -\frac{1}{2} + (n-1)\frac{1}{2} = -1 + \frac{n}{2}. \tag{4.1.17}$$

Again, applying the preconditioner for $H^1(\Omega)$ then leads to $\gamma(n) = \frac{n}{2}$ as witnessed in Table 4.1. A theoretical proof for the value $\gamma(1) = \frac{1}{2}$ of the Laplace operator in one dimension can be found in [39].

Remark 4.12 *According to the results of Example 4.11 and (4.1.15), we should expect no specific convergence rate in 2D* ($n = 2$) *for the operator* $F(u) = -\Delta u + u^3$, *since* $\gamma = \min(1,3) = 1$ *leads to* $s^\star = 0$. *But the experimental results shown in Section 4.6 suggest an effective value of* $\gamma = 3$ ($s^\star = 1$) *in this case, as if the linear part was absent. In [11], it was shown that the best approximation rate* $s^\star$ *for linear operators and piecewise polynomial wavelets can be larger than what (4.1.15) promises for the correct value of* γ. *But formally, the prerequisites of Theorem 4.8 are not satisfied because the Laplace operator has a lower decay exponent value than the nonlinear part.*

We now turn to a short discussion of the iterative solution algorithms.

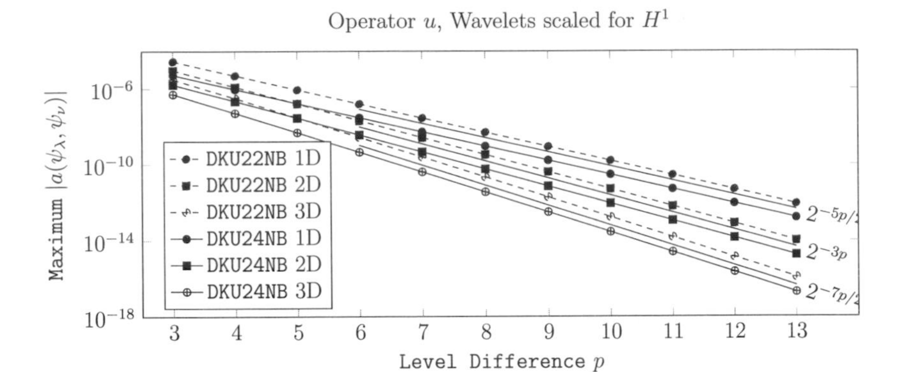

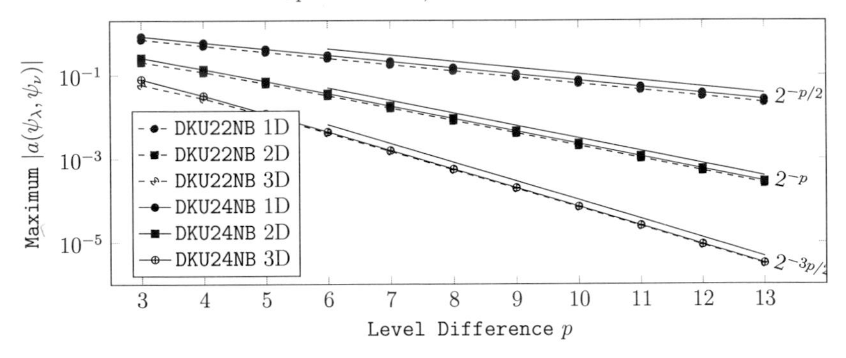

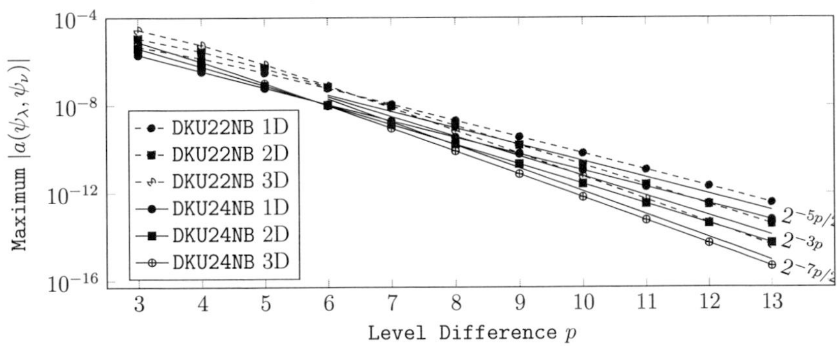

Figure 4.1: Maximum values of the bilinear forms $a(u, v) = \int u\, v\, d\mu$, $a(u, v) = \int \nabla u \cdot \nabla v\, d\mu$ and $a(u, v) = \int u^3\, v\, d\mu$ in 1D for a fixed wavelet $u = \psi_\lambda$ and all wavelets $v \in \{\psi_\nu\}$ on level $|\nu| = |\lambda| + p$ for different wavelet types and dimensions. The wavelet index λ was in each case on level $|\lambda| = 4$, position $\mathbf{k}_i = 4$ and type $\mathbf{e}_i = 1$ for $i = 1, 2, 3$.

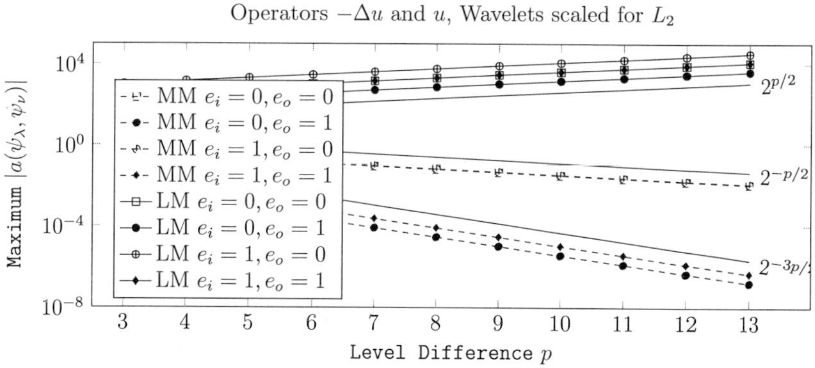

Operators $-\Delta u$ and u, Wavelets scaled for L_2

Figure 4.2: Maximum values of the bilinear forms $a(u, v) = \int \nabla u \cdot \nabla v d\mu$ and $a(u, v) = \int u v d\mu$ for all DKU22NB wavelets $u = \psi_\lambda$ on level $|\lambda| = 4$ and all wavelets $v \in \{\psi_\nu\}$ on level $|\nu| = |\lambda| + p$ for different function types $e_i, e_o \in \{0, 1\}$.

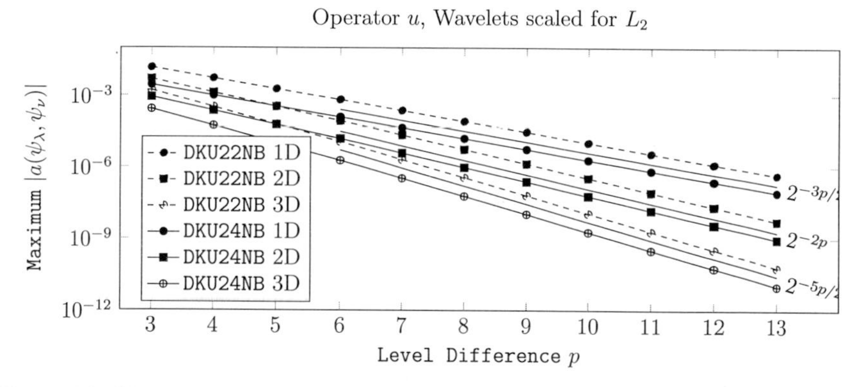

Figure 4.3: Maximum values of the bilinear forms $a(u, v) = \int u v d\mu$ for a fixed wavelet $u = \psi_\lambda$ and all wavelets $v \in \{\psi_\nu\}$ on level $|\nu| = |\lambda| + p$ for different wavelet types and dimensions. The wavelet index λ was in each case on level $|\lambda| = 4$, position $\mathbf{k}_i = 4$ and type $\mathbf{e}_i = 1$ for $i = 1, 2, 3$. Here, the preconditioner was not applied, i.e., the wavelets were scaled according to L_2, not H^1.

	Operator u	Operator $-\Delta u$	Operator u^3
n = 1	5/2	1/2	5/2
n = 2	3	1	3
n = 3	7/2	3/2	7/2

Table 4.1: Observed values of the decay parameter γ for different PDE operators $H^1 \to (H^1)'$ for wavelets with polynomial exactness order $d = 2$.

4.2 Richardson Iteration

The Richardson solver, see Algorithm 4.4, corresponds to the iteration scheme (4.1.2) with $\mathbf{C}_i := \alpha\mathbf{I}$. Although this step operator is practically very easy to compute, there are a lot of parameters in the algorithm that need to be chosen in accordance with the convergence proof from [33]. This includes the step size parameter $\alpha \in \mathbb{R}$ and the number of solver steps K before a coarsening of the iterands is necessary. For example, the explicit number of $K \in \mathbb{N}$ depends nonlinearly on the step parameter α, the spectral radius (4.2.5) and the constant C of (3.3.4), which implicitly turns up in the coarsening step.

Remark 4.13 *To avoid excessive iterations in the inner loop, it is possible to exit before the K steps have been executed once the residual in line 17 has fallen below $\sim \varepsilon_i$. One has to ensure that the constant involved here is small enough, since a premature exit from the inner iteration will result in worse convergence rates. The same can happen if the target tolerance ε is not very small, because then, in the coarsening step Algorithm 21, the factor $\frac{2}{5}$ can lead to a important coefficients being coarsed. In this case, the factor $\frac{2}{5}$ can be set to a lower value, e.g. $\frac{1}{5}$, which can reestablish quick convergence.*

Our version of the Richardson iteration therefore shows only the general structure of the algorithm, we discuss the convergence properties of Algorithm 4.4 for linear and nonlinear PDEs in the next sections. In [33], it was shown, that, with sufficient assumptions to the choice of the parameters, the Richardson iteration converges:

Theorem 4.14 *The iterates $\mathbf{u}^i$ produced by Algorithm 4.4 satisfy in each step*

$$\|\mathbf{u}^\star - \mathbf{u}^i\|_{\ell_2} \leq \varepsilon_i,$$

so that, in particular, $\|\mathbf{u}^\star - \mathbf{u}_\varepsilon\|_{\ell_2} \leq \varepsilon$. The number $K \in \mathbb{N}$ is uniformly bounded, independent of ε and the data. Moreover, if **`RESIDUAL`** *is $s^\star$-sparse (Theorem 4.7) for some $s^\star > 0$ and if $\mathbf{u}^\star \in \mathcal{A}^s_{tree}$ for some $s < s^\star$, then holds*

$$\# \operatorname{supp} \mathbf{u}_\varepsilon \lesssim \varepsilon^{-1/s} \|\mathbf{u}^\star\|^{1/s}_{\mathcal{A}^s_{tree}},$$
$$\|\mathbf{u}_\varepsilon\|_{\mathcal{A}^s_{tree}} \lesssim \|\mathbf{u}^\star\|_{\mathcal{A}^s_{tree}},$$

where the constants depend only on s as $s \to s^\star$. The number of operations needed to compute $\mathbf{u}_\varepsilon$ is proportional to $\varepsilon^{-1/s}\|\mathbf{u}^\star\|^{1/s}_{\mathcal{A}^s_{tree}}$.

Restating the proof would entail citing the very technical details, the interested reader can either look it up in [33] or find a variant of the proof in [147]. We will just briefly state some of the guiding factors in the determination of the parameters of the algorithm. The number of steps K before a coarsening operation is necessary is shown to be given by the number

$$K := \min\left\{ k \in \mathbb{N} \,|\, \beta\,(\omega_k\,\hat{\rho} + 4)\,\hat{\rho}^{k-1} \leq \frac{1}{2(1+2\,C)} \right\}, \tag{4.2.1}$$

with C of (3.3.4), ω_k a value of a summable sequence $\sum \omega_i = 1$, $\beta = \frac{1}{\alpha\, c_\Psi^2\, c_A}$ for (semi-)linear operators F with c_A from (1.4.13) and $\hat{\rho}$ is the maximum of the parameter $\overline{\rho}$ and

Algorithm 4.4 Adaptive computation of the solution $\mathbf{F}(\mathbf{u}) = \mathbf{f}$ up to accuracy $\varepsilon > 0$ by Richardson Iteration for an initial guess $\mathbf{v}$.

1: **procedure** SOLVER_RICHARDSON(ε, $\mathbf{F}$, $\mathbf{f}$, $\mathbf{v}$) $\rightarrow \mathbf{u}_\varepsilon$
2: // Init
3: $i \leftarrow 0$ ▷ Step counter
4: $\mathbf{u}^0 \leftarrow \mathbf{v}$ ▷ Start vector
5: Choose $K \in \mathbb{N}$ ▷ Coarsening counter
6: Choose $\alpha \in \mathbb{R}_+$ ▷ Step size
7: Choose $\overline{\rho} \in (0,1)$ ▷ Reduction factor
8: $\varepsilon_0 \leftarrow \|\mathbf{F}(\mathbf{u}^0) - \mathbf{f}\|_{\ell_2}$ ▷ Initial Error, computed on static index set $S(\mathbf{u}^0)$
9: $\varepsilon_1 \leftarrow \frac{1}{2}\varepsilon_0$
10: // Loop
11: **for** $i = 1, 2, \ldots$ **do**
12: // Error Reduction
13: $\eta_0 \leftarrow \varepsilon_i$ ▷ Stage dependent error
14: $\hat{\mathbf{u}}^{i,0} \leftarrow \mathbf{u}^i$
15: **for** $l = 1, \ldots, K$ **do**
16: $\eta_l \leftarrow \overline{\rho}\,\eta_{l-1}$
17: $\mathbf{r}_{\eta_l} \leftarrow$ RESIDUAL($\eta_l, \mathbf{F}, \mathbf{f}, \hat{\mathbf{u}}^{i,l-1}$) ▷ Compute residual using Algorithm 4.3
18: $\hat{\mathbf{u}}^{i,l} \leftarrow \hat{\mathbf{u}}^{i,l-1} + \alpha\,\mathbf{r}_{\eta_l}$ ▷ Perform Richardson Iteration
19: **end for**
20: // Coarsening
21: $\mathbf{u}^{i+1} \leftarrow$ TREE_COARSE($\frac{2}{5}\varepsilon_i, \hat{\mathbf{u}}^{i,K}$) ▷ Coarse the iterand using Algorithm 3.3
22: **if** $\varepsilon_i \leq \varepsilon$ **then**
23: $\mathbf{u}_\varepsilon \leftarrow \mathbf{u}^{i+1}$
24: **break** ▷ Exit loop
25: **end if**
26: // Error Reduction: When residual is small enough
27: **if** $\|\mathbf{r}_{\eta_K}\|_{\ell_2} \lesssim \varepsilon_i$ **then**
28: $\varepsilon_{i+1} \leftarrow \frac{1}{2}\varepsilon_i$ ▷ Set target accuracy for the next iteration step
29: **end if**
30: **end for**
31: **return** $\mathbf{u}_\varepsilon$
32: **end procedure**

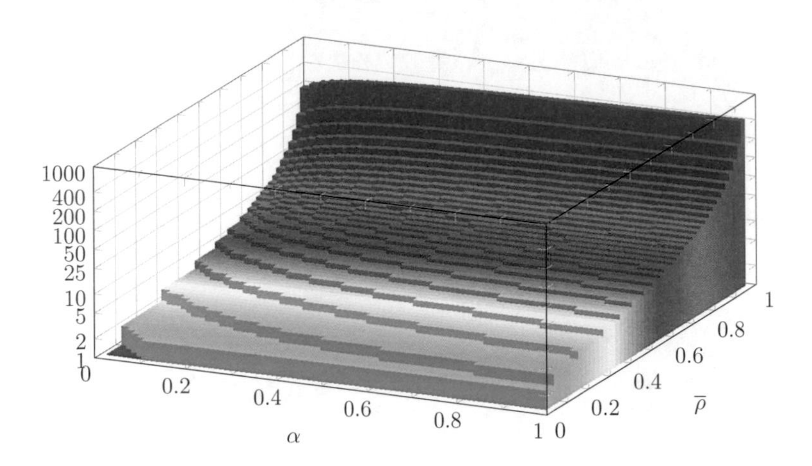

Figure 4.4: Values of K by (4.2.2) for $\alpha = 0.02, \ldots, 0.98$ and $\overline{\rho} = 0.02, \ldots, 0.98$ both with increments of 0.02. Note the exponentially scaled z-axis. While the value of K depends formally on both parameters, the parameter α only has a minor influence. When the value of $\overline{\rho}$ approaches 1, the value increases exponentially with a super-linear exponent. The color in the diagram only depends on the value of $\overline{\rho}$ to mark same values.

the true reduction rate ρ of (4.1.4). In applications, one probably does not know all the values of the of above constants with great precision and calculating them would be very cumbersome. It would be extremely difficult to accurately determine the numerical values of all the involved constants. Instead, one applies heuristics and uses meaningful guesses for most of them.
The only real parameter in (4.2.1), that is the one whose values is not given by the operators or wavelets at hand, is the value of the Richardson step parameter $\alpha \in \mathbb{R}$. The value of α usually has to be determined through trial and error, since it has to be chosen small enough for convergence but as high as possible for high convergence speeds.

Example 4.15 *Setting the right hand side to* $1/10$, *i.e.,* $C = 2$, *and assuming* $\hat{\rho} = 4/5$, $\omega_i = 2^{-i}$ *and* $c_\Psi^2 c_A = 1/100$, *then* $K = 39$.

It was shown in [147], that, when an error reduction of $\frac{1}{10}$ for the inner iterations is sought, the number K can be expressed in the case of **linear** operators as

$$K := \min\left\{k \in \mathbb{N} \,|\, \overline{\rho}^k + \alpha\, k\, \overline{\rho}^{k-1} \leq \frac{1}{10}\right\}, \tag{4.2.2}$$

which has the advantage of only depending on the known parameters α and $\overline{\rho}$. This simplifies the calculation so much that we use the above definition also for semilinear operators. Comparing (4.2.1) and (4.2.2), it is clear that both inequalities have, w.r.t. the unknown variable k, almost the same structure. A plot of the calculated values can be found in Figure 4.4.

4.2.1 Linear Operators

For this section, we will assume that a linear operator $\mathbf{A} : \ell_2 \to \ell_2$ is **symmetric positive definite**. This assumption makes the error analysis much easier, but it is not unmotivated. First, many operators are intrinsically symmetric, because they are based upon symmetric and positive definite **bilinear forms**, which itself is often rooted in **coercivity** (1.4.11). In the following, for $\mathbf{A}$ s.p.d., we denote by

$$\Sigma(\mathbf{A}) := \{\lambda \in \mathbb{R} \,|\, \mathbf{I} - \lambda\mathbf{A} \text{ is singular }\}, \tag{4.2.3}$$

the spectrum and its biggest element

$$\sigma(\mathbf{A}) := \max\{\lambda \in \mathbb{R} \,|\, \lambda \in \Sigma(\mathbf{A})\}, \tag{4.2.4}$$

is called the **spectral radius** of $\mathbf{A}$. Analogously to finite dimensional matrices, the Richardson iteration converges if the step size α is so small, that the step operator is a contraction, i.e.,

$$\sigma(\mathbf{I} - \alpha\mathbf{A}) = \max\{|1 - \lambda_{\max}(\alpha\mathbf{A})|, |1 - \lambda_{\min}(\alpha\mathbf{A})|\} < 1. \tag{4.2.5}$$

Since the generalized eigenvalues of $\mathbf{A}$ are all positive, $\lambda_{\max}(\alpha\mathbf{A}) < 2$ is a necessary and sufficient condition for (4.2.5). Thus, the Richardson scheme converges for $0 < \alpha < \frac{2}{\lambda_{\max}(\mathbf{A})}$ and the best choice is

$$\alpha_{\text{best}} := \frac{2}{\lambda_{\min}(\mathbf{A}) + \lambda_{\max}(\mathbf{A})}. \tag{4.2.6}$$

One can then show for the error in step n to hold

$$\|\mathbf{u}^\star - \mathbf{u}^i\|_{\ell_2} \leq \frac{1}{c_A\, c_\Psi^2}\|\mathbf{R}(\mathbf{u}^i)\|_{\ell_2}, \tag{4.2.7}$$

with the constants of (1.4.13) and (2.2.17). In particular, the scheme converges for the start vector $\mathbf{u}^0 := \mathbf{0} \in \ell_2$.

4.2.2 Semilinear Operators

As was explained in Section 1.5.2, symmetry and positive definiteness is replaced with **monotonicity** for semilinear operators $F = A + G$. For given starting point $\mathbf{u}^0 \in \ell_2(\mathcal{I})$ holds

$$\|\mathbf{u}^0 - \mathbf{u}^\star\|_{\ell_2} \leq c_A^{-1}\|\mathbf{R}(\mathbf{u}^0)\|_{\ell_2} =: \delta_0, \tag{4.2.8}$$

and this iteration converges for step size

$$0 < \alpha < \frac{2}{C_A + \hat{C}(\delta_0)}, \tag{4.2.9}$$

with again C_A being the constant from (1.4.13). Specifically, it holds for such α,

$$\varrho(\alpha) := \max\left\{|1 - c_A\,\alpha|, |1 - \alpha(C_A + \hat{C}(\delta_0))|\right\} < 1,$$

and then the following error reduction estimate is true:

$$\|\mathbf{u}^\star - \mathbf{u}^i\|_{\ell_2} \leq \varrho(\alpha)\|\mathbf{u}^\star - \mathbf{u}^{i-1}\|_{\ell_2}. \tag{4.2.10}$$

Of course, estimating this range of admissible step sizes (4.2.9) is again just as realistic as the determination of (4.2.6) in applications: It is theoretically possible, but requires very precise knowledge of all properties of the involved operators and the wavelet bases.
To avoid the determination of the step size parameter α, we take a quick look at the Gradient Iteration next.

4.3 Gradient Iteration

The greatest shortcoming of the Richardson Methods is the determination of the optimal step size α_{best} (4.2.6), since its value is often not directly computable, because the numerical properties of the operator are usually not known exactly. By dumping the assumption of a constant step parameter $\alpha \in \mathbb{R}$ for all steps, the Gradient Iteration, also known as Steepest Descent Method, computes the optimal step parameter for a linear operator $\mathbf{A}$ by the rule

$$\alpha_{\mathbf{A}} := \alpha_{\mathbf{A}}(\mathbf{u}) := \frac{(\mathbf{R}(\mathbf{u}))^T\,\mathbf{R}(\mathbf{u})}{(\mathbf{R}(\mathbf{u}))^T\,\mathbf{A}\,\mathbf{R}(\mathbf{u})}. \tag{4.3.1}$$

This scheme features the same convergence speed as the Richardson scheme with (4.2.6), but needs a second application of the operator $\mathbf{A}$. It converges for all $\mathbf{u}^0 \in \ell_2$ and the error in step n can be shown to abide by

$$\|\mathbf{u}^\star - \mathbf{u}^i\|_{\mathbf{A}} \leq \left(\frac{\kappa_2(\mathbf{A}) - 1}{\kappa_2(\mathbf{A}) + 1}\right)^i \|\mathbf{u}^\star - \mathbf{u}^0\|_{\mathbf{A}}. \tag{4.3.2}$$

Since the $\mathbf{A}$-norm is equivalent to the ℓ_2-norm by Theorem 2.25, this result is similar to (4.2.7). In our adaptive setting, it is theoretically possible for two adaptive vectors to be non-empty, but their scalar product to be zero. This cannot happen here since the support of the result of the computation Algorithm 3 is assumed to be a superset of the input vector $\mathbf{r} = \mathbf{R}(\mathbf{u})$, i.e., $S(\mathbf{A}\,\mathbf{r}) \supseteq S(\mathbf{r})$. Thus, the value computed in Algorithm 4 is well-defined unless $\mathbf{r} = \mathbf{0}$. This could feasibly only happen if the right hand side of the equation $\mathbf{F}(\mathbf{u}) =$ is also exactly zero. Nevertheless, it should be checked that the return value of Algorithm 4.5 is not infinite.
According to [33], for a semilinear operator $\mathbf{F} = \mathbf{A} + \mathbf{G}$, the steepest descent direction is still given by the residual $\mathbf{R}(\mathbf{u}^i)$ and to obtain the optimal step size α, one has to minimize the function

$$g(\alpha) = (\mathbf{f} - \mathbf{A}(\mathbf{u} + \alpha\,\mathbf{r}) - \mathbf{G}(\mathbf{u} + \alpha\,\mathbf{r}))^T\mathbf{r}. \tag{4.3.3}$$

Solving this equation can be done using **Newton's method** using $g'(\alpha) = -\mathbf{r}^T(\mathbf{A} + D\mathbf{G}(\mathbf{u} + \alpha\,\mathbf{r}))\mathbf{r}$. Since $g(\alpha_{\mathbf{A}}) = -\mathbf{G}(\mathbf{u} + \alpha_{\mathbf{A}}\mathbf{r})^T\mathbf{r}$, the linear value (4.3.1) can either be used as a starting value for the Newton's method, or, if the semilinear does not operator "behave" very nonlinear, even as the actual step parameter. The adaptive wavelet theory and algorithm is very similar to the details given in Section 4.2. In fact, both solvers are

covered by the same theoretical results in [33]. Instead of reiterating the results already cited in Section 4.2, we will focus here on the the difference in implementation.
This algorithm is then executed before line 18 in Algorithm 4.4 and the return value used for the step size α in the update of the iteration variable $\hat{\mathbf{u}}^{i,l}$. In order for the two extra operator applications to pay off w.r.t. execution times, convergence must be faster than the Richardson solver in actual applications. This will be tested in Section 4.6.
To improve on the overall complexity of these solution methods, we discuss the potentially more effective, but also more involved, Newton's method.

Algorithm 4.5 Adaptive computation of the optimal step size parameter for the Gradient Iteration for the operator $\mathbf{F}$ and the residual $\mathbf{r}$.

1: **procedure** SOLVER_GRADIENT_ITERATION_STEP_SIZE(ε, $\mathbf{F}$, $\mathbf{u}$, $\mathbf{f}$, $\mathbf{r}$) $\to \alpha$
2: $\quad \mathbf{w} \leftarrow$ APPLY($\varepsilon/5, \mathbf{F}, \mathbf{r}$) $\quad\triangleright$ APPLY stands for one of the algorithms presented,
3: $\quad\quad$ whichever is appropriate for this $\mathbf{F}$ or $D\mathbf{F}$
4: $\quad \alpha_0 \leftarrow \dfrac{\mathbf{r}^T\mathbf{r}}{\mathbf{r}^T\mathbf{w}}$
5: $\quad$ **for** $i = 0, 1, \ldots$ **do**
6: $\quad\quad \Delta\alpha_i \leftarrow \mathbf{r}^T$ (RHS($\varepsilon/2, \mathbf{f}$) $-$ APPLY($\varepsilon/2, \mathbf{F}, \mathbf{u} + \alpha_{i+1}\mathbf{r}$)) /
7: $\quad\quad\quad \left(-\mathbf{r}^T\text{APPLY}(\varepsilon/2, D\mathbf{F}(\mathbf{u} + \alpha_{i+1}\mathbf{r}), \mathbf{r})\right)$
8: $\quad\quad \alpha_{i+1} \leftarrow \alpha_i - \Delta\alpha_i$ $\quad\triangleright$ $\alpha_{i+1} \leftarrow \alpha_i - (g'(\alpha_i))^{-1}g(\alpha_i)$
9: $\quad\quad$ **if** $\Delta\alpha_i < 10^{-3}$ **then exit loop**
10: $\quad\quad$ **end if**
11: $\quad$ **end for**
12: $\quad$ **return** α_i
13: **end procedure**

4.4 Newton's Method

The Newton iteration $\mathbf{C}_i := D\mathbf{R}(u^i)^{-1}$ is traditionally executed in two steps, that is

$$\text{Solve} \quad D\mathbf{R}(\mathbf{u}^i)\mathbf{w}^i = -\mathbf{R}(\mathbf{u}^i), \tag{4.4.1}$$

$$\text{Step} \quad \mathbf{u}^{i+1} = \mathbf{u}^i + \mathbf{w}^i, \tag{4.4.2}$$

where the inversion in the first step is only done **approximately**, using a solver for linear equations. The implementation of these two lines in Algorithm 4.6 follows the general scheme Section 4.1.1 closely. Just as for Algorithm 4.4, the constant C denotes the constant from (3.3.4). The other constant $\widetilde{\omega}$ will be explained in Section 4.4.1.
The Algorithm leaves two details unspecified: the determination of the accuracy η_i for which to solve subproblem (4.4.1) and the algorithm with which this solution is computed. We discuss the first topic now and refer to Section 4.4.2 for the second one.
The whole discussion of convergence of Algorithm 4.6 comes down to proving that, if the tolerances η_i are chosen appropriately, then a sufficiently good initial vector $(\mathbf{u}^0)^T\Psi = u^0$, i.e., $u^0 \in B_\delta(u^\star)$ for some δ, will produce iterates for which it also holds that $(\mathbf{u}^i)^T\Psi = u^i \in B_\delta(u^\star)$. The initial condition $u^0 \in B_\delta(u^\star)$ has to be verified independently, but the conclusion from this condition w.r.t. the iterates can be assured by the following limits:

Fixing a positive constant $\beta < 1$, and assuming a $\delta > 0$ is chosen to ensure that

$$\delta < \min\left\{\frac{c_\Psi^3}{(1+2C)\,C_\Psi^3\,\omega}, \frac{\beta\,c_\Psi}{(1+2C)\,\widetilde{\omega}}\right\}, \tag{4.4.3}$$

then the condition

$$\eta_i \leq \eta_0 < \frac{\delta}{2(1+2C)\,C_\Psi}, \qquad \text{for all } n = 1, 2, \ldots, \tag{4.4.4}$$

implies that $u^i \in B_\delta(u^\star)$ for all subsequent iterations. Moreover, if

$$\eta_i \leq \frac{\varepsilon_i(\beta - (1+2C)\,\widetilde{\omega}\,\varepsilon_i)}{1+2C} = \varepsilon_i \frac{\beta}{1+2C} - \varepsilon_i^2 \widetilde{\omega}, \qquad \text{for all } n = 1, 2, \ldots, \tag{4.4.5}$$

then one has for $\hat{\eta}_i$ defined in Line 13,

$$\varepsilon_{i+1} = (1+2C)\,\hat{\eta}_i \leq \beta\,\varepsilon_i, \qquad \text{for all } n = 1, 2, \ldots. \tag{4.4.6}$$

Using these restrictions, [33] arrive at a similar result as Theorem 4.14:

Theorem 4.16 *Assuming that the restrictions (4.4.3)-(4.4.5) hold, then Algorithm 4.6 terminates after finitely many steps and produces a finitely supported* $\mathbf{u}_\varepsilon$ *satisfying*

$$\|\mathbf{u}^\star - \mathbf{u}_\varepsilon\|_{\ell_2} \leq \varepsilon.$$

Algorithm 4.6 Adaptive computation of the solution $\mathbf{F}(\mathbf{u}) = \mathbf{f}$ up to accuracy $\varepsilon > 0$ by Newton's Method for an initial guess $\mathbf{v}$.

1: **procedure** SOLVER_NEWTON(ε, $\mathbf{F}$, $\mathbf{f}$, $\mathbf{v}$) $\rightarrow \mathbf{u}_\varepsilon$
2: // Init
3: $i \leftarrow 0$ ▷ Step counter
4: $\mathbf{u}^0 \leftarrow \mathbf{v}$ ▷ Start vector
5: $\varepsilon_0 \leftarrow \|\mathbf{F}(\mathbf{u}^0) - \mathbf{f}\|_{\ell_2}$ ▷ Initial Error, computed on static index set $S(\mathbf{u}^0)$
6: // Loop
7: **for** $i = 1, 2, \ldots$ **do**
8: // Error Reduction
9: Choose η_i ▷ Stage dependent error
10: $\mathbf{w}_{\eta_i} \leftarrow$ SOLVE(η_i, $D\mathbf{R}(\mathbf{u}^i)$, $-\mathbf{R}(\mathbf{u}^i)$) ▷ Compute (4.4.1)
11: $\hat{\mathbf{u}}^{i+1} \leftarrow \mathbf{u}^i + \mathbf{w}_{\eta_i}$ ▷ Perform (4.4.2)
12: // Coarsening
13: $\hat{\eta}_i \leftarrow \widetilde{\omega}\,\varepsilon_i^2 + \eta_i$
14: $\mathbf{u}^{i+1} \leftarrow$ TREE_COARSE($2\,C\,\hat{\eta}_i$, $\hat{\mathbf{u}}^{i+1}$) ▷ Coarse the iterand using Algorithm 3.3
15: **if** $\varepsilon_i \leq \varepsilon$ **then**
16: $\mathbf{u}_\varepsilon \leftarrow \mathbf{u}^{i+1}$
17: **break** ▷ Exit loop
18: **end if**
19: // Error Estimation
20: $\varepsilon_{i+1} \leftarrow (1+2C)\,\hat{\eta}_i$ ▷ Set target accuracy for the next iteration step
21: **end for**
22: **return** $\mathbf{u}_\varepsilon$
23: **end procedure**

The second part of Theorem 4.14, the estimation of the work complexity, requires a detailed knowledge of the solution produced by the not yet determined solver in Line 10. We discuss the details of the solution methods employed to solve (4.4.1) after looking at the necessary requirements of our (nonlinear) PDE operators for the Newton scheme to be applicable.

Remark 4.17 *Since by Remark 1.54 it holds* $D\mathbf{A} = \mathbf{A}$ *for linear operators, a single step of the Newton scheme would result in the exact solution. But first, the solution of* $(D\mathbf{A})\mathbf{v} = \mathbf{f}$ *has to be computed, which is the original problem. Therefore, it does not make much sense to use Newton's scheme on linear problems.*

4.4.1 Semilinear Operators

Considering **semilinear** equations (1.5.2) $F = A + G$, the nonlinear term complicates analysis of the schemes. Simple continuity (1.4.10) is replaced by the notion of **stability** (1.5.8), which is transferred to the wavelet setting as

$$\|\mathbf{G}(\mathbf{u}) - \mathbf{G}(\mathbf{v})\|_{\ell_2} \leq \hat{C}_G(\max\{\|\mathbf{u}\|_{\ell_2}, \|\mathbf{v}\|_{\ell_2}\})\|\mathbf{u} - \mathbf{v}\|_{\ell_2}, \tag{4.4.7}$$

where $\hat{C}_G(\cdot) := C_\Psi^2\, C_G(C_\Psi\, \cdot)$ with the constant C_Ψ from (2.2.17). For the theoretical proof of convergence, two more assumptions are required:
Recall $R(v) := F(v) - f$, then holds $DR(v) = DF(v)$. There exists an open ball $\mathcal{U} \subset \mathcal{H}$ for which it holds:

(N1) The Fréchet derivative (1.5.6) $DR(v) : \mathcal{H} \to \mathcal{H}'$ is an isomorphism and there exists $\omega > 0$ such that for all $v \in \mathcal{U}$ and $y \in \mathcal{H}$ such that $v + y \in \mathcal{U}$ holds

$$\|(DR(v))^{-1}(DR(v+y) - DR(v))y\|_{\mathcal{H}} \leq \omega\|y\|_{\mathcal{H}}^2. \tag{4.4.8}$$

(N2) There exists a solution $u^\star \in \mathcal{U}$ and an initial guess $u^0 \in \mathcal{U}$ such that for the ω of (N1) holds

$$\|u^\star - u^0\|_{\mathcal{H}} \leq \delta < \frac{2}{\omega}, \quad \text{and} \quad B_\delta(u) \subseteq \mathcal{U}. \tag{4.4.9}$$

With these assumptions, it was shown in [60,61] that the iterates arising from the Newton iteration fulfill

$$\|u^\star - u^i\|_{\mathcal{H}} < \delta \quad \text{for } i \in \mathbb{N}_0, \qquad \text{and} \lim_{i\to\infty} \|u^\star - u^i\|_{\mathcal{H}} = 0. \tag{4.4.10}$$

Also, the convergence is **locally** (in $\mathcal{U}$) **quadratic**, i.e.,

$$\|u^\star - u^{i+1}\|_{\mathcal{H}} \leq \frac{\omega}{2}\|u^\star - u^i\|_{\mathcal{H}}^2, \quad \text{for all } i \in \mathbb{N}_0. \tag{4.4.11}$$

This translates into the wavelet domain as

$$\|\mathbf{u}^\star - \mathbf{u}^{i+1}\|_{\ell_2} \leq \widetilde{\omega}\|\mathbf{u}^\star - \mathbf{u}^i\|_{\ell_2}^2, \quad \text{for all } i \in \mathbb{N}_0, \qquad \text{with } \widetilde{\omega} := \frac{C_\Psi^2\omega}{2c_\Psi}, \tag{4.4.12}$$

with the constants c_Ψ,C_Ψ from (2.2.17).

Example 4.18 *For a linear isomorphic operator, both (4.4.8) and (4.4.9) are obviously true. For nonlinear operators, both assumptions have to be verified. This was done generally in [33] for semilinear operators of type (1.5.2) with properties (1.5.7) and (1.5.9) as assumed herein.*
More specifically, a proof for our case $G(u) := u^3(x)$ *can be found in [147]. In this case holds* $DG(z)(\cdot) : \mathcal{H} \to \mathcal{H}'$ *with* $DG(z) := 3\,z^2$*, and in wavelet coordinates* $D\mathbf{G}(\mathbf{z}) := ((\langle \psi_\lambda, 3z^2\psi_\mu \rangle))_{\lambda,\mu}$.
Furthermore, the operator $DF(z) := DG(z) + A$ *is symmetric positive definite: For the linear part* A*, this follows from coercivity (1.4.11). The nonlinear part* $DG(z)$ *is positive semi-definite by monotonicity (1.5.9):*

$$\begin{aligned}
\langle v, DG(z)(v) \rangle &= \lim_{h\to 0} \frac{1}{h} \langle v, F(z + h\,v) - F(z) \rangle\,, \\
&= \lim_{h\to 0} \frac{1}{h^2} \langle z + h\,v - z, F(z + h\,v) - F(z) \rangle\,, && w := z + h\,v, \\
&= \lim_{h\to 0} \frac{1}{h^2} \langle w - z, F(w) - F(z) \rangle \geq 0, && \text{for all } z, v \in \mathcal{H}.
\end{aligned}$$

In the sum, $DF(z)$*, is thus positive definite. For this operator* $G(u) := u^3(x)$*, the bilinear form based upon the Fréchet derivative* $DG(z)(\cdot)$ *can be interpreted as a* **(square-) weighted mass matrix***, since the bilinear form evaluates for fixed* z *to*

$$a_z(u,v) := \int_\Omega 3z^2\, u\, v\, d\mu, \qquad \text{for } u, v \in \mathcal{H}. \tag{4.4.13}$$

The application of the operator A_z *in the wavelet context was discussed in Section 3.5.4.*

4.4.2 Solving the Inner System

For simplicity, we will assume that $D\mathbf{R}(\mathbf{z}) = \mathbf{A} + D\mathbf{F}(\mathbf{z})$ is symmetric positive definite. This is the case for our nonlinearity $F(u) = u^3$, so this is not a big restriction, but simplifies the solution process considerably.

Adaptive Solution Method

This "adaptive solution method" refers to the original proposed method in [33]. Here, the subproblem (4.4.1) is treated as a separate adaptive linear equation with operator

$$\mathbf{Q}(\mathbf{z})(\cdot) : \ell_2(\mathcal{I}) \to \ell_2(\mathcal{I}), \quad \mathbf{Q}(\mathbf{z})(\mathbf{w}) := \mathbf{A}\mathbf{w} + (D\mathbf{G}(\mathbf{z}))(\mathbf{w}), \tag{4.4.14}$$

and right hand side

$$\mathbf{g}(\mathbf{z}) := \mathbf{F}(\mathbf{z}) - \mathbf{f} \in \ell_2(\mathcal{I}), \tag{4.4.15}$$

both for a fixed $\mathbf{z} \in \ell_2(\mathcal{I})$. Although $\mathbf{z} \in \ell_2(\mathcal{I})$ is a constant in this setting, this is not true for $D\mathbf{F}(\mathbf{z})$ and $\mathbf{F}(\mathbf{z})$ until a tolerance for evaluating the functions is chosen. Since the subproblem (4.4.1) is by construction a **linear problem**, linear solvers can be employed if an appropriate **residual** evaluation procedure like Algorithm 4.3 is available.

Remark 4.19 *In order to apply the linear operator theory of Section 4.2, it must be shown that the operator* $\mathbf{Q}(\cdot,\cdot)$ *satisfies assumptions (3.3.13) and (3.3.14) for Algorithm 3.4. This only concerns the* **nonlinear part** $G(\cdot)$, *as it changes under the derivative, unlike* A. *If the nonlinear part* $G(\cdot)$ *satisfies (1.5.11) for some* r *up to* $s^\star \geq 1$, *then* $DG(\cdot)(\cdot)$ *trivially satisfies the same growth condition for* $r-1$ *in the first argument and* 1 *in the second argument up to* $s^\star - 1$.

Since both the operator and the right hand side depend on another argument in a non-trivial way, a separate procedure is warranted and was presented in [33]. This algorithm uses the finite constant $\hat{C}$ which fulfills

$$\|D\mathbf{R}(\mathbf{u})(\mathbf{u}^\star - \mathbf{u}) + \mathbf{R}(\mathbf{u})\|_{\ell_2} \leq \hat{C}\|\mathbf{u}^\star - \mathbf{u}\|_{\ell_2}^2, \tag{4.4.16}$$

which is easily proven by Taylor expansion, and assumes $\|\mathbf{u}^\star - \mathbf{u}\|_{\ell_2} \leq \xi$ to be so small, that $\eta/4 > \hat{C}\xi^2$. This is necessary to ensure convergence of the solution of the approximately solved linear subproblem towards the solution of the original nonlinear problem.

Algorithm 4.7 Adaptive computation of the residual $\mathbf{r} := D\mathbf{F}(\mathbf{z})(\mathbf{v}) - (\mathbf{F}(\mathbf{z}) - \mathbf{f})$ for a proper tree $\mathbf{v}$ up to accuracy $\varepsilon > 0$.

1: **procedure** RESIDUAL_NEWTON_SUB(ε, $\mathbf{F}$, $\mathbf{f}$, $\mathbf{z}$, $\mathbf{v}$) $\to \mathbf{r}_\varepsilon$
2: $\quad\mathbf{w} \leftarrow$ APPLY($\varepsilon/2, D\mathbf{F}(\mathbf{z}), \mathbf{v}$) $\quad\triangleright$ APPLY stands for one of the algorithms presented,
3: $\quad$ whichever is appropriate for this **DF**
4: $\quad\mathbf{y} \leftarrow$ APPLY($\varepsilon/4 - \hat{C}\zeta^2, \mathbf{F}, \mathbf{z}$) $\quad\triangleright$ Application of operator $\mathbf{F}$ to $\mathbf{z}$ with
5: $\quad$ whichever algorithm works for F
6: $\quad\mathbf{z} \leftarrow$ RHS($\varepsilon/4, \mathbf{f}$) $\quad\triangleright$ Right hand side $\mathbf{f}$ can be constructed
7: $\quad$ like any normal function
8: $\quad\mathbf{r}_\varepsilon \leftarrow \mathbf{w} - (\mathbf{y} - \mathbf{z})$ $\quad\triangleright$ Do not delete canceled out values,
9: $\quad$ then $\mathbf{r}_\varepsilon$ is a proper tree
10: $\quad$**return** $\mathbf{r}_\varepsilon$
11: **end procedure**

This variant of Algorithm 4.3 is still $s^\star$**-sparse**:

Theorem 4.20 *Let* $s^\star := \frac{2\gamma - n}{2n}$, *where* γ *is the parameter of (3.3.14) of* $D\mathbf{F}(\mathbf{z})(\cdot)$ *and* n *is the dimension, then* $\mathbf{r}_\varepsilon :=$ *RESIDUAL_NEWTON_SUB*$(\varepsilon, \mathbf{F}, \mathbf{f}, \mathbf{z}, \mathbf{v})$ *is* $s^\star$-**sparse**, *i.e.,*

$$\begin{aligned}\#\operatorname{supp}\mathbf{r}_\varepsilon &\lesssim \varepsilon^{-1/s}\left(1 + \|\mathbf{v}\|_{\mathcal{A}^s_{tree}}^{1/s} + \|\mathbf{u}^\star\|_{\mathcal{A}^s_{tree}}^{1/s} + \|\mathbf{z}\|_{\mathcal{A}^s_{tree}}^{1/s}\right), \\ \|\mathbf{r}_\varepsilon\|_{\mathcal{A}^s_{tree}} &\lesssim 1 + \|\mathbf{v}\|_{\mathcal{A}^s_{tree}} + \|\mathbf{u}^\star\|_{\mathcal{A}^s_{tree}} + \|\mathbf{z}\|_{\mathcal{A}^s_{tree}},\end{aligned}$$

as long as $\|\mathbf{z}\|_{\mathcal{A}^s_{tree}} \lesssim 1$.

The proof is based on the properties of the involved algorithms and the arguments given in Remark 4.19. It is also shown in [33], that, in addition to (4.4.5), the **stage dependent error** η_i cannot be chosen too small, but must obey

$$\eta_i \gtrsim \frac{\hat{C}}{\rho^K}\varepsilon_i^2, \tag{4.4.17}$$

as the target accuracy for Algorithm 4.4 with ρ from (4.1.4) and K from (4.2.1). The admissible range for η_i in accordance with the upper limit (4.4.5) and this lower limit allows for

- **linear convergence** by choosing $\eta_i \sim \varepsilon_i$,
- **quadratic convergence** by choosing $\eta_i \sim \varepsilon_i^2$.

But as always with Newton's method, quadratic convergence is only possible when $\mathbf{u}^i$ is already close to the exact solution, i.e., (4.4.12) holds.
Altogether, under these assumptions, the second part to Theorem 4.16, the complexity results from [33], can be stated as follows:

Theorem 4.21 *If $D\mathbf{R}(\mathbf{z})(\cdot)$ fulfills (3.3.13) and (3.3.14), and if the exact solution $\mathbf{u}^*$ belongs to $\mathcal{A}^s_{tree}$ for some $s < s^\star = (2\gamma - n)/(2n)$, the output $\mathbf{u}_\varepsilon$ of Algorithm 4.6 has the following properties:*

$$\# \operatorname{supp} \mathbf{u}_\varepsilon \lesssim \varepsilon^{-1/s} \|\mathbf{u}^\star\|^{1/s}_{\mathcal{A}^s_{tree}},$$
$$\|\mathbf{u}_\varepsilon\|_{\mathcal{A}^s_{tree}} \lesssim \|\mathbf{u}^\star\|_{\mathcal{A}^s_{tree}}.$$

The number of operations needed to compute $\mathbf{u}_\varepsilon$ is proportional to $\varepsilon^{-1/s} \|\mathbf{u}^\star\|^{1/s}_{\mathcal{A}^s_{tree}}$.

This result mirrors exactly the properties of the algorithm put forth in Theorem 4.14

Solving on a Fixed Grid

An alternative to the full-blown adaptive strategy of the previous paragraph is fixing the given by the right hand side, i.e., $\mathcal{T}' := S(\mathbf{R}(\mathbf{u}^i))$, and then solve (4.4.1) for a solution vector $\mathbf{w}^i$ on the same set $\mathcal{T}'$. Since the computation of $D\mathbf{R}(\mathbf{u}^i)$ would involve Algorithm 3.4, we can assume $\mathcal{T} := S(\mathbf{u}^i) \subset \mathcal{T}'$ and thus the operator $D\mathbf{R}(\mathbf{u}^i)$ is well-defined on $\mathcal{T}'$.
There are two advantages to this approach, i.e., using a fixed support for all ℓ_2–vectors:

1. First, one avoids the determination or estimation of the involved constants needed to adaptively solve the subproblem (4.4.1), as was presented in the previous paragraph.
2. The second advantage is the possibility of employing standard full-grid solver, e.g., the **CG** method, see [59].

Both reasons simplify the implementing process of this method compared to the adaptive solution method considerably. The main disadvantage of this setting is simply that no convergence theory exists in this case.
Another noteworthy detail to address here is the **coarsening step** in line 13 of Algorithm 4.6: By definition, working on a fixed set of wavelet coefficients, the number of wavelet coefficients cannot grow during the execution of the **CG**. That means there is really no incentive to apply the coarsening step after the solution of the inner problem has been determined. Instead, as in Algorithm 4.4, the coarsening should be applied after the set of coefficients $\mathcal{T}'$ has been allowed to grow for a number of iterations.
This approach was first tried and implemented in [147].

Before putting all these algorithms to the test, we go into a few details of the implementation not discussed above.

4.5 Implementational Details

We present here a few "short cuts", i.e., deviations from the theoretical results meant to simplify implementation and program flow.
Technical details about the implementation can be found in Appendix B.

4.5.1 The Maximum Level

Restricting the maximum level $J \geq j_0$ seems like a counter-intuitive idea in the context of adaptive wavelet methods. After all, one of the design principles of adaptive methods is the ability to refine elements wherever needed and to any extent necessary. But in reality, there is no such thing as an infinite number (of levels) in computer programs. The most common approach is to initialize all possible levels in an adaptive storage container (explained in Section B.3) on creation of the object. Alternatively, it is possible to create more levels as needed, e.g., when wavelet indices are being inserted, thus creating the illusion of an infinite number of levels. This strategy only works up to the maximum number of levels representable by the data type used for the levels. The naive implementation using an `(unsigned) integer` would allow billions of levels but wastes memory, since, as is explained further down, the number cannot safely grow to even 400. In case of the bit field wavelet coefficients of Section B.3, which are designed for most efficient memory use, the number of levels never exceeds a few dozen.
In light of the preconditioning values of Theorem 2.18, it is not advisable to allow more than 324 levels since much higher levels would construct values in the preconditioner that are not representable using `double` floating point numbers (and would thus be computed to be zero, see [81]). Actually, numerical instabilities would be encountered much earlier, probably already around 50 levels, at which point values close to the machine precision `eps` are created frequently.
So, in summary, there is no such thing as an infinite number of levels. Restricting the number of levels arbitrarily actually has a number of advantages:

- The absolute number of wavelet coefficients becomes bounded by $\#\Delta_J^{\square} = \mathcal{O}\left(2^{nJ}\right)$ of (2.4.7). This effectively limits the complexity of all involved algorithms, e.g., Algorithm 4.4, and allows extremely speed-up computations.

- For linear (sub-)problems, the methods of Section 2.5 can be employed. This is only possible if the effective level is so low that full vectors can be created within the memory constraints of the employed computer. This approach enables one to employ well-known proven standard algorithms, especially for the inverses of Riesz operators as in Section 3.5.3.

Of course, the main disadvantage of this approach is the loss of accuracy that can be achieved when using only a small number of levels. But with sufficiently smooth functions, this effect is not very constrictive.
Once the accuracy of a computed solution can no longer be improved upon, this vector can be used as a starting point for a calculation involving additional levels.

4.5.2 Starting Solver

Related to the above idea is the notion to use different solution methods, depending on the accuracy sought. For a linear problem, a direct solver, i.e., **QU** decomposition, can be chosen if the maximum levels $J \approx j_0$. This particular approach is, of course, not directly applicable to **nonlinear** problems. But nevertheless, it is still possible to compute a good starting vector for a general iterator solver in this setting. To this end, we simply restrict the maximum level J of the wavelet discretization to, e.g., j_0+1 and let an iterative solver run until the **discretization error** is reached.

4.5.3 Increasing the Decay Parameter γ

The exact value of the decay parameter γ of (3.3.14) as given by theory is not necessarily the optimal value in application. During operator application, the value of γ regulates the size of the predicted tree by (3.3.19). The inverse nature of the dependence means that doubling the value of γ means roughly that the predicted tree will only be refined locally to half the depth. Because of the exponential growth of the influence sets (3.3.18), only increasing γ by 1 effectively means cutting the growth of the tree in half. Since most of the computational complexity, as seen in Figure 3.13, stems from the computation of the polynomial representation and the decomposition of the wavelet coefficient values, increasing γ is a very good way to avert long program runtimes. We usually use $\gamma = 7$ for our experiments. As Figure 4.17 shows, the increased value is regrettably, but expectedly, not translated into a higher approximation rate (4.1.15). Otherwise arbitrarily high convergence rates could be achieved by raising the value of γ. At least the increased value of γ does not lower the observed approximation rate.

4.5.4 Caching Data

Since Algorithm 4.1 has to be called very often where the only difference is the value of the error tolerance ε. To speed up the computation of the coarsed tree, it is advantageous to assemble and save the tree given by (3.3.5) and supply this in conjunction with $\mathbf{f} \in \ell_2$ to Algorithm 3.3. As explained in Remark 3.26, the modified error functional $\widetilde{e}(\lambda)$ of (3.3.7) is a local information and can be quickly calculated for all $\lambda \in \mathcal{T}$ and thus needs not be precomputed.
Similarly, the quadratic polynomial $3z^2$ of operator (3.5.19) needs only be computed once per call to the solution algorithm of the linear subproblem given by (4.4.14) with right hand side (4.4.15). As long as (3.5.21) holds, no recomputation of the weight polynomial is necessary. If, after a coarsening step within the inner solver, (3.5.21) does no longer hold, the easiest way to ensure it is simply to insert all wavelet coefficients $S(\mathbf{z})$ into the coarsed vector again.

4.5.5 Zero Subtrees in Differences of Vectors

Two vectors are subtracted at many places in different algorithms, e.g., Algorithm 4.3. Subtraction could yield zero subtrees, i.e., subtrees that contain only almost-zero values (exact zero values cannot be expected). Hence, an erase of zero valued elements and a downstream tree reconstruction can improve computation times in some occasions. In case of the residual, convergence would then entail smaller vector sizes.

This procedure is not advisable if the size of the support of the tree is of paramount importance to subsequent algorithms, e.g., as in Section 3.5.4.

4.5.6 Estimating The Constants

In the implementation, we tried not to insert arbitrary constants in the equations or estimates where constants such as (2.2.31) appear. However, this cannot be always avoided to guarantee accuracy or convergence of the numerical schemes in all circumstances. A different case arises in situation like (4.1.14), where a term is estimated by splitting it in two parts. In theory, the constants can be distributed in any way compliant with the theoretical proof. In applications, however, how the individual values are chosen can make a big influence on the performance or accuracy of the numerical algorithm. Figuring out the best values is usually only possible by trial and error, unless the constants are known exactly or at least approximately.

4.5.7 About Runtimes

The target accuracy (3.2.16) and the results like Theorem 4.8 establish that the complexity of acquiring a solution is proportional to N^{-s}, where s is some fixed constant. It also means that the computation of a solution of up to accuracy $\frac{1}{2}\varepsilon$ from a solution of accuracy ε will let the vector grow according to

$$N_{\varepsilon/2} \sim \left(\frac{\varepsilon}{2}\right)^{-1/s} = 2^{1/s}\varepsilon^{-1/s} \sim 2^{1/s} N_\varepsilon,$$

Iterating this line of thought, starting from some value ε_0 then follows after M halving periods,

$$\varepsilon = 2^{-M}\varepsilon_0, \quad \Longrightarrow \quad N_\varepsilon \sim 2^{M/s} N_{\varepsilon_0},$$

The runtime of each step w.r.t. the previous one grows exponentially by a factor of $2^{1/s}$. Just as with the pyramid scheme of the fast wavelet transform (2.1.33), viewing from the perspective of the last period then holds for all intermediate steps,

$$N_\varepsilon \sim 2^{-1/s} N_{2\varepsilon} \sim 2^{-2/s} N_{4\varepsilon} \sim \ldots \sim 2^{-M/s} N_{\varepsilon_0},$$

which means the complexity of the last step is always as large as the complexity of **all the previous steps combined**:

$$\sum_{k=1}^{M} 2^{-k/s} N_{2^k\varepsilon} = \frac{1-2^{-M/s}}{2^{1/s}-1} N_\varepsilon \leq \frac{1}{2^{1/s}-1} N_\varepsilon.$$

Thus, the **overall runtime**, as seen from the last step, is linear, but the witnessed overall runtime starting from ε_0 grows exponentially, i.e.,

$$\sum_{k=0}^{M} 2^{k/s} N_{\varepsilon_0} = \frac{2^{(M+1)/s}-1}{2^{1/s}-1} \sim 2^{M/s} N_{\varepsilon_0}.$$

Lost in the last equivalence is the factor attributed to the sum of all the addends, e.g., for $s = 1$ it would be a factor of 2. In such an application, this means the computation until an error of, e.g., $\varepsilon_1 = 5 \times 10^{-3}$ could take only a few seconds, but the continued computation until $\varepsilon_2 = 1 \times 10^{-4}$ is reached can take longer by a factor of $2\,\varepsilon_1/\varepsilon_2 = 100$, increasing the computation times from mere seconds to several minutes. If the runtime until ε_1 is reached is already in the range of a few minutes, this means the overall runtime could be in the range of hours. A plot of the runtimes taken from the examples of the next section can be found in Figure 4.5. It is therefore essential have high approximation rates $s^\star$ (4.1.15) and to use efficient **preconditioners** enabling high convergence rates for iterative solvers, which means fewer steps and thus operator applications have to be executed.

Next, we put the algorithms of this section to use on a nonlinear PDE.

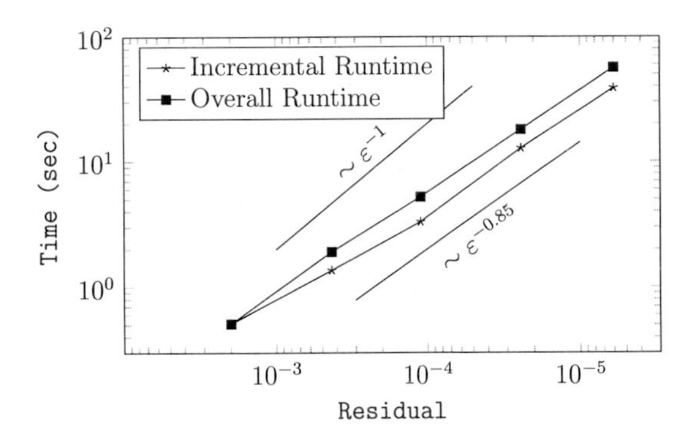

Figure 4.5: Overall and individual runtimes of the examples discussed in Section 4.6.2 using `MP244NB-SMD-SMOrth` wavelets. Note the reverse direction of the abscissa. The approximation rate seen in Figure 4.17 suggest $s^\star = 1$, the slope of which is given here as the black continuous line. Here, it seems as if the rate is slightly lower than $s^\star = 1$, which is probably attributed to overhead in the initialization of data structures and increased complexity due to the handling of the high number of wavelet coefficients in the unordered associative containers, see Section B.3.

4.6 A 2D Example Problem

To illustrate the effect of the individual components laid out in the previous section 4.1 – 4.4, we compare the costs of solving the following nonlinear equation in weak form (1.5.3):

$$-\Delta u(x) + u^3(x) = \left\{ \begin{array}{ll} \frac{3}{2} - \frac{4}{3}x_1, & \frac{1}{4} \leq x_1 \leq \frac{5}{8} \wedge \frac{5}{8} \leq x_2 \leq \frac{3}{4}, \\ \frac{3}{2} - \frac{4}{3}x_2, & \frac{5}{8} \leq x_1 \leq \frac{3}{4} \wedge \frac{3}{8} \leq x_2 \leq \frac{5}{8}, \end{array} \right\}, \quad \text{in } \Omega, \qquad (4.6.1)$$

$$\frac{\partial u}{\partial \nu} = 0, \quad \text{on } \partial\Omega.$$

The construction of the right hand side data vector will be performed using Algorithm 4.2. Several result vectors of the constructions can be seen in Figure 4.7. We are going to compare the solution algorithms presented in the previous sections to determine which computes the solution more quickly and how to choose the parameters to optimize the execution times.

Remark 4.22 *The reference solution used in the following experiments was obtained numerically by letting the Richardson solver run until the residual had reached a value of* 10^{-7}. *We used the same adaptive data for the right hand side and solution data in all experiments, adjusted to the different wavelets. One just has to make sure the data is applicable, e.g., same functions used in the primal or dual wavelet expansions, and adapt it to the employed preconditioner or basis transformation.*

A rastered reference solution plot and the wavelet coefficient distribution can be found in Figure 4.6. We will now present numerical experiments to qualitatively judge and quantitatively determine the best solution strategy for the above problem.

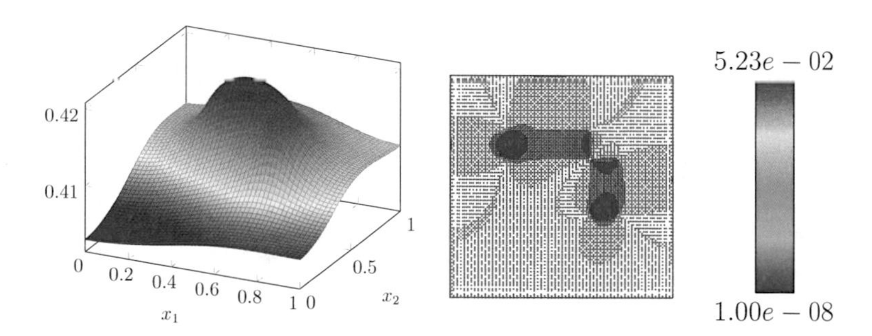

Figure 4.6: Plot of the computed reference solution using DKU22NB wavelets and its coefficient distribution (scatter diagram). The adaptive solution vector consists of roughly 1.5 million wavelet coefficients. The polynomial plot was generated by rastering with 50 points in each direction because the number of piecewise polynomials given by all these wavelets cannot be plotted accurately simultaneously. For the same reason the scatter diagram was filtered to only display values greater than 10^{-8} and levels up to 7, because otherwise the whole diagram would have been covered in dots and no distribution would have been recognizable. This is a limitation of the plotting means.

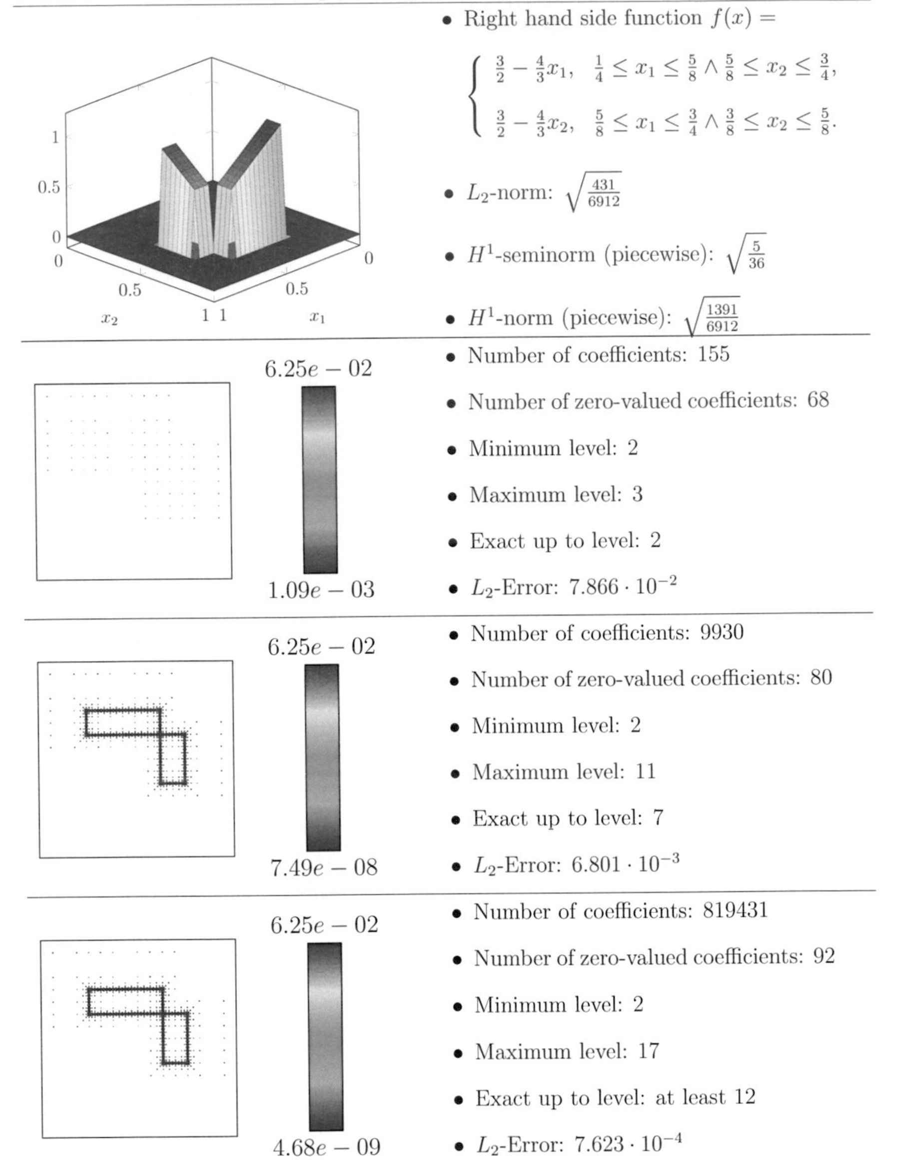

Figure 4.7: A piecewise polynomial right hand side function and the adaptive vectors of dual expansion coefficients computed by Algorithm 4.2 using DKU-22 wavelets. As the error computed by (4.1.12) decreases, the number of levels which are exactly computed, i.e., which contain all non-zero wavelet coefficients of a reference vector computed by full-grid methods up to level 12, grows.

4.6.1 Solving with Richardson Iteration

This section is entirely about using Algorithm 4.4 to solve (4.6.1). We start by examining the influence of some parameters on the Richardson iteration and then move to more general topics, e.g., preconditioning.

The parameter ρ

The first parameter we want to discuss is $\overline{\rho}$, which the theory of [33] says can be chosen freely in the range of $(0, 1)$. As such, we expect the overall solution process to be mostly independent of $\overline{\rho}$. Fixing otherwise all parameters and only changing the value of $\overline{\rho}$, the results depicted in Table 4.3 show that the influence on the solution process of this parameter is in fact negligible, except when approaching 1. Neither the execution time nor the total number of steps fluctuates greatly. The only influence it seems to have is being able to shift the amount of internal and external steps. Not depicted is the memory usage, which is hard to capture exactly, but it is usually higher for higher values of K, as then more wavelet coefficients can be accumulated before the vectors are being coarsened. Therefore it is advisable to choose a value of $\overline{\rho}$ with K in the middle range, e.g., $\overline{\rho} = 0.9$.

The Preconditioner

The efficiency of the Richardson iteration depends greatly on the value of the step parameter α, which depends on the spectral condition of the operator by (4.2.6) and (4.2.9). For the operator $A(u) := -\Delta u + u$, which leads to the bilinear form exactly representing the H^1-norm, it is clear that (2.2.33) is the better choice than (2.2.15). Further improvements were made by the orthogonal basis transformation (2.3.17) and this was demonstrated numerically in [122]. Although the operator $F(u) := -\Delta u + u^3$ is also an isomorphism from H^1 to its dual $(H^1)'$, it is not clear if the same preconditioning strategies will have the same effects.
To judge the preconditioning, we solve our example problem for $\mathbf{D}_1$, $\mathbf{D}_a$ and $\mathbf{D}_{\{O,a\}}$ ($\mathbf{D}_a$ with (2.3.17)) using the same settings except for the step size parameter α, which is chosen so the iteration converges most quickly. The maximum values of α, given in Table 4.2, were here found experimentally. These values indicate that the condition number of the operator $F(u) := -\Delta u + u^3$ is lowest with the configuration `MP244NB-SMD-SMOrth`, especially it should be much lower than the condition number when using `DKU22NB-P2`. The iteration studies presented in Figures 4.8 – 4.10 show clearly that the type of precondi-

Wavelet	`DKU22NB`	`DKU24NB`	`MP244NB`
`P2`	1.1×10^{-1}	8×10^{-2}	1×10^{-1}
`SMD`	5.3×10^{-1}	5.5×10^{-1}	6.4×10^{-1}
`SMD-SMOrth`	5.3×10^{-1}	5.5×10^{-1}	6.4×10^{-1}

Table 4.2: Maximum values of the Richardson step size parameter α for different preconditioners. See Section A for details on the names of the wavelet configurations.

$\overline{\rho}$	K	Max. Red. $\overline{\rho}^K$	Richardson Steps	Overall Steps	Time	$\|\mathbf{u} - \mathbf{u}^*\|_{\ell_2}$
0.1	3	1×10^{-3}	692	2076	589s	0.352
0.2	3	8×10^{-3}	692	2076	568s	0.352
0.3	4	8.1×10^{-3}	517	2067	638s	0.357
0.4	5	1.024×10^{-2}	414	2066	588s	0.357
0.5	7	7.812×10^{-3}	296	2065	649s	0.358
0.6	9	1.007×10^{-2}	231	2066	604s	0.357
0.7	14	6.782×10^{-3}	149	2065	616s	0.358
0.75	18	5.637×10^{-3}	117	2066	654s	0.357
0.8	24	4.722×10^{-3}	88	2066	552s	0.357
0.85	34	3.983×10^{-3}	63	2066	595s	0.357
0.9	56	2.738×10^{-3}	39	2066	604s	0.357
0.95	131	1.207×10^{-3}	18	2067	627s	0.357
0.99	845	2.049×10^{-3}	5	2066	548s	0.357

Table 4.3: Execution statistics for Algorithm 4.4 for varying values of $\overline{\rho}$. The values of K were determined according to (4.2.2). The Richardson scheme was then run to compute the residual to an accuracy of 10^{-3}. The last column shows the error $\|\mathbf{u} - \mathbf{u}^*\|_{\ell_2}$ of the computed solution to the reference solution depicted in Figure 4.6. The number of Richardson iterations varies inversely proportional to K, but the execution time and the overall number of steps, which includes counting the internal steps between coarsenings, remains roughly constant. This holds even when $\overline{\rho}$ gets very close to 1, where K grows exponentially (see Figure 4.4), because the inner iteration can exit prematurely if the residual has already fallen sufficiently, see Remark 4.13.

tioning has a tremendous influence on the speed of convergence. The preconditioners $\mathbf{D}_1$ of (2.2.15) and $\mathbf{D}_a$ of (2.2.33) exhibit for all wavelet type the lowest asymptotic rates in the Richardson iteration. In all cases, the rate of exponential decay of the residual and the ℓ_2-error are equal. As shown in [122], the basis transformation (2.3.17) effectively annihilates the influence of the coarsest level Ψ_{j_0-1} on the condition number of the operator $A(u) := -\Delta u + u$. Although not specifically constructed for the operator of (4.6.1), the stiffness matrix orthogonalizing basis transformation are very effective for the overall solution construction process here. So effective indeed, that the solution process is sped up by a factor of 20 in steps in this case. In terms of execution times, the speedup is even higher, bringing computations that take hours down to minutes and (in conjunction with other implementational devices of Section 4.5) into the realm of "real time computations", where the computation can be finished within fractions of seconds, see Table 4.5. The almost constant per-wavelet computations times in all cases show that there are no "hidden costs", i.e., the runtime is solely proportional to the number of treated wavelet coefficients. It should be noted that `DKU22NB-SMD-SMOrth` finished about twice as fast as `MP244NB-SMD-SMOrth`, which is simply a consequence of the larger support of the `MP244NB` wavelet compared to the `DKU22NB` wavelet, e.g., see the wavelet constructions presented in Section A.1.4 and Section A.1.6.

Iteration Histories and Approximation Order

A more detailed analysis of the best cases can be found in Figure 4.11. Here, the **internal steps** of the Richardson iteration are also displayed. The effect of the coarsening of the iterand can be observed in the jumping up of the residual value and dropping down of the number of wavelet coefficients contained in the adaptive vector. It is noticeable that the number of elements in the iterand grows steadily with increasing accuracy of the solution. After each coarsening step, the vector fills up again the quicker the more steps have already been executed.
There are two reasons the iterand $\hat{\mathbf{u}}^{i,l}$ in Algorithm 17 grows in size: In Algorithm 4.3 a tree prediction in done within Algorithm 3.8 on the current iterate and a tree coarsening is executed on the right hand side vector within Algorithm 4.1. As the stage dependent error $\eta_l := \varepsilon_i \overline{\rho}^l$ with $\overline{\rho} < 1$ takes on smaller values in each step and this value is used as tolerances for the tree prediction and coarsening, both algorithms will produce larger trees each time as a result. Here, at some point, the tolerance will be so small that the tree coarsening of the right hand side will simply return the whole vector. As seen in Figure 4.7, the right hand side vector with an error smaller than 10^{-3} contains almost 10^6 wavelet coefficients. The most important, i.e., of highest absolute value, coefficients are usually within this tree and not on the leaves. The tree prediction algorithm will then, after several iterations, stagnate and insert no more new elements.
An important conclusion should thus be that the right hand side data should be very sparse to ensure sparseness in all the iterands throughout the whole execution of the Richardson scheme.
Lastly, in Figure 4.17, the approximation order, i.e., the relation of the error ε and the number of wavelet coefficients N, is displayed for all the $\mathbf{D}_{\{O,a\}}$ cases. In this diagram, it can be observed that the accuracy in all cases drops proportional to N^{-1}, which is expected for $\gamma = 3$ as determined in Table 4.1 for the nonlinear operator G by (4.1.15). It seems that the linear operator A in this application seems to (quite unexpectedly, see Remark 4.12) exhibit a value $\gamma \geq 3$ of its own.

4.6.2 Solving with Gradient Iteration

Since the gradient iteration is so similar to the Richardson iteration, we only compare the `SMD-SMOrth` configurations. The iteration histories can be found in Figure 4.12. For these, the step size parameter was chosen according to (4.3.3) as depicted in Algorithm 4.5. The effort to compute the optimal step size value pays off as the iteration now converges uniformly and reaches the required residual accuracy after several steps for all wavelet types. As the value of the internal tolerances η_l only attains a value of $0.9^9 \approx 0.38$ times the starting tolerance ε, the number of active wavelet coefficients does not grow over the course of the iterations. This effect can also be seen in Figure 4.11, where the number of wavelet coefficients also stays constant at $\#\mathcal{N}_0(\mathcal{T})$ for the first few steps. A plot of the approximation rate would thus be meaningless, as this diagram would only constitute of a vertical line. The runtimes given in Table 4.5 show that this solver finishes first in the `SMD-SMOrth` configurations of all the evaluated solvers. This is not true for the other configurations and the runtimes indicate that in those cases the determination of the ideal step size takes too long, which makes the Gradient iteration inefficient, although it takes fewer steps to reach the target accuracy than the Richardson iteration.

4.6.3 Solving with Newton Iteration

The main advantage of the Newton iteration compared to the Richardson solver is the potential quadratic convergence, if the iterand comes into the neighborhood $\mathcal{U} \subset \mathcal{H}$ close to the solution.

Quadratic Convergence

As laid out in Section 4.4, one has to choose $\eta_i \sim \varepsilon_i^2$ in $\mathcal{U}$ to achieve quadratic convergence. Determining whether the iterand has entered the neighborhood $\mathcal{U}$ poses a problem similar to determining the optimal step size parameter α in the Richardson. The most practical approach to start the phase of quadratic convergence is when the value of the residual $\|\mathbf{r}\|_{\ell_2}$ drops below a predefined limit. If started before the residual falls below a certain value, the target tolerance $\eta_i \sim \varepsilon_i^2$ grows too small for the solution of the subproblem to really be computable to a precision of η_i. In this case, the Richardson solver does not converge properly and spends a large number of iterations without the value of the residual dropping below a certain threshold. This prolongs the computation time of the Newton solver considerably and makes the use of this solver unfeasible. In our studies, the most reliable threshold value was 10^{-3}. As this means $\eta_i \lesssim 10^{-6}$, which is probably much smaller than the desired target tolerance $\varepsilon = 10^{-4}$, this would entail computations to very high precision and thus great computational complexity. Therefore, the local tolerance η_i should be limited to the value of the target tolerance ε. We can then only expect one more step in the Newton solver, since an error reduction of a factor of $10 = 10^{-3}/10^{-4}$ should be accomplished within one step with quadratic convergence.

CG on Static Grid vs. Adaptive Solver

The Newton iteration histories where the inner system is solved using CG on the inner grid can be found in Figure 4.13 to Figure 4.15. Again, as in the case of the Richardson solver in Section 4.6.1, the convergence speed greatly depends on the employed preconditioner. It is obvious again that the basis transformation (2.3.17) effectively cuts the complexity of solving the equation by at least one order of magnitude, and in the best case by several orders of magnitudes. We spare the same iteration histories for all preconditioners for the Richardson scheme as the inner solver, the results are qualitatively the same. The iteration histories for the `SMD-SMOrth` configurations and the Richardson solver on the inner grid can be found in Figure 4.16. For the subproblem, the Richardson step size parameter ϱ slightly rounded down values of the ones given in Table 4.2 were chosen.
In the diagrams Figure 4.16, the quadratic convergence phase can hardly be distinguished. In experiments, disabling the quadratic convergence assumption will reduce the time necessary to decrease the residual from 10^{-3} to 10^{-4}, although it increases the number of steps needed. It simply seems a fact that it is computationally advantageous to execute a few steps to just decrease the residual by a factor of $\frac{1}{2}$ several times instead of trying

to aim for a factor of $\frac{1}{10}$ or lower directly.
Turning to Table 4.6, it is fairly obvious that the full adaptive treatment of the Richardson solver for the solution of the subproblem entails a higher computational cost. As can be seen in Figure 4.16 this even holds although the convergence rate is higher compared to the respective cases depicted in Figure 4.13 to Figure 4.15.

Approximation Order

The approximation order, i.e., the relation of the error ε and the number of wavelet coefficients N, is displayed in Figure 4.17. As stated before in Section 4.6.1, the accuracy in all cases drops proportional to N^{-1}, see Section 4.6.1 and the comments therein.

4.6.4 Conclusions

In the previous sections, we have presented many theoretical estimates including theoretical constants, representing properties of a wide range of operators and wavelets. It is very time-consuming and often also numerically difficult to determine the values of all of these constants to a high accuracy. Therefore, in practical applications, one often uses approximations for some of these constants and focuses on the generic behavior, e.g., squaring a value in each step. For the solution process, it is very important to optimize these involved values, especially the coarsening tolerances. Otherwise, the convergence towards the solution might slow down considerably or the solution might never be attained at all. From the data displayed in the previous sections, we can draw the following conclusions:

- Because of the outstanding performance improvements of the `SMD-SMOrth` configurations, we will use only these configurations from now on.
- In the best cases, all solvers can both output a solution within the time frame of a second. The per-wavelet computational costs shown in Tables 4.4–4.6 remained roughly constant at $\sim 10^5/s$ in all cases, hence no solver is more efficient than any other.
- The quadratic convergence of the Newton iteration is, theoretically, the quickest way to compute a solution to a high accuracy. In practice, however, the quadratic tolerances $\eta_i \sim \varepsilon_i^2$ entail such a great incline in computational complexity that the advantage does not pay off in faster execution times.

In summary, it is most important to choose a well-conditioned wavelet configuration, but there is no such thing as a "black box" solver in this setting. The closest to this comes the gradient iteration of Section 4.3, which gives good results with only minimal manual tuning. In summary, manual optimization of the solver parameters is always required and often has a high impact on the overall performance.

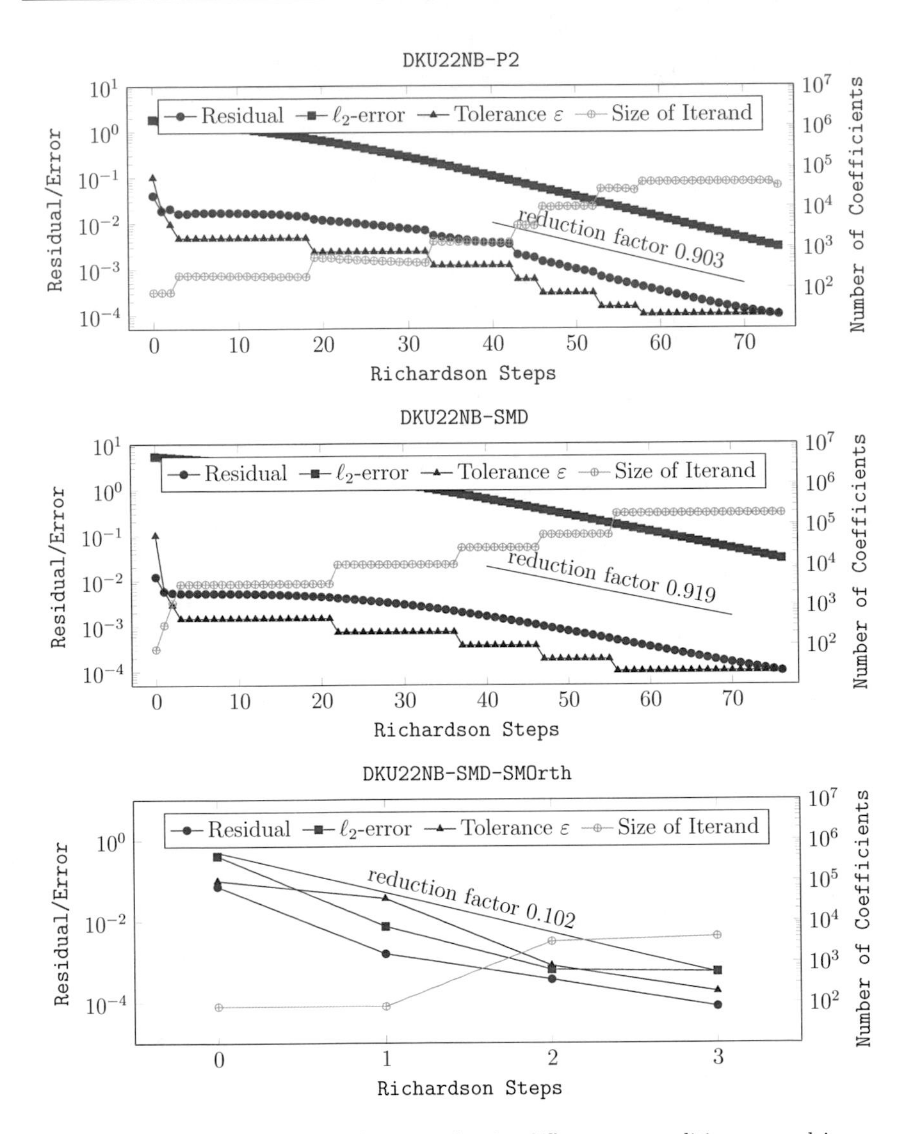

Figure 4.8: Richardson iteration histories for the different preconditioners used in conjunction with the DKU22NB wavelet. The plots depict the norm of the residual $\|\mathbf{F}(\mathbf{u})-\mathbf{f}\|_{\ell_2}$, the error $\|\mathbf{u}-\mathbf{u}^*\|_{\ell_2}$ (scale on the left side) and the number of coefficients $\#S(\mathcal{T})$ (scale on the right side). On the x-axis, we plot only the "outer" steps of the Richardson iteration, i.e., without the K inner steps. Each step thus corresponds to (at most) K operator applications and one coarsening of the iterands. The full history, including the inner steps, of the case DKU22NB-SMD-SMOrth can be found in Figure 4.11.

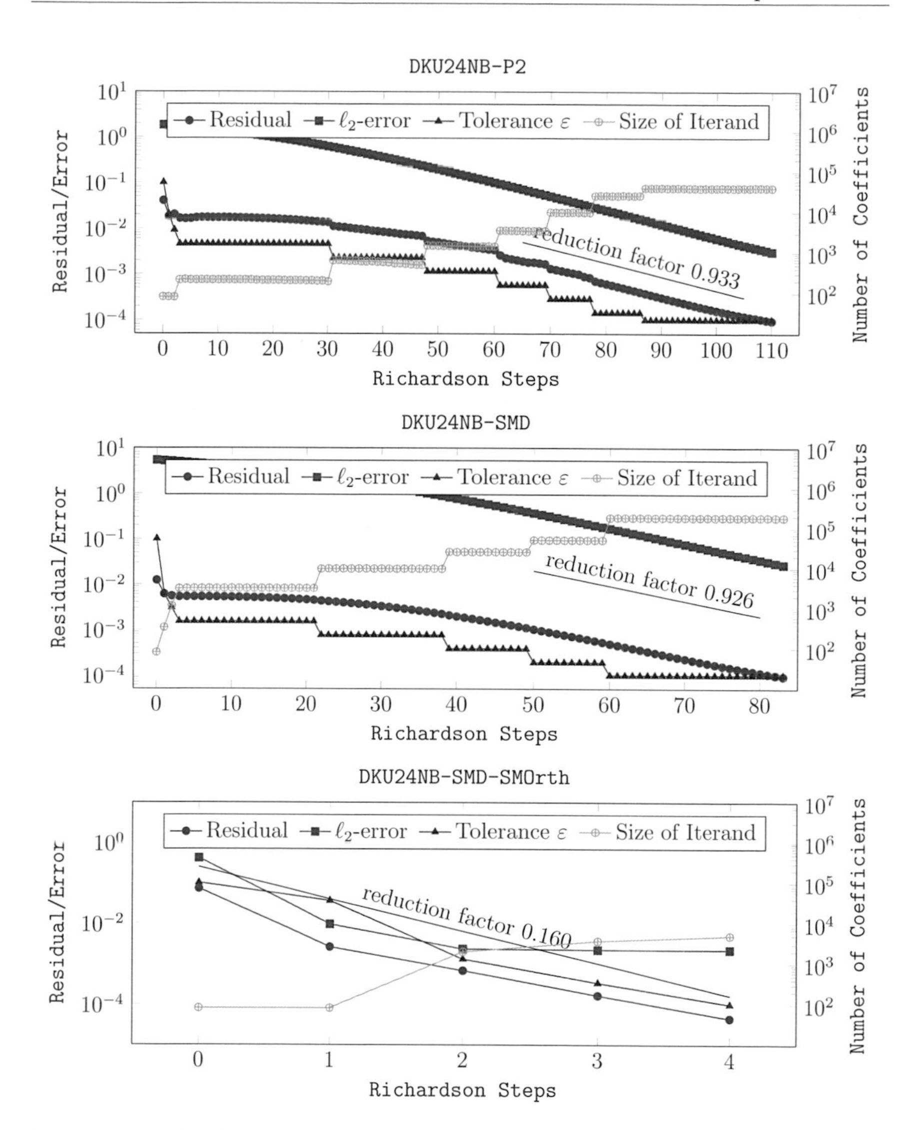

Figure 4.9: Richardson iteration histories for the different preconditioners used in conjunction with the DKU24NB wavelet. The plots depict the norm of the residual $\|\mathbf{F}(\mathbf{u})-\mathbf{f}\|_{\ell_2}$, the error $\|\mathbf{u}-\mathbf{u}^*\|_{\ell_2}$ (scale on the left side) and the number of coefficients $\#S(\mathcal{T})$ (scale on the right side). On the x-axis, we plot only the "outer" steps of the Richardson iteration, i.e., without the K inner steps. Each step thus corresponds to (at most) K operator applications and one coarsening of the iterands. The full history, including the inner steps, of the case DKU24NB-SMD-SMOrth can be found in Figure 4.11.

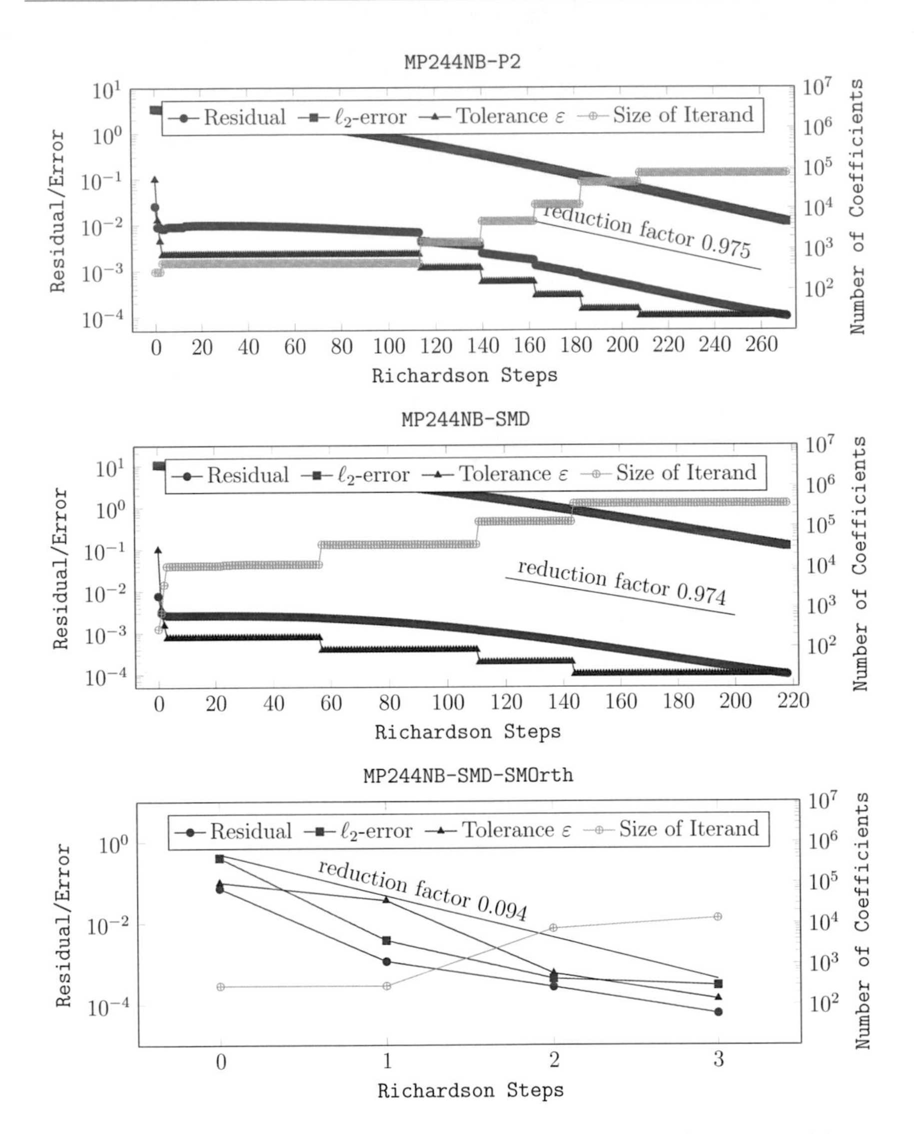

Figure 4.10: Richardson iteration histories for the different preconditioners used in conjunction with the `MP244NB` wavelet. The plots depict the norm of the residual $\|\mathbf{F}(\mathbf{u})-\mathbf{f}\|_{\ell_2}$, the error $\|\mathbf{u}-\mathbf{u}^*\|_{\ell_2}$ (scale on the left side) and the number of coefficients $\#S(\mathcal{T})$ (scale on the right side). On the x-axis, we plot only the "outer" steps of the Richardson iteration, i.e., without the K inner steps. Each step thus corresponds to (at most) K operator applications and one coarsening of the iterands. The full history, including the inner steps, of the case `MP244NB-SMD-SMOrth` can be found in Figure 4.11.

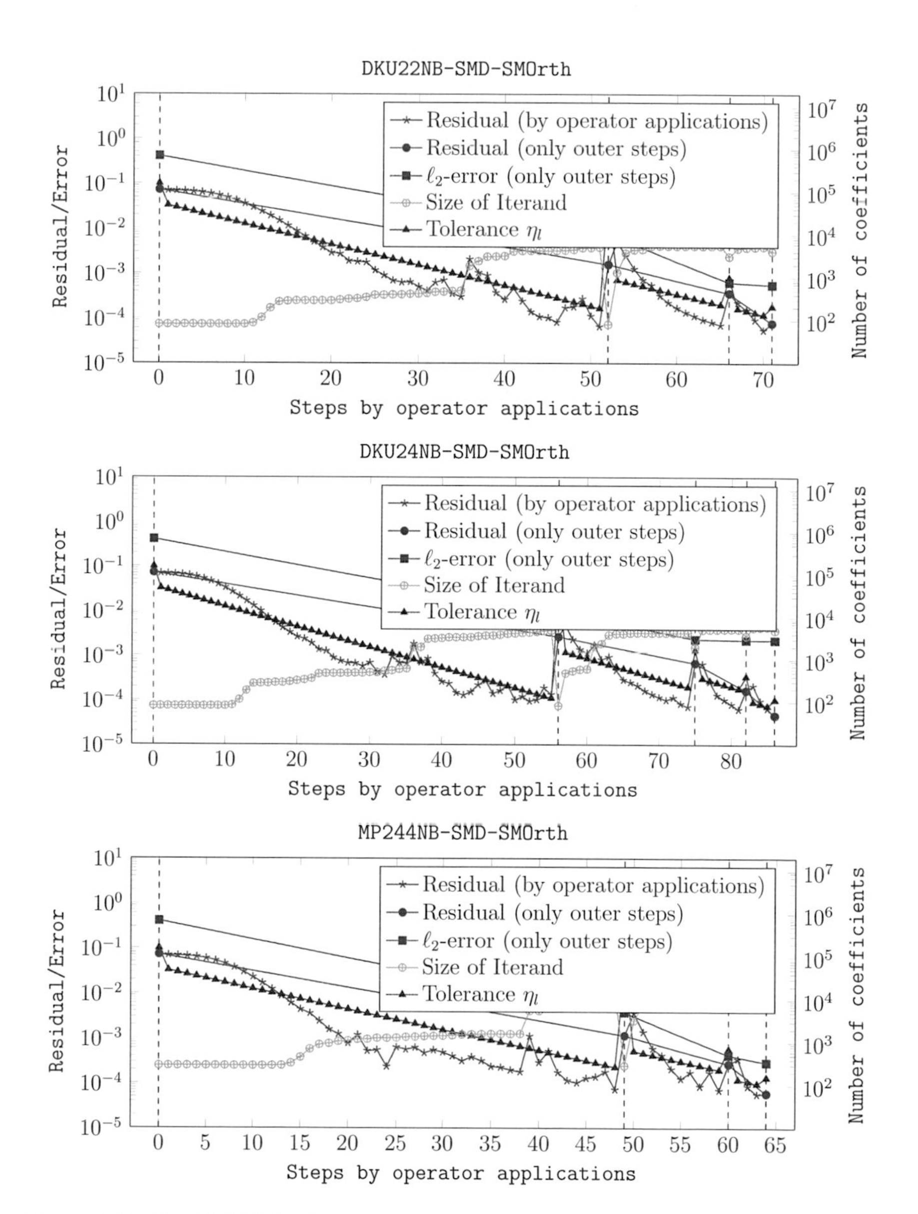

Figure 4.11: Detailed Richardson iteration histories for all wavelets in conjunction with the $\mathbf{D}_{\{O,a\}}$ preconditioner.

Richardson Runtime

Wavelet	DKU22NB		DKU24NB	
Preconditioner	Overall	Per Wavelet	Overall	Per Wavelet
P2	6.46×10^{2}	7.97×10^{-6}	1.26×10^{3}	1.15×10^{-5}
SMD	3.98×10^{3}	1.12×10^{-5}	6.34×10^{3}	1.59×10^{-5}
SMD-SMOrth	1.45	6.63×10^{-6}	2.36	1.01×10^{-5}

Wavelet	MP244NB	
Preconditioner	Overall	Per Wavelet
P2	1.91×10^{4}	1.62×10^{-5}
SMD	7.54×10^{4}	2.10×10^{-5}
SMD-SMOrth	3.93	1.22×10^{-5}

Table 4.4: Overall runtime and runtime in seconds per treated wavelet coefficient of the Richardson solver for the different Wavelet configurations. These times were measured from individual runs of the program with all options disabled that do not contribute to the computation of the solution directly. The comparison of the increase in runtime per coefficient in case of the more involved `SMD` preconditioner compared to the `P2` preconditioner hints at the cost of the more involved computations needed to determine the value of (2.2.32) for each wavelet coefficient. Since the runtime and number of wavelet coefficients of the `SMD-SMOrth` are so small, statistical fluctuations impact this configuration the most and thus the per wavelet coefficient values have the highest associated uncertainty.

Gradient Iteration Runtime

Wavelet	DKU22NB		DKU24NB	
Preconditioner	Overall	Per Wavelet	Overall	Per Wavelet
P2	4.44×10^{3}	9.34×10^{-6}	7.86×10^{3}	1.39×10^{-5}
SMD	1.33×10^{4}	1.13×10^{-5}	1.91×10^{4}	1.67×10^{-5}
SMD-SMOrth	9.2×10^{-2}	1.53×10^{-5}	9.7×10^{-2}	1.61×10^{-5}

Wavelet	MP244NB	
Preconditioner	Overall	Per Wavelet
P2	1.61×10^{5}	2.02×10^{-5}
SMD	2.55×10^{5}	2.14×10^{-5}
SMD-SMOrth	8.3×10^{-1}	4.10×10^{-5}

Table 4.5: Overall runtime and runtime per treated wavelet coefficient in seconds of the Gradient iteration for the different Wavelet configurations. The general comments of Table 4.4 also apply here.

Newton Runtime

Wavelet	DKU22NB		DKU24NB	
Preconditioner	Overall	Per Wavelet	Overall	Per Wavelet
Newton (CG) Runtime				
P2	5.98×10^1	7.31×10^{-6}	1.00×10^2	1.09×10^{-5}
SMD	4.26×10^2	8.66×10^{-6}	4.12×10^2	1.16×10^{-5}
SMD-SMOrth	1.30	7.06×10^{-6}	2.35	1.05×10^{-5}
Newton (Richardson) Runtime				
P2	3.95×10^3	8.04×10^{-6}	1.05×10^4	1.30×10^{-5}
SMD	1.31×10^3	7.75×10^{-6}	2.07×10^3	1.16×10^{-5}
SMD-SMOrth	2.50	6.92×10^{-6}	8.40	1.09×10^{-5}

Wavelet	MP244NB	
Preconditioner	Overall	Per Wavelet
Newton (CG) Runtime		
P2	1.50×10^3	1.41×10^{-5}
SMD	7.56×10^3	1.44×10^{-5}
SMD-SMOrth	4.01	1.28×10^{-5}
Newton (Richardson) Runtime		
P2	9.64×10^4	1.44×10^{-5}
SMD	1.88×10^4	1.30×10^{-5}
SMD-SMOrth	1.07×10^1	1.37×10^{-5}

Table 4.6: Overall runtime and runtime per treated wavelet coefficient in seconds of the Newton solver for the different Wavelet configurations. The name of the inner (linear) subproblem solver is given in parentheses in the title. It is directly obvious that the adaptive treatment of the linear subproblem entails an increase in needed runtime. Also, the larger support of the $\widetilde{d} = 4$ wavelets leads to an increase in complexity per treated wavelet index. The motivation to use smoother wavelet configurations is of course the benefit of higher theoretical approximation orders which should translate into higher convergence orders. The general comments of Table 4.4 also apply here.

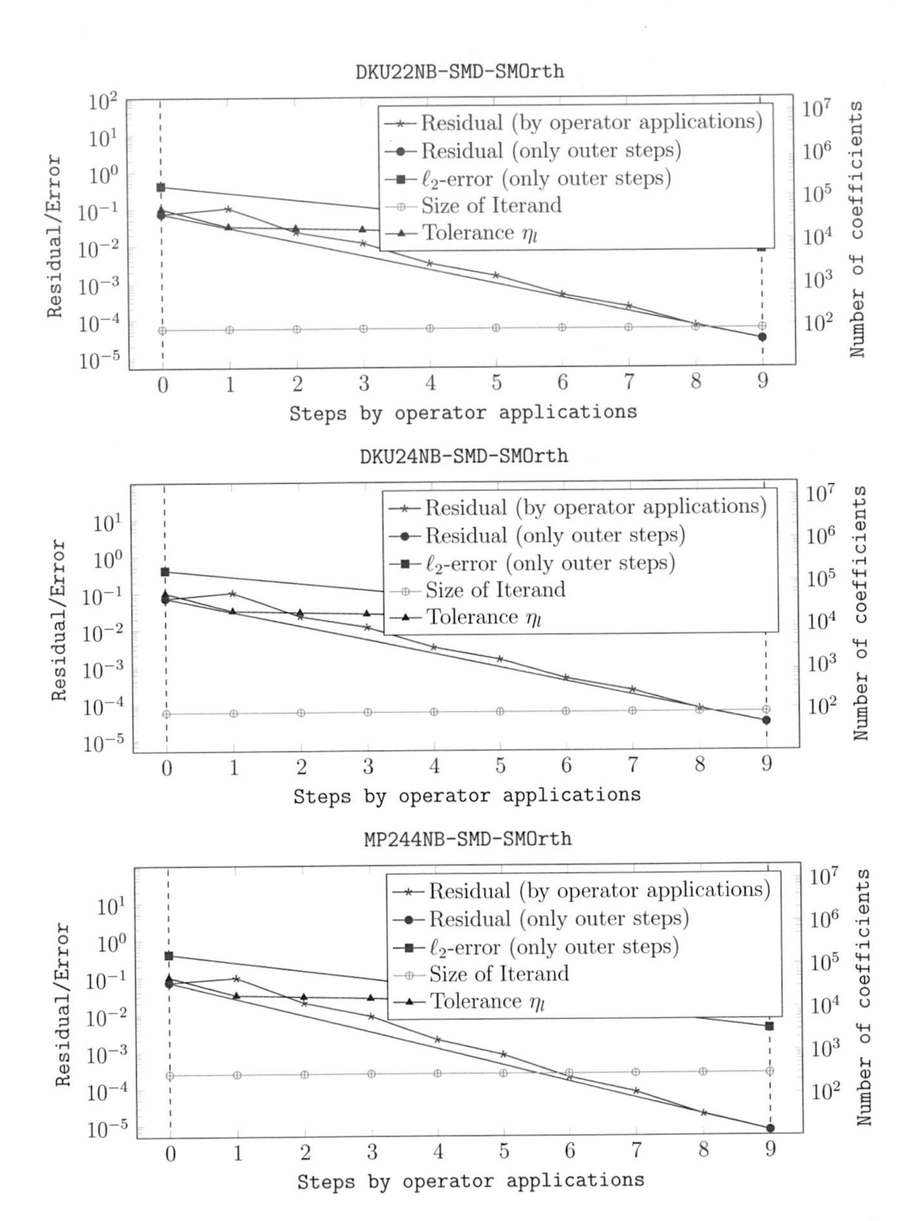

Figure 4.12: Detailed Gradient Iteration histories for all wavelets in conjunction with the $\mathbf{D}_{\{O,a\}}$ preconditioner.

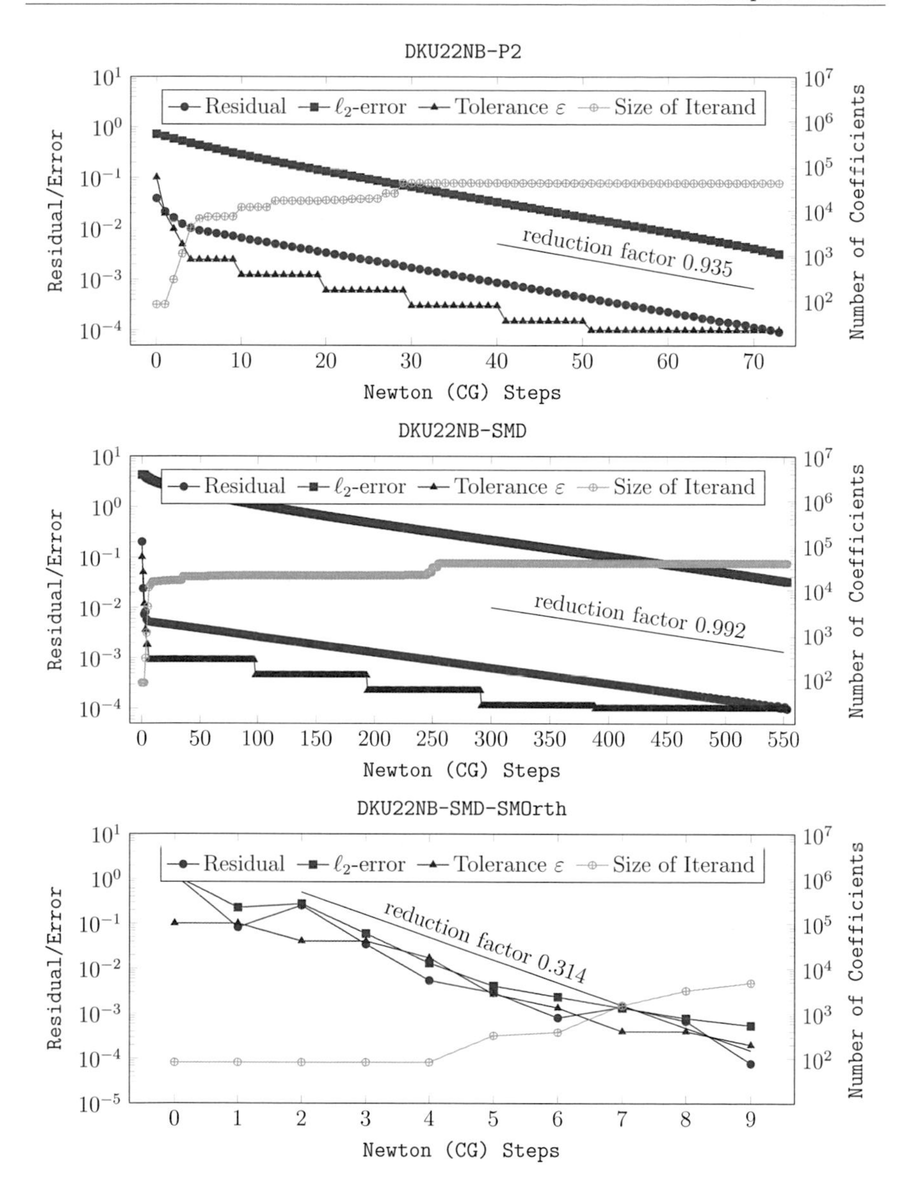

Figure 4.13: Newton iteration histories with CG as subproblem solver for the different preconditioners used in conjunction with the DKU22NB wavelet. The plots depict the norm of the residual $\|\mathbf{F}(\mathbf{u}) - \mathbf{f}\|_{\ell_2}$, the error $\|\mathbf{u} - \mathbf{u}^*\|_{\ell_2}$ (scale on the left side) and the number of coefficients $\#S(\mathcal{T})$ (scale on the right side). On the x-axis, we plot the steps of the Newton iteration.

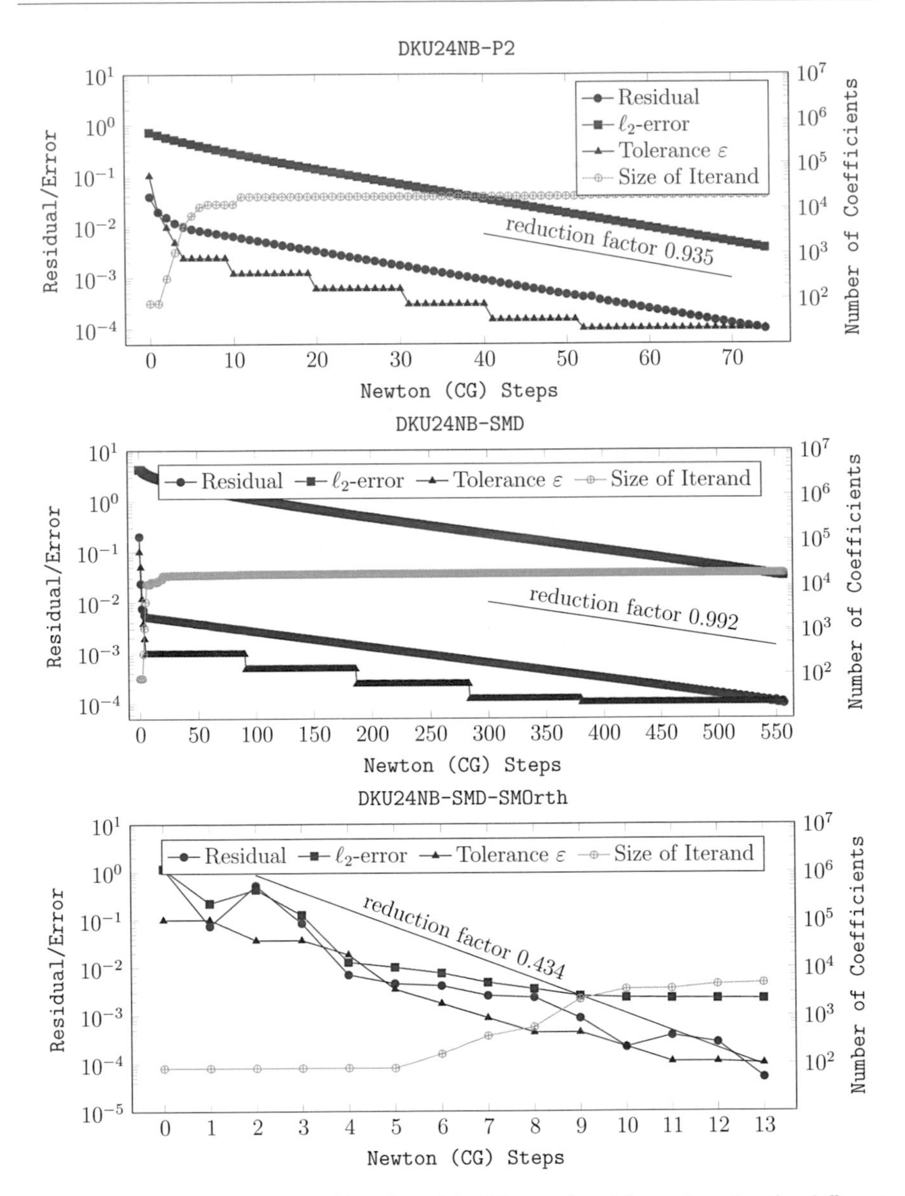

Figure 4.14: Newton iteration histories with CG as subproblem solver for the different preconditioners used in conjunction with the DKU24NB wavelet. The plots depict the norm of the residual $\|\mathbf{F}(\mathbf{u}) - \mathbf{f}\|_{\ell_2}$, the error $\|\mathbf{u} - \mathbf{u}^*\|_{\ell_2}$ (scale on the left side) and the number of coefficients $\#S(\mathcal{T})$ (scale on the right side). On the x-axis, we plot the steps of the Newton iteration.

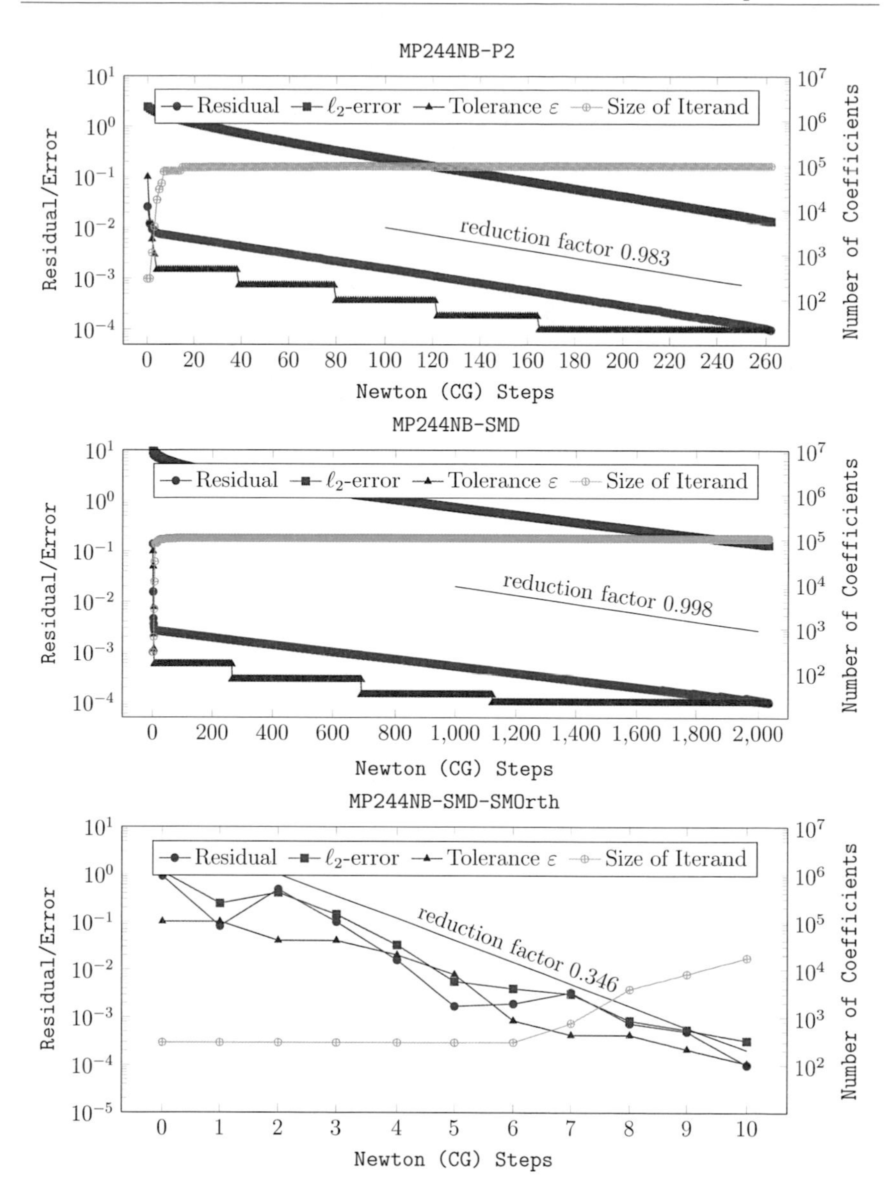

Figure 4.15: Newton iteration histories with CG as subproblem solver for the different preconditioners used in conjunction with the MP244NB wavelet. The plots depict the norm of the residual $\|\mathbf{F}(\mathbf{u}) - \mathbf{f}\|_{\ell_2}$, the error $\|\mathbf{u} - \mathbf{u}^*\|_{\ell_2}$ (scale on the left side) and the number of coefficients $\#S(\mathcal{T})$ (scale on the right side). On the x-axis, we plot the steps of the Newton iteration.

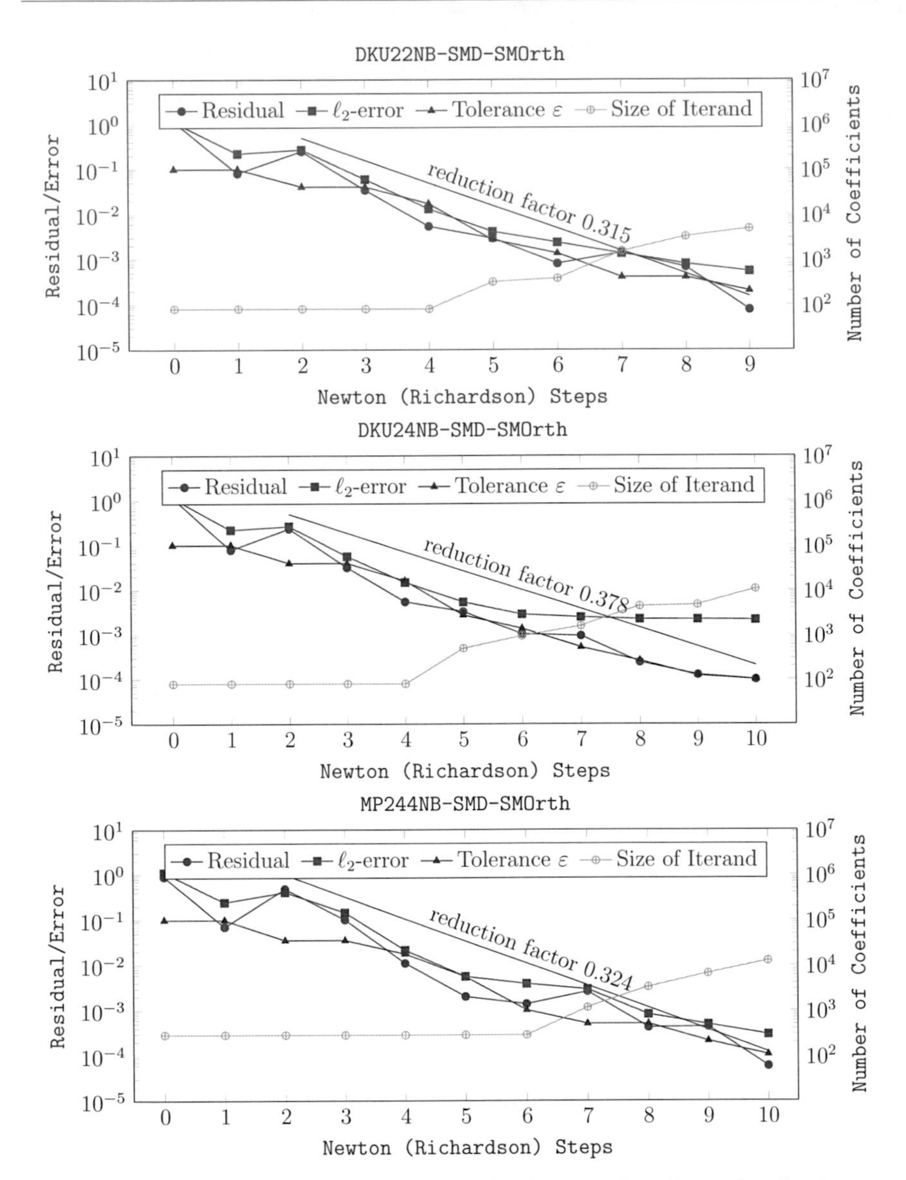

Figure 4.16: Newton iteration histories with Richardson as subproblem solver for the preconditioner $\mathbf{D}_{\{O,a\}}$ used in conjunction with the DKU22NB, DKU24NB and MP244NB wavelets. The plots depict the norm of the residual $\|\mathbf{F}(\mathbf{u}) - \mathbf{f}\|_{\ell_2}$, the error $\|\mathbf{u} - \mathbf{u}^*\|_{\ell_2}$ (scale on the left side) and the number of coefficients $\#S(\mathcal{T})$ (scale on the right side). On the x-axis, we plot the steps of the Newton iteration.

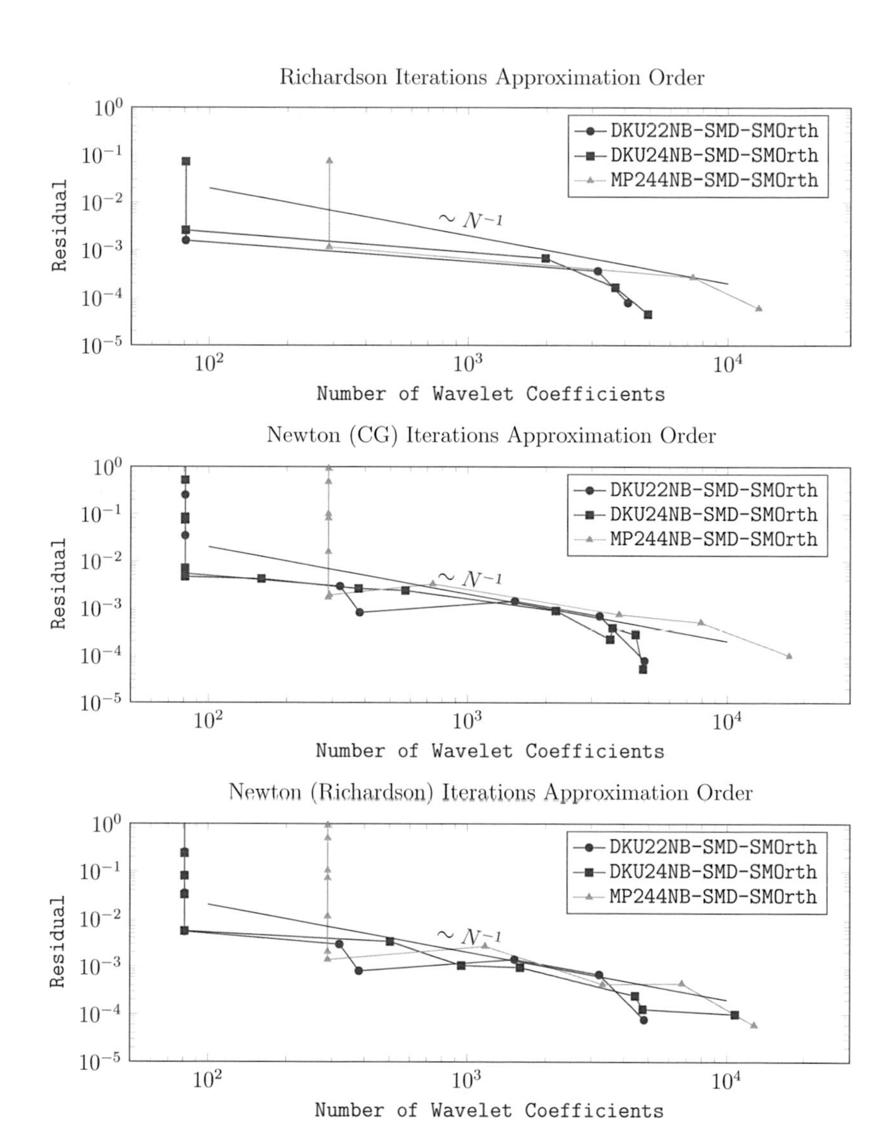

Figure 4.17: Approximation order for the different wavelets using preconditioner $\mathbf{D}_{\{O,a\}}$. On the abscissa, the number of wavelet coefficients after each coarsening is plotted against the residual for the coarsed vector on the ordinate axis.

4.7 A 3D Example Problem

As the three-dimensional example PDE, we study the nonlinear partial differential equation,

$$\begin{aligned} -\Delta u(x) + u^3(x) &= f, \quad && \text{in } \Omega, \\ \frac{\partial u}{\partial \nu} &= 0, \quad && \text{on } \partial\Omega, \end{aligned} \tag{4.7.1}$$

where the right hand side function is given by the piecewise polynomial

$$f(x_1,x_2,x_3) := \begin{cases} 0, & 0 < x_1 \le \frac{1}{2}, \; 0 < x_2 \le \frac{1}{2}, \; 0 < x_3 \le \frac{1}{2}, \\ 1, & \frac{1}{2} < x_1 < 1, \; 0 < x_2 \le \frac{1}{2}, \; 0 < x_3 \le \frac{1}{2}, \\ 2, & 0 < x_1 \le \frac{1}{2}, \; \frac{1}{2} < x_2 < 1, \; 0 < x_3 \le \frac{1}{2}, \\ 3, & \frac{1}{2} < x_1 < 1, \; \frac{1}{2} < x_2 < 1, \; 0 < x_3 \le \frac{1}{2}, \\ 4, & 0 < x_1 \le \frac{1}{2}, \; 0 < x_2 \le \frac{1}{2}, \; \frac{1}{2} < x_3 < 1, \\ 5, & \frac{1}{2} < x_1 < 1, \; 0 < x_2 \le \frac{1}{2}, \; \frac{1}{2} < x_3 < 1, \\ 6, & 0 < x_1 \le \frac{1}{2}, \; \frac{1}{2} < x_2 < 1, \; \frac{1}{2} < x_3 < 1, \\ 7, & \frac{1}{2} < x_1 < 1, \; \frac{1}{2} < x_2 < 1, \; \frac{1}{2} < x_3 < 1. \end{cases} \tag{4.7.2}$$

Since this function is smooth on each patch but not at the boundaries, we expect most of the wavelet coefficients to aggregate on the hyperplanes $(x_1, 0, 0)$, $(0, x_2, 0)$ and $(0, 0, x_3)$. This setup is sparse in three dimensions, but the discontinuities cover the whole of the three two-dimensional hyperplanes. The three-dimensional scatter diagram of the wavelet coefficients computed by Algorithm 4.2 can be seen in Figure 4.18. Although the distribution seems perfectly symmetrical, the height of the jump at the interfaces is different depending on which plane is examined. By construction, crossing the intersection of the hyperplane x_i, $i = 0, 1, 2$ entails a jump of 2^i. As such, the values of the constructed expansion vector are not equally weighted, there is a slight tendency of the coefficients constructed by Algorithm 4.2 towards the hyperplane $(0, 0, x_3)$. But this effect is hardly recognizable in the diagram.
In these tests, we solve the equation (4.7.1) up to $\varepsilon = 1 \times 10^{-3}$. The references in each case are solutions computed up to $\varepsilon = 1 \times 10^{-5}$. A density plot of the pointwise evaluated computed solution can be seen in Figure 4.19.

4.7.1 Solving with Richardson Iteration

The experimentally found step sizes used can be seen in Table 4.7. The Richardson solver proved very effective in this case, its simplicity being the predominant reason. Computing in dimension $n = 3$ entails a substantially higher runtime, simply because there are at least an order of magnitude more wavelet coefficients in any vector, even by just considering the number of all roots $\mathbb{N}_0^n$ (3.2.2) which grows exponentially in n by the setup (2.4.7).
In Figure 4.24, the approximation rates are displayed for all wavelet constructions. In this diagram, it can be observed that the accuracy in all cases drops proportional to $N^{-2/3}$, which is expected for $\gamma = \frac{7}{2}$ and $n = 3$ as determined in Table 4.1 for the nonlinear operator G by (4.1.15).

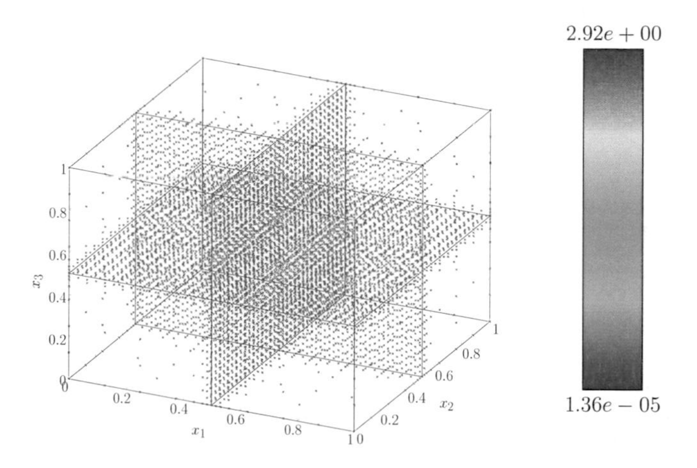

Figure 4.18: Wavelet coefficients of the right hand side (4.7.2) expanded in `DKU22NB-SMD-SMOrth` wavelets. Due to the high dimension $n = 3$, only the coefficients of up to level $J = 5$ are presented. But this limitation also makes it possible to recognize the structure of the distribution more easily. Except for the coarsest level coefficients, the wavelet coefficients are gathered at the three hyperplanes $(x_1, 0, 0)$, $(0, x_2, 0)$ and $(0, 0, x_3)$. The above diagram contains $N = 9483$ markers for individual coefficients, the actual vector used in the experiments contains several million wavelet coefficients.

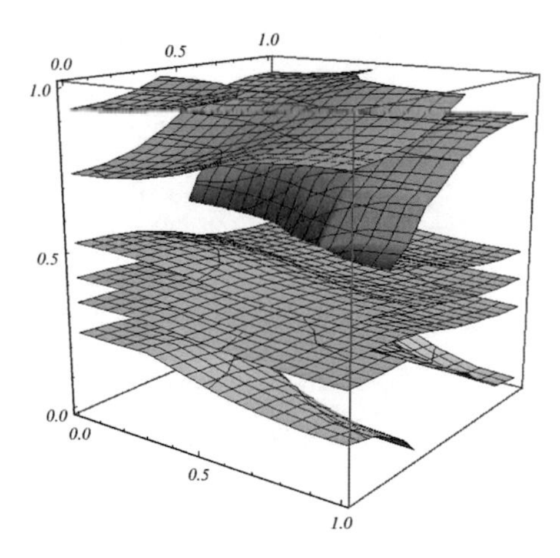

Figure 4.19: Solution of (4.7.1). The plot shows the isosurfaces of the pointwise evaluation of the polynomial represented by the vector of wavelet coefficients.

Wavelet	DKU22NB	DKU24NB	MP244NB
SMD-SMOrth	2.5×10^{-1}	2.5×10^{-1}	2.6×10^{-1}

Table 4.7: Maximum values of the Richardson step size parameter α for different preconditioners. See Section A for details on the names of the wavelet configurations.

4.7.2 Solving with Gradient Iteration

Again, the main difference to the prior solver is the computation of the optimal step size α in each step. In the two-dimensional example, this computation was not very costly to execute and the overall solution process benefited greatly. But here, each operator application can take several minutes and computing the optimal step size can thus take dozens of minutes in each case, depending on how many steps are needed to solve (4.3.3). Since the incurred overhead of computing α is countermanded by the speedup in convergence, it is not advisable to use this solver when compared to the Richardson solver.

4.7.3 Solving with Newton Iteration

Using CG as the subproblem solver proves to be a very effective way to solve the inner linearized system in the light of the high number of wavelet coefficients in this example. The downside of this solver is a rather unsteady convergence history (Figure 4.22) and the expected approximation rate (Figure 4.25) is just barely recognizable. On the upside, the solutions were acquired very quickly in each case, the runtimes documented in Table 4.9 are among the lowest for this example.
The overall runtimes for the `MP244NB` wavelet are in every case more than an order higher than for the `DKU` wavelets. The relative (per wavelet coefficients) runtimes show that a factor of ~ 2 is attributed to the wavelet, but the absolute values on the abscissa in Figure 4.25 show that the vector contain at least ten times as many coefficients.
In the case the Gradient Iteration is used as the internal solver, the approximation rate is much more recognizable, but the execution times grow extensively. Especially the first few Newton steps take a very long time, although the accuracy in these cases is not lower than 0.1. In fact, for the first 11 of the 16 overall steps, over 80% of the runtime is needed. Ignoring this initial phase, the performance using the `MP244NB` wavelet is en par with the `DKU244NB` wavelet. Once a residual smaller than the initial tolerance 0.1 is achieved, the last 5 steps execute (on average) in accordance with the expected approximation rate. It thus seems a good strategy to precede the Newton methods by a few steps of another solver, e.g., the Richardson scheme, to compute a better starting vector.

4.7.4 Conclusions

In three dimensions, a sparse right hand side can contain a non-sparse subset of a two dimensional manifold. This entails a large number of non-zero wavelet coefficients. Although the solution algorithms behave in accordance with the predictions when it comes to approximation rates and therefore the number of wavelet coefficients within a working copy vector, the high number of wavelet coefficients entail a runtime that increases the solution process from mere seconds in the best cases of Section 4.6 to several minutes at least here. Since the complexity estimates of Theorem 4.8 guarantee that the runtime stays proportional to the number of computed wavelet coefficients, the only practical way to speed up these computations considerably is by parallelization techniques, which are not a subject of this work, see Section 6.2.

As a solution process for a single nonlinear PDE is now established, we turn to boundary value problems based upon said nonlinear PDEs.

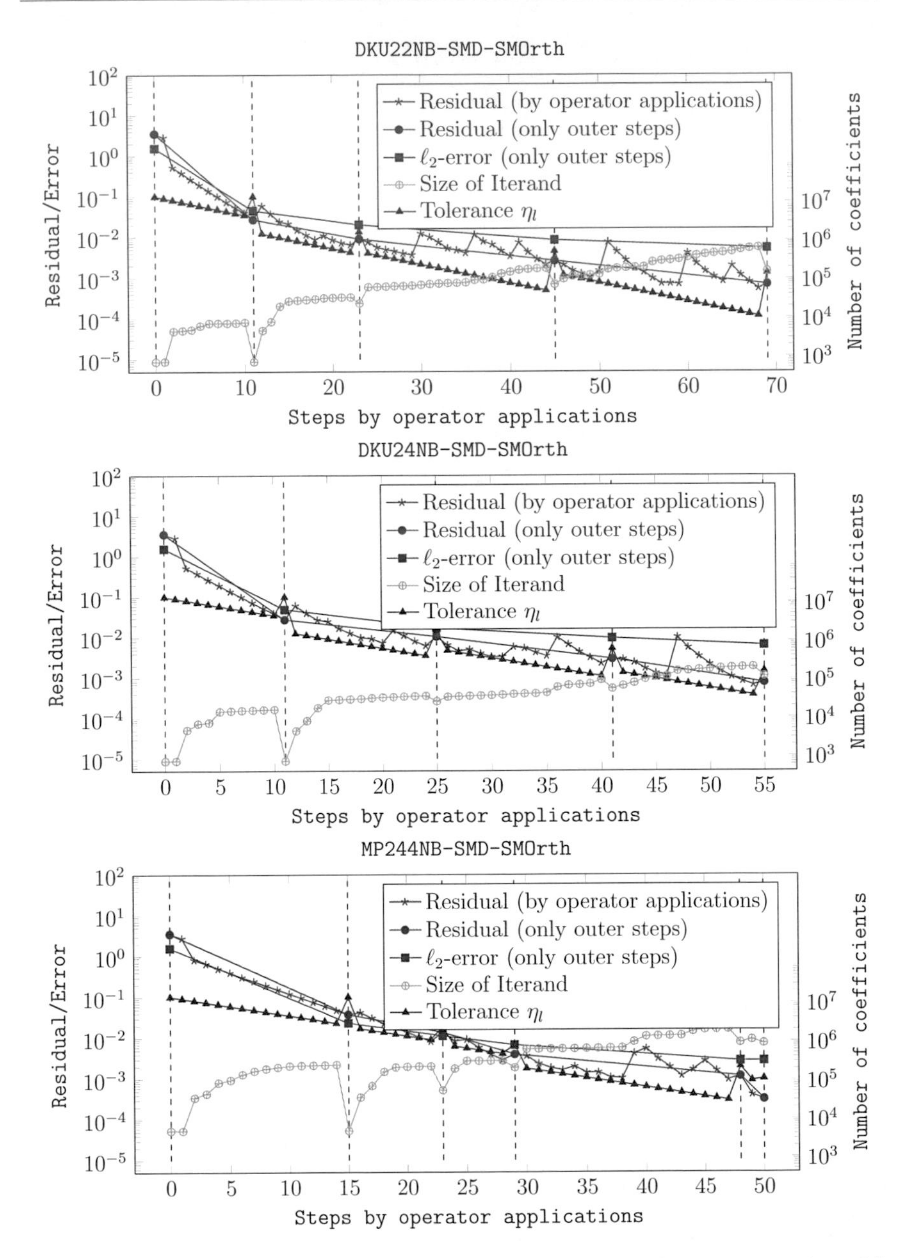

Figure 4.20: Detailed Richardson Iteration histories for all wavelets in conjunction with the $\mathbf{D}_{\{O,a\}}$ preconditioner.

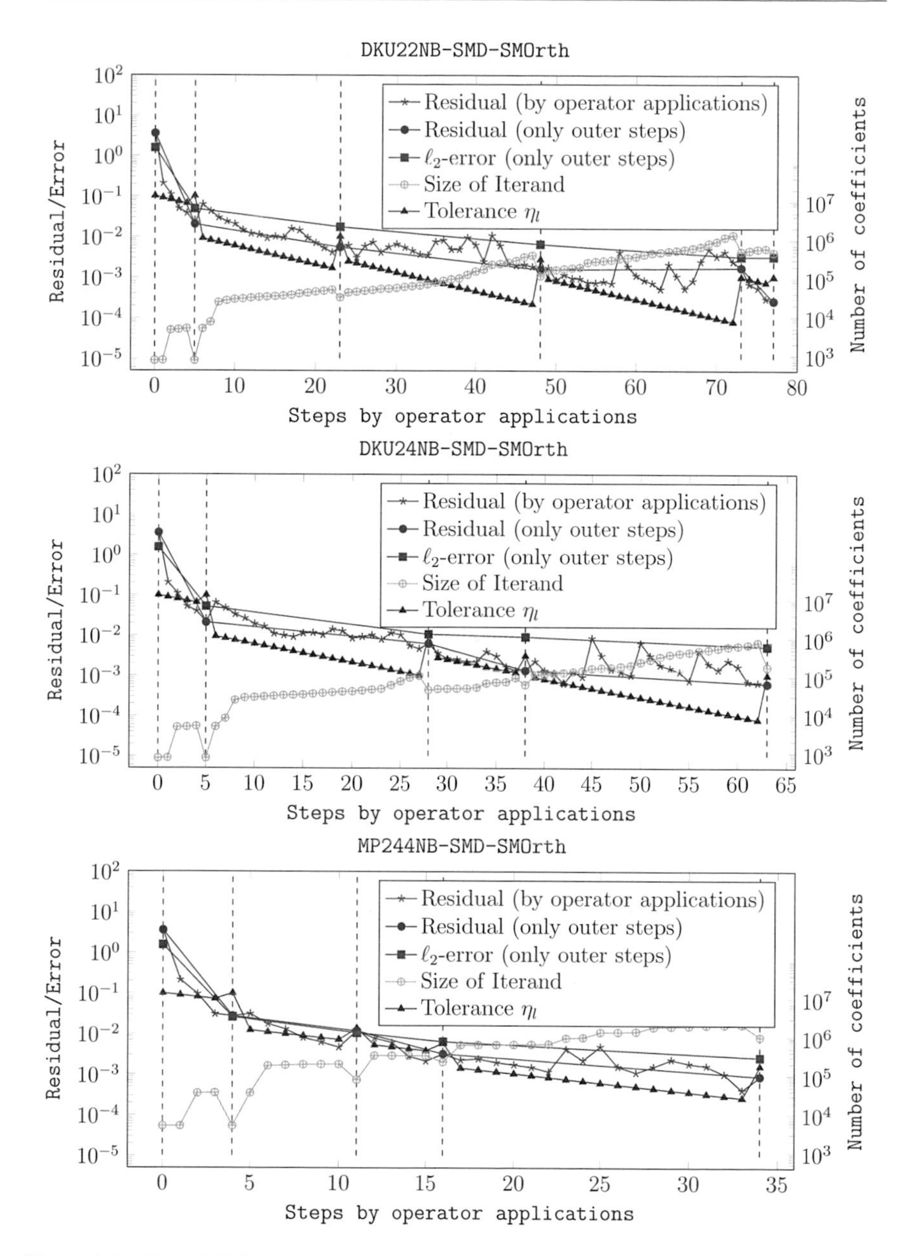

Figure 4.21: Detailed Gradient Iteration histories for all wavelets in conjunction with the $\mathbf{D}_{\{O,a\}}$ preconditioner.

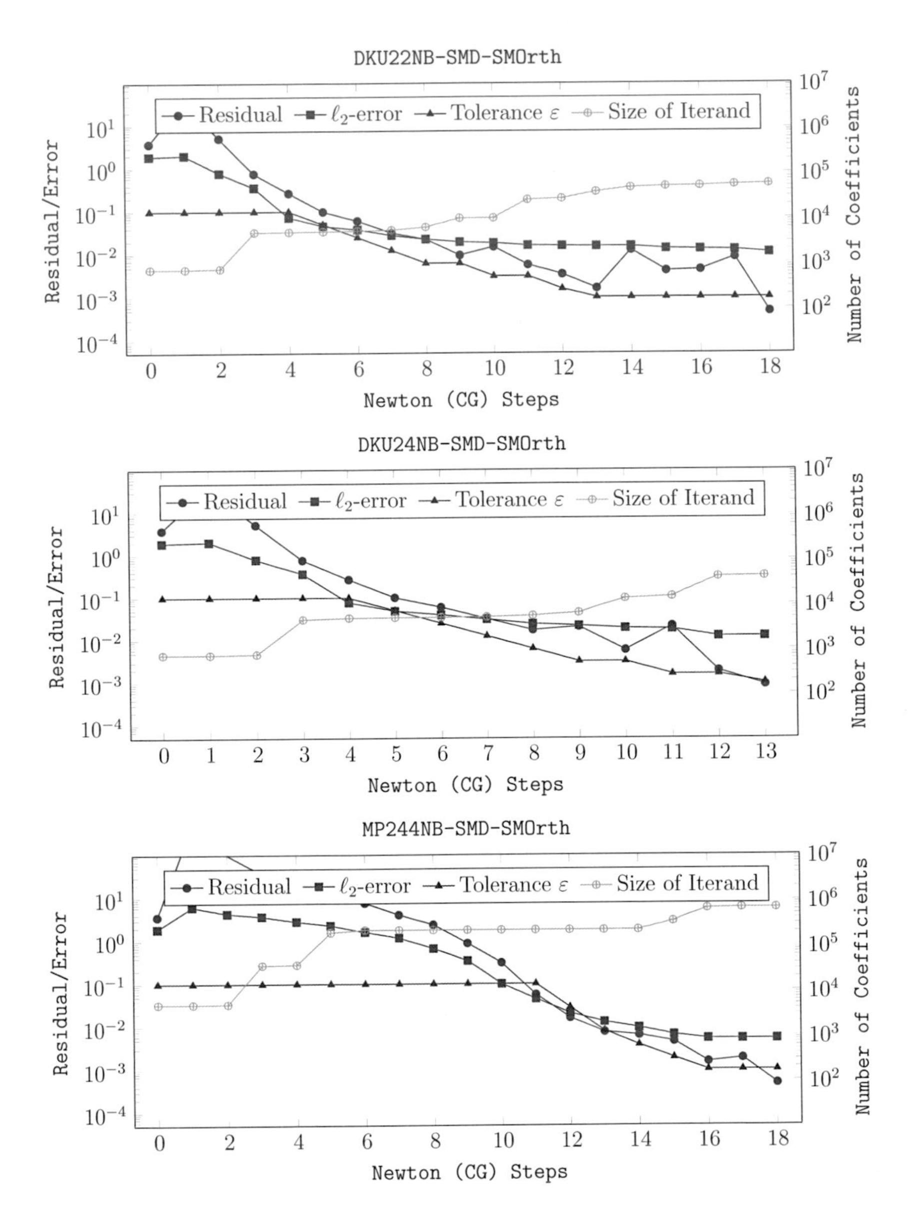

Figure 4.22: Newton Iteration histories using the CG solver for the linearized subproblem for all wavelets in conjunction with the $\mathbf{D}_{\{O,a\}}$ preconditioner.

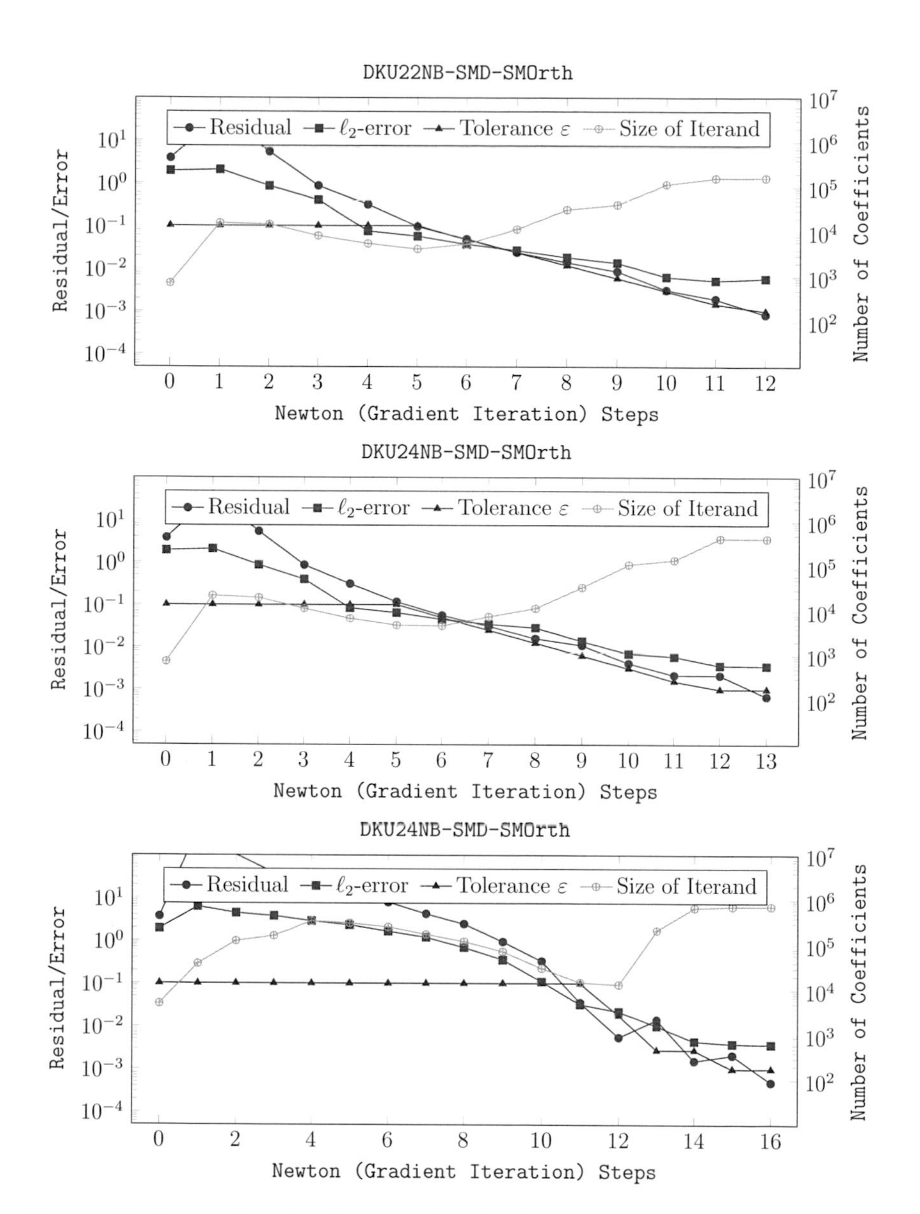

Figure 4.23: Newton Iteration histories using the Gradient solver for the linearized subproblem for all wavelets in conjunction with the $\mathbf{D}_{\{O,a\}}$ preconditioner.

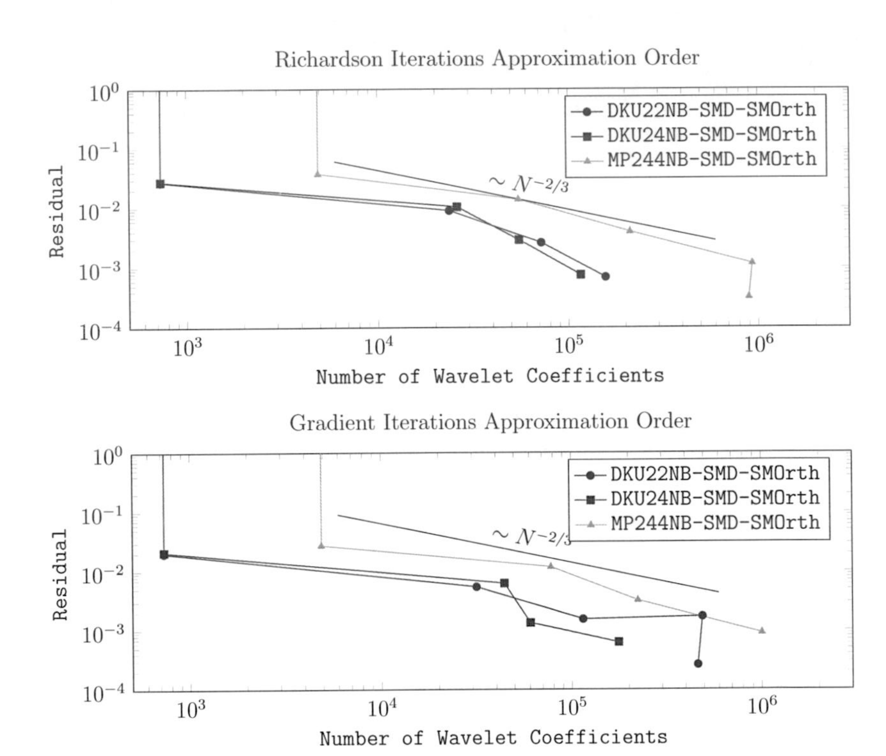

Figure 4.24: Approximation order for the different wavelets using preconditioner $\mathbf{D}_{\{O,a\}}$. On the abscissa, the number of wavelet coefficients after each coarsening is plotted against the residual for the coarsed vector on the ordinate axis.

Richardson and Gradient Iteration Runtime

Wavelet	DKU22NB		DKU24NB		MP244NB	
Prec.	Overall	Per Wav.	Overall	Per Wav.	Overall	Per Wav.
Richardson Runtime						
SMD-SMOrth	5.87×10^2	6.07×10^{-5}	3.03×10^2	9.04×10^{-5}	4.26×10^3	1.41×10^{-4}
Gradient Iteration Runtime						
SMD-SMOrth	1.38×10^4	7.69×10^{-5}	1.39×10^4	1.40×10^{-4}	3.91×10^4	1.89×10^{-4}

Table 4.8: Overall runtime and runtime per treated wavelet coefficient in seconds of the Richardson and Gradient Iteration solver for the different Wavelet configurations. The general comments of Table 4.4 also apply here.

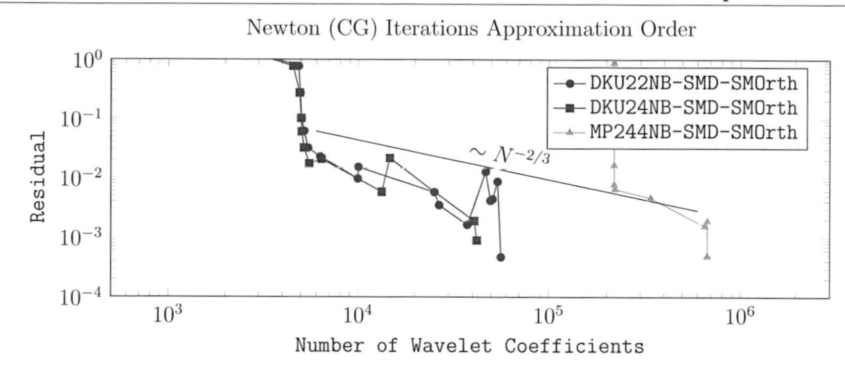

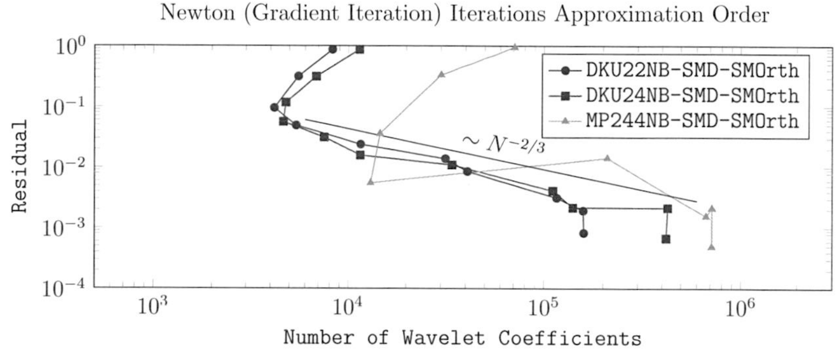

Figure 4.25: Approximation order for the different wavelets using preconditioner $\mathbf{D}_{\{O,a\}}$. On the abscissa, the number of wavelet coefficients after each coarsening is plotted against the residual for the coarsed vector on the ordinate axis.

Wavelet	DKU22NB		DKU24NB		MP244NB	
Prec.	Overall	Per Wav.	Overall	Per Wav.	Overall	Per Wav.
Newton (CG) Runtime						
SMD-SMOrth	1.34×10^{2}	4.36×10^{-5}	8.72×10^{1}	6.63×10^{-5}	4.97×10^{3}	1.67×10^{-4}
Newton (Gradient) Runtime						
SMD-SMOrth	1.12×10^{4}	6.22×10^{-5}	4.34×10^{4}	1.32×10^{-4}	2.49×10^{5}	1.73×10^{-4}

Table 4.9: Overall runtime and runtime per treated wavelet coefficient of the Newton solver for the different Wavelet configurations. The name of the inner (linear) subproblem solver is given in parentheses in the title. Again, using an adaptive solver for the linear subproblem entails an increase in runtime. Also, the larger support of the $\widetilde{d} = 4$ wavelets leads to an increase in complexity per treated wavelet index. The general comments of Table 4.4 also apply here.

5 Boundary Value Problems as Saddle Point Problems

This chapter gives a short introduction into **saddle point problems** and especially to formulate **elliptic boundary value problems** as saddle point problems.
The theoretical framework of this chapter is generally based upon [47,99], which itself build upon the theory put forth in [9] as the **Lagrange multiplier** method. Here, the space of test functions is chosen not to incorporate any Dirichlet boundary conditions. Instead, these are supposed to be attained by the Lagrangian multipliers only where needed. In the wavelet setting, this method allows for optimal preconditioning and adaptive methods offer automatic **stabilization** which voids the necessity of the **LBB** condition mentioned in Section 2.5.2. Adaptive wavelet methods for saddle point problems have been developed only recently, the main works being [32, 40, 42, 55]. The following excerpt is based upon these papers. Also, our main motivation to employ **Uzawa** algorithms is based upon our previous numerical results, see [122].

5.1 Saddle Point Problems

Consider Hilbert spaces $\mathcal{H}$ and $\mathcal{V}$ with their dual spaces $\mathcal{H}'$ and $\mathcal{V}'$ together with their respective dual forms $\langle\cdot,\cdot\rangle_{\mathcal{H}\times\mathcal{H}'}$, $\langle\cdot,\cdot\rangle_{\mathcal{V}\times\mathcal{V}'}$. We define the product Hilbert space $\mathcal{Z}$ and its dual $\mathcal{Z}'$ as

$$\mathcal{Z} := \mathcal{H} \times \mathcal{V}', \quad \mathcal{Z}' := \mathcal{H}' \times \mathcal{V}. \tag{5.1.1}$$

This definition of $\mathcal{Z}$ involving $\mathcal{V}'$ instead of $\mathcal{V}$ is more convenient for the specification considered later. The $\mathcal{Z}$-inner product will be given by

$$(\cdot,\cdot)_{\mathcal{Z}} := (\cdot,\cdot)_{\mathcal{H}} + (\cdot,\cdot)_{\mathcal{V}'}, \tag{5.1.2}$$

inducing the canonical norm on $\mathcal{Z}$ as

$$\left\| \begin{pmatrix} v \\ q \end{pmatrix} \right\|_{\mathcal{Z}}^2 := \|v\|_{\mathcal{H}}^2 + \|q\|_{\mathcal{V}'}^2, \qquad \text{for all } v \in \mathcal{H}, q \in \mathcal{V}'. \tag{5.1.3}$$

5.1.1 The Linear Case

Suppose $a(\cdot,\cdot) : \mathcal{H} \times \mathcal{H} \to \mathbb{R}$ is a continuous bilinear form with some constant $\alpha_2 > 0$ such that

$$|a(v,w)| \le \alpha_2 \|v\|_{\mathcal{H}} \|w\|_{\mathcal{H}}, \qquad \text{for all } v, w \in \mathcal{H}, \tag{5.1.4}$$

and likewise for $b(\cdot,\cdot) : \mathcal{H} \times \mathcal{V}' \to \mathbb{R}$ holds with some other constant $\beta_2 > 0$,

$$|b(v,q)| \le \beta_2 \|v\|_{\mathcal{H}} \|q\|_{\mathcal{V}'}, \qquad \text{for all } v \in \mathcal{H}, q \in \mathcal{V}'. \tag{5.1.5}$$

The problem to solve is as follows: Given $f \in \mathcal{H}'$ and $g \in \mathcal{V}$, find $y \in \mathcal{H}$ such that,

$$\left\{ \begin{array}{rrcll} \text{solve} & a(y,v) & = & \langle v, f\rangle, & \text{for all } v \in \mathcal{H}, \\ \text{under the constraint} & b(y,w) & = & \langle g, w\rangle, & \text{for all } w \in \mathcal{V}'. \end{array} \right. \tag{5.1.6}$$

In this form, the problem formulation is actually not tractable: If the first equation has a unique solution, it either satisfies the second equation or not. To get a well-defined problem, it has to be reposed as constrained minimization problem. Under these assumptions on the bilinear forms, it is well known (see [17, 19, 67]) that the solution $y \in \mathcal{H}$ of the first equation of (5.1.6) is the unique minimizer of the quadratic (and thus convex) functional

$$J(u) = \frac{1}{2}a(u,u) - \langle u, f \rangle . \tag{5.1.7}$$

The proper problem formulation is then

$$\left\{ \begin{array}{rcll} \text{minimize} \quad J(y) & = & \frac{1}{2}a(y,y) - \langle y, f \rangle , & \\ \text{for all } y \text{ satisfying} \quad b(y,w) & = & \langle g, w \rangle , & \text{for all } w \in \mathcal{V}'. \end{array} \right. \tag{5.1.8}$$

The above minimization problem can be solved by appending the linear constraints of the second equation in (5.1.6) by means of a **Lagrangian multiplier** $q \in \mathcal{V}'$ to the functional (5.1.7), by defining

$$K(u,q) := J(u) + b(u,q) - \langle g, q \rangle , \tag{5.1.9}$$

and solving the system of equations resulting from the necessary (and in this case also sufficient) minimization conditions:

$$\partial K(v,p) = 0 \quad \Longleftrightarrow \quad \partial_u K(v,p) = 0 \wedge \partial_q K(v,p) = 0.$$

Explicitly calculating these derivatives leads to the following reformulation of the above problem: Given $(f,g) \in \mathcal{Z}'$, find $(y,p) \in \mathcal{Z}$ such that

$$\left\langle \begin{pmatrix} v \\ p \end{pmatrix}, L_A \begin{pmatrix} y \\ p \end{pmatrix} \right\rangle_{\mathcal{Z}\times\mathcal{Z}'} := \left\{ \begin{array}{lcl} a(y,v) + b(v,p) & = & \langle v, f \rangle_{\mathcal{H}\times\mathcal{H}'} \\ b(y,p) & = & \langle g, p \rangle_{\mathcal{V}\times\mathcal{V}'} \end{array} \right. , \quad \text{for all } v \in \mathcal{H} \text{ and } p \in \mathcal{V}', \tag{5.1.10}$$

which is written shortly using the operator definitions (1.4.12) and (3.6.2) as

$$L_A \begin{pmatrix} y \\ p \end{pmatrix} := \begin{pmatrix} A & B' \\ B & 0 \end{pmatrix} \begin{pmatrix} y \\ p \end{pmatrix} = \begin{pmatrix} f \\ g \end{pmatrix}. \tag{5.1.11}$$

Loosely speaking, the extra degree of freedom that $p \in \mathcal{V}'$ provides assures to find a solution $y \in \mathcal{H}$ with $B\,y = g$ such that $A\,y = f - B'\,p$, i.e., the right hand side has been altered to accommodate for the constrained solution space.

Remark 5.1 *When $a(\cdot,\cdot)$ is symmetric positive definite, the solution (y,p) of (5.1.10) solves the extremal problem*

$$\inf_{u\in\mathcal{H}} \sup_{q\in\mathcal{V}'} \left(\frac{1}{2}a(u,u) - \langle f, u \rangle_{\mathcal{H}'\times\mathcal{H}} + b(u,q) - \langle u, q \rangle_{\mathcal{V}\times\mathcal{V}'} \right).$$

and satisfies

$$K(y,q) \leq K(y,p) \leq K(u,p), \qquad \textit{for all } u \in \mathcal{H}, q \in \mathcal{V}',$$

which explains the designation **saddle point problem.**

The operator L_A therefore maps $\mathcal{Z}$ into $\mathcal{Z}'$, and this even bijectively under the following conditions: Important for the theoretical assertions is the relation of the operator A to the **kernel** of B, i.e.,

$$\ker B := \{v \subset \mathcal{H} \,|\, b(v,q) = 0 \text{ for all } q \in \mathcal{V}'\} \subset \mathcal{H}, \tag{5.1.12}$$

Another prerequisite is given by the **inf-sup condition**, i.e., let there be some constant $\beta_1 > 0$ such that holds

$$\inf_{q\in\mathcal{V}'} \sup_{v\in\mathcal{H}} \frac{b(v,q)}{\|v\|_{\mathcal{H}}\|q\|_{\mathcal{V}'}} \geq \beta_1. \tag{5.1.13}$$

The inf-sup condition (5.1.13) means that the range of the operator B is closed in $\mathcal{V}'$. The following theorem from [21, 100] combines these two and some other properties to show the existence of a unique solution to (5.1.10).

Theorem 5.2 *Let the linear operator A be invertible on* $\ker B$, *i.e., for some constant $\alpha_1 > 0$ holds,*

$$\begin{aligned} \inf_{v\in\ker B} \sup_{w\in\ker B} \frac{\langle Av, w\rangle_{\mathcal{H}'\times\mathcal{H}}}{\|v\|_{\mathcal{H}}\|w\|_{\mathcal{H}}} &\geq \alpha_1, \\ \inf_{v\in\ker B} \sup_{w\in\ker B} \frac{\langle A'v, w\rangle_{\mathcal{H}'\times\mathcal{H}}}{\|v\|_{\mathcal{H}}\|w\|_{\mathcal{H}}} &\geq \alpha_1, \end{aligned} \tag{5.1.14}$$

and let the **inf-sup condition** *(5.1.13) hold for B.*
Then there exists a unique solution $(y,p)^T \in \mathcal{Z}$ to (5.1.10) for all $(f,g)^T \in \mathcal{Z}'$. That is, the operator L_A of (5.1.11) is an isomorphism, and one has the norm equivalence

$$c_{L_A} \left\| \begin{pmatrix} v \\ q \end{pmatrix} \right\|_{\mathcal{Z}} \leq \left\| L_A \begin{pmatrix} v \\ q \end{pmatrix} \right\|_{\mathcal{Z}'} \leq C_{L_A} \left\| \begin{pmatrix} v \\ q \end{pmatrix} \right\|_{\mathcal{Z}}, \quad \textit{for all } (v,q)^T \in \mathcal{Z}, \tag{5.1.15}$$

where the constants $c_{L_A}, C_{L_A} > 0$ are given as

$$\begin{aligned} c_{L_A} &:= \left(\frac{1}{\alpha_1\beta_1}\left(1+\frac{\alpha_2}{\alpha_1}\right) + \max\left\{ \frac{2}{\alpha_1^2}, \frac{1}{\beta_1^2}\left(1+\frac{\alpha_2}{\alpha_1}\right)^2 + \left(\frac{\alpha_2}{\beta_1^2}\left(1+\frac{\alpha_2}{\alpha_1}\right)\right)^2 \right\} \right)^{-1/2}, \\ C_{L_A} &:= \sqrt{2(\alpha_2^2+\beta_2^2)}. \end{aligned}$$

Remark 5.3 *The first prerequisite (5.1.14) is trivially fulfilled if A is invertible on all $\mathcal{H} \supset$* $\ker B$. *This is assured by (5.1.4) with* **ellipticity** *(1.4.11), which gives the property (1.4.14).*
The second assumption, the inf-sup condition (5.1.13), is also trivially satisfied if B is **surjective**, *i.e.,* $\operatorname{range} B = \mathcal{V}'$ *or equivalently* $\ker B' = \{0\}$.

The reason the invertibility of the operator A is only needed on $\ker B$ stems from the homogeneous problem formulation $g \equiv 0$. Then, it is obvious that any solution y must fulfill $y \in \ker B$. The inhomogeneous problem $g \neq 0$ can be reduced to the homogeneous case by the **Brezzi splitting trick**: Considering $y = y_g + y_0$ with $B\,y_g = g$ and $y_0 \in \ker B$, the inhomogeneous equations simplify to a homogeneous problem in the unknown y_0: $A\,y_0 = f - A\,y_g$. This trick therefore hinges on the **linearity** of the operator A.

The Schur Complement

In the following, we will assume the linear operator A to be invertible, and that also the prerequisites of Theorem 5.5 holds. Under these conditions, the solution to (5.1.26) can be expressed analytically using only the operators A and B and the right hand side data. The **Schur complement** is the operator

$$S := B\,A^{-1}\,B', \quad S : \mathcal{V}' \to \mathcal{V}, \tag{5.1.16}$$

which is inherits the monotonicity properties of A^{-1}, if B is surjective because of

$$\begin{aligned}\langle \widetilde{u}_1 - \widetilde{u}_2, S(\widetilde{u}_1) - S(\widetilde{u}_2)\rangle &= \left\langle \widetilde{u}_1 - \widetilde{u}_2, B\,A^{-1}(B'\widetilde{u}_1) - B(A^{-1}\,B'\widetilde{u}_2)\right\rangle, \qquad \text{for all } \widetilde{u}_1, \widetilde{u}_2 \in \mathcal{V}',\\ &= \left\langle B'\widetilde{u}_1 - B'\widetilde{u}_2, A^{-1}(B'\widetilde{u}_1) - A^{-1}(B'\widetilde{u}_2)\right\rangle,\\ &=: \left\langle \widetilde{v}_1 - \widetilde{v}_2, A^{-1}(\widetilde{v}_1) - A^{-1}(\widetilde{v}_2)\right\rangle.\end{aligned}$$

In particular, when A is an **isomorphism**, A^{-1} then inherits the coercivity property (1.4.11) from A, and so does S. We can then use (5.1.16) to rewrite (5.1.26) by eliminating y as

$$S\,p = B\,A^{-1}\,f - g. \tag{5.1.17}$$

The above remarks show that S is invertible and can be brought to the other side in this equation. Substituting p into the first equation of (5.1.26) leads to an explicit representation of y as

$$y = A^{-1}(f - B'\,S^{-1}(B\,A^{-1}f - g)). \tag{5.1.18}$$

The importance of the Schur complement goes beyond theoretical transformations, it also determines the efficiency of the **Uzawa** algorithm of Section 5.3.
If the operator A induces a norm on the space $\mathcal{H}$, then the Schur complement can be used to define an **energy norm** of the operator L_A on the space $\mathcal{Z}$ as

$$\left\|\begin{pmatrix} v\\ q\end{pmatrix}\right\|_{L_A}^2 := \|v\|_A^2 + \|q\|_S^2 = \langle v, A\,v\rangle_{\mathcal{H}\times\mathcal{H}'} + \langle q, S\,q\rangle_{\mathcal{V}\times\mathcal{V}'}\,. \tag{5.1.19}$$

This follows quickly from under the above conditions, i.e., if B is bounded and the inf-sup condition (5.1.13) holds, then

$$\|q\|_S \sim \|q\|_{\mathcal{V}'},$$

see [42] for a proof. Since already holds $\|\cdot\|_A \sim \|\cdot\|_{\mathcal{H}}$ by (1.4.13), the equivalence (5.1.19) follows.
It should be said that the Schur complement is never computed explicitly, as the operator A^{-1} is never set up exactly.

5.1.2 The Semilinear Case

This excerpt is based upon [67, 149]. In the following, we add a nonlinear term $\xi(\cdot)$ to the symmetric and coercive bilinear form $a(\cdot,\cdot)$. This function ξ interpreted as a function

$\xi : \mathbb{R} \to \mathbb{R}$ shall be continuous and nondecreasing with $\xi(0) = 0$. The setting is therefore in accordance with Section 1.5. Also, we assume to possess knowledge of a non-negative, **convex** and lower semicontinuous functional $j : \mathcal{H} \to \mathbb{R}$, i.e.,

$$\liminf_{v \to v_0} j(v) \geq j(v_0), \qquad \text{for all } v, v_0 \in \mathcal{H}, \tag{5.1.20}$$

which satisfies

$$Dj(v; w) = \langle w, \xi(v) \rangle, \qquad \text{for all } v, w \in \mathcal{H}.$$

Simply stated: $Dj \equiv \xi$. Operators with this property are called **potential operators** and the potential is given by

$$j(u) := \int_0^1 \langle \xi(t\,u), u \rangle \, dt. \tag{5.1.21}$$

The function $\xi(\cdot)$ is convex on $\mathcal{H}$, if and only if the functional $j(\cdot) : \mathcal{H} \to \mathbb{R}$ is monotone, see [149]. It then follows, because of the symmetry of $a(\cdot, \cdot)$, that the task to find $y \in \mathcal{H}$ for given $f \in \mathcal{H}'$ for which holds

$$a(y, v) + \langle v, \xi(y) \rangle = \langle v, f \rangle, \qquad \text{for all } v \in \mathcal{H}, \tag{5.1.22}$$

are the **Euler equations** of the functional

$$J(u) = \frac{1}{2} a(u, u) + j(u) - \langle u, f \rangle, \tag{5.1.23}$$

which means that for the solution $u \in \mathcal{H}$ of (5.1.22) holds $J(u) \leq J(v)$, for all $v \in \mathcal{H}$. We can then again consider the constrained minimization problem: Given $f \in \mathcal{H}'$ and $g \in \mathcal{V}$, find $y \in \mathcal{H}$ such that,

$$\left\{ \begin{array}{rrcl} \text{minimize} & J(y) & = & \frac{1}{2} a(y, y) + j(y) - \langle y, f \rangle, \\ \text{for all } y \text{ satisfying} & b(y, w) & = & \langle g, w \rangle, \qquad \text{for all } w \in \mathcal{V}'. \end{array} \right. \tag{5.1.24}$$

The optimality conditions with the Lagrangian multiplier $p \in \mathcal{V}'$ then read

$$\left\{ \begin{array}{lcll} a(y, v) + \langle v, \xi(y) \rangle + b(v, p) & = & \langle v, f \rangle, & \text{for all } v \in \mathcal{H}, \\ b(y, w) & = & \langle g, w \rangle, & \text{for all } w \in \mathcal{V}'. \end{array} \right. \tag{5.1.25}$$

Denoting the operator of the form $\xi(\cdot)$ by $G(\cdot)$ as in (1.5.2) and setting $F := A + G$, then the above equations can be written as

$$L_F \begin{pmatrix} y \\ p \end{pmatrix} := \begin{pmatrix} F & B' \\ B & 0 \end{pmatrix} \begin{pmatrix} y \\ p \end{pmatrix} = \begin{pmatrix} f \\ g \end{pmatrix}. \tag{5.1.26}$$

Remark 5.4 *If the operator F is not symmetric, the problem (5.1.6) is not given as the optimality conditions of the saddle point problem formulation in Remark 5.1. Therefore, the term saddle point problem is not exactly fitting here for problem (5.1.24), but is often used regardless.*

Theorem 5.5 *Let the semilinear operator F be invertible on $M := \{v \in \mathcal{H} \,|\, B\,v = g\}$ and let otherwise hold the hold the assumptions of Theorem 5.2 on M and Theorem 1.57 for the nonlinear part. Then has the equation (5.1.26) a unique solution $(y,p) \in \mathcal{H} \times \mathcal{V}'$ for any $(f,g) \in \mathcal{H}' \times \mathcal{V}$. It also fulfills the a priori estimates*

$$\|y\|_{\mathcal{H}} \le \frac{1}{\alpha_1}\left(1 + \frac{\beta_2}{\beta_1}\right)\|f\|_{\mathcal{H}'} + \frac{\alpha_2'\,\beta_2}{\alpha_1\,\beta_1^2}\|g\|_{\mathcal{V}}, \tag{5.1.27}$$

$$\|p\|_{\mathcal{H}} \le \frac{1}{\beta_1}\|f\|_{\mathcal{H}'} + \frac{\alpha_2'}{\beta_1^2}\|g\|_{\mathcal{V}}, \tag{5.1.28}$$

with $\alpha_2' := \alpha_2 + C_G(\|y\|_{\mathcal{H}})$ as in (1.5.10).

Proof:

- Existence:
  By Theorem 25.C of [149], the functional $J : M \subseteq \mathcal{H} \to \mathbb{R}$ has a minimum in M, if M is a nonempty bounded closed convex set in $\mathcal{H}$ and $J(\cdot)$ is weakly lower semicontinuous on M.
  The set $M \subset \mathcal{H}$ is nonempty if B is surjective. Since B is linear, M is obviously convex and closed. It holds $\|y\| \lesssim \|g\|$, for any $y \in M$, which will also be used for the estimates. The weakly lower semicontinuity is assumed for the nonlinear part $j(\cdot)$ and follows from ordinary continuity for the other terms. Thus, $J(\cdot)$ has a minimizer on M.
  Any local minimum of $J(\cdot)$ in M now satisfies equations (5.1.25): Let $y \in M$ be the minimum, then, because the set M is convex, it holds for and any $v \in M$ and $t \in [0,1]$,

  $$J(y) \le J(y + t(v-y)) \quad \Longrightarrow \quad 0 \le \frac{J(y + t(v-y)) - J(y)}{t}.$$

  Expanding the terms and taking the limit $t \to 0$ gives

  $$0 \le a(y, v-y) + \langle \xi(y), v-y\rangle - \langle f, v-y\rangle\,,$$

  which means, because v is arbitrary and $v - y \in \ker B$,

  $$0 \le a(y,w) + \langle \xi(y), w\rangle - \langle f, w\rangle\,, \qquad \text{for all } w \in \ker B.$$

  From this follows equality and thus $F(y) - f \in (\ker B)^0 \subset \mathcal{H}'$. Since B is injective, B' is surjective. Thus, there exists a $p \in \mathcal{V}'$ for that holds

  $$-B'\,p = F(y) - f \quad \Longrightarrow \quad F(y) + B'\,p - f = 0 \in \mathcal{H}',$$

  which shows (5.1.25).

- Uniqueness:
  Let $(y_1,p_1),(y_2,p_2)$ be two different solutions, i.e., at least $y_1 \ne y_2$ or $p_1 \ne p_2$. Subtracting both sets of equations then yields

  $$\left\{\begin{array}{lcll} a(y_1 - y_2, v) + \langle v, \xi(y_1) - \xi(y_2)\rangle + b(v, p_1 - p_2) & = & \langle v, 0\rangle\,, & \text{for all } v \in \mathcal{H}, \\ b(y_1 - y_2, w) & = & \langle 0, w\rangle\,, & \text{for all } w \in \mathcal{V}'. \end{array}\right.$$

From the second equation, we can infer $y_1 - y_2 \in \ker B$. For the first equation thus follows for $v = y_1 - y_2$,

$$a(y_1 - y_2, y_1 - y_2) + \langle y_1 - y_2, \xi(y_1) - \xi(y_2)\rangle + 0 = 0,$$

and together with (1.4.11), which here is used simply as $a(y_1 - y_2, y_1 - y_2) > 0$,

$$\langle y_1 - y_2, \xi(y_1) - \xi(y_2)\rangle < 0,$$

which is a contradiction to the monotonicity assumption (1.5.9) of $\xi(\cdot)$. Thus, $y_1 = y_2$. Reevaluating the first equation above for $y_1 = y_2$ then leads to

$$b(v, p_1 - p_2) = 0, \qquad \text{for all } v \in \mathcal{H},$$

which can only hold for $p_1 = p_2$.

- Estimates:
  The calculation of the upper bounds are mainly just applying the assumptions. The only deviation is the use of estimate

$$\|B\,y\|_{\mathcal{V}} = \|g\|_{\mathcal{V}} \quad \Longrightarrow \quad \|y\|_{\mathcal{H}} \leq \frac{1}{\beta_1}\|g\|_{\mathcal{V}},$$

  which requires the closed range theorem, see [21,22]. Then easily follows

$$\beta_1\|p\| \leq \|B'p\| = \|f - F(y)\| \leq \|f\| + \alpha_2'\|y\| \leq \|f\| + \frac{\alpha_2'}{\beta_1}\|g\|,$$

  which yields (5.1.28). Similarly for y,

$$\alpha_1\|y\|^2 \leq a(y,y) \leq |\,\langle f, y\rangle - \langle B'p, y\rangle\,| \leq \|y\|\|f\| + \beta_2\|p\|\|y\|,$$

  from which, together with the estimate for $\|p\|$, follows (5.1.27).

■

Of course the estimate (5.1.27) is only properly usable if the value $C_G(\|y\|)$ within α_2' can be estimated a priori, too. As the function $C(\cdot)$ is only generally described as being positive and nondecreasing, this would only be possible under more specific assumptions on the nonlinearity.

Remark 5.6 *In the linear case, better a priori estimates, given by the constant c_{L_A} in Theorem 5.2, are achieved again by the Brezzi splitting trick: By acquiring estimates for $y = y_g + y_0$ independently for the components $y_0 \in \ker B$ and $y_g \in (\ker B)^\perp$, several instances of the bilinear form $b(\cdot,\cdot)$ can be made to vanish by restricting the first equation (5.1.6) on* $\ker B$. *This approach does not work for the here considered nonlinear operators.*

By principally the same proof as for Theorem 5.2, the operator L_F is stable in the sense of (1.5.8) with the constant,

$$C_{L_F}(\{(y_1,p_1)^T, (y_2,p_2)^T\}) := \sqrt{2\,(C_F^2(\max\{\|y_1\|_{\mathcal{H}}, \|y_2\|_{\mathcal{H}}\}) + \beta_2^2)}. \tag{5.1.29}$$

Another simple way to derive the upper estimates for the operator L_F is to note the structure

$$L_F = L_A + \begin{pmatrix} G & 0 \\ 0 & 0 \end{pmatrix}, \tag{5.1.30}$$

and use the triangle inequality. In particular, it follows for the Fréchet derivative by Remark 1.54

$$D\,L_F = L_A + \begin{pmatrix} D\,G(z) & 0 \\ 0 & 0 \end{pmatrix}, \tag{5.1.31}$$

from which immediately follows for the upper constant of (1.5.7) $C_{z,L_F} := C_{L_A} + C_{z,G}$ with the constants from Theorem 5.2 and (1.5.7) for the nonlinear operator. By the same reasoning as used here and in the proof of Theorem 1.57 follows $c_{z,L_A} := c_{L_A}$.

The Reduced Equation

In the **nonlinear** case, the reduced equation (5.1.17) reads

$$B\,F^{-1}(f - B'\,p) = g, \tag{5.1.32}$$

which cannot be expressed using the Schur complement (5.1.16) because of the nonlinearity of F^{-1}. The Uzawa algorithm presented in Section 5.3 is a **Richardson** iteration on the equation

$$\widetilde{S}(B,F,f)(p) = g, \quad \text{with} \quad \widetilde{S}(B,F,f)(p) := B\,F^{-1}(f - B'\,p). \tag{5.1.33}$$

The nonlinearity of this operator does not only come from the operator F^{-1}, but also from the translation $f - B'\,p$. This could considerably impair the convergence properties in actual applications.

We will now examine how PDE Based Boundary Value Problems fit in the abstract saddle point problem.

5.2 PDE Based Boundary Value Problems

In the following, let $\Omega \subseteq \square \subset \mathbb{R}^n$ be a domain bounded by the cube $\square = \square^n = (0,1)^n$. We assume Ω has a Lipschitz continuous boundary $\partial\Omega \in C^{0,1}$ and $\Gamma \subseteq \partial\Omega$ is a subset of $\partial\Omega$ with strictly positive surface measure. For example, if $\Omega = \square^n := (0,1)^n$, then $\Gamma \subseteq \partial\Omega$ may be an edge or a face of this cube. We then consider the following general **boundary value problem**: For given functions $f_\Omega \in (H^1(\Omega))'$ and $g \in H^{1/2}(\Gamma)$, search $y \in H^1(\square)$ satisfying

$$\begin{aligned} -\Delta y + \xi(y) &= f_\Omega, && \text{in } \Omega, \\ y &= g, && \text{on } \Gamma, \\ \nabla y \cdot \mathbf{n} &= 0, && \text{on } \partial\Omega \setminus \Gamma, \end{aligned} \tag{5.2.1}$$

where $\mathbf{n} = \mathbf{n}(x)$ is the outward normal at any point $x \in \Gamma$. Here, the operator $\xi(\cdot)$ potentially nonlinear, e.g., $\xi(y) = y^3$, but could also be linear, e.g., $\xi(y) = y$. In

order to apply the theory from Section 5.1, we transform the above problem into a weak formulation, expressing the equations by **(bi-)linear forms**. In the view of Section 5, we choose specifically the spaces $\mathcal{H} = H^1(\Omega)$ and $\mathcal{V} = H^{1/2}(\Gamma)$ as Sobolev spaces on the bounded domains bounded in $\mathbb{R}^n$. We can formulate the partial differential equation (5.2.1) by the theory of Section 1.4.2 as follows:
Given $f_\Omega \in (H^1(\Omega))'$ and $g \in H^{1/2}(\Gamma)$, find the solution $y \in H^1(\Omega)$, which solves

$$\begin{cases} \text{solve} \;\; a_\Omega(y,v) &= \langle v, f_\Omega \rangle_\Omega, \qquad \text{for all } v \in H^1_{0,\Gamma}(\Omega), \\ \text{satisfying} \qquad y|_\Gamma &= g, \end{cases} \tag{5.2.2}$$

where the space of the test functions is defined as

$$H^1_{0,\Gamma}(\Omega) := \left\{ v \in H^1(\Omega) \,|\, v|_\Gamma = 0 \right\}. \tag{5.2.3}$$

Here the **Neumann boundary conditions** on $\Gamma \subset \partial\Omega$ were incorporated into the weak formulation and need not be stated explicitly any more, i.e., they are natural boundary conditions, see Section 1.4.2 for details. The **Dirichlet boundary condition** for Γ cannot be integrated into the solution space unless $g \equiv 0$. The form $a_\Omega(\cdot,\cdot)$ is given by Section 1.4.2 as

$$a_\Omega(v,w) := \int_\Omega (\nabla v \cdot \nabla w + \xi(v)w)\, d\mu. \tag{5.2.4}$$

The (possibly) inhomogeneous boundary conditions on Γ are missing from the (bi-)linear form and cannot generally incorporated into the test space, i.e., they are essential boundary conditions. The next section explains how we deal with these.

5.2.1 The Fictitious Domain–Lagrange Multiplier Approach

Including the essential boundary conditions in our weak formulation, a standard approach was introduced in [9] and is known as the **Lagrange multiplier method**. We will describe it here in the context of the **fictitious domain** scheme.
The motivation for the fictitious domain setup is the idea of solving a PDE on a complex domain $\Omega \subset \square$ by embedding it into a simple domain $\square$ and solving the problem on the larger domain. Under the right constraints, the solution of computed solution restricted to the domain Ω is then the sought solution. In this setting, we call $\square \subset \mathbb{R}^n$ the **fictitious domain**. This approach thus enlarges the range of problems from simple domains $\square$ to complex domains Ω.

Remark 5.7 *The fictitious domain procedure bears more difficulties when $\Gamma \neq \partial\Omega$. Especially when enforcing mixed Neumann and Dirichlet boundary conditions on different parts of the boundary $\partial\Omega$ this approach can cause problem because of decreased regularity at the interfaces, see [85]. Therefore, for the fictitious domain assertions $\Omega \subsetneq \square$, assume $\Gamma = \partial\Omega$. Regardless, the* **Lagrange Multiplier Approach** *is our preferred approach to deal with the inhomogeneous boundary conditions of (5.2.1).*

By **extending** the problem from the domain Ω onto the fictitious domain $\square$, we first have to discuss under which conditions the problem stays well-defined. Let the right hand side $f_\Omega \in (H^1(\Omega))'$ be expanded onto the cube $\square$ such that

$$f := f_\square \in (H^1(\square))', \quad f_\Omega = f_\square|_\Omega. \tag{5.2.5}$$

Moreover, extend the form $a_\Omega(\cdot,\cdot)$ to $\Box$ by defining

$$a(v,w) := a_\Box(v,w) := \int_\Box (\nabla v \cdot \nabla w + \xi(v)w)\, d\mu. \tag{5.2.6}$$

Depending on the (constant) coefficients within (5.2.6), one has to make sure the (bi-)linear form remains uniformly positive definite and/or symmetric, whatever characteristics the original (bi-)linear form possessed.
To generate a solution that satisfies the original (Dirichlet) boundary conditions, these have to be enforced as the essential boundary conditions using the **trace operators** from Section 1.2.2. The problem setup could require different boundary conditions on parts of the boundary $\partial\Omega$ which would require several different trace operators, see [86,87] for an example problem.
To express these essential Dirichlet boundary conditions, we employ the **trace operator** of (1.2.25), i.e.,

$$\gamma_0 v = v|_\Gamma.$$

The trace is well-defined for any $v \in H^1(\Omega)$ since Γ is Lipschitzian as a subset of $\partial\Omega$ and thus holds $\gamma_0 v \in H^{1/2}(\Gamma)$, cf. Theorem 1.21. We define the **bilinear form** $b(\cdot,\cdot)$ by setting

$$b(v,q) := \langle q, \gamma_0 v\rangle_{(H^{1/2}(\Gamma))' \times H^{1/2}(\Gamma)} = \int_\Gamma v|_\Gamma\, q\, ds, \qquad \text{for } v \in H^1(\Omega), q \in (H^{1/2}(\Gamma))', \tag{5.2.7}$$

which is well-defined because of the above remarks. The weak reformulation of (5.2.1) thus takes the form (5.1.8) (or (5.1.24)): Given $f \in (H^1(\Box))'$ and $g \in H^{1/2}(\Gamma)$, find $y \in H^1(\Box)$ such that,

$$\left\{ \begin{array}{rrcll} \text{minimize} & J(y) & = & \frac{1}{2} a_\Box(y,y) - \langle y, f\rangle, & \\ \text{for all } y \text{ satisfying} & b(y,w) & = & \langle g, w\rangle, & \text{for all } w \in (H^{1/2}(\Gamma))', \end{array} \right. \tag{5.2.8}$$

which then leads to the optimality conditions

$$\left\{ \begin{array}{lcll} a_\Box(y,v) + b(v,p) & = & \langle v, f\rangle, & \text{for all } v \in H^1(\Box), \\ b(y,w) & = & \langle g, w\rangle, & \text{for all } w \in (H^{1/2}(\Gamma))', \end{array} \right. \tag{5.2.9}$$

with $p \in (H^{1/2}(\Gamma))'$ In this formulation, the essential boundary conditions are not enforced in $H^1(\Box)$, but appended by the **Lagrangian multiplier** $p \in (H^{1/2}(\Omega))'$.

Remark 5.8 *In the view of Remark 5.1, the* **saddle point problem** *is given by: For $f \in (H^1(\Omega))'$ and $g \in H^{1/2}(\Gamma)$, find the solution of*

$$\inf_{v \in H^1(\Omega)} \sup_{q \in (H^{1/2}(\Gamma))'} \frac{1}{2} a_\Box(v,v) - \langle f, v\rangle_{(H^1(\Omega))' \times H^1(\Omega)} + b(v,q) - \langle g, q\rangle_{H^{1/2}(\Gamma) \times (H^{1/2}(\Gamma))'}. \tag{5.2.10}$$

However, a consequence of the nature of the saddle point problem, is that the operator L_A is **indefinite**. *This means that to actually solve (5.1.11), we have to use different iterative solvers than for system with positive definite operators. The most well-known algorithms for such indefinite symmetric systems are* **Uzawa**-*type algorithms, see Section 5.3 for the implementation details.*

By design, this technique allows for a decoupling of the differential operator from the boundary constraints. Consequentially, changing boundary conditions or changing boundaries can be treated by updating the right hand side g or by adapting the trace operator γ_0 to a new domain. Since these actions only involve a lower dimensional manifold, this can be done relatively easy compared to the cost a change of the domain Ω would induce. Specifically, changes to the domain Ω excluding Γ have no effect on the setup, as long as the domain is still bounded by the same fictitious domain $\square$. The topology of the fictitious domain is obviously chosen to be as simple as possible to allow for an easy setup and evaluation of the (bi-)linear form (5.2.6).

Remark 5.9 *In the case of the* **Dirichlet problem**, *the Lagrange multipliers of the solution $p \in (H^{1/2}(\Gamma))'$ can be shown to be the conormal derivative of y at Γ, $p = \mathbf{n} \cdot \nabla y$. This is often interpreted as the* **stress** *of the solution at the boundary.*

We can now employ the theory from Section 5.1 and derive the optimality conditions of (5.2.10). This yields the following reformulation of our elliptic boundary value problem: Given $(f, g) \in (H^1(\Omega))' \times H^{1/2}(\Gamma)$, find $(y, p) \in H^1(\Omega) \times (H^{1/2}(\Gamma))'$ such that holds

$$\begin{aligned} a_\square(y, v) + b(v, p) &= \langle f, v \rangle_{(H^1(\Omega))' \times H^1(\Omega)}, && \text{for all } v \in H^1(\Omega), \\ b(y, q) &= \langle g, q \rangle_{(H^{1/2}(\Gamma))' \times H^{1/2}(\Gamma)}, && \text{for all } q \in (H^{1/2}(\Gamma))'. \end{aligned} \tag{5.2.11}$$

As formulated in the abstract setting from Section 5.1, we can write (5.2.11) in operator form (5.1.11) or (5.1.26), depending on $\xi(\cdot)$. Hence, the operators A, B and B' are defined by their roles as functionals $A\,v \in (H^1(\Omega))'$ and $B'\,q \in (H^1(\Omega))'$ acting on elements of the space $H^1(\Omega)$.

Remark 5.10 *Note that this operator A defined by (1.4.12) is* **self-adjoint**, *i.e., $A' = A$ if the bilinear form (5.2.6) is symmetric.*

It remains to answer the question whether the solution y on $\square$ is really the solution y_Ω when restricted to the domain Ω. This can be answered positively if $\Gamma = \partial\Omega$, see [66, 68]. In case $\Gamma \subset \partial\Omega$, this is no longer automatically valid and still an open question in general. In fact this depends on the way the right hand side extension is constructed, cf. [116].

5.2.2 The Case $\Omega = \square$, $\Gamma = |$

Let $I = (0, 1)$ and $\Omega = \square = (0, 1)^n \subset \mathbb{R}^n$ for a fixed $n \geq 2$. We focus here on the case $n = 2$, since it is easiest to visualize. This domain has a piecewise smooth boundary $\partial\Omega$, in particular $\partial\Omega \in C^{0,1}$.
The task is now to find a solution to the following boundary value problem

$$\begin{aligned} -\Delta y + \xi(y) &= f, && \text{in } \Omega, \\ y &= g, && \text{on } \Gamma, \\ \nabla y \cdot \mathbf{n} &= 0, && \text{on } \partial\Omega \setminus \Gamma, \end{aligned} \tag{5.2.12}$$

where $\mathbf{n} = \mathbf{n}(x)$ is the outward normal at any point $x \in \Gamma$. We assume our Dirichlet boundary Γ to be one of the sides of the hypercube $\square$. In total, there then are $2n$ trace operators, i.e., in dimension $1 \leq i \leq n$ onto the faces at $k \in \{0, 1\}$, the trace operators w.r.t. to the domains

$$\Gamma_{i,k} := \{x = (x_1, \ldots, x_n) \in \mathbb{R}^n \mid x_i = k, 0 \leq x_1, \ldots, x_{i-1}, x_{i+1}, \ldots x_n \leq 1\}. \tag{5.2.13}$$

These two faces $k \in \{0, 1\}$ for $i = 1$ shall be designated Γ_W (west) and Γ_E (east) respectively, see the diagram for the case $n = 2$:

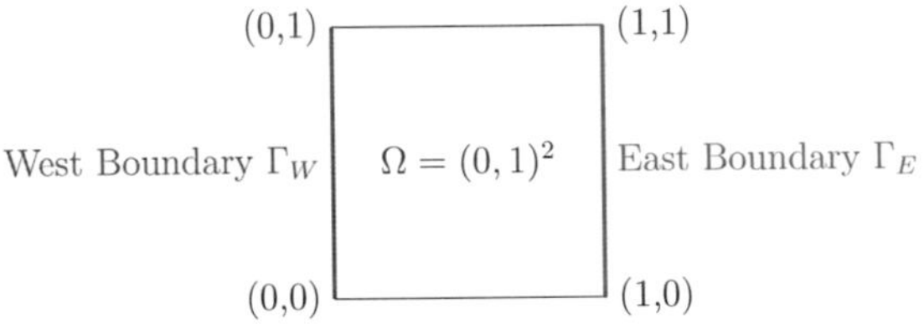

(0,1) (1,1)

West Boundary Γ_W $\Omega = (0,1)^2$ East Boundary Γ_E

(0,0) (1,0)

The choices of the Dirichlet boundary edges are completely arbitrary and the symmetry of the domain permits immediate transfer of any results to the respective boundary value problems with Γ_N (north) and Γ_S (south) boundaries. This argument obviously also applies to higher dimensions $n > 2$.
We can infer from Section 1.2.2 that the trace operators

$$\gamma_W : H^1(\Omega) \to H^{1/2}(\Gamma_W), \quad v \mapsto v|_{\Gamma_W}, \tag{5.2.14}$$

$$\gamma_E : H^1(\Omega) \to H^{1/2}(\Gamma_E), \quad v \mapsto v|_{\Gamma_E}, \tag{5.2.15}$$

are well-defined. In the following, we fix one operator and refer to it with the symbol $\gamma \in \{\gamma_E, \gamma_W\}$.
The Dirichlet boundary $\Gamma \in \{\Gamma_E, \Gamma_W\}$ shall be uniquely determined by this operator. The bilinear form of the trace operator γ is defined by (5.2.7) and the bilinear form of the PDE by (5.2.6).
Thus, all that needs to be checked are the assumptions of Theorem 5.2 and Theorem 5.5 in each case. The boundedness and coercivity of the bilinear form $a(\cdot, \cdot)$ have already been discussed in Section 1.4.1. The trace operator details were given in Section 1.2.2. Remaining are the details of the nonlinear terms.

Example 5.11 *For the nonlinearity $\xi(u) := u^3$, define*

$$\Xi(t) := \int_0^t \xi(\tau) d\tau,$$

then the potential functional $j : H^1(\Omega) \to \mathbb{R}$ is given by

$$j(u) := \int_\Omega \Xi(u) d\mu = \frac{1}{4} \int_\Omega u^4 d\mu.$$

From the construction, it is obvious that for the Fréchet derivative of $j(u)$ holds

$$j'(u; v) = \int_\Omega \Xi'(u; v) d\mu = \langle \xi(u), v \rangle.$$

The functional $j(\cdot)$ is obviously non-negative with $j(0) = 0$ and (lower semi-)continuous. Convexity follows simply by linearity of the integral and the property $t^4 \leq t$ for $t \in [0, 1]$.

Example 5.12 *The square-weighted operator* $\xi_z(u) := 3z^2u$ *is linear in* u. *Its functional is given by*

$$j(u) := \frac{3}{2}\int_\Omega z^2u^2 d\mu,$$

which is (lower semi-)continuous and convex by the same reasoning as in the previous example. The non-negativity holds because of the squaring within the integral, but it can hold that $j(u) = 0$ *for* $u \neq 0$ *if* $y = 0$ *on a domain* $\Lambda \subset \Omega$ *of positive measure. Note that the trace spaces* Γ *are of measure zero, i.e.,* $\mu(\Gamma) = 0$.

Since our operator equations are thus in accordance with the theory of Section 5.1.1 and Section 5.1.2, we discuss the wavelet methods for these problems next.

5.2.3 Wavelet Discretization

We now employ the wavelet theory from Section 2 to our problem. With the background of Section 2, we have assured the existence of biorthogonal wavelet bases Ψ^1_Ω, $\widetilde{\Psi}^1_\Omega$ and $\Psi^{1/2}_\Gamma$, $\widetilde{\Psi}^{1/2}_\Gamma$ for the spaces $H^1(\Omega)$, $H^{1/2}(\Gamma)$ and their duals $(H^1(\Omega))'$,$(H^{1/2}(\Gamma))'$, such that the norm equivalences (2.2.13), (2.2.14) hold for the required ranges. For all multi-dimensional domains, the **isotropic** wavelet constructions from Section 2.4.3 are used.
Hence, in accordance with the notation introduced in Corollary 2.23, a wavelet basis for $\mathcal{Z} := H^1(\Omega) \times (H^{1/2}(\Gamma))'$ is given by $\left(\Psi^1_\Omega, \widetilde{\Psi}^{1/2}_\Gamma\right)^T$ with the index set $\ell_2(\mathcal{I}_\Omega) \times \ell_2(\mathcal{I}_\Gamma)$. Since writing $\ell_2(\mathcal{I}_\Omega) \times \ell_2(\mathcal{I}_\Gamma)$ is too cumbersome every time, we will shorten it to ℓ_2, especially on the norm. Likewise, a basis for the dual space $\mathcal{Z}' = (H^1(\Omega))' \times H^{1/2}(\Gamma)$ is $\left(\widetilde{\Psi}^1_\Omega, \Psi^{1/2}_\Gamma\right)^T$. We can expand the right hand side $(f,g)^T \in \mathcal{Z}'$ in these scaled wavelet bases as

$$(f,g)^T = \left(\mathbf{f}^T\widetilde{\Psi}_\Omega, \mathbf{g}^T\widetilde{\Psi}_\Gamma\right)^T, \quad \text{with } \mathbf{f} \in \ell_2(\mathcal{I}_\Omega), \mathbf{g} \in \ell_2(\mathcal{I}_\Gamma). \tag{5.2.16}$$

The solution vector $(y,p)^T \in \mathcal{Z}$ has an analogous expansion

$$(y,p)^T = \left(\mathbf{y}^T\Psi_\Omega, \mathbf{p}^T\widetilde{\Psi}_\Gamma\right)^T, \quad \text{with } \mathbf{y} \in \ell_2(\mathcal{I}_\Omega), \mathbf{p} \in \ell_2(\mathcal{I}_\Gamma). \tag{5.2.17}$$

By Section 2.2.4, the discretized infinite-dimensional operator $\mathbf{L_F}$ from (5.1.26) is now given by

$$\mathbf{L_F}\begin{pmatrix}\mathbf{y}\\ \mathbf{p}\end{pmatrix} := \begin{pmatrix}\mathbf{F} & \mathbf{B}^T\\ \mathbf{B} & \mathbf{0}\end{pmatrix}\begin{pmatrix}\mathbf{y}\\ \mathbf{p}\end{pmatrix} = \begin{pmatrix}\mathbf{f}\\ \mathbf{g}\end{pmatrix}. \tag{5.2.18}$$

In the linear case $\mathbf{F} \equiv \mathbf{A}$, where $\mathbf{A}$ is an isomorphism (2.2.28), the following result holds.

Lemma 5.13 *Let the operator* $\mathbf{A} : \ell_2 \to \ell_2$ *be an isomorphism. Then* $\mathbf{L_A} : \ell_2(\mathcal{I}_\Omega) \times \ell_2(\mathcal{I}_\Gamma) \to \ell_2(\mathcal{I}_\Omega) \times \ell_2(\mathcal{I}_\Gamma)$, *given by*

$$\mathbf{L_A}\begin{pmatrix}\mathbf{y}\\ \mathbf{p}\end{pmatrix} = \begin{pmatrix}\mathbf{A} & \mathbf{B}^T\\ \mathbf{B} & \mathbf{0}\end{pmatrix}\begin{pmatrix}\mathbf{y}\\ \mathbf{p}\end{pmatrix} = \begin{pmatrix}\mathbf{f}\\ \mathbf{g}\end{pmatrix}, \tag{5.2.19}$$

is also an **isomorphism**, *i.e.*,

$$\left\| \mathbf{L}_{\mathbf{A}} \begin{pmatrix} \mathbf{v} \\ \mathbf{q} \end{pmatrix} \right\|_{\ell_2} \sim \left\| \begin{pmatrix} \mathbf{v} \\ \mathbf{q} \end{pmatrix} \right\|_{\ell_2}, \qquad \text{for all } (\mathbf{v}, \mathbf{q}) \in \ell_2(\mathcal{I}_\Omega) \times \ell_2(\mathcal{I}_\Gamma). \tag{5.2.20}$$

The constants in these norm equivalences only depend on the constants c_{L_A}, C_{L_A} *from Theorem 5.2 and the constants in the norm equivalences (2.2.13) and (2.2.14) for the wavelet bases* Ψ^1_Ω, $\widetilde{\Psi}^1_\Omega$ *and* $\Psi^{1/2}_\Gamma$, $\widetilde{\Psi}^{1/2}_\Gamma$.

Proof: The proof is imminent from the details of Section 2.2.4 and Theorem 5.2, see also [99]. ∎

Remark 5.14 *When using discretizations based upon full grids as in Section 2.5, the above lemma says that the spectral condition* $\kappa_2(\mathbf{L}_{\mathbf{A}})$ *is* **uniformly** *bounded and an iterative solution scheme can produce a solution in optimal time and complexity, i.e., both are linear in the number of unknowns, see [45, 46]. But this presupposes the* **stability** *of the discretizations, i.e., the boundedness of the error when going from an infinite wavelet expansion to a finite one, see Section 2.5.2.*

The coefficient vectors $\mathbf{f}, \mathbf{g}$ are calculated from functions $(f, g) \in \mathcal{Z}'$ by the primal and dual expansions (2.1.55), (2.1.56), i.e.,

$$\mathbf{f} = \langle f, \Psi^1_\Omega \rangle \, \widetilde{\Psi}^1_\Omega, \qquad \mathbf{g} = \left\langle g, \widetilde{\Psi}^{1/2}_\Gamma \right\rangle \Psi^{1/2}_\Gamma. \tag{5.2.21}$$

Remark 5.15 *The details of the application of the operator* $F : \mathcal{H} \to \mathcal{H}'$ *in wavelet coordinates were presented in Section 3.4.6 and numerical results were provided in Section 4.6. The application of the linear trace operator* $B : \mathcal{H} \to \mathcal{V}$ *and its adjoint* $B' : \mathcal{V}' \to \mathcal{H}'$ *based upon the bilinear form (3.6.1) was discussed in Section 3.6.*

The following section deals with the adaptive numerical solution of equations of the above types.

5.3 Adaptive Solution Methods

Now we turn to acquiring a solution of the system of equations (5.2.19) numerically.

5.3.1 The Normalized Equation

The standard approach to solve any non-symmetric indefinite (and non-square) equation $\mathbf{A}\,\mathbf{x} = \mathbf{f}$ is to minimize the functional

$$J(\mathbf{u}) := \frac{1}{2} \|\mathbf{A}\,\mathbf{u} - \mathbf{f}\|^2,$$

which leads to the symmetric positive definite (and square) operator equation

$$\widetilde{\mathbf{A}}\,\mathbf{u} := \mathbf{A}^T \mathbf{A}\,\mathbf{u} = \mathbf{A}^T \mathbf{f}. \tag{5.3.1}$$

The main problem when employing a simple iterative scheme, e.g., Algorithm 4.4 and Algorithm 4.6, on the normalized equation is, that the condition number of the operator $\mathbf{A}^T\mathbf{A}$ is approximately squared compared to the single operator $\mathbf{A}$, i.e.,

$$\kappa_2(\widetilde{\mathbf{A}}) \sim \kappa_2(\mathbf{A})^2,$$

which means more iteration steps are necessary to converge to a target tolerance ε. Also, since the application of the operator $\widetilde{\mathbf{A}}$ takes at least twice as much computational work, the experienced runtime of the solvers can be much larger than for the original equation. Therefore, the normalized equations are usually only considered if no other iterative solver can be applied.

5.3.2 A Positive Definite System

An interesting approach to solve the linear boundary value problem was proposed in [32], based upon ideas put forth in [20]. There, the indefinite saddle point system (5.1.11) is multiplied for $0 < \gamma < c_A$ (2.2.27) with

$$\begin{pmatrix} \gamma^{-1}\mathbf{I} & \mathbf{0} \\ \gamma^{-1}\mathbf{B} & -\mathbf{I} \end{pmatrix},$$

which transforms it into a positive definite system in the inner product

$$\left[\begin{pmatrix} \mathbf{y} \\ \mathbf{p} \end{pmatrix}, \begin{pmatrix} \mathbf{v} \\ \mathbf{q} \end{pmatrix}\right] := \mathbf{v}^T\left(\mathbf{A} - \gamma\mathbf{I}\right)\mathbf{y} + \mathbf{q}^T\mathbf{p}.$$

This would allow standard solvers which could exhibit fast convergence speeds. Especially promising is the prospect of not having to solve the inner system explicitly. The disadvantage is that the value of γ below the upper limit c_A has to be determined. The spectral condition of the transformed system depends (roughly) inversely on $c_A - \gamma$, so γ has to be chosen as large as possible.

5.3.3 Uzawa Algorithms

The general details of this section can be found in [19], amongst others. Uzawa algorithm are iterative solvers for **saddle point problems** (5.2.19). These algorithms generally have a lower complexity than a general solver applied to the normalized equation (5.3.1) for (5.2.19). They can be derived as simple solvers applied to the reduced equation (5.1.17). Conceptually, the Uzawa algorithms can be written as

$$\left.\begin{array}{rcl} \mathbf{A}\,\mathbf{y}^i & = & \mathbf{f} - \mathbf{B}^T\mathbf{p}^{i-1}, \\ \mathbf{p}^i & = & \mathbf{p}^{i-1} + \alpha^i(\mathbf{B}\,\mathbf{y}^i - \mathbf{g}), \end{array}\right\} i = 1, 2, \ldots. \tag{5.3.2}$$

The step size parameter $\alpha^i \in \mathbb{R}$ has to be chosen small enough to ensure convergence depending on the spectral properties of $\mathbf{A}$ and $\mathbf{B}$. Since it also determines the convergence speed, α^i must be chosen as large as possible in applications. It is known that the upper bound for α^i is given by the norm of the **Schur complement** (5.1.16) as

$$\alpha^i < \frac{2}{\|\mathbf{B}\mathbf{A}^{-1}\mathbf{B}^T\|_{\ell_2}}. \tag{5.3.3}$$

Assuming the vector $\mathbf{y}^i$ is calculated analytically yields the representation for the residual error

$$\left\| \mathbf{L_A} \begin{pmatrix} \mathbf{y}^i \\ \mathbf{p}^i \end{pmatrix} - \begin{pmatrix} \mathbf{f} \\ \mathbf{g} \end{pmatrix} \right\|_{\ell_2} = \|\mathbf{B}\,\mathbf{y}^i - \mathbf{g}\|_{\ell_2} = \|\boldsymbol{\mu}^i\|_{\ell_2}, \tag{5.3.4}$$

in every step, where the **defect** is defined as

$$\boldsymbol{\mu}^i := \mathbf{g} - \mathbf{B}\,\mathbf{y}^i. \tag{5.3.5}$$

In the case where $\mathbf{A}$ is a symmetric positive definite operator, the optimal step size is in each step given by

$$\alpha^i = \frac{(\boldsymbol{\mu}^i, \boldsymbol{\mu}^i)}{(\boldsymbol{\mu}^i, \mathbf{B}\,\mathbf{A}^{-1}\,\mathbf{B}^T\,\boldsymbol{\mu}^i)} = \frac{(\boldsymbol{\mu}^i, \boldsymbol{\mu}^i)}{((\mathbf{B}^T\,\boldsymbol{\mu}^i), \mathbf{A}^{-1}\,(\mathbf{B}^T\,\boldsymbol{\mu}^i))}. \tag{5.3.6}$$

This follows by considering the **Steepest Descent Method** of Section 4.3 applied to equation (5.1.17). In the full-grid implementation, this value can be computed without extra operator applications by a simple trick: The solution component $\mathbf{A}^{-1}\mathbf{f}$ can be computed once and taken out of subsequent steps. The update of the solution component $\mathbf{y}$ is then based upon the relation,

$$\frac{\mathbf{y}^i - \mathbf{y}^{i-1}}{\alpha^{i-1}} = \frac{1}{\alpha^{i-1}} \left(\mathbf{A}^{-1}(\mathbf{f} - \mathbf{B}^T\,\mathbf{p}^{i-1}) - \mathbf{A}^{-1}(\mathbf{f} - \mathbf{B}^T\,\mathbf{p}^{i-2}) \right) = -\mathbf{A}^{-1}(\mathbf{B}^T\boldsymbol{\mu}^{i-1}),$$

which is exactly the term in (5.3.6). Interpreting the relation the other way, it could be used to determine a value for α^i in terms of α^{i-1}. In the adaptive case, this no longer works as the right hand side $\mathbf{f}$ in different steps is determined approximately and thus does not cancel out. Evaluating the right hand side in every step with a very high accuracy instead so that the above relation still holds would defeat the purpose of the adaptive procedure.
Even better convergence properties can be obtained by employing **conjugate directions**, which is a very effective strategy for conventional discretizations using wavelets, see [122]. In the adaptive wavelet setting, the convergence speed might increase, i.e., fewer steps are required to obtain a solution, but the overhead for computing the optimal step size might increase the overall runtime. Especially when an inverse operator application is involved as it is here, simpler iterations can have a better performance than more complicated schemes.

An Adaptive Uzawa Algorithm

The first publication of this algorithm in the wavelet domain for linear problems was [40]. The Uzawa algorithm is designed as a Richardson iteration on the reduced system (5.1.17). But instead of the simple application of a single operator, one has to solve the first equation in (5.3.2) and then evaluate the second equation. We treat the Uzawa algorithm here in the framework of Algorithm 4.4, for a complete description see [40, 42, 147].

Remark 5.16 *In [40], the authors applied a* **Riesz operator** $\mathbf{R}_{\mathcal{V}'} : \mathcal{V}' \to \mathcal{V}$ *(2.2.34) to map the defect* $\boldsymbol{\mu}^i \in \mathcal{V}'$ *(5.3.5) from* $\mathcal{V}' = (H^{1/2}(\Gamma))'$ *to* $\mathcal{V} = H^{1/2}(\Gamma)$ *to make it* *more compatible as an update for the Lagrangian multiplier* $\mathbf{p}^i \in \mathcal{V}$. *In their setup, the trace operator mapped into* $\mathcal{V}'$, *which made this necessary. In our setup, by the Sobolev space embedding relations (1.2.9), this is not necessary, but it could still be beneficial in practice. However, we have not found any speed up in numerical experiments but the additional complexity for computing the application of the Riesz Operator increased runtimes measurably (although only by a few percent of the overall runtime). For this reason, we did not incorporate the Riesz operator* $\mathbf{R}_{\mathcal{V}} : \mathbf{V} \to \mathbf{V}'$ *into the algorithm here.*

To calculate $\mathbf{y}^i$ adaptively, the solvers of Section 4 are employed, we just have to formulate an algorithm to evaluate the right hand side, i.e., to compute $\mathbf{q}_\varepsilon$ satisfying

$$\|\mathbf{q}_\varepsilon - (\mathbf{f} - \mathbf{B}^T \mathbf{p}^{i-1})\|_{\ell_2} \leq \varepsilon. \tag{5.3.7}$$

Lemma 5.17 *Let proper trees* $\mathbf{f}, \mathbf{p}^{i-1} \in \ell_2$ *and* $\varepsilon > 0$ *be given. Assuming a routine that computes a vector* $\mathbf{q}_1 \in \ell_2$ *such that*

$$\|\mathbf{q}_1 - \mathbf{B}^T \mathbf{p}^{i-1}\|_{\ell_2} < \varepsilon/2,$$

for any $\varepsilon > 0$. *Then Algorithm 5.1 computes a vector satisfying (5.3.7).*

Proof: With the property (4.1.7) of Algorithm 4.1, the estimate is immediate:

$$\begin{aligned} \|\mathbf{q}_\varepsilon - (\mathbf{f} - \mathbf{B}^T \mathbf{p}^{i-1})\|_{\ell_2} &\leq \|\mathbf{q}_2 - \mathbf{q}_1 - (\mathbf{f} - \mathbf{B}^T \mathbf{p}^{i-1})\|_{\ell_2} \\ &\leq \|\mathbf{q}_2 - \mathbf{f}\|_{\ell_2} + \|\mathbf{q}_1 - \mathbf{B}^T \mathbf{p}^{i-1}\|_{\ell_2} \leq \varepsilon/2 + \varepsilon/2 = \varepsilon. \end{aligned}$$

■

Plugging this right hand side into Algorithm 4.3 then gives the convergence results of Theorem 4.8 for the first equation of (5.3.2) since $\mathbf{A}$ is assumed to be an operator of the form considered in Section 4.
Algorithm 5.1 enables us to compute the **defect** (5.3.5), which is, up to the error made in the approximate inversion of the operator $\mathbf{A}$, equal to the **residual** because of

$$\left\| \mathbf{L}_\mathbf{A} \begin{pmatrix} \mathbf{y}^i \\ \mathbf{p}^i \end{pmatrix} - \begin{pmatrix} \mathbf{f} \\ \mathbf{g} \end{pmatrix} \right\|_{\ell_2} = \|\mathbf{A}\,\mathbf{y}^i + \mathbf{B}^T \mathbf{p}^i - \mathbf{f}\|_{\ell_2} + \|\mathbf{B}\,\mathbf{y}^i - \mathbf{g}\|_{\ell_2} \leq \varepsilon_i + \|\boldsymbol{\mu}^i\|_{\ell_2}, \tag{5.3.8}$$

Algorithm 5.1 Construct a vector $\mathbf{q}_\varepsilon$ which that satisfies $\|\mathbf{q}_\varepsilon - (\mathbf{f} - \mathbf{B}^T \mathbf{p})\| \leq \varepsilon$ for $\varepsilon > 0$.

1: **procedure** SOLVER_UZAWA_SUB_RHS($\varepsilon, \mathbf{f}, \mathbf{B}^T, \mathbf{p}$) $\to \mathbf{q}_\varepsilon$
2: // Apply Adjoint of Trace Operator
3: $\mathbf{q}_1 \leftarrow$ APPLY($\varepsilon/2, \mathbf{B}^T, \mathbf{p}$) ▷ APPLY stands for the routine
4: applicable to this operator $\mathbf{B}^T$
5: // Coarsening of $\mathbf{f}$
6: $\mathbf{q}_2 \leftarrow$ RHS($\varepsilon/2, \mathbf{f}$)
7: $\mathbf{q}_\varepsilon \leftarrow \mathbf{q}_2 - \mathbf{q}_1$
8: **return** $\mathbf{q}_\varepsilon$
9: **end procedure**

where ε_i denotes the step-dependent error to which the first equation was solved. By setting this tolerance to a lower value than the target tolerance ε to which the overall problem is to be solved, it is reasonable to use the norm of the defect only as the residual stopping criterion.

Lemma 5.18 *Let proper trees* $\mathbf{f}, \mathbf{g}, \mathbf{q} \in \ell_2$ *and* $\varepsilon > 0$ *be given. Assuming a routine that computes a vector* $\boldsymbol{\mu}_2 \in \ell_2$ *such that*

$$\|\boldsymbol{\mu}_2 - \mathbf{B}\,\mathbf{y}\|_{\ell_2} < \frac{\varepsilon}{3},$$

then Algorithm 5.2 computes a vector satisfying

$$\left\|\boldsymbol{\mu}_\varepsilon - \left(\mathbf{g} - \mathbf{B}\left(\mathbf{A}^{-1}(\mathbf{f} - \mathbf{B}^T\mathbf{q})\right)\right)\right\|_{\ell_2} \lesssim \varepsilon. \tag{5.3.9}$$

Proof: The proof of the estimate follows standard lines: Let $\widetilde{\mathbf{y}}$ be the exact solution of $\mathbf{A}^{-1}(\mathbf{f} - \mathbf{B}^T\,\mathbf{q})$. Then the promise of `SOLVE` is

$$\|\mathbf{y} - \widetilde{\mathbf{y}}\|_{\ell_2} \le \frac{\varepsilon}{3\,\beta_2}.$$

The estimate (5.3.9) can then be written as

$$\begin{aligned}\|\boldsymbol{\mu}_\varepsilon - (\mathbf{g} - \mathbf{B}\,\widetilde{\mathbf{y}})\|_{\ell_2} = \|\boldsymbol{\mu}_1 - \mathbf{g} - (\boldsymbol{\mu}_2 - \mathbf{B}\,\widetilde{\mathbf{y}})\|_{\ell_2} &\le \|\boldsymbol{\mu}_1 - \mathbf{g}\|_{\ell_2} + \|\boldsymbol{\mu}_2 - \mathbf{B}\,\widetilde{\mathbf{y}}\|_{\ell_2} \\ &\le \varepsilon/3 + \|\boldsymbol{\mu}_2 - \mathbf{B}\,\widetilde{\mathbf{y}}\|_{\ell_2}.\end{aligned}$$

For the second term follows

$$\|\boldsymbol{\mu}_2 - \mathbf{B}\,\widetilde{\mathbf{y}}\|_{\ell_2} \le \|\boldsymbol{\mu}_2 - \mathbf{B}\,\mathbf{y}\|_{\ell_2} + \|\mathbf{B}\,\mathbf{y} - \mathbf{B}\,\widetilde{\mathbf{y}}\|_{\ell_2} \lesssim \varepsilon/3 + \beta_2\|\mathbf{y} - \widetilde{\mathbf{y}}\|_{\ell_2} \le \varepsilon/3 + \frac{\beta_2\,\varepsilon}{3\,\beta_2} = 2\,\varepsilon/3.$$

Thus holds (5.3.9). ■

The constants skipped in the above estimates depend only on the **Riesz Stability** constants given in (2.2.31). Since this only affects the solution of the first equation, it should suffice to multiply the target tolerance $\frac{\varepsilon}{3\,\beta_2}$ by another constant factor.

Remark 5.19 *Within the overall Uzawa algorithm, it is sensible to use the result* $\mathbf{y}^{i-1}$ *of the previous step as starting vector for the next call to* ***SOLVE****. But since* $\mathbf{y}^i$ *is not created from* $\mathbf{y}^{i-1}$ *by the addition of an offset, the vector* $\mathbf{y}^i$ *will not require coarsening like the normal step vectors of Section 4.1.1.*

Algorithm 5.2 Adaptive computation of the defect $\boldsymbol{\mu}_\varepsilon$ up to accuracy $\varepsilon > 0$ for the vector $\mathbf{p}$.

1: **procedure** SOLVER_UZAWA_RESIDUAL($\varepsilon, \mathbf{A}, \mathbf{B}, (\mathbf{f}, \mathbf{g}), \mathbf{p}) \to \boldsymbol{\mu}_\varepsilon$
2: // Init
3: $\mathbf{y} \leftarrow \mathbf{0}$
4: $\boldsymbol{\mu}_1 \leftarrow \mathbf{0}$, $\boldsymbol{\mu}_2 \leftarrow \mathbf{0}$ ▷ Start vectors
5: // First Equation
6: $\mathbf{y} \leftarrow$ SOLVE($\frac{\varepsilon}{3\,\beta_2}, \mathbf{A}, \mathbf{f} - \mathbf{B}^T\mathbf{p}$) ▷ Solve $\mathbf{A}\,\mathbf{y} = \mathbf{f} - \mathbf{B}^T\mathbf{p}$ by whatever algorithm
7: can be employed.
8: The right hand side is evaluated by Algorithm 5.1
9: // Residual
10: $\boldsymbol{\mu}_1 \leftarrow$ RHS($\varepsilon/3, \mathbf{g}$)
11: $\boldsymbol{\mu}_2 \leftarrow$ APPLY($\varepsilon/3, \mathbf{B}, \mathbf{y}$) ▷ APPLY stands for the routine
12: applicable to this operator $\mathbf{B}$
13: $\boldsymbol{\mu}_\varepsilon \leftarrow \boldsymbol{\mu}_1 - \boldsymbol{\mu}_2$
14: **return** $\boldsymbol{\mu}_\varepsilon$
15: **end procedure**

5.3.4 Convergence Properties – The Linear Case

The subject of the adaptive Uzawa algorithm based upon linear PDEs in the wavelet context has been discussed in [40,42] and we quote their result in this section. Since their results deal with linear operators, i.e., matrices, the decay property (3.3.14) was not yet in use. Instead, the essential property required of operators is **compressibility**:

Definition 5.20 [$s^\star$-Compressibility]
A matrix $\mathbf{C}$ *is called* $s^\star$-**compressible** *if for any* $0 < s < s^\star$ *and* $j \in \mathbb{N}$ *there exists a real summable sequence* $(\alpha_j)_{j=0}^\infty$ *and a matrix* $\mathbf{C}_j$ *with* $\mathcal{O}(2^j)$ *non-zero values in every row and column satisfying*

$$\|\mathbf{C} - \mathbf{C}_j\|_{\ell_2 \to \ell_2} \leq \alpha_j 2^{-j\,s}.$$

The set of all such matrices is denoted $\mathcal{C}_{s^\star}$.

As was pointed out in [31], a **local** (3.1.1) linear operator satisfying (3.3.14) for some $\gamma > \frac{n}{2}$ is in $\mathcal{C}_{s^\star}$ for $s^\star = \frac{2\gamma - n}{2n}$. This is exactly the best approximation rate (4.1.15) and this hints at further results. A direct characterization of the class $\mathcal{C}_{s^\star}$ given in [31] concerns the values of the matrix itself.

Definition 5.21 [Quasi-Sparseness]
A matrix $\mathbf{B} = (b_{\lambda,\mu})$ *is called* **quasi-sparse** *if for* $\sigma > n/2$ *and* $\beta > n$ *holds*

$$|b_{\lambda,\mu}| \lesssim 2^{-\sigma||\lambda|-|\mu||} \left(1 + 2^{\min(|\lambda|,|\mu|)} \operatorname{dist}(S_\lambda, S_\mu)\right)^{-\beta}.$$

If the operator is local, i.e., $\operatorname{dist}(S_\lambda, S_\mu) = 0$ for all non-zero matrix entries, only $\sigma > n/2$ is necessary for quasi-sparseness. Then, the matrix $\mathbf{B}$ is $s^\star$-compressible with $s^\star = \frac{n}{2} - \frac{1}{2}$ (Prop. 3.5 in [31]).

Proposition 5.22 *The matrix generated by any trace operator $B : H^1(\Omega) \to H^{1/2}(\Gamma)$ is $\mathcal{C}_{s^\star}$ with (at least) $s^\star = \frac{1}{2}$.*

Proof: By the general bilinear form definition (3.6.4) holds for properly scaled wavelets

$$|b_{\lambda,\mu}| = \left|b\left(2^{-|\lambda|}\,\psi_\lambda^\Omega, 2^{|\mu|/2}\,\widetilde{\psi}_\mu^\Gamma\right)\right| = \left|\int_\Gamma 2^{-|\lambda|}(\psi_\lambda^\Omega)|_\Gamma\, 2^{|\mu|/2}\widetilde{\psi}_\mu^\Gamma ds\right| = 2^{-|\lambda|}\,2^{|\mu|/2}\left|\int_\Gamma (\psi_\lambda^\Omega)|_\Gamma\,\widetilde{\psi}_\mu^\Gamma ds\right|.$$

Expressing the trace $(\psi_\lambda^\Omega)|_\Gamma$ in the wavelet base Ψ^Γ and noting (2.1.14), it follows

$$|b_{\lambda,\mu}| = 2^{-|\lambda|}\,2^{|\mu|/2}\left|\int_\Gamma 2^{|\lambda|/2}\sum_{\lambda'} c_{\lambda'}\psi_{\lambda'}^\Gamma\,\widetilde{\psi}_\mu^\Gamma ds\right| \leq 2^{-|\lambda|/2}\,2^{|\mu|/2}\sum_{\lambda'}\left|\int_\Gamma c_{\lambda'}\psi_{\lambda'}^\Gamma\,\widetilde{\psi}_\mu^\Gamma ds\right|$$
$$\lesssim 2^{-(|\lambda|-|\mu|)/2},$$

where we used **biorthogonality** $(\mathcal{B})$(2.1.54) in the last step. ∎

Remark 5.23 *In reality, the value $s^\star$ is much larger, especially for the trace operators discussed in Section 3.6.1. The matrix $\mathbf{B}^T$ is in this case extremely sparse, the number of non-zero entries per row is not only bounded by 2^j but often by 1. Were it not that the traces of isotropic wavelets can be single scale functions that have to be decomposed which generates more non-zero values on lower levels, the matrix $\mathbf{B}^T$ would essentially be a bloated diagonal matrix with many rows containing only zero values. Experimentally determined values can be seen in Figure 5.1. These results show that the prerequisite $\sigma > n/2$ of Definition 5.21 is satisfied, although just by an ε.*

The importance of the quasi-sparseness is that there is a complete adaptive scheme based upon the adaptive application of such operators. These results can be found in [31,32,55], among others. In that setting, the adaptive vectors are **not trees**, the characterization is based solely on the error of an N-term approximation of order N^{-s}. Specifically, the following subspace of ℓ_2 is employed:

Definition 5.24 [Weak ℓ_τ-Space]
For any vector $\mathbf{v} \in \ell_2(\mathcal{I})$, let $\mathbf{v}'_n$ be the n-largest value of $|\mathbf{v}_m|$ and $\mathbf{v}'$ be the vector of the rearranged coordinates (3.2.14). The space of all sequences $\mathbf{v} \in \ell_2$ for which the **decreasing rearrangement** $\mathbf{v}'$ *satisfies*

$$\|\mathbf{v}\|_{\ell_\tau^\omega} := \|\mathbf{v}\|_{\ell_2} + \sup_{n\geq 1} n^{1/\tau}\mathbf{v}'_n < \infty, \tag{5.3.10}$$

for some $0 < \tau < 2$ is called the **weak ℓ_τ-space** ℓ_τ^ω.

In fewer words, the values of $\mathbf{v}$ can be rearranged so that the values decay like $N^{-1/\tau}$. The elements of this space can be characterized by the **best N-term approximation** $\sigma_N(\mathbf{v})$ (3.2.13) and the space $\mathcal{A}^s$ defined analogously to Definition 3.18 is exactly ℓ_τ^ω for

$$\frac{1}{\tau} = s + \frac{1}{2}, \quad s > 0.$$

For linear PDEs, results analogous to any given in Section 4 have been established in the norms ℓ^{ω}_{τ} and $\mathcal{A}^s$. Since $\mathcal{A}^s_{\text{tree}}$ is a subspace of $\mathcal{A}^s$, the only defining characteristic is the tree structure, the results given in the norm ℓ^{ω}_{τ} are still valid when considering only vectors with tree structure. The following result from [40] shows that the overall expectations for the Uzawa algorithm are consistent with the results from Section 4.1.4.

Theorem 5.25 *For matrices* $\mathbf{A}$, $\mathbf{B} \in \mathcal{C}_{s^\star}$ *computes the adaptive Uzawa algorithm an approximate solution* $(\mathbf{y}_\varepsilon, \mathbf{p}_\varepsilon)$ *for any* $\varepsilon > 0$ *satisfying*

$$\|\mathbf{y} - \mathbf{y}_\varepsilon\| \lesssim \varepsilon, \quad \|\mathbf{p} - \mathbf{p}_\varepsilon\| \lesssim \varepsilon.$$

If $\mathbf{y}, \mathbf{p} \in \ell^{\omega}_{\tau}$, *i.e.,* $y, p \in \mathcal{A}^s$ *for some* $s < s^\star$, *for* $\dfrac{1}{\tau} < s^\star + \frac{1}{2}$, *then hold the estimates*

$$\|\mathbf{y}_\varepsilon\|_{\ell^{\omega}_{\tau}} \lesssim \|\mathbf{u}\|_{\ell^{\omega}_{\tau}}, \quad \|\mathbf{p}_\varepsilon\|_{\ell^{\omega}_{\tau}} \lesssim \|\mathbf{p}\|_{\ell^{\omega}_{\tau}},$$

and the size of the vectors are bound by

$$\#\operatorname{supp}\mathbf{y}_\varepsilon \lesssim \|\mathbf{u}\|^{1/s}_{\ell^{\omega}_{\tau}} \varepsilon^{-1/s}, \quad \#\operatorname{supp}\mathbf{p}_\varepsilon \lesssim \|\mathbf{p}\|^{1/s}_{\ell^{\omega}_{\tau}} \varepsilon^{-1/s}.$$

Moreover, the computational work needed to compute $\mathbf{y}_\varepsilon$, $\mathbf{p}_\varepsilon$ *is also of the order* $\varepsilon^{-1/s}$.

In particular, for $0 < s < s^\star$, the asymptotically optimal approximation and complexity rates are achieved. But the maximum value $s^\star$ given by the properties of the operators $\mathbf{A}, \mathbf{B} \in \mathcal{C}_{s^\star}$ can only be achieved if the solution (y, p) is smooth enough.

Remark 5.26 *In our algorithm, the solution of the first equation in (5.3.2) is not used as an update for* $\mathbf{y}^i$ *but it is* $\mathbf{y}^{i+1}$. *This setup is necessary for the nonlinear operator but could limit the approximation speed to the value of the used operator, even if the operator is linear.*

Besov Regularity

The last step is identifying when the linear boundary value problem admits solutions in $\mathbf{u}, \mathbf{p} \in \ell^{\omega}_{\tau}$. The problem is generally in the same category as the problems of Section 4.1.4, we can expect the same regularity $u \in B^{m+s\,n}_{\tau}(L_\tau(\Omega))$, for sufficiently smooth right hand sides $f \in H^{-s}(\Omega)$, $s > -1/2$. For the trace $\gamma_0 u$ of such a function $u \in B^{m+s\,n}_{\tau}(L_\tau(\Omega))$ with $\frac{1}{\tau} < s + \frac{1}{2}$ now holds by the trace theorems in [2],

$$\begin{aligned} &\gamma_0 u \in B^{m+s\,n-1/\tau}_{\tau}(L_\tau(\Gamma)), \\ \Longrightarrow\quad &\gamma_0 u \in B^{m+s\,n-(s+1/2)}_{\tau}(L_\tau(\Gamma)), \qquad \text{because } -\frac{1}{\tau} > -(s+\frac{1}{2}), \\ \Longrightarrow\quad &\gamma_0 u \in B^{(m-1/2)+s(n-1)}_{\tau}(L_\tau(\Gamma)). \end{aligned}$$

Since Γ is a $n-1$-dimensional manifold, this smoothness index has exactly the known pattern for $m' := m-1/2$ and $n' := n-1$. By Theorem 1.21, it also holds $\gamma_0 u \in H^{m-1/2}(\Gamma)$, so that by Proposition 3.20 the expansion coefficients of the trace $\gamma_0 u$ are in $\mathcal{A}^s_{\text{tree}}$ for $\frac{1}{\tau} < s + \frac{1}{2}$. Therefore, again for sufficiently smooth right hand side data, the same

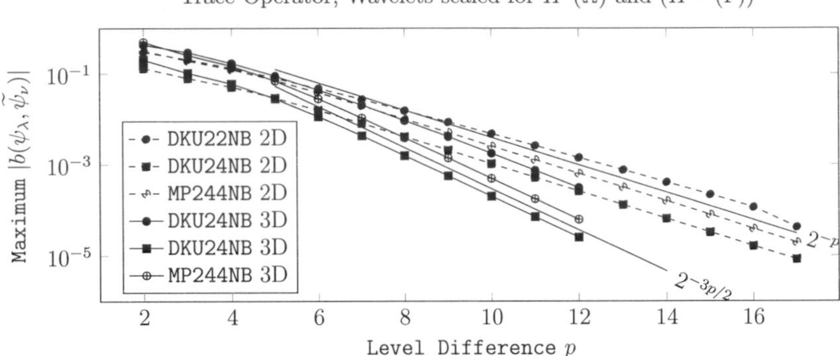

Figure 5.1: Maximum of the bilinear form values $|b(\psi_\lambda, \widetilde{\psi}_\nu)|$ for all $|\nu| = j_0$ and $|\lambda| = |\nu|+p$. The 2D values seem to decay exactly with rate 1 but the decay rate is slightly increasing for higher levels. Only considering the last 4 data points, the decay rate is ≈ 1.01 for the DKU24NB and MP244NB wavelets. In each case, the higher regularity of the dual wavelets make the decay rates approach the threshold $n/2$ quicker.

approximation rates should be expected for both solution components y and p in the Uzawa algorithm.

There are, as of yet, no comparable results concerning the semilinear case discussed in Section 5.1.2. This is of course an important future research topic, especially if more involved problems are considered. For instance, control problems involving semilinear PDEs as discussed in Section 6.2.

To test these algorithms, we now solve a few example problems numerically. The standard case of boundary value problem is given by the setting of Section 5.2.2. Within this, the operator $\xi(u)$ is varied to be either $\xi(u) = u$ or $\xi(u) = u^3$. The trace operator, which determines the trace boundary $\Gamma \subset \partial\Omega$, is usually the restriction to one coordinate axis.

5.4 A 2D Linear Boundary Value Example Problem

As our first example problem, we seek to solve the boundary value problem,

$$\begin{aligned} -\Delta y + a_0\, y &= 1, && \text{in } \Omega, \\ y &= 0, && \text{on } \Gamma, \\ \nabla y \cdot \mathbf{n} &= 0, && \text{on } \partial\Omega \setminus \Gamma, \end{aligned} \tag{5.4.1}$$

with a constant $a_0 \geq 0$, $\Omega = (0,1)^2$ and $\Gamma \equiv \Gamma_E$. This problem has the unique solution, see [122],

$$y^\star_{a_0}(x_1, x_2) := \frac{1}{a_0} - \frac{\cosh(\sqrt{a_0}x_1)}{a_0 \cosh(\sqrt{a_0})}, \tag{5.4.2}$$

which converges for $a_0 \to 0$ to the solution,

$$y_0^\star(x_1, x_2) := \frac{1}{2}(1 - x_1^2). \tag{5.4.3}$$

As these functions are globally smooth, we expect an almost uniform distribution of wavelet coefficients to be generated by any solution algorithm. Also, in such a case, **coarsening** is not only unnecessary, it is outright detrimental to the convergence, i.e., since there are no superfluous wavelet coefficients and all are equally significant, deletion of wavelet coefficients will always decrease the precision of the vector. The ideal tree $S(\mathbf{y})$ is in this case not a "thin" and "deep" one, i.e., few wavelet coefficients on all levels, but a "fat" and "wide" one, i.e., many or all wavelet coefficients on very few levels. Since the curvature on $y_{a_0}^\star$ gets larger for $x_2 \to 1$, we can expect a slightly higher concentration of wavelet coefficients there than at $x_1 \to 0$, but the effect should be small. Also, since the solution (5.4.2) does not depend on x_2, i.e., it is **constant** w.r.t. x_2, we expect no wavelet coefficients of **type** $\mathbf{e}_2 = 1$, i.e., only the (root) single scale function coefficients in direction x_2 will be needed.

5.4.1 Numerical Results

The reference solution here was computed using full-grid techniques on level $J = 10$. This means the expected accuracy of the reference is in the range of $2^{-10} \sim 1 \times 10^{-3}$ by Lemma 1.47, which is observed as the lower limit for the y component in Figures 5.4 to 5.6. The above mentioned wavelet coefficient distribution can be seen in Figure 5.2. In all cases, the target tolerance is $\varepsilon = 1 \times 10^{-4}$. The inner equation is solved using the Gradient iteration outlined in Section 4.3 using the optimal step size (5.3.6). The maximum values of the step size parameter α for the Uzawa algorithm can be found in Table 5.1. Coarsening was done with an extra factor of 0.01, because higher values hindered convergence. The approximation rates of the vectors can be found in Figure 5.7. The rate of the component y is very close to 1 which corresponds well with the rates in Section 4.6. The rate of the preferred variable p is well higher, which is probably due to the linear operator and smooth right hand side $g \equiv 0$.

Wavelet	DKU22NB	DKU24NB	MP244NB
SMD-SMOrth	0.19	0.22	0.51

Table 5.1: Maximum values of the Uzawa step size parameter α for different preconditioners. See Section A for details on the names of the wavelet configurations.

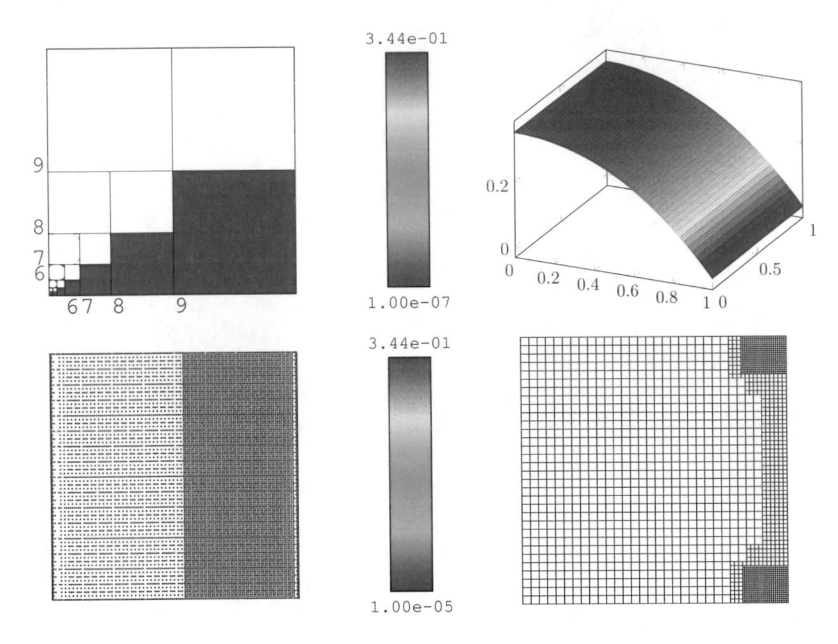

Figure 5.2: Plot of the solution $y_1^\star$ of (5.4.1) and the isotropic wavelet coefficient diagrams. Details of the diagrams' structure can be found in Appendix A.2. The minimum displayed values for the diagrams were chosen to maximize recognizability of the distribution patterns. Since these diagrams were generated from full-grid data structures, all coefficients are present and not enforcing the lower limit would result in plots completely covered in data points. The wavelet vector was similarly thresholded prior to the generation of the grid to emphasize its structure.

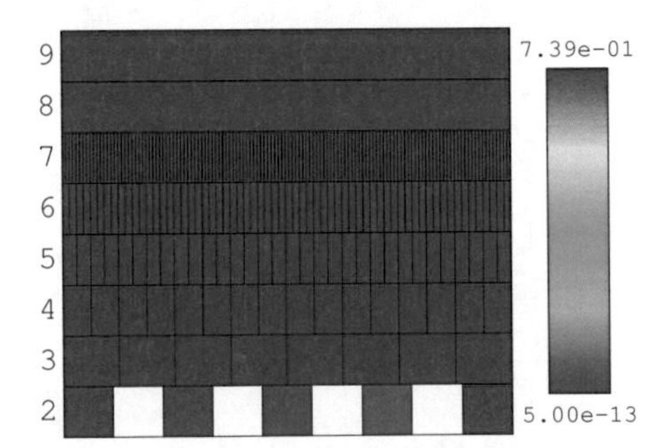

Figure 5.3: Diagram of the Lagrangian multiplier p for the solution $y_1^\star$ of (5.4.2). The data was generated using full-grid data structures. The values on the coarsest level are not symmetrically distributed because of the basis transformation (2.3.22).

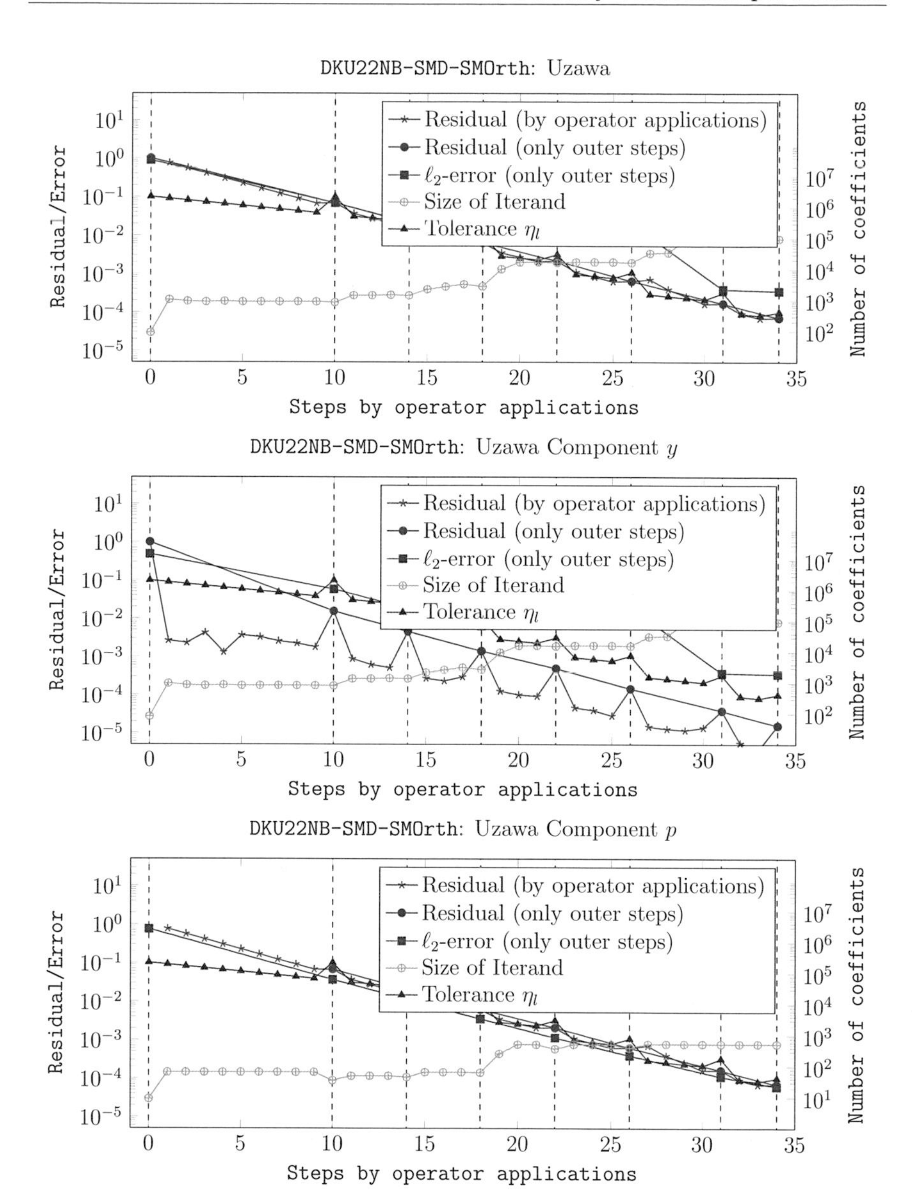

Figure 5.4: Detail iteration histories for the Uzawa algorithm. The first diagram shows the overall values, the other two diagrams deal with the components y and p, respectively. Since the starting vectors contain zero values only and $g = 0$, the residual in the first step for p is exactly zero, which is not displayed on the logarithmic scale.

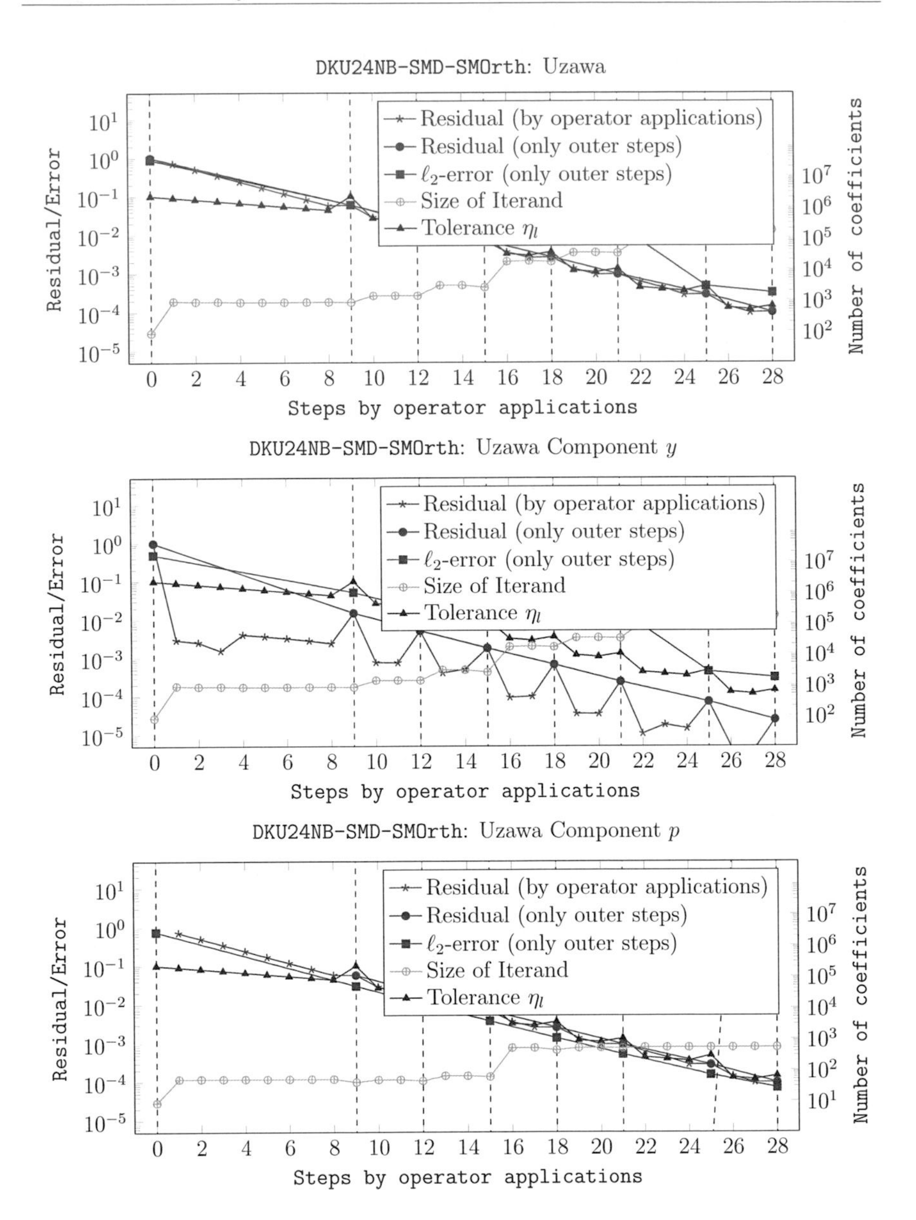

Figure 5.5: Detail iteration histories for the Uzawa algorithm. The first diagram shows the overall values, the other two diagrams deal with the components y and p, respectively. Since the starting vectors contain zero values only and $g = 0$, the residual in the first step for p is exactly zero, which is not displayed on the logarithmic scale.

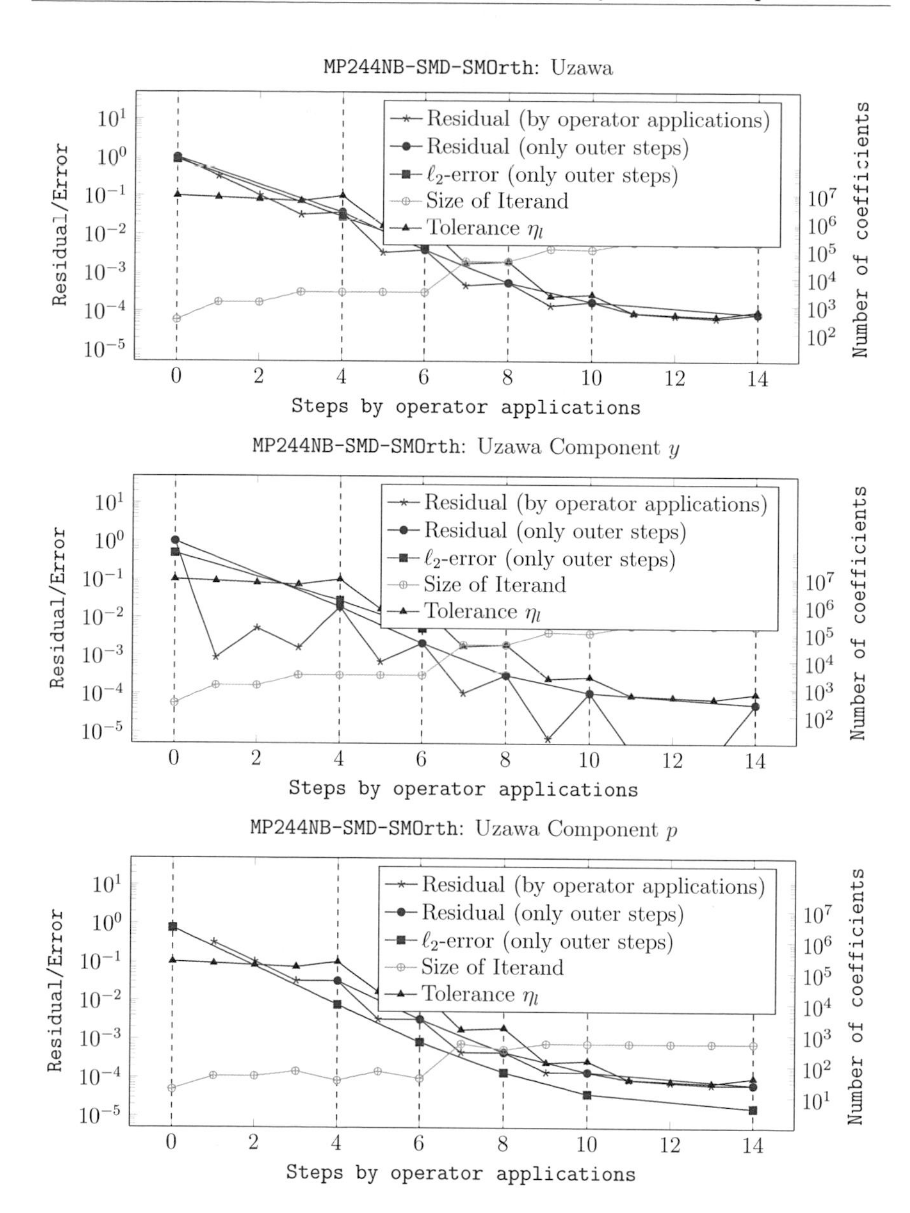

Figure 5.6: Detail iteration histories for the Uzawa algorithm. The first diagram shows the overall values, the other two diagrams deal with the components y and p, respectively. Since the starting vectors contain zero values only and $g = 0$, the residual in the first step for p is exactly zero, which is not displayed on the logarithmic scale.

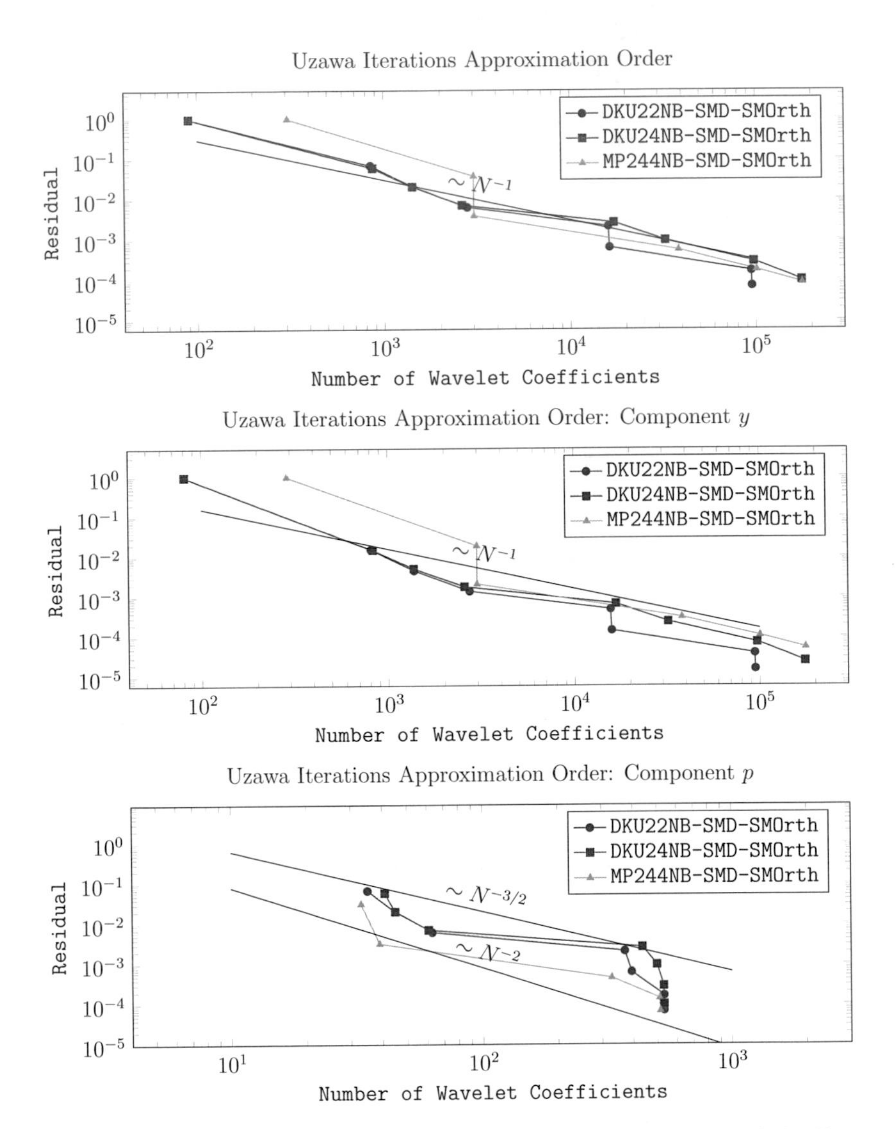

Figure 5.7: Approximation order for the overall and individual vectors of the Uzawa algorithm for the different wavelets using preconditioner $\mathbf{D}_{\{O,a\}}$. On the abscissa, the number of wavelet coefficients after each coarsening is plotted against the residual for the coarsed vector on the ordinate axis.

Wavelet	DKU22NB		DKU24NB	
Preconditioner	Overall	Per Wavelet	Overall	Per Wavelet
SMD-SMOrth	5.81×10^2	1.56×10^{-5}	1.48×10^3	2.15×10^{-5}

Wavelet	MP244NB	
Preconditioner	Overall	Per Wavelet
SMD-SMOrth	5.76×10^2	2.17×10^{-5}

Table 5.2: Runtimes of the Uzawa solver for different preconditioners.

5.5 A 2D Linear PDE and a Boundary on a Circle

The afore discussed boundary value problems are "good-natured" in the sense, that one does not expect complications from the discretization of the trace operator. This is a consequence of the trace space Γ being a side of the tensor product domain Ω. Instead, we want to digress shortly and again consider the linear boundary value problem,

$$\begin{aligned} -\Delta y + y &= 1, \qquad \text{in } \Omega, \\ y &= 0, \qquad \text{on } \partial\Omega, \end{aligned} \tag{5.5.1}$$

where the domain is a **disk**,

$$\Omega := \left\{ x \in \mathbb{R}^2 \mid \|x - c\|_2 \leq R \right\}. \tag{5.5.2}$$

By the theory of Section 5.2.1, this is equivalent of embedding Ω into a square domain $\square$ and enforcing the boundary conditions by a trace operator on the disk's boundary, i.e.,

$$\Gamma := \left\{ x \in \mathbb{R}^2 \mid \|x - c\|_2 = R \right\}. \tag{5.5.3}$$

For this example, we center the disk at the point $c = (\frac{1}{2}, \frac{1}{2}) \in \mathbb{R}^2$ and set the radius to $R = \frac{1}{2}$, the fictitious domain is thus given by $\square = (0,1)^2$. This problem has been considered before in the wavelet context, see [47]. One of the difficulties of this problem using standard discretization techniques lies in the compliance of the **LBB condition**, see Section 2.5.2. The discretized saddle point system $\mathbf{L}$ is ill-conditioned, if the discretization level ℓ on the trace space Γ is too high w.r.t. the discretization level j on the domain $\square$: If $\ell > j+1$, then the iterative solver needs significantly more steps than in the case $\ell \leq j+1$, which indicates a violation of the LBB condition. But a fine mesh on the trace space is necessary to achieve the required accuracy to ensure the boundary conditions. In the adaptive case, this problem should not arise.
Since Γ is a closed circle, we use one-dimensional **periodic** wavelets for the discretization. The mapping $\kappa : [0,1) \to \Gamma \subset \square$ is given by

$$\kappa(t) := \begin{pmatrix} \kappa_1(t) \\ \kappa_2(t) \end{pmatrix} := \begin{pmatrix} c_1 + R\cos(2\pi t) \\ c_2 + R\sin(2\pi t) \end{pmatrix}. \tag{5.5.4}$$

The trace operator bilinear form (3.6.1) then takes the form

$$b(v, \widetilde{q}) = \int_\Gamma v|_\Gamma \, \widetilde{q} \, d\mu = \int_0^1 v(\kappa(t)) \, \widetilde{q}(t) \, dt, \qquad \text{for all } v \in \Psi^\square, \widetilde{q} \in \widetilde{\Psi}^\Gamma, \tag{5.5.5}$$

where $\Psi^\square$, $\widetilde{\Psi}^\Gamma$ denote primal and dual wavelet bases on $\square$ and Γ, respectively. The bilinear form cannot be evaluated straightforwardly because the trace $\psi_\lambda(\kappa(\cdot))$ is, unlike ψ_λ, not a piecewise polynomial. The approach to calculate the above terms undertaken in [47] was the **refineable integrals** scheme of [52], which, by a series of operators approximating the integral, converges towards the exact value. Since here we only want to do a short example, we use a simpler approximation strategy than that of [52], one that can be easily implemented using our existing tool set.

5.5.1 Application of the Trace Operator in Wavelet Coordinates

Instead of approximating the integrals, we approximate the trace $\psi_\lambda(\kappa(t))$ by the primal wavelets Ψ^Γ and evaluate the integral exactly. To this end, the trace of each function $\psi_\lambda^\square$ intersecting the circle Γ is discretized by the **single scale** generators Φ_J^Γ for a high level J. This expansion is then adaptively decomposed using Algorithm 3.2 giving the wavelet vector $\mathbf{d} \in \ell_2$. By biorthogonality $(\mathcal{B})$(2.1.54) and linearity of (5.5.5) then follows

$$b\left(\sum_\lambda d_\lambda \psi_\lambda^\Gamma, \widetilde{\psi}_\nu^\Gamma\right) = \sum_\lambda d_\lambda \int_\Gamma \psi_\lambda^\Gamma \widetilde{\psi}_\nu^\Gamma \, d\mu = \sum_\lambda d_\lambda \delta_{\lambda,\nu} = d_\nu, \qquad \text{for fixed wavelet index } \nu.$$

The result vector $\psi_\lambda^\Omega|_\Gamma$ is thus directly given by the decomposed wavelet expansion vector $\mathbf{d}$.

Remark 5.27 *This resulting set of wavelet coefficients could also be used as a starting point for the refineable integrals scheme. Having identified the relevant wavelet coefficients this way, determination of the values of the bilinear form could be delegated to the algorithm of [52].*

The application of the trace operator $\mathbf{B}$ to a vector $\mathbf{y} \in \ell_2$ is then straightforward: Sum up all result vectors until a specified tolerance $\varepsilon > 0$ is met. Alternatively, sum up all individual vectors and coarse the result appropriately to obtain a tree of minimal size with an accuracy still smaller than ε.
If the discretization is done on a high enough level $J \gg j_0$, then the computed values should be at least accurate to **discretization error** of the maximum level used. To this end, it can also be prudent to set an admissible highest level for the adaptive structures, see Section 4.5.1, lower than J. This approach can be easily implemented by considering the following details:

- Because of (2.1.16), only a finite number of wavelets $\psi_\lambda^\square$ on each level intersect with the circle Γ. On the coarsest level j_0, all wavelets can be checked for intersecting with Γ. On the higher levels $j > j_0$, only the wavelets intersecting the supports of the wavelets on the previous levels have to be considered. This set can be easily computed using the **Support Operator** of Remark 3.55.

- Whether a function crosses the circle Γ can easily be determined in the **piecewise polynomial** representation (3.4.3). A **support cube** $\square \subset S_\lambda$ (3.4.4) intersects Γ if at least one vertex lies inside the disk encircled by Γ and at least one lies outside.

- Only those single scale functions of Φ_J^Γ have to be considered that actually lie within the support of $\psi_\lambda^\square(\kappa(\cdot))$. By determining the support cubes $\square \subset S_\lambda$ and the points (x_1, x_2) of intersection with Γ, then $t = \kappa^{-1}(x_1, x_2)$ gives the minimum and maximum locations $k_1, k_2 \in \Delta_J$ between which the support of $\psi_\lambda^\square(\kappa(\cdot))$ falls. Thus, in general only $k_2 - k_1 + 1$ function evaluations are necessary.

The set of wavelet coefficients determined in this manner can be seen in Figure 5.8.

5.5.2 Numerical Results

The following numerical example was using `DKU22NB-SMD-SMOrth` wavelets on the domain $\square$ and `DKU22PB-SMD-SMOrth` wavelets on the boundary Γ. The Uzawa algorithm was executed with a target tolerance of $\varepsilon = 1 \times 10^{-3}$. Diagrams and plots of the solution are given in Figure 5.9 and Figure 5.10. As expected, most of the wavelet coefficients of the solution component y are in the vicinity of the trace boundary Γ. Since the problem has rotational symmetry for every multiple angle of $\frac{\pi}{2}$, the wavelet coefficients are evenly distributed according to this symmetry. As anticipated, the plot of the state component y constrained to Γ shows that the boundary conditions on Γ are fulfilled up to the target tolerance 1×10^{-3}. For this, only the coarsest three levels of wavelet coefficients for the vector p are necessary, as seen in its wavelet diagram.
The iteration histories of the vectors can be found in Figure 5.12. The reference solution was computed on a maximum level of $J = 7$ by a `QU` decomposition, its accuracy is bounded by the discretization error of $\sim 10^{-2}$. With just a few data points, a good estimate of the approximation rate depicted in Figure 5.11 is very difficult. It seems the approximation rate of component p is slightly larger than the rate of component y.

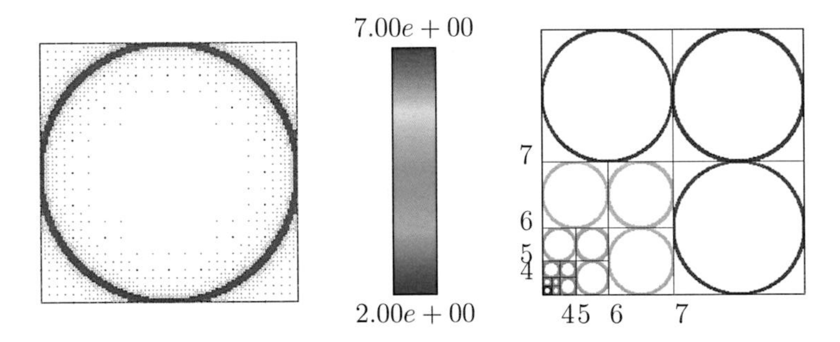

Figure 5.8: Computed wavelet coefficients intersecting with the circle centered around $(\frac{1}{2}, \frac{1}{2})$ with radius $R = \frac{1}{2}$. The coefficients are colored according to their level $|\lambda|$. With increasing level, the coefficients' support reduce in size and thus have to be centered closer towards the circle itself. The number of wavelet coefficients intersecting any point on Γ on each level is bounded uniformly by a constant.

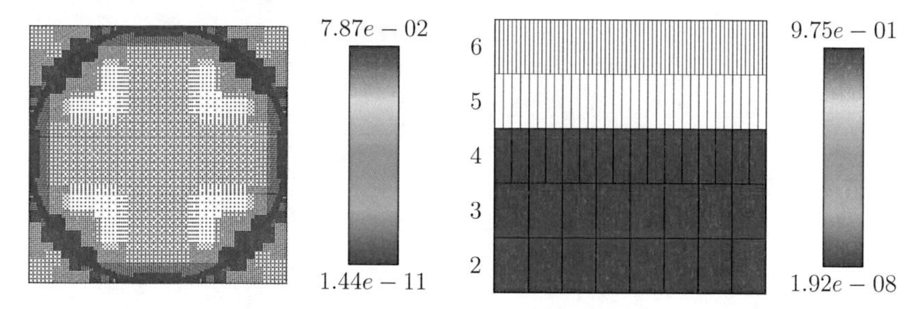

Figure 5.9: Solution of (5.5.1) solved to the accuracy $\varepsilon = 1 \times 10^{-3}$. The left diagram shows the wavelet coefficients distribution of the y component, the right diagram the wavelet coefficients of the p component. The values on coarsest level of p are not symmetrically distributed because of the basis transformation (2.3.22). Without it, the values follow the periodic nature of the circle w.r.t. the domain $\square$.

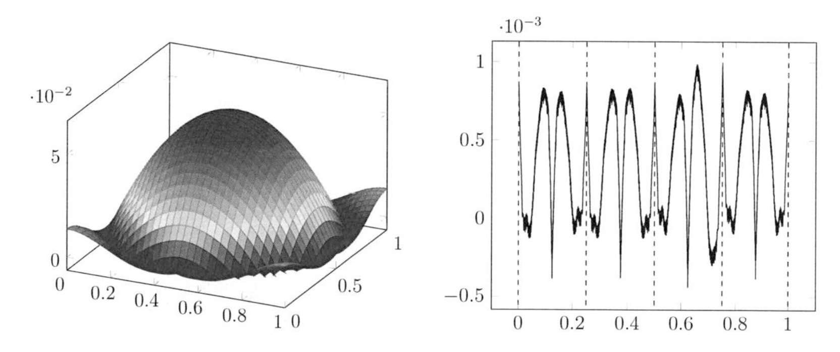

Figure 5.10: Solution of (5.5.1) solved to the accuracy $\varepsilon = 1 \times 10^{-3}$. The left plot shows the y component, the right plot the trace $y|_\Gamma$. The dashed lines indicate the parameter values where the circle touches the middle of the edges of the square domain $\square$.

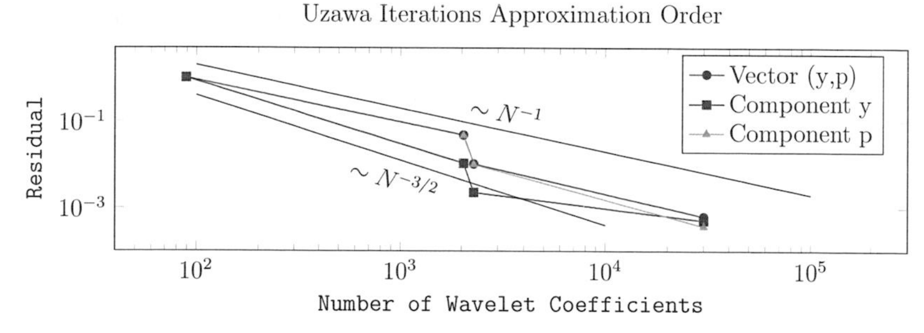

Figure 5.11: Approximation orders of the overall vector (y, p) and the individual components. The y component is discretized in `DKU22NB-SMD-SMOrth` wavelets, the p component uses `DKU22PB-SMD-SMOrth` wavelets.

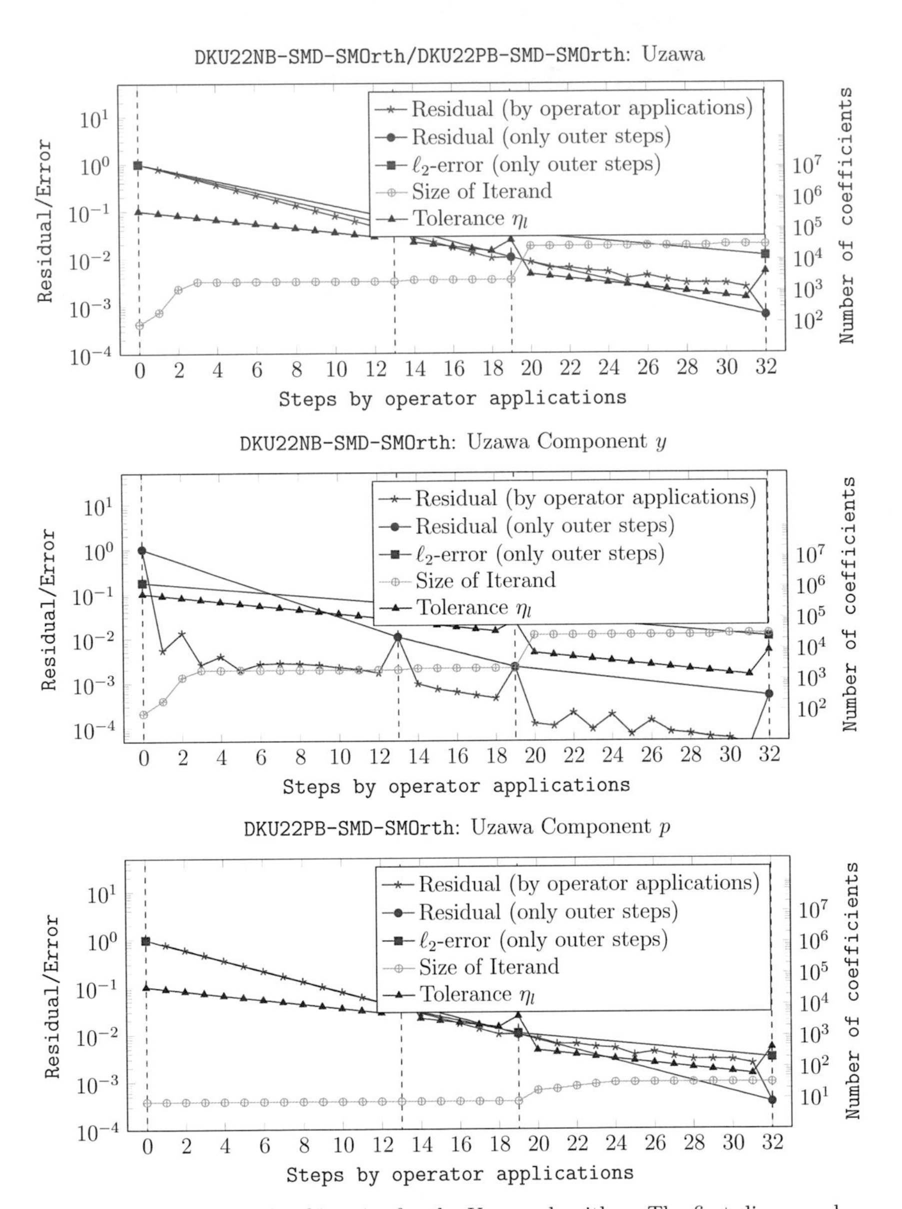

Figure 5.12: Detail iteration histories for the Uzawa algorithm. The first diagram shows the overall values, the other two diagrams deal with the components y and p, respectively. Since the starting vectors contain zero values only and $g = 0$, the residual in the first step for p is exactly zero, which is not displayed on the logarithmic scale.

5.6 A 2D Nonlinear Boundary Value Example Problem

As our nonlinear two-dimensional example problem, we seek to solve the boundary value problem,

$$\begin{aligned} -\Delta y + y^3 &= f, && \text{in } \Omega, \\ y &= 0, && \text{on } \Gamma, \\ \nabla y \cdot \mathbf{n} &= 0, && \text{on } \partial\Omega \setminus \Gamma, \end{aligned} \tag{5.6.1}$$

with $\Omega := (0,1)^2$ and $\Gamma \equiv \Gamma_E$. The right hand side function f is again given as in Section 4.6. This problem formulation has the advantage of also being able to be formulated as a single nonlinear PDE. Since the support of the right hand side f is far enough away from the trace boundary Γ, the problem (5.6.1) can be solved using the tensor product of `DKU22NZB` and `DKU22NB` wavelets, enforcing the homogeneous boundary conditions on Γ in the space of test functions. This allows a verification at least of the first component y of the computed solutions to (5.6.1). The solution computed in this way can be seen in Figure 5.13. The wavelet expansion vector of the second component p can then be recovered by using the system equations (5.2.19) and computing

$$\mathbf{p} = (\mathbf{B}\,\mathbf{B}^T)^{-1}\,\mathbf{B}\,\left(\mathbf{B}^T\mathbf{p}\right) = (\mathbf{B}\,\mathbf{B}^T)^{-1}\,\mathbf{B}\,(\mathbf{f} - \mathbf{F}(\mathbf{y}))\,. \tag{5.6.2}$$

But if the support of the dual wavelets associated to the boundary Γ_E is so large that it intersects with the support of the right hand side f, then the computed solution $\mathbf{y}$ would be slightly different and this shortcut does not work.

5.6.1 Numerical Results

In all numerical experiments, the target tolerance is $\varepsilon = 1 \times 10^{-4}$. The inner equation is solved using the Gradient iteration outlined in Section 4.3 using the optimal step size (5.3.6). The maximum values of the step size parameter α for the Uzawa algorithm can be found in Table 5.3. Figures 5.14–5.15 show the wavelet coefficient distribution of the computed solutions for the `DKU22NB-SMD-SMOrth` wavelets. The iteration histories for all wavelet configurations can be found in Figures 5.16–5.18. The approximation rates in Figure 5.19 show a comparable and even higher rate than in the linear example of Section 5.4. The relative runtimes presented in Table 5.4 show that the Uzawa solver adds only little complexity to the inner solver when comparing the runtimes with the results of Table 4.5. But the absolute runtimes show that a large number of steps is necessary until the target tolerance is met, which is a result of the small admissible step sizes used in the Uzawa algorithm.

Wavelet	`DKU22NB`	`DKU24NB`	`MP244NB`
`SMD-SMOrth`	5.3×10^{-2}	4.7×10^{-2}	4.4×10^{-2}

Table 5.3: Maximum values of the Uzawa step size parameter α for different preconditioners. See Section A for details on the names of the wavelet configurations.

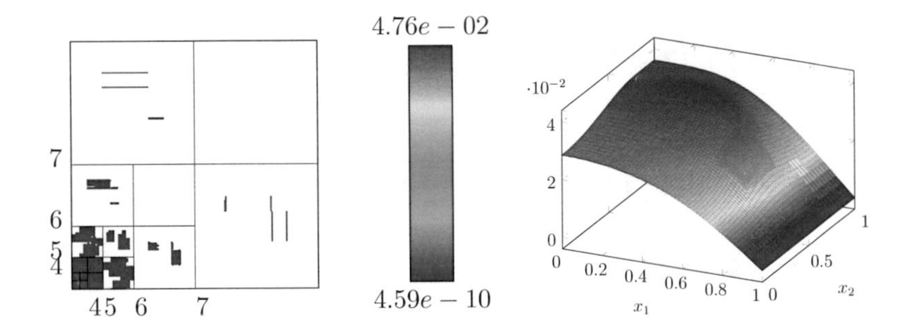

Figure 5.13: Plot of the solution and the isotropic coefficient diagram of the solution y to the nonlinear PDE $-\Delta y + y^3 = f$ with right hand side of (4.6.1) and zero valued boundary conditions on Γ_E. We used the solver of Section 4.3 with a target tolerance $\varepsilon = 1 \times 10^{-4}$.

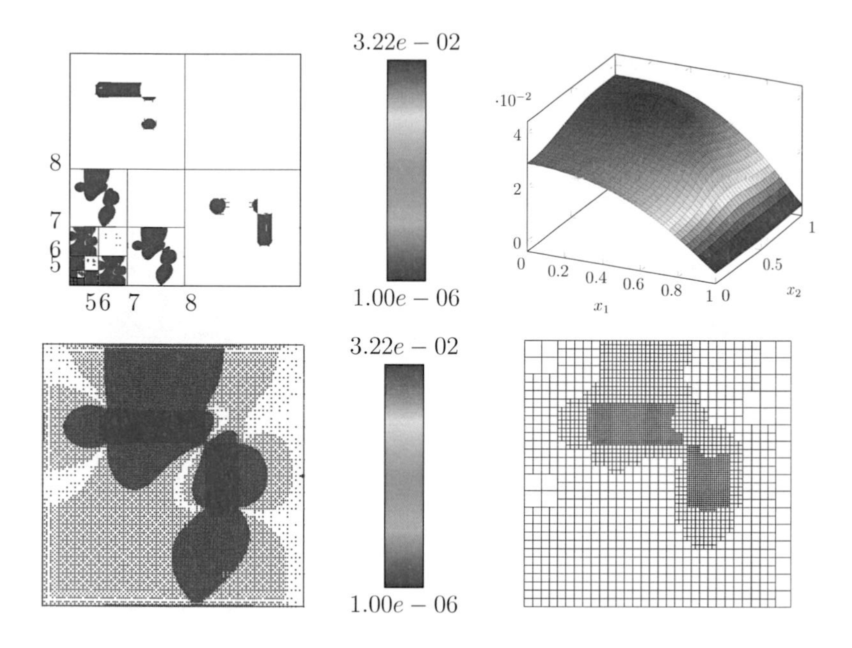

Figure 5.14: Plots of the solution component y of (5.6.1) for the `DKU22NB-SMD-SMOrth` wavelets and the isotropic coefficient diagrams. Wavelet coefficients with values below 1×10^{-6} were omitted to maximize recognizability of the distribution patterns. The plot of the function was rastered using 32 points in every direction.

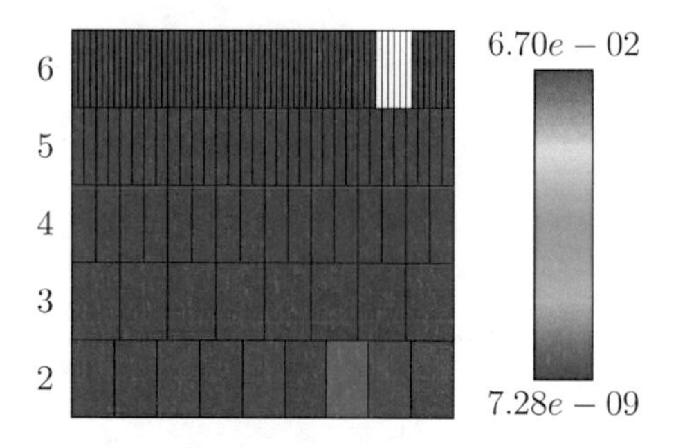

Figure 5.15: Diagram of the Lagrangian multiplier p for the solution part p of (5.6.1). The values on the coarsest level are not symmetrically distributed because of the basis transformation (2.3.22).

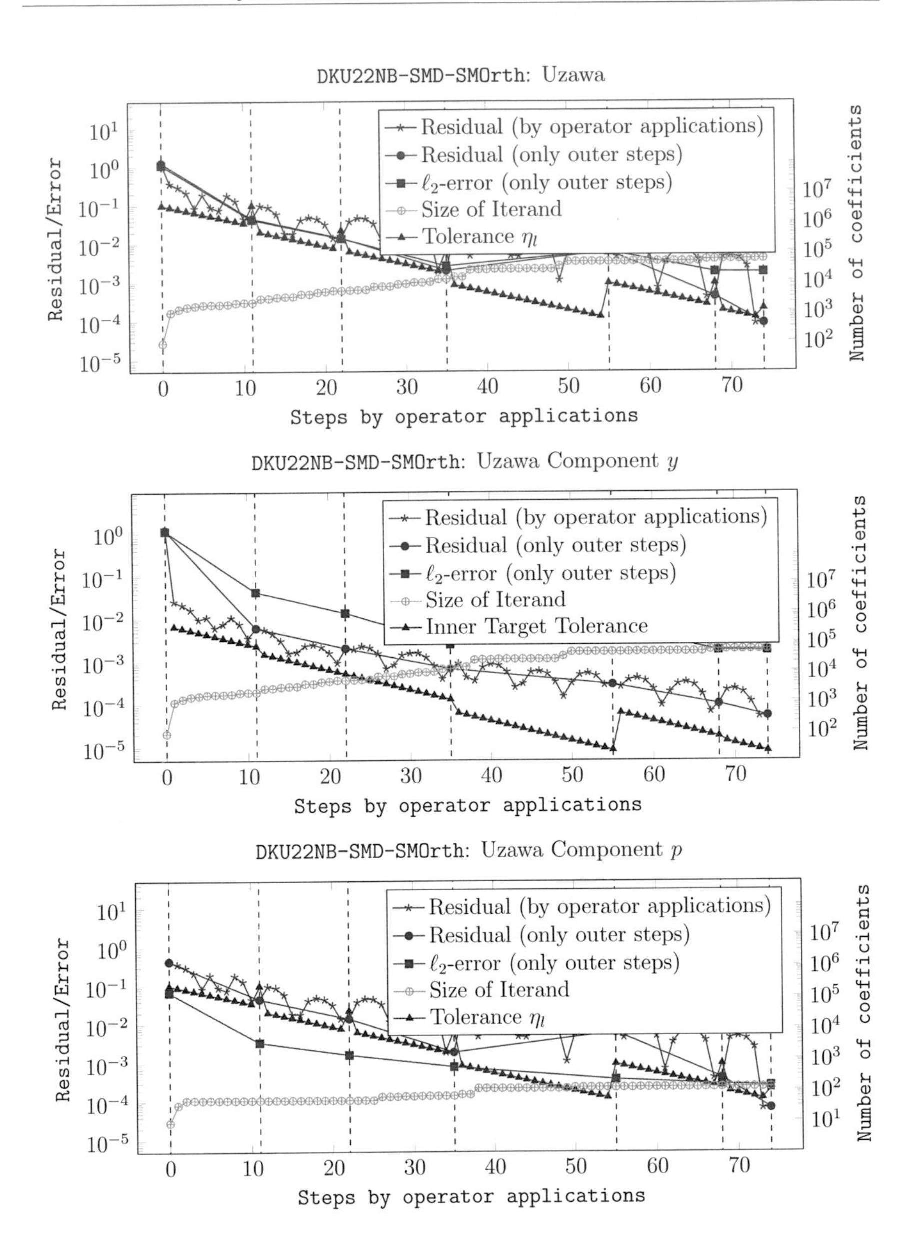

Figure 5.16: Detail iteration histories for the Uzawa algorithm. The first diagram shows the overall values, the other two diagrams deal with the components y and p, respectively.

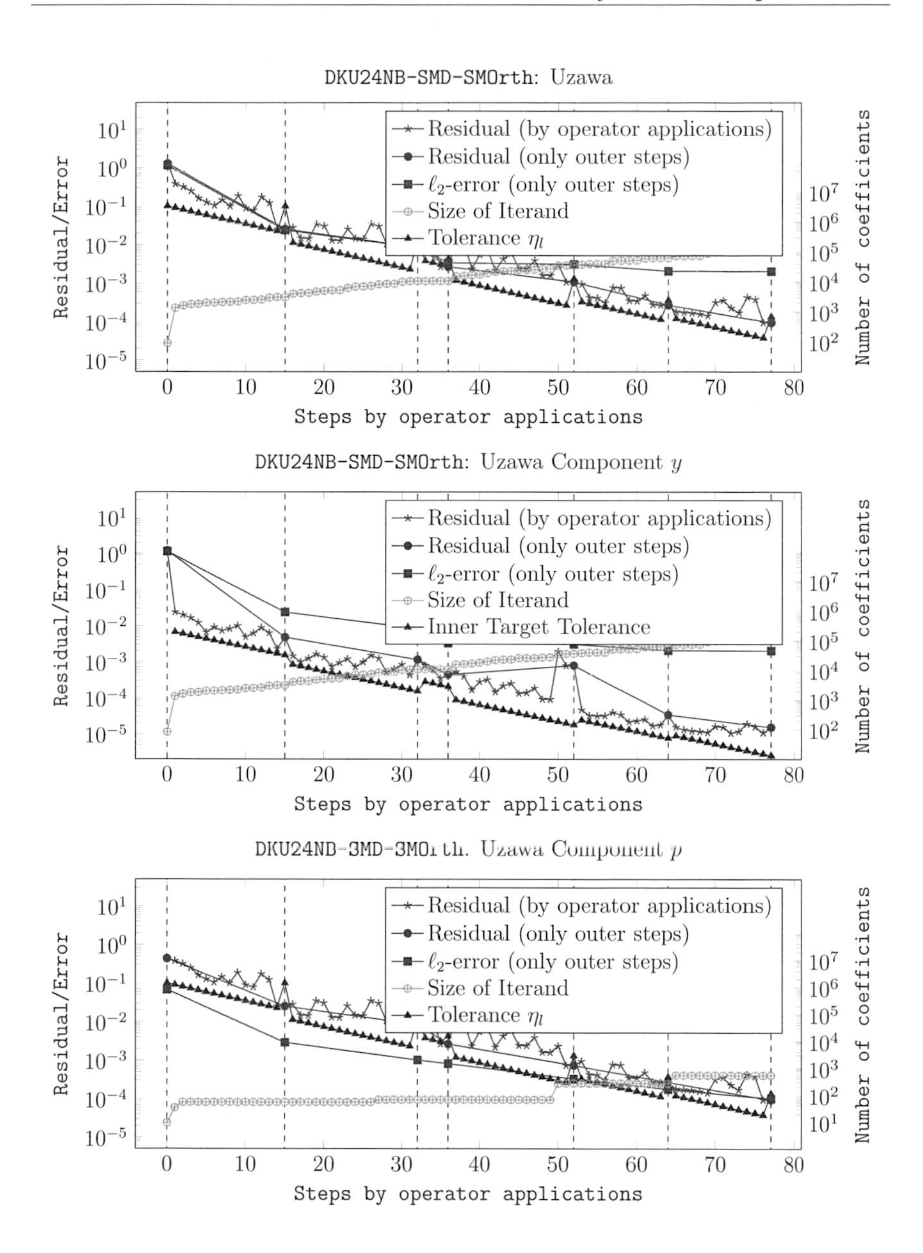

Figure 5.17: Detail iteration histories for the Uzawa algorithm. The first diagram shows the overall values, the other two diagrams deal with the components y and p, respectively.

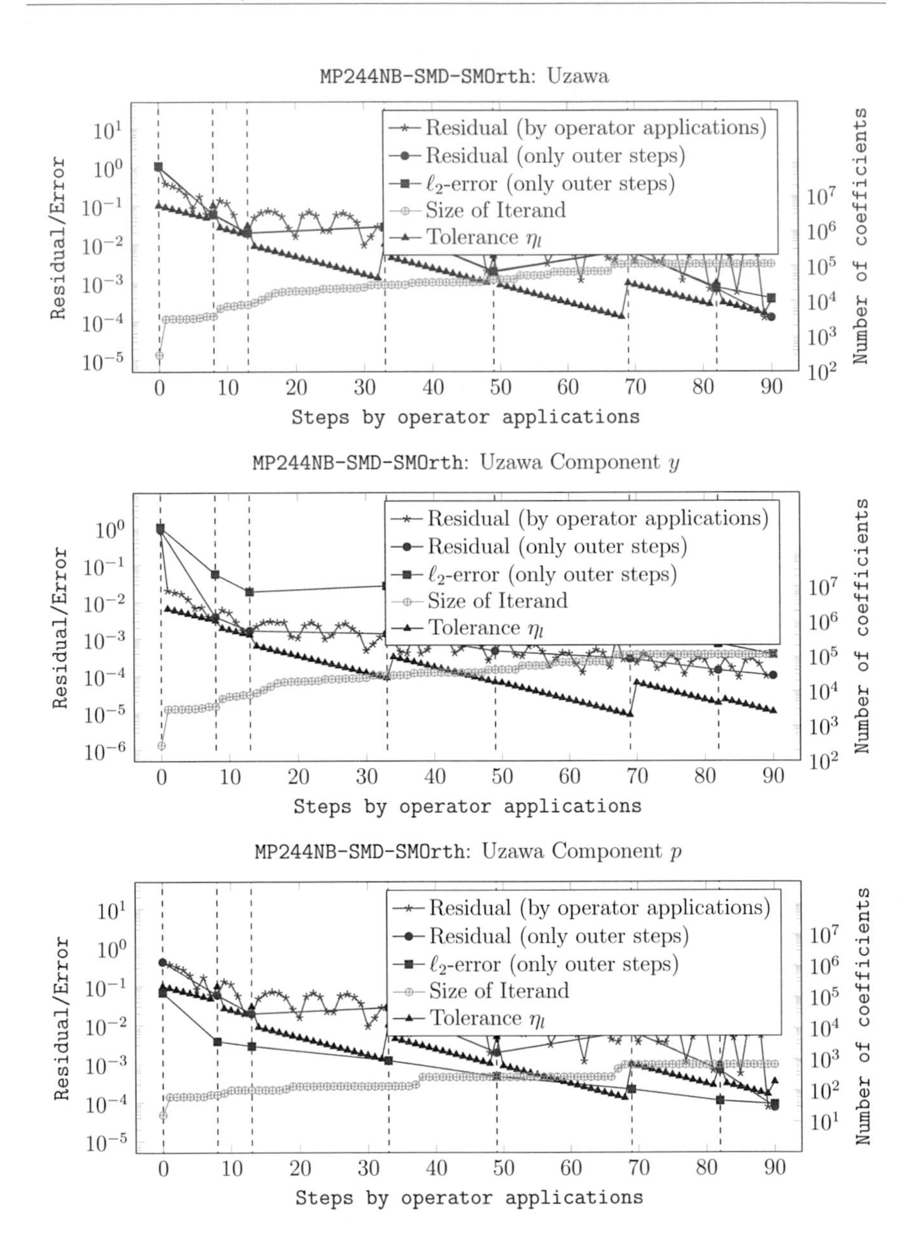

Figure 5.18: Detail iteration histories for the Uzawa algorithm. The first diagram shows the overall values, the other two diagrams deal with the components y and p, respectively.

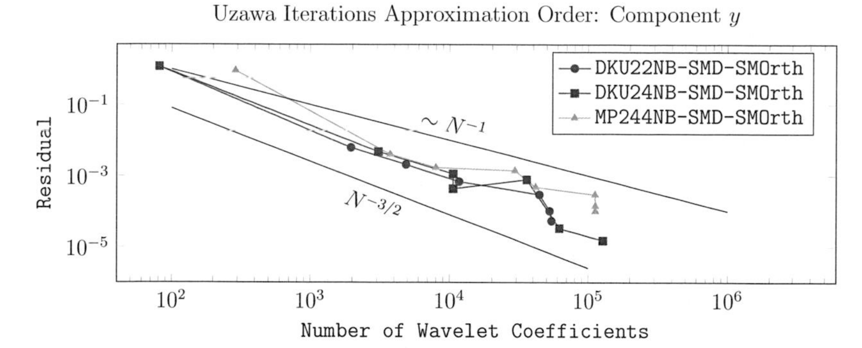

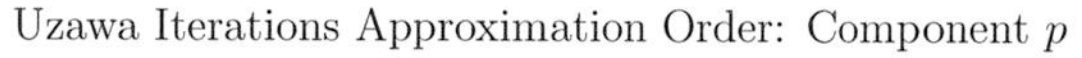

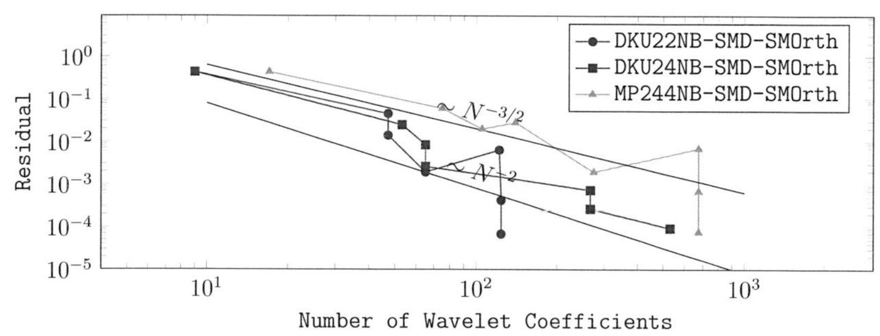

Figure 5.19: Approximation orders of the overall vector (y, p) and the individual components.

Wavelet	DKU22NB		DKU24NB	
Preconditioner	Overall	Per Wavelet	Overall	Per Wavelet
SMD-SMOrth	5.87×10^{3}	8.55×10^{-6}	1.14×10^{4}	1.46×10^{-5}

Wavelet	MP244NB	
Preconditioner	Overall	Per Wavelet
SMD-SMOrth	2.38×10^{4}	1.28×10^{-5}

Table 5.4: Runtimes of the Uzawa solver for different preconditioners.

5.7 A 3D Nonlinear Boundary Value Example Problem

As our nonlinear three-dimensional example problem, we seek to solve the boundary value problem,

$$\begin{aligned} -\Delta y + y^3 &= f, && \text{in } \Omega, \\ y &= 0, && \text{on } \Gamma, \\ \nabla y \cdot \mathbf{n} &= 0, && \text{on } \partial\Omega \setminus \Gamma, \end{aligned} \tag{5.7.1}$$

with $\Omega = (0,1)^3$ and $\Gamma \equiv \Gamma_R := \{(1, x_2, x_3) \,|\, 0 \leq x_2, x_3 \leq 1\}$. The right hand side function f is not given as in Section 4.7, because of the high number of wavelet coefficients which entails a long runtime. A reference solution for y can be computed by solving the nonlinear PDE using wavelets with zero boundary conditions on Γ_R, as in Section 5.6. But because the right hand side f has support on the whole domain Ω, formula (5.6.2) cannot be employed. Hence, we only use a reference for the component y in our experiments. Because of the runtimes of the experiments conducted in Section 4.7, we use the Richardson solver Section 4.2 for computing the solution of the PDE.

5.7.1 Numerical Results

In all numerical experiments, the target tolerance is $\varepsilon = 1 \times 10^{-2}$. The decay parameter (3.3.14) was increased to $\gamma = 11$ and the number of internal steps in the Uzawa algorithm, i.e., the number of steps before the internal tolerance was reset and a coarsening was applied, was set to 10 to achieve lower runtimes. The exact Influence set (3.3.18) was used for solving the nonlinear subproblem because this improved the convergence properties. The maximum values of the step size parameter α for the Uzawa algorithm can be found in Table 5.5. For the right hand side f, a constructed wavelet expansion vector, which can be seen plotted on Figure 5.20, was used. This vector was generated by inserting two random coefficients on a high level and this vector was made into an expanded tree (3.2.5). The values of the individual wavelet coefficients were then randomized but chosen proportional to $\sim 2^{-|\lambda|}$, i.e., exponentially decaying by level. The right hand side vectors of the individual wavelet configurations, e.g., `DKU22NB` and `DKU24NB`, are thus not describing the same function. The computed solutions are thus also not describing the same function, but the similarities in the right hand side functions should suffice for a qualitative comparison of the Uzawa solver. As can be seen in Figure 5.22, the structure of the y function is very complicated and it does not represent a classical analytical function. The wavelet coefficient distribution given in Figure 5.21 shows that a high spatial concentration of wavelet coefficients can also be found at positions where the right hand side f does not have any wavelet coefficients, an effect that is expected for nonlinear operators. The distribution of the wavelet coefficients of the solution component p on the boundary Γ_R in Figure 5.27 shows no specific patterns, only a few coordinates with a higher concentration of coefficients. Figures 5.23–5.25 show the iteration histories of the Uzawa solver using the step sizes given in Table 5.5. It is again clearly recognizable that higher smoothness of the dual wavelets has a very positive impact on the efficiency of the Uzawa algorithm in terms of runtimes and computational complexity. The values of the approximation rates exhibited in the diagrams of Figure 5.26 are at least as high as in the 3D example of Section 4.7. The comment about the runtimes in Section 5.6 applies again when comparing the values in Table 5.6 to the values given in Table 4.8.

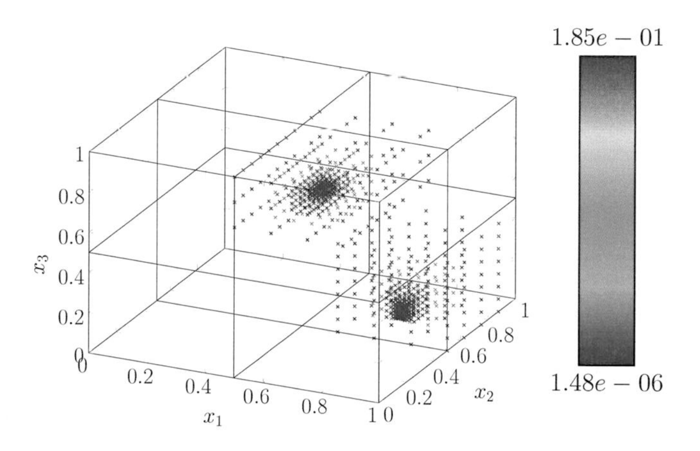

Figure 5.20: Scatter plot of the right hand side f of (5.7.1) for the DKU22NB wavelets.

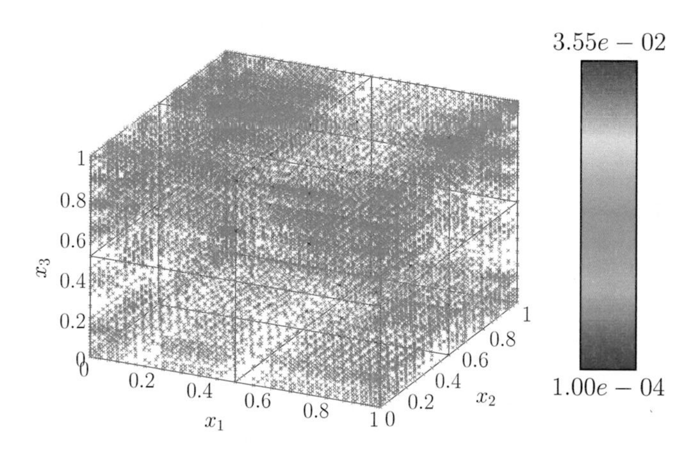

Figure 5.21: Scatter plot of the wavelet coefficients of the solution y of (5.7.1) for the DKU22NB wavelets. The boundary Γ_R is the right facing side of the cube.

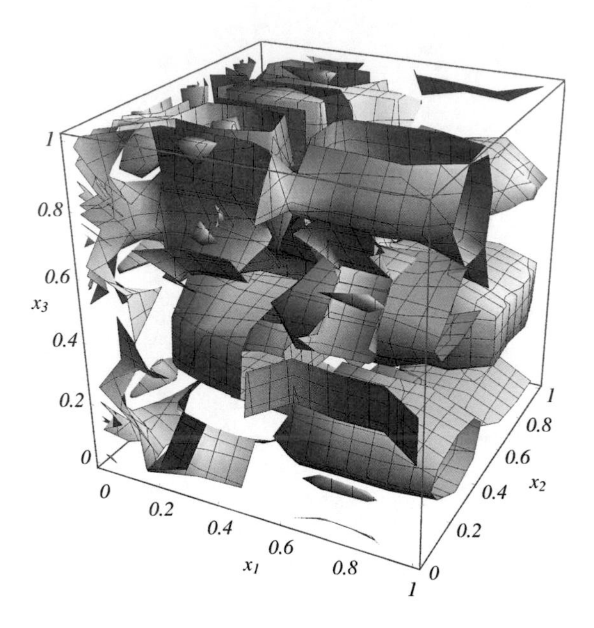

Figure 5.22: Contour plot of the solution y of (5.7.1) for the DKU22NB wavelets. The boundary Γ_R is the right facing side of the cube.

Wavelet	DKU22NB	DKU24NB	MP244NB
SMD-SMOrth	2.3×10^{-2}	2.5×10^{-2}	3.0×10^{-2}

Table 5.5: Maximum values of the Uzawa step size parameter α for different preconditioners. See Section A for details on the names of the wavelet configurations.

Wavelet	DKU22NB		DKU24NB	
Preconditioner	Overall	Per Wavelet	Overall	Per Wavelet
SMD-SMOrth	8.47×10^{5}	6.60×10^{-5}	1.97×10^{5}	1.09×10^{-4}

Wavelet	MP244NB	
Preconditioner	Overall	Per Wavelet
SMD-SMOrth	2.04×10^{5}	1.32×10^{-4}

Table 5.6: Runtimes of the Uzawa solver for different wavelet types and preconditioners.

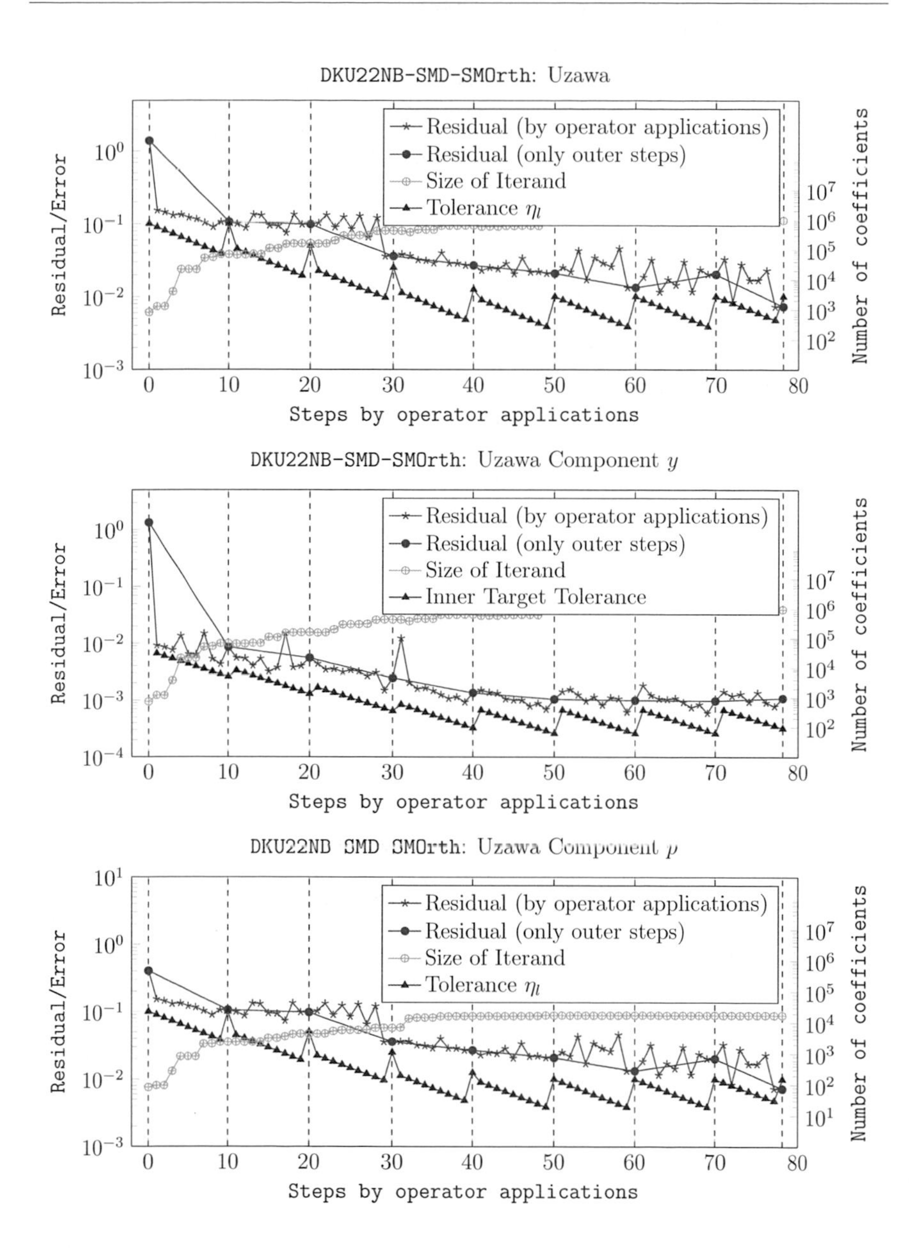

Figure 5.23: Detail iteration histories for the Uzawa algorithm. The first diagram shows the overall values, the other two diagrams deal with the components y and p, respectively.

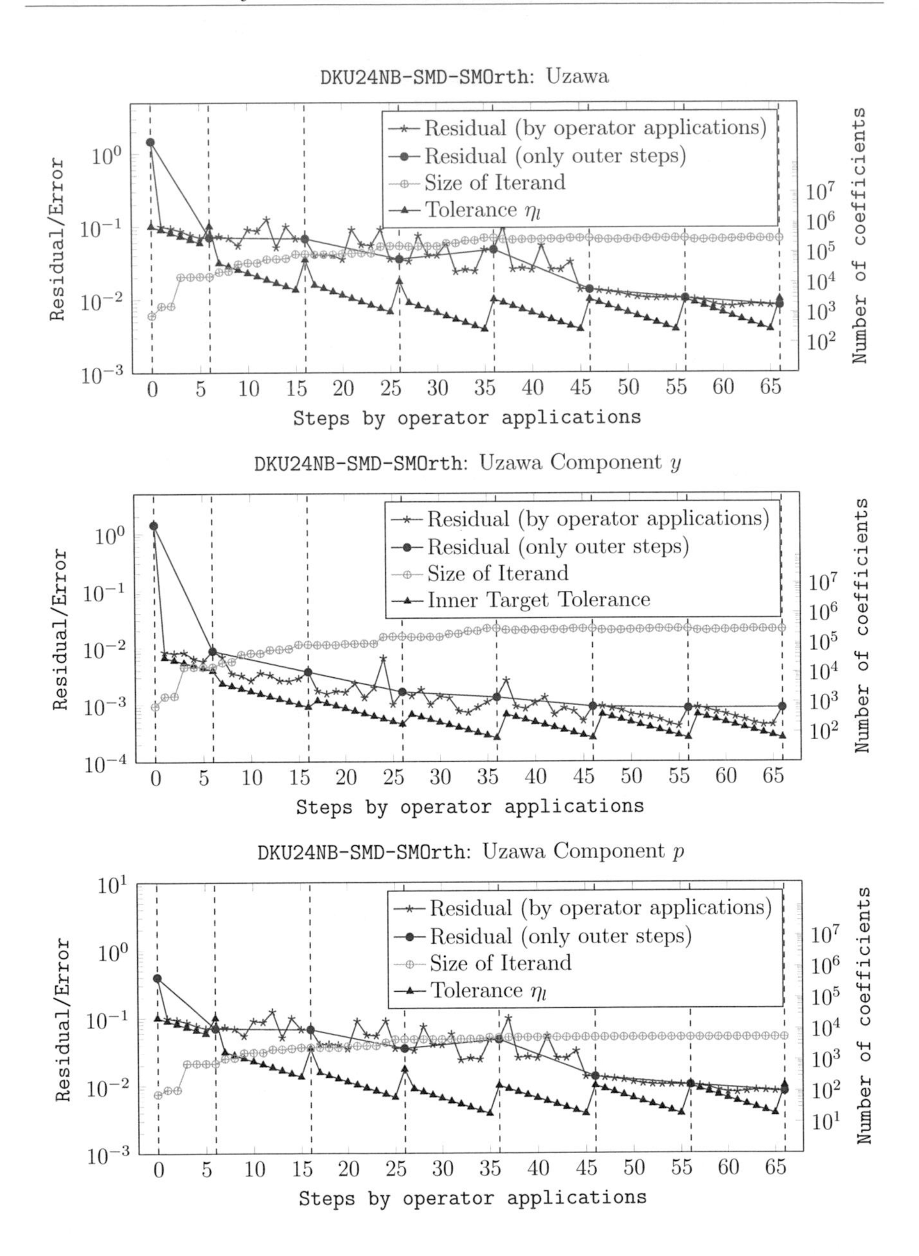

Figure 5.24: Detail iteration histories for the Uzawa algorithm. The first diagram shows the overall values, the other two diagrams deal with the components y and p, respectively.

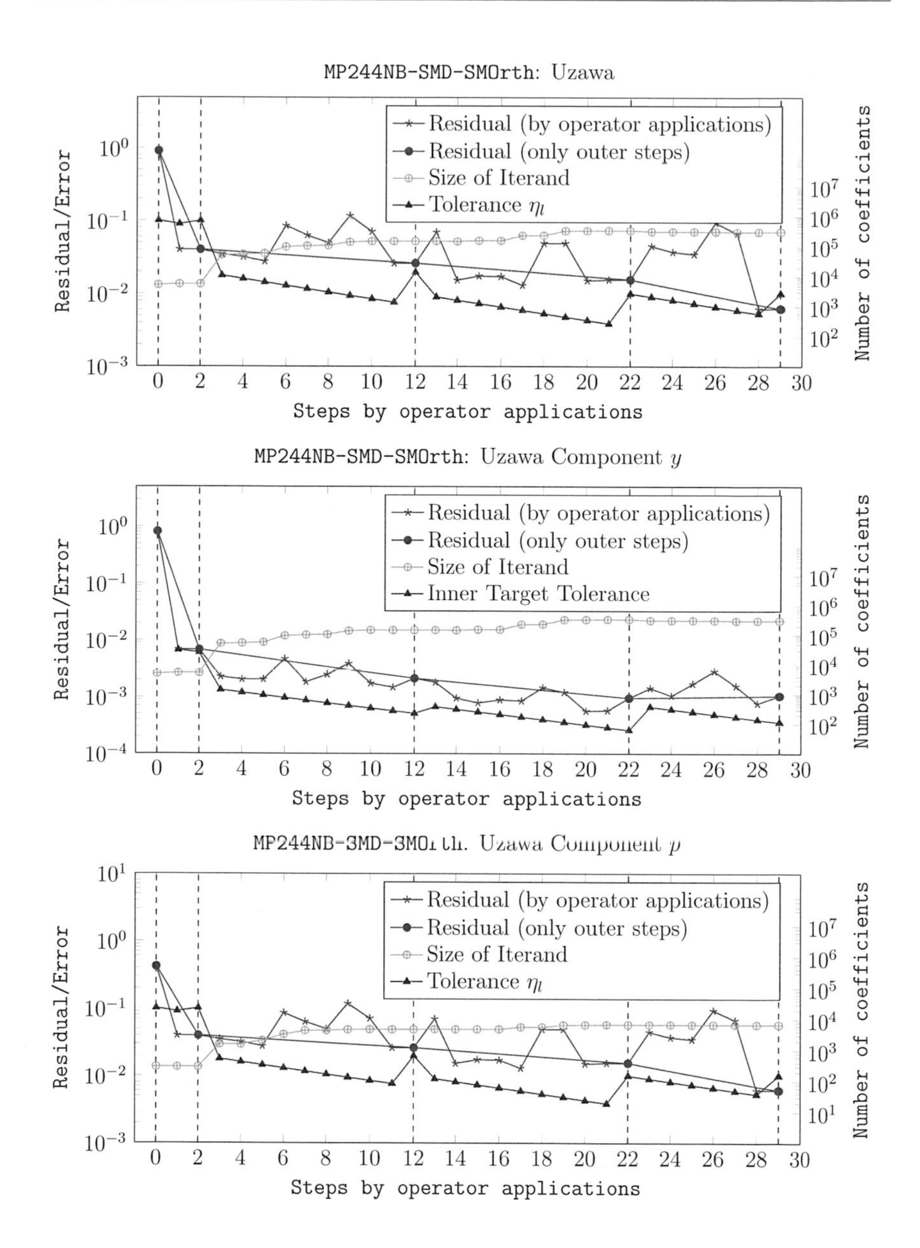

Figure 5.25: Detail iteration histories for the Uzawa algorithm. The first diagram shows the overall values, the other two diagrams deal with the components y and p, respectively.

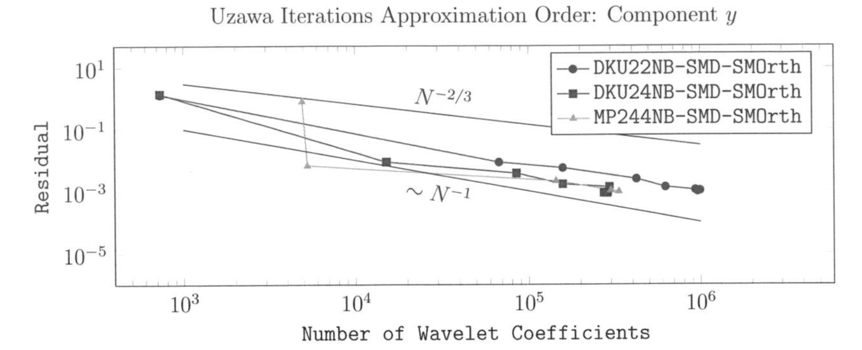

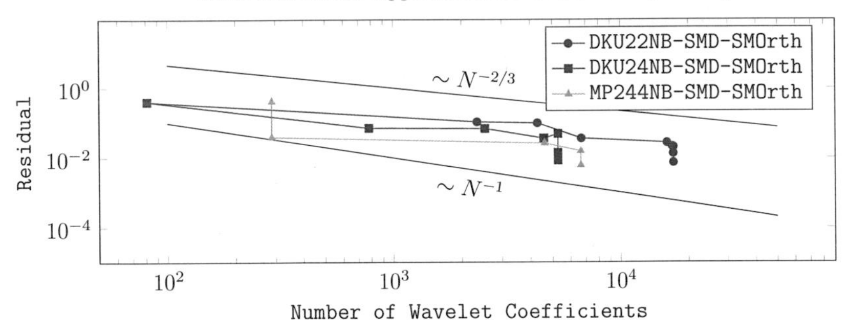

Figure 5.26: Approximation orders of the overall vector (y, p) and the individual components.

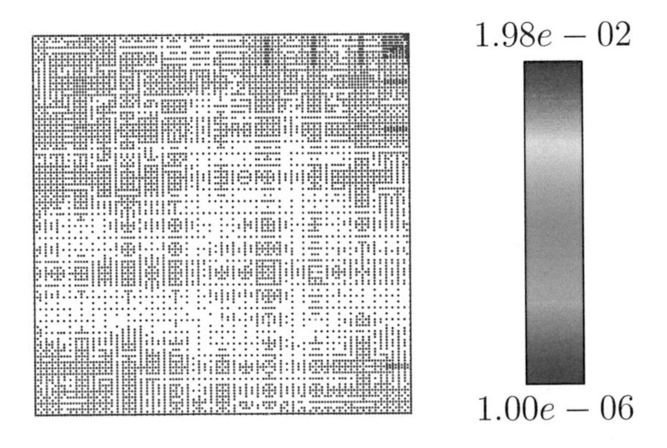

Figure 5.27: Scatter plot of the wavelet coefficients of the solution component p on Γ_R of (5.7.1) for the DKU22NB wavelets. The lower limit of the value of the wavelet coefficients in this plot was set to 1×10^{-6}.

6 Résumé and Outlook

6.1 Conclusions

In this thesis I have presented the numerical realization of the theoretical framework on adaptive wavelet methods put forth in [32–34]. The code works for generic tensor product domains, can employ different wavelet configurations in different spatial directions and offers high customizability using compile and command line options.

In [115] we showed that a tree structured set of wavelet indices can be represented using local polynomials on a partition of standard cubes of the underlying domain. Linear and nonlinear local operators can then be applied piecewise on this set of disjoint cells and from this the wavelet coefficients of the target tree can be reconstructed. Both transformations and the operator can be proven to be applicable in linear time w.r.t. the number of wavelet coefficients, the numerical examples clearly demonstrate the linear complexity. Extending the family of operators, I implemented the means to adaptively apply linear operators given only by the values of the bilinear form and trace operators for several two- and three-dimensional domains.

The runtimes to solve a simple two dimensional nonlinear PDE in the wavelet context was, using the code of [147], in the realms of hours and could even take more than a day, see [137]. My code can accomplish the same task within minutes and sometimes even fractions of a second, in part owed to transformations in the wavelet bases lowering the absolute spectral condition number and in part to optimizations in the program code. The measured approximation rates and complexity estimates in example problems agree very well with the theoretical predictions, thus confirming the soundness of the overall adaptive wavelet method.

For boundary value problems based upon linear PDEs, theoretical results of complexity estimates and convergence rates extending those presented for PDEs are available. These results are also confirmed in my numerical experiments. In these computations, because of the implicit inverse operator application, dual wavelets with higher smoothness yield better performance. This can be understood by interpreting the inverse operator as an approximation to the dual operator, as is the case with the mass matrix. The adaptive data structures also allow easy set up of algorithms for trace operators for domains not parallel to Cartesian coordinate axes, enabling the handling of more complex domains, especially those encountered in real life problems. In the case of nonlinear boundary value problems, a theoretical proof has net yet been proven. The numerical results are encouraging, hinting at results similar to the linear case. A possible remedy to the long runtimes is a scheme similar to the positive definite system explained in Section 5.3.2. Since in this setting the inner system is not solved explicitly, it can possibly deliver the solution much faster.

In summary, the numerical experiments confirm the theoretical predictions perfectly, from the linear complexity of operators to uniformly bounded condition numbers due to preconditioners to the approximation rates when computing the solutions of semilinear PDEs. Still, there are some instances to improve the results given in this thesis. Most importantly, a more effective algorithm to solve semilinear saddle point problem has to be devised. Although the linear complexity of the adaptive Uzawa algorithm is clearly demonstrated and the approximation rates are high, the step size parameters have to be chosen relatively small to compute the result in just a few iteration steps. This is the main obstacle on the way to more complex problems based upon nonlinear boundary value problems.

6.2 Future Work

6.2.1 Control Problems Governed by Nonlinear Elliptic PDEs

It was originally envisioned for this work to cover PDE constrained nonlinear elliptic control problems with Dirichlet boundary control, i.e, problems of the form

$$\text{Minimize} \qquad J(y,u) = \frac{1}{2}\|y - y_{\Gamma_y}\|^2_{H^s(\Gamma_y)} + \frac{\omega}{2}\|u\|^2_{H^t(\Gamma)},$$

$$\text{under the constraint} \qquad \begin{aligned} -\Delta y + y^3 &= f \quad \text{in } \Omega, \\ y &= u \quad \text{on } \Gamma \subset \partial\Omega, \end{aligned}$$

with given right hand side $f \in (H^{-1}(\Omega))'$, target function $y_{\Gamma_y} \in H^s(\Gamma_y)$ for $\Gamma_y \subset \partial\Omega$ and some $\omega > 0$. The natural Sobolev smoothness parameters in this case are the values $s, t = \frac{1}{2}$, this fact follows directly from the solution theory of the PDE and the trace theorems. Considering the necessary and sufficient conditions of this problem, one has to solve a system of (non-)linear boundary value problems.
The most important future work in this regard is the development of a solver for the nonlinear boundary value problem that can determine a solution faster than the Uzawa algorithms employed in Section 5. This goal could be achieved by some other solution method as discussed above, combined with parallelization strategies to lower execution times.

6.2.2 Parallelization Strategies

A very promising (yet highly complex to implement and execute) approach to speed up numerical computations is **parallelization**. That is, utilizing multiple arithmetic computing units at the same time.

As a first step, assuming that the compiler can identify code segments that can be executed in parallel, i.e., when consecutive computations do not depend on each others results, a **data parallelization** approach may be considered (see also Section B.1.1). This technique is particularly efficient when the same operation, e.g., a multiplication or an addition, has to be applied to each element in a vector of data values. Given that proper optimizations are enabled during the compilation process, such concurrent computations are often automatically detected by modern compilers. Practically however, this is only feasible if the data, to which the operation is applied, occupies consecutive memory cells. Although this is not true for the unordered containers employed in the software developed herein (see Section B.3), this technique is applied at numerous places in the code.

The next avenue which may be explored is the so called **(auto-)parallelization**. This involves executing repetitive code segments, e.g., a matrix multiplication, on multiple CPU cores simultaneously. This can be most easily accomplished using `OpenMP` directives, where the compiler sets up and dictates the complete code to (i) start different threads; and (ii) combine the results after each thread has finished its work, see [120].

The new `C++11` and `C++14` standards [141], also incorporate support for language based, and thus portable, asynchronous computations, i.e., thread enabled parallelization techniques. These new programming language features potentially allow one to harness the complete computational power of all the cores in a CPU at once, but require very little time and effort on the side of the developer. Such measures should be implemented as soon as the software support by the **compiler** developers is tried and tested.
If applied to the software developed herein, this technique would allow the parallel execution of the individual phases of the adaptive operator application detailed in Section 3.4.3. In particular, of great parallelization interest is the **operator application** Algorithm 3.8. This is due to the fact that this procedure is at the heart of all solution algorithms and is by far the most computationally involved within these schemes. By the results given in Figure 3.13, it is clear that the highest speedup can be achieved by parallelization of Algorithms 3.5 and 3.7, as these steps make up the majority of the complexity of the adaptive operator application. In effect, these algorithms consider the implementation of the **linear** operator (3.4.19) and its transpose. Exactly because these operators are linear, it is possible to compute the application in parallel and then combine the results, basically by adding up the components. It is worth underlining that this step requires synchronization of the data for the parallel program flows ("threads"); concurrent writes to a single variable or memory address can easily lead to race conditions and undefined behavior, i.e., erroneous data values being generated. On the other hand, synchronization is far from trivial. It entails a certain amount of computational complexity which in turn may incur a high computational overhead and hence a low parallel efficiency. However, the need for data synchronization can be minimized in the framework developed herein as the adaptive vector structures use independent data containers for each level. This means that an easy parallelization approach could be to assign to each thread one level of the process.

Lastly, if the combined computing power of the multiple cores in a modern CPU is not sufficient, **full parallelization** can be considered, i.e., running multiple instances of a program simultaneously on different computers not sharing a common memory space. In this setup, the uniform distribution and synchronization of data over a network is of utmost importance and a great difficulty. A software specifically designed to provide this function is called `MPI` [118]. Parallelization using `MPI` would require a refactoring of the source code, as in this setup the parallel nature of the program flow must be put first in the development process.
For such a fully parallelized program, a main task is **load-balancing**, i.e., distributing the data among all compute nodes in such a way that a comparable amount of complexity is executed by each node while at the same time minimizing the network traffic required for synchronization among different nodes. Dynamic load-balancing algorithms for structures based upon trees are already established [25], which could be a good starting point for a parallel adaptive wavelet code.

6.2.3 Parabolic Partial Differential Equations

As it was pointed out in Section 1.2.5, there are inherently **anisotropic** tensor product spaces. Such spaces can turn up in the context of parabolic or Schrödinger-type equations as test- and trial spaces by using full **space-time weak formulations**, see [113]. Classical approaches like the **method of lines** [132] or **Rothe's method** [84], which are essentially time marching methods, only look at a single point in time and then successively take a step towards the next one. The space-time weak formulation can treat the whole problem holistically, i.e., the explicit time dependency is eliminated, so that one can discretize in space and time simultaneously, as if time was an extra dimension in space. This in turn means that no time stepping is required and a **Petrov-Galerkin** (where the test and trial spaces are different) solution can be calculated directly in space and time.

A full space-time weak formulation can thus bring the full power of the adaptive wavelet theory into these kinds of problems.

Example 6.1 *Consider an abstract evolution equation*

$$\frac{du(t)}{dt} + A(t)u(t) = g(t) \quad \text{in } \mathcal{H}', \qquad u(0) = u_0 \quad \text{in } \mathcal{V}, \tag{6.2.1}$$

for $t \in [0,T]$ a.e., initial condition $u_0 \in \mathcal{V}$ and an operator $A(t) : \mathcal{H} \to \mathcal{H}'$ on the Hilbert space **Gelfand triple** *$\mathcal{H} \hookrightarrow \mathcal{V} \cong \mathcal{V}' \hookrightarrow \mathcal{H}'$. The right hand side is a mapping $g(\cdot) : [0,T] \to \mathcal{H}'$ which needs to satisfy certain smoothness criteria. These are given by* **Sobolev-Bochner spaces**, *see [6, 56], e.g., $g \in L_2(0,T;\mathcal{H}')$.*
To arrive at the space-time weak formulation, the equation (6.2.1) is tested using a function $v \in L_2(0,T;\mathcal{H})$ and integrated over $t \in [0,T]$. By additionally taking the initial condition into account, we arrive at the variational problem of finding a solution

$$u \in X := L_2(0,T;\mathcal{H}) \cap H^1(0,T;\mathcal{H}') \tag{6.2.2}$$

such that

$$b(u,v) = f(v), \qquad \text{for all } v = (v_1,v_2) \in Y := L_2(0,T;\mathcal{H}) \times \mathcal{V}, \tag{6.2.3}$$

with $b(\cdot,\cdot) : X \times Y \to \mathbb{C}$ defined by

$$b(u,(v_1,v_2)) := \int_0^T \left(\left\langle \tfrac{du(t)}{dt}, v_1(t) \right\rangle + \langle A(t)u(t), v_1(t) \rangle \right) dt + \langle u(0), v_2 \rangle \,, \tag{6.2.4}$$

and right hand side $f(\cdot) : Y \to \mathbb{C}$ given by

$$f(v) := \int_0^T \langle g(t), v_1(t) \rangle \; dt + \langle u_0, v_2 \rangle \,. \tag{6.2.5}$$

It was shown in [133] that the operator $B : X \to Y'$ defined by $(Bu)(v) := b(u,v)$ is boundedly invertible for bounded and coercive linear operators $A(t)$. Stability results are presented in [5, 112].

A slightly different formulation to (6.2.1) was given in [29], where the initial condition is incorporated as a natural boundary condition instead of an essential one by applying integration by parts to the first term.
In both formulations, the trial- and test spaces are **Sobolev-Bochner spaces**, respectively intersections of them. It is well known, see [6], that Sobolev-Bochner spaces are

isometric to Hilbert tensor product spaces, so that one can identify the spaces of (6.2.2) and (6.2.3) as

$$X \cong L_2(0,T) \otimes \mathcal{H} \cap H^1(0,T) \otimes \mathcal{H}', \qquad Y \cong (L_2(0,T) \otimes \mathcal{H}) \times \mathcal{V}. \tag{6.2.6}$$

That is, here one actually needs to deal with anisotropic tensor products which implies that one needs to use anisotropic bases for discretization in order to obtain conforming subspaces. It was shown in [72,133] that wavelets of sufficiently high order d, $\widetilde{d}$ (to cover the full smoothness range required) can be rescaled in such a way that they form Riesz-bases for intersections of tensor product spaces. This is a crucial detail, as the **optimal convergence rate** for this problem depends on the type of wavelet basis construction used. Assuming $\mathcal{H} = H^m(\Omega)$ and wavelets of orders d_t and d_x for the time and the n space dimensions, then, for sufficiently smooth functions, the best possible approximation rate with an **isotropic wavelet basis** is

$$r^{\text{iso}} := \min(d_t - 1, \frac{d_x - m}{n}). \tag{6.2.7}$$

In contrast, the best possible approximation rate when using **anisotropic tensor product wavelet basis** is

$$r^{\text{ani}} := \min(d_t - 1, d_x - m), \tag{6.2.8}$$

which is independent of the dimension n. The kind of inverse dependency of the dimension n as in (6.2.7) is known as **the curse of dimensionality**. So one can circumvent the curse of dimensionality here by using anisotropic tensor product wavelets instead of an isotropic construction, thus ensuring a high optimal convergence rate for all space dimensions. A more detailed description can be found in [133], see also the references therein. A good introduction and numerical experiments with isotropic wavelets can be found in [137].
These techniques could also serve as a basis for adaptive wavelet methods for control problems based upon parabolic PDEs as devised in [74]. Furthermore, an extension to control problems with elliptic, parabolic and stochastic PDEs, as developed in [103], is a prospective future application.

A Wavelet Details

In our software, we have implemented a number of wavelet constructions with varying boundary types and polynomial orders. It is possible to choose different wavelet types for different dimensions, even at runtime. The list of one-dimensional wavelet types implemented include:

`HAAR11PB`	Haar Wavelets (orthogonal), $d = 1$, $\widetilde{d} = 1$, $j_0 = 1$, periodic boundary.
`DKU22NB`	DKU Wavelets, $d = 2$, $\widetilde{d} = 2$, $j_0 = 3$, boundary adapted.
`DKU22ZB`	DKU Wavelets, $d = 2$, $\widetilde{d} = 2$, $j_0 = 3$, zero boundary.
`DKU22PB`	DKU Wavelets, $d = 2$, $\widetilde{d} = 2$, $j_0 = 3$, periodic boundary.
`DKU22NZB`	DKU Wavelets, $d = 2$, $\widetilde{d} = 2$, $j_0 = 3$, left ($x = 0$): boundary adapted, right ($x = 1$): zero boundary.
`DKU22ZNB`	DKU Wavelets, $d = 2$, $\widetilde{d} = 2$, $j_0 = 3$, left ($x = 0$): zero boundary, right ($x = 1$): boundary adapted.
`DKU24NB`	DKU Wavelets, $d = 2$, $\widetilde{d} = 4$, $j_0 = 3$, boundary adapted.
`CB24NB`	DKU Wavelets with scaling (A.1.14), $d = 2$, $\widetilde{d} = 4$, $j_0 = 3$, boundary adapted.
`MP244NB`	Primbs Wavelets, $d = 2$, $\widetilde{d} = 4$, $j_0 = 4$ boundary adapted.
`DKU33PB`	DKU Wavelets, $d = 3$, $\widetilde{d} = 3$, $j_0 = 3$, periodic boundary.

Since we only need boundary adapted wavelets in the course of this work, we present only the details of these wavelets in the next section.

Preconditioner and Basis Configurations

We use the following textual abbreviations to identify specific wavelet configurations.

`DKU22NB-P2`	The wavelet details can be found in Section A.1.4, the preconditioner in (2.2.15).
`DKU22NB-SMD`	Same wavelets, but with preconditioner (2.2.33).
`DKU22NB-SMD-SMOrth`	Additionally to the above, the orthogonal basis transformation (2.3.17) is employed, resulting in the setup given in (2.3.23).

The names for the wavelet types `DKU24` of Section A.1.5 and `P244` of Section A.1.6 are constructed in the same manner. If the boundary conditions are omitted anywhere in this document in the names of wavelets, it can be assumed to be of type `NB`.

A.1 Boundary Adapted Wavelets on the Interval $(0,1)$

Here we briefly show all the necessary data needed to use a specific wavelet in a numerical implementation. The theoretical background can be found in Section 2.3. More details can be found in [122] and in the references given in the specific sections.

The dual wavelets and generators are here not known explicitly and in particular, are not piecewise polynomials and thus cannot be plotted directly. Their existence is assured by Theorem 2.32 which suffices for our purpose. However, by (2.1.60) and (2.1.62), we can project the space $S(\widetilde{\Psi}_j)$ onto the space $S(\Phi_J)$, thus approximating the dual wavelets with piecewise linear functions. Obviously this method is inexact, but choosing $J \gg j_0$ and then plotting only dual functions on level j_0 leads to a sufficiently accurate visualization.

A.1.1 Boundary Adapted Hat Functions $d = 2$

Hat Functions consist of piecewise linear polynomials. The base can be defined for all $j_0 \geq 0$, and it holds

$$\#\Delta_j = 2^j + 1, \quad \#\nabla_j = 2^j. \tag{A.1.1}$$

The **hat function** is given as

$$\varphi_{j,k}(x) := \begin{cases} 2^j x - (k-1), & 2^j x, \in [k-1, k), \\ -2^j x + (k+1), & 2^j x, \in [k, k+1], \\ 0, & \text{otherwise.} \end{cases} \tag{A.1.2}$$

The primal **generators** on level $j \geq j_0$ are given by

$$\phi_{j,k} = 2^{j/2} \begin{cases} \varphi^-_{j,k+1}, & k = 0, \\ \varphi_{j,k}, & k = 1, \ldots, 2^j - 1, \\ \varphi^+_{j,k-2}, & k = 2^j, \end{cases} \tag{A.1.3}$$

where $\varphi^+_{j,k}$ is the first branch of the hat function (with positive slope) and $\varphi^-_{j,k}$ the second branch (with negative slope). The **hat function** is known to be refinable, and in the form of (2.1.10), the **mask** is given by

$$\phi_{j,k} = \frac{1}{2\sqrt{2}}\phi_{j+1,2k-1} + \frac{1}{\sqrt{2}}\phi_{j+1,2k} + \frac{1}{2\sqrt{2}}\phi_{j+1,2k+1}$$

which we abbreviate as

$$\left\{ \frac{1}{2\sqrt{2}}, \frac{1}{\sqrt{2}}, \frac{1}{2\sqrt{2}}; \quad -1 \right\} \quad \text{for } k = 1, \ldots, \#\Delta_j - 2, \tag{A.1.4}$$

identifying the last number as the **offset** in the position to be added to the index $k \to 2k$ to the first function. Traversing the mask from left to right, the position is increased by one, the offset always added. It is easy to verify that the boundary generators are refinable with masks

$$\begin{aligned} &\left\{ \frac{1}{\sqrt{2}}, \frac{1}{2\sqrt{2}}; \quad 0 \right\} && \text{for } k = 0, \\ &\left\{ \frac{1}{2\sqrt{2}}, \frac{1}{\sqrt{2}}; \quad -1 \right\} && \text{for } k = \#\Delta_j - 1. \end{aligned} \tag{A.1.5}$$

Since the function are piecewise linear, it follows $d = 2$ for the parameter of $(\mathcal{P})$(2.2.3). The sparsity of $\mathbf{M}_{j,0}$ is thus obvious: we have at most three non-zero values per column. It is well known (see [53]), that piecewise linear and globally continuous wavelets are contained in the Sobolev space H^s up to $s < 3/2$, or, by Definition 2.17 we have $\gamma = 3/2$.

Tree Structure

For the tensor product constructions, we need to define two mappings: A mapping between $\Phi_j \leftrightarrow \Phi_{j+1}$ and one between $\Phi_j \leftrightarrow \Psi_j$.
Because the number of single scale functions $\#\Delta_j$ (A.1.1) grows slightly slower than 2, not every function can have exactly two children. One just needs to decide how to adapt the tree relation at the boundaries. We opted for a relation which is based on the tree structure of **hat functions with zero boundary conditions**. This means we handle the boundary adapted indices independent of the other indices.
For each index $\mu = (j,k,0), k \in \Delta_j$ the set of children $\mathcal{C}(\mu)$ is set to be

$$\mathcal{C}(\mu) := \begin{cases} \{(j+1,0,0)\}, & \text{for } k = 0, \\ \{(j+1,2k-1,0),(j+1,2k,0)\}, & \text{for } k = 1,\ldots,\lfloor \#\Delta_j/2 \rfloor - 1, \\ \{(j+1,2k-1,0),(j+1,2k,0), & \text{for } k = \lfloor \#\Delta_j/2 \rfloor, \\ \qquad (j+1,2k+1,0)\}, & \\ \{(j+1,2k,0),(j+1,2k+1,0)\}, & \text{for } k = \lfloor \#\Delta_j/2 \rfloor + 1,\ldots,\#\Delta_j - 2, \\ \{(j+1,\#\Delta_{j+1}-1,0)\}, & \text{for } k = \#\Delta_j - 1. \end{cases} \tag{A.1.6}$$

For the same level mapping $\Phi_j \leftrightarrow \Psi_j$ one has the problem that $\#\Delta_j = \#\nabla_j + 1$, thus the number of parents is actually one less than the number of children. At least one parent can thus have no child. For symmetry reasons, it should either be the middle index $(j, \lfloor \#\Delta_j/2 \rfloor, 0)$ or the two boundary indices $(j,0,0)$ and $(j, \#\Delta_j - 1, 0)$. For the boundary value problems of Section 5, where the trace operator works on a side of the domain $(0,1)^n$, it seems more appropriate to choose the second variant, i.e., that the children of the boundary adapted functions are itself boundary adapted functions.

$$\mathcal{C}(\mu) := \begin{cases} \{(j,k,1)\}, & \text{for } k = 0,\ldots,\lfloor \#\Delta_j/2 \rfloor - 1, \\ \emptyset, & \text{for } k = \lfloor \#\Delta_j/2 \rfloor, \\ \{(j,k+1,1)\}, & \text{for } k = \lfloor \#\Delta_j/2 \rfloor + 1,\ldots,\#\Delta_j - 1, \end{cases} \tag{A.1.7}$$

Depictions of these tree structures can be found in Figure A.1. The standard wavelet tree structure (3.2.8) can be found in Figure A.2. This choice should stimulate the selection of indices of boundary adapted functions in Algorithm 3.4.

A.1.2 Boundary Adapted Dual Generators $\widetilde{d} = 2$

The dual generators are, except for a constant factor of $\frac{1}{\sqrt{2}}$, given by the masks

$$
\begin{aligned}
\left\{\frac{5}{4}, \frac{3}{2}, \frac{-1}{8}; \quad 0\right\} \quad & \text{for } k = 0, \\
\left\{\frac{-1}{8}, \frac{1}{4}, \frac{13}{8}, \frac{1}{2}, \frac{-1}{4}; \quad -2\right\} \quad & \text{for } k = 1, \\
\left\{\frac{-1}{4}, \frac{1}{2}, \frac{3}{2}, \frac{1}{2}, \frac{-1}{4}; \quad -2\right\} \quad & \text{for } k = 2, \ldots, \#\Delta_j - 3, \\
\left\{\frac{-1}{4}, \frac{1}{2}, \frac{13}{8}, \frac{1}{4}, \frac{-1}{8}; \quad -2\right\} \quad & \text{for } k = \#\Delta_j - 2, \\
\left\{\frac{-3}{4}, \frac{3}{2}, \frac{5}{4}; \quad -2\right\} \quad & \text{for } k = \#\Delta_j - 1.
\end{aligned}
\tag{A.1.8}
$$

Plots of these functions can be found in Figure A.3.
The values of the dual mass matrix $a(\widetilde{\phi}_{j,k}, \widetilde{\phi}_{j,m})$ are (entries numbered for m from left to

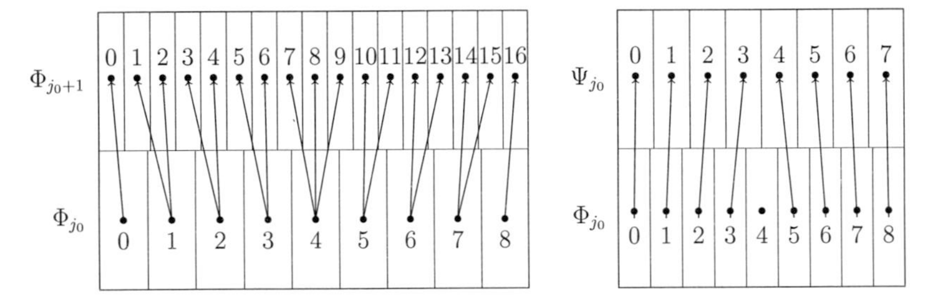

Figure A.1: Visualization of the tree structure (A.1.6) on the left and (A.1.7) on the right. The location index k is given in each rectangle, the basis type is given next to each row.

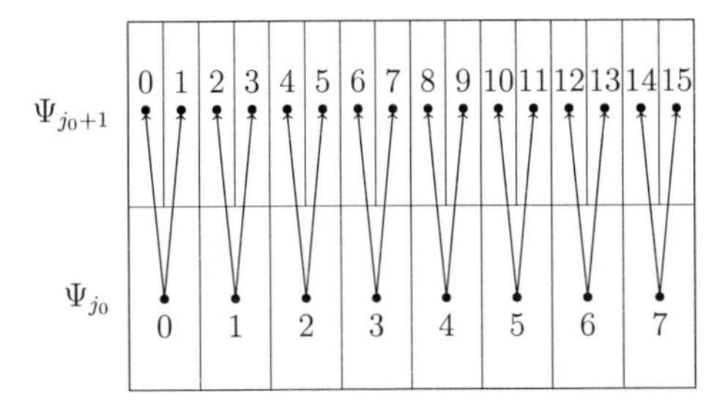

Figure A.2: Visualization of the tree structure (3.2.8). The location index k is given in each rectangle, the basis is given next to each row.

right):

$$\begin{aligned}
&\left\{\frac{17}{4}, \frac{-49}{32}, \frac{19}{48}, \frac{1}{96}\right\} && \text{for } k=0, m=0,\dots,3,\\
&\left\{\frac{-49}{32}, \frac{77}{32}, \frac{-221}{288}, \frac{71}{576}, \frac{1}{288}\right\} && \text{for } k=1, m=0,\dots,4,\\
&\left\{\frac{19}{48}, \frac{-221}{288}, \frac{77}{36}, \frac{-67}{96}, \frac{1}{8}, \frac{1}{288}\right\} && \text{for } k=2, m=0,\dots,5,\\
&\left\{\frac{1}{96}, \frac{71}{576}, \frac{-67}{96}, \frac{77}{36}, \frac{-67}{96}, \frac{1}{8}, \frac{1}{288}\right\} && \text{for } k=3, m=0,\dots,6,\\
&\left\{\frac{1}{288}, \frac{1}{8}, \frac{-67}{96}, \frac{77}{36}, \frac{-67}{96}, \frac{1}{8}, \frac{1}{288}\right\} && \text{for } k=4,\dots,\#\Delta_j-5, m=k-3,\dots,m=k+3\\
&&& \text{(A.1.9)}\\
&\left\{\frac{1}{288}, \frac{1}{8}, \frac{-67}{96}, \frac{77}{36}, \frac{-67}{96}, \frac{71}{576}, \frac{1}{96}\right\} && \text{for } k=\#\Delta_j-4, m=\#\Delta_j-7,\dots,\#\Delta_j-1,\\
&\left\{\frac{1}{288}, \frac{1}{8}, \frac{-67}{96}, \frac{77}{36}, \frac{-221}{288}, \frac{19}{48}\right\} && \text{for } k=\#\Delta_j-3, m=\#\Delta_j-6,\dots,\#\Delta_j-1,\\
&\left\{\frac{1}{288}, \frac{71}{576}, \frac{-221}{288}, \frac{77}{32}, \frac{-49}{32}\right\} && \text{for } k=\#\Delta_j-2, m=\#\Delta_j-5,\dots,\#\Delta_j-1,\\
&\left\{\frac{1}{96}, \frac{19}{48}, \frac{-49}{32}, \frac{17}{4}\right\} && \text{for } k=\#\Delta_j-1, m=\#\Delta_j-4,\dots,\#\Delta_j-1.
\end{aligned}$$

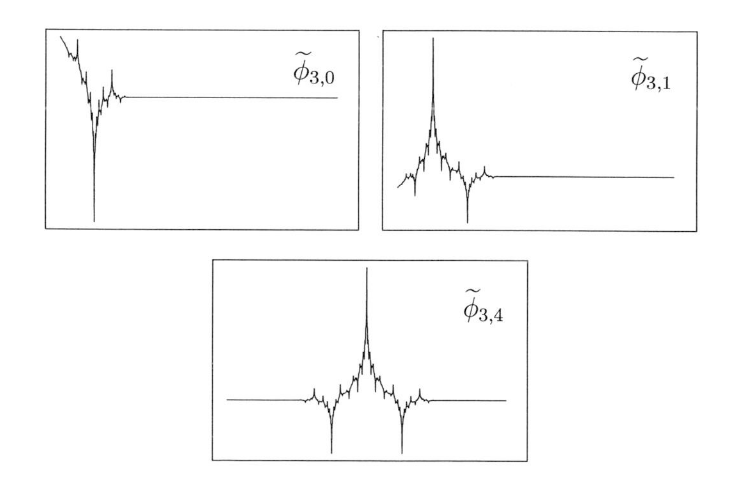

Figure A.3: Dual generators of type DKU22NB, projected onto $S(\Phi_{10})$. This means every function is approximated by piecewise linear functions of support $\approx 2^{-10}$. The recurring function is $\widetilde{\phi}_{3,4}$, the others are boundary adapted special cases.

A.1.3 Boundary Adapted Dual Generators $\widetilde{d} = 4$

The dual generators are, except for a constant factor of $\frac{1}{\sqrt{2}}$, given by the masks

$$
\begin{aligned}
\left\{\frac{93}{64},\frac{35}{32},\frac{-5}{16},\frac{-15}{32},\frac{15}{64};\quad 0\right\} &\quad \text{for } k=0,\\
\left\{\frac{-241}{768},\frac{241}{384},\frac{245}{192},\frac{105}{128},\frac{-93}{256},\frac{-3}{32},\frac{3}{64};\quad -2\right\} &\quad \text{for } k=1,\\
\left\{\frac{41}{384},\frac{-41}{192},\frac{-13}{96},\frac{31}{64},\frac{187}{128},\frac{19}{32},\frac{-1}{4},\frac{-3}{32},\frac{3}{64};\quad -4\right\} &\quad \text{for } k=2,\\
\left\{\frac{-5}{256},\frac{5}{128},\frac{1}{64},\frac{-9}{128},\frac{-67}{256},\frac{19}{32},\frac{45}{32},\frac{19}{32},\frac{-1}{4},\frac{-3}{32},\frac{3}{64};\quad -6\right\} &\quad \text{for } k=3,\\
\left\{\frac{3}{64},\frac{-3}{32},\frac{-1}{4},\frac{19}{32},\frac{45}{32},\frac{19}{32},\frac{-1}{4},\frac{-3}{32},\frac{3}{64};\quad -4\right\} &\quad \text{for } k=4,\dots,\#\Delta_j-5,
\end{aligned}
\tag{A.1.10}
$$

$$
\begin{aligned}
\left\{\frac{3}{64},\frac{-3}{32},\frac{-1}{4},\frac{19}{32},\frac{45}{32},\frac{19}{32},\frac{-67}{256},\frac{-9}{128},\frac{1}{64},\frac{5}{128},\frac{-5}{256};\quad -4\right\} &\quad \text{for } k=\#\Delta_j-4,\\
\left\{\frac{3}{64},\frac{-3}{32},\frac{-1}{4},\frac{19}{32},\frac{187}{128},\frac{31}{64},\frac{-13}{96},\frac{-41}{192},\frac{41}{384};\quad -4\right\} &\quad \text{for } k=\#\Delta_j-3,\\
\left\{\frac{3}{64},\frac{-3}{32},\frac{-93}{256},\frac{105}{128},\frac{245}{192},\frac{241}{384},\frac{-241}{768};\quad -4\right\} &\quad \text{for } k=\#\Delta_j-2,\\
\left\{\frac{15}{64},\frac{-15}{32},\frac{-5}{16},\frac{35}{32},\frac{93}{64};\quad -4\right\} &\quad \text{for } k=\#\Delta_j-1.
\end{aligned}
$$

The mask length is $9 \approx 2\widetilde{d}$, which is a direct consequence of (2.3.3) combined with (2.32(i),(iii)). Plots of these functions can be seen in Figure A.4.

The values of the dual mass matrix $a(\widetilde{\phi}_{j,k}, \widetilde{\phi}_{j,m})$ are (entries numbered for m from left to right):

$$
\begin{aligned}
\{a_{0,0},\dots,a_{0,7}\} &\quad \text{for } k=0, m=0,\dots,7,\\
\{a_{0,1},a_{1,1},\dots,a_{1,6},a_1,a_0\} &\quad \text{for } k=1, m=0,\dots,8,\\
\{a_{0,2},a_{1,2},a_{2,2},\dots,a_{2,6},a_2,a_1,a_0\} &\quad \text{for } k=2, m=0,\dots,9,\\
\{a_{0,3},a_{1,3},a_{2,3},a_{3,3},\dots,a_{3,6},a_3,\dots,a_0\} &\quad \text{for } k=3, m=0,\dots,10,\\
\{a_{0,4},a_{1,4},a_{2,4},a_{3,4},a_7,\dots,a_0\} &\quad \text{for } k=4, m=0,\dots,11,\\
\{a_{0,5},\dots,a_{3,5},a_6,a_7,\dots,a_0\} &\quad \text{for } k=5, m=0,\dots,12,\\
\{a_{0,6},\dots,a_{3,6},a_5,a_6,a_7,\dots,a_0\} &\quad \text{for } k=6, m=0,\dots,13,\\
\{a_{0,7},a_1,\dots,a_7,\dots,a_0\} &\quad \text{for } k=7, m=0,\dots,14,\\
\{a_0,\dots,a_7,\dots,a_0\} &\quad \text{for } k=8,\dots,\#\Delta_j-9, m=k-7,\dots,k+7,
\end{aligned}
\tag{A.1.11}
$$

$$
\begin{array}{rl}
\{a_0, \dots, a_7, \dots, a_1, a_{0,7}\} & \text{for } k = \#\Delta_j - 8, m = \#\Delta_j - 15, \dots, \#\Delta_j - 1, \\
\{a_0, \dots, a_7, a_6, a_5, a_{3,6}, a_{2,6}, a_{1,6}, a_{0,6}\} & \text{for } k = \#\Delta_j - 7, m = \#\Delta_j - 14, \dots, \#\Delta_j - 1, \\
\{a_0, \dots, a_7, a_6, a_{3,5}, a_{2,5}, a_{1,5}, a_{0,5}\} & \text{for } k = \#\Delta_j - 6, m = \#\Delta_j - 13, \dots, \#\Delta_j - 1, \\
\{a_0, \dots, a_7, a_{3,4}, \dots, a_{0,4}\} & \text{for } k = \#\Delta_j - 5, m = \#\Delta_j - 12, \dots, \#\Delta_j - 1, \\
\{a_0, \dots, a_3, a_{3,6}, \dots, a_{3,3}, a_{2,3}, a_{1,3}, a_{0,3}\} & \text{for } k = \#\Delta_j - 4, m = \#\Delta_j - 11, \dots, \#\Delta_j - 1, \\
\{a_0, a_1, a_2, a_{2,6}, \dots, a_{2,2}, a_{1,2}, a_{0,2}\} & \text{for } k = \#\Delta_j - 3, m = \#\Delta_j - 10, \dots, \#\Delta_j - 1, \\
\{a_0, a_1, a_{1,6}, \dots, a_{1,1}, a_{0,1}\} & \text{for } k = \#\Delta_j - 2, m = \#\Delta_j - 9, \dots, \#\Delta_j - 1, \\
\{a_{0,7}, \dots, a_{0,0}\} & \text{for } k = \#\Delta_j - 1, m = \#\Delta_j - 8, \dots, \#\Delta_j - 1,
\end{array}
$$

with the following definitions:

$$a_{0,0} := \frac{6665765124563151867639088393}{1836742536536418841421414400}, \qquad a_{0,1} := \frac{-180522950635150021475645737399}{165306828288277695727927296000},$$

$$a_{0,2} := \frac{2138552910114357514699300337}{5903815296009917704568832000}, \qquad a_{0,3} := \frac{-33216135061795320582043458053}{330613656576555391455854592000},$$

$$a_{0,4} := \frac{5790765461324751880387}{399885889154850068889600}, \qquad a_{0,5} := \frac{57625368402443}{46634638088601600},$$

$$a_{0,6} := \frac{2447177}{474392070400}, \qquad a_{0,7} := \frac{21321}{3795136563200},$$

$$a_{1,1} := \frac{728885767926046009515034156247}{360669443538060427042750464000}, \qquad a_{1,2} := \frac{-75344609883948280847203025159}{122449502435761256094760960000},$$

$$a_{1,3} := \frac{1703303494242923977691255987933}{9918409697296661743675637760000}, \qquad a_{1,4} := \frac{-87635150175456101315267}{2665905927699000459264000},$$

$$a_{1,5} := \frac{733332973252793}{559615657063219200}, \qquad a_{1,6} := \frac{4563943311}{18975682816000},$$

$$a_{2,2} := \frac{9102770208630960638350027475101}{4959204848648330871837818880000}, \qquad a_{2,3} := \frac{-1261514908182928217865367195607}{2479602424324165435918909440000},$$

$$a_{2,4} := \frac{7759709295780780783436367}{59982883373227510333440000}, \qquad a_{2,5} := \frac{-101159974306543711}{4197117427974144000},$$

$$a_{2,6} := \frac{20441015537}{10673821584000},$$

$$a_{3,3} := \frac{118177743629794844436577567139}{66790637692233412415324160000}, \qquad a_{3,4} := \frac{-1762462517518769957236217}{3635326265044091535360000},$$

$$a_{3,5} := \frac{116527049496700033}{932692761772032000}, \qquad a_{3,6} := \frac{-6249298745239}{256171718016000},$$

$$a_0 := \frac{21321}{18975682816000}, \qquad a_1 := \frac{4873033}{4743920704000},$$

$$a_2 := \frac{13834149977}{56927048448000}, \qquad a_3 := \frac{40856239771}{21347643168000},$$

$$a_4 := \frac{-12498464767253}{512343436032000}, \qquad a_5 := \frac{592985214967}{4743920704000},$$

$$a_6 := \frac{-82631485669781}{170781145344000}, \qquad a_7 := \frac{56427081576079}{32021464752000}.$$

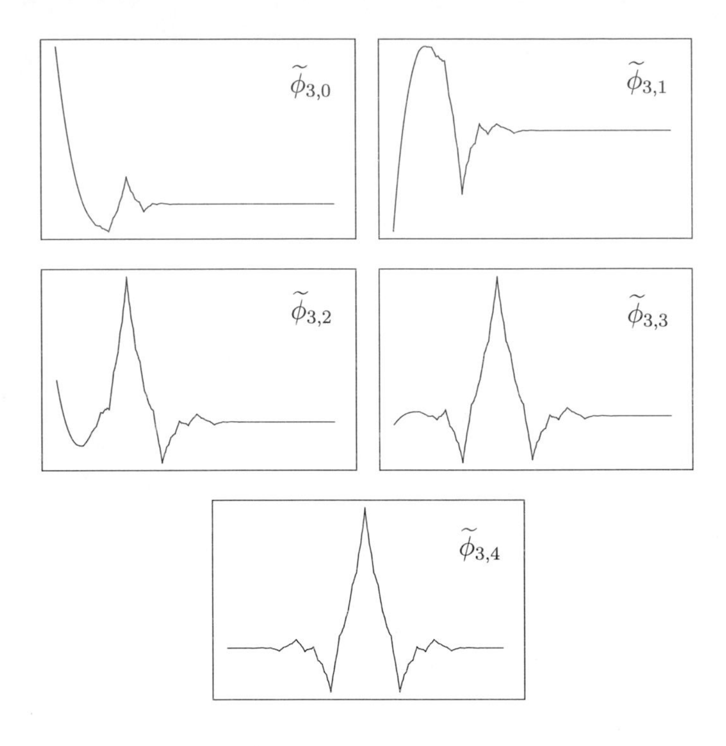

Figure A.4: Dual generators for $\widetilde{d} = 4$, projected onto $S(\Phi_{10})$. This means every function is approximated by piecewise linear functions of support $\approx 2^{-10}$. The recurring function is $\widetilde{\phi}_{3,4}$, all others are boundary adapted special cases.

A.1.4 Boundary Adapted Wavelets $d = 2$, $\widetilde{d} = 2$, $j_0 = 3$ (DKU)

The following data is taken from [51]. The primal wavelet refinement coefficients can be scaled just as the coefficients in Section A.1.5. Their values are

$$\begin{aligned}
&\left\{\frac{3}{4}, \frac{-7}{16}, \frac{3}{8}, \frac{-11}{16}, \frac{1}{4}, \frac{1}{8}; \quad 0\right\} && \text{for } k = 0, \\
&\left\{\frac{-1}{4}, \frac{-1}{2}, \frac{3}{2}, \frac{-1}{2}, \frac{-1}{4}; \quad -1\right\} && \text{for } k = 1, \ldots, \#\nabla_j - 2, \\
&\left\{\frac{1}{8}, \frac{1}{4}, \frac{-11}{16}, \frac{3}{8}, \frac{-7}{16}, \frac{3}{4}; \quad -3\right\} && \text{for } k = \#\nabla_j - 1.
\end{aligned} \tag{A.1.12}$$

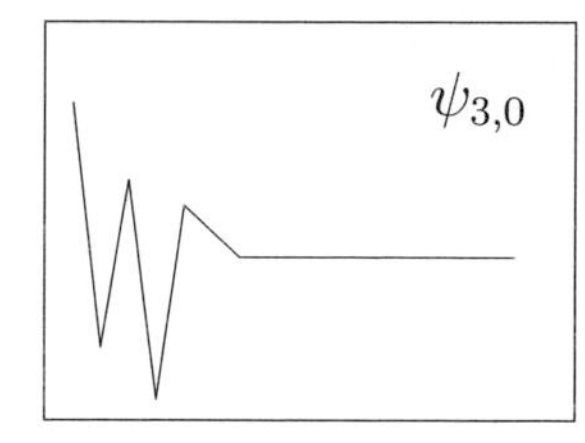

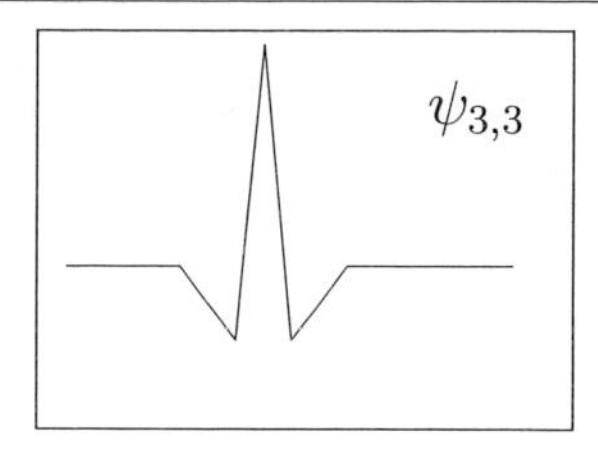

Figure A.5: Wavelets of type DKU22NB, consisting of piecewise linear polynomials. The recurring functions is $\psi_{3,3}$, the other is the boundary adapted special case.

Dual Wavelets

The refinement masks of the dual wavelets are

$$\begin{aligned}
\{1,-2,1;\quad 0\} &\quad \text{for } k=0,\\
\left\{\frac{1}{2},-1,0,1,\frac{-1}{2};\quad -2\right\} &\quad \text{for } k=1,\\
\left\{\frac{-1}{2},1,\frac{-1}{2};\quad 0\right\} &\quad \text{for } k=2,\ldots,\#\nabla_j-3,\\
\left\{\frac{-1}{2},1,0,-1,\frac{1}{2};\quad 0\right\} &\quad \text{for } k=\#\nabla_j-2,\\
\{1,-2,1;\quad 0\} &\quad \text{for } k=\#\nabla_j-1.
\end{aligned} \tag{A.1.13}$$

The dual wavelets can be seen in Figure A.6. Note that in this case holds $\widetilde{\gamma} > 0.6584$ (see [36]).

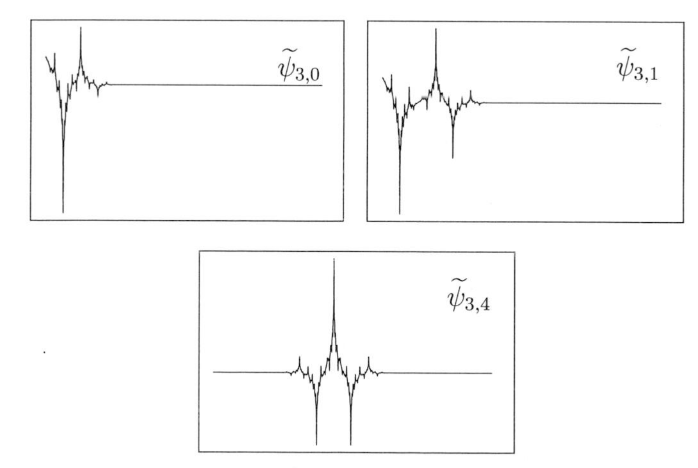

Figure A.6: Dual wavelets of type DKU22NB, projected onto $S(\Phi_{10})$. This means every function is approximated by piecewise linear functions of support $\approx 2^{-10}$. The recurring functions is $\widetilde{\psi}_{3,4}$, the others are boundary adapted special cases.

A.1.5 Boundary Adapted Wavelets $d = 2$, $\widetilde{d} = 4$, $j_0 = 3$ (DKU)

The following results were first established in [50,51] and further investigated in [23] by using only rational numbers in the construction process, which avoids rounding errors. During the latter construction, it became obvious, that a scaling parameter $r \neq 0$ can be chosen arbitrarily. This parameter scales the primal and dual wavelets w.r.t. the primal and dual single scale bases respectively. The mask coefficients of the primal wavelets are scaled by r, the mask coefficients of the dual wavelets by r^{-1}. This scaling can therefore affect computations by having an impact on the absolute value of the condition numbers of the finite discretized systems. There are two values of r that have proved to be of value in applications and we will denote them by

$$r_{DKU} := 1\,, \quad r_{CB} := \sqrt{2}\,. \tag{A.1.14}$$

Setting $r = r_{DKU}$ leads to the refinement matrices of [49,51] and setting $r = r_{CB}$ is the construction from [23]. With this construction, two wavelets have to be adapted to the interval boundary at either side of $(0,1)$. There are therefore five types of wavelets, one is translated in its position and makes up most of the basis:

$$\begin{aligned}
&\left\{\frac{5}{16},\frac{-185}{384},\frac{139}{192},\frac{-73}{128},\frac{13}{96},\frac{23}{384},\frac{-1}{64},\frac{-1}{128};\quad 0\right\} && \text{for } k=0,\\
&\left\{\frac{15}{32},\frac{-45}{256},\frac{-105}{128},\frac{345}{256},\frac{-31}{64},\frac{-53}{256},\frac{9}{128},\frac{9}{256};\quad -2\right\} && \text{for } k=1,\\
&\left\{\frac{3}{64},\frac{3}{32},\frac{-1}{4},\frac{-19}{32},\frac{45}{32},\frac{-19}{32},\frac{-1}{4},\frac{3}{32},\frac{3}{64};\quad -3\right\} && \text{for } k=2,\dots,\#\nabla_j-3,\\
&\left\{\frac{9}{256},\frac{9}{128},\frac{-53}{256},\frac{-31}{64},\frac{345}{256},\frac{-105}{128},\frac{-45}{256},\frac{15}{32};\quad -3\right\} && \text{for } k=\#\nabla_j-2,\\
&\left\{\frac{-1}{128},\frac{-1}{64},\frac{23}{384},\frac{13}{96},\frac{-73}{128},\frac{139}{192},\frac{-185}{384},\frac{5}{16};\quad -5\right\} && \text{for } k=\#\nabla_j-1.
\end{aligned} \tag{A.1.15}$$

We here show the masks for $r = 2$, which conveniently leaves only rational numbers.

Dual Wavelets

The refinement masks of the dual wavelets are

$$\begin{aligned}
&\{1,-2,1;\quad 0\} && \text{for } k=0,\\
&\{\tfrac{1}{2},-1,0,1,\tfrac{-1}{2};\quad 0\} && \text{for } k=1,\\
&\{\tfrac{-1}{2},1,\tfrac{-1}{2};\quad 0\} && \text{for } k=2,\dots,\#\nabla_j-3,\\
&\{\tfrac{-1}{2},1,0,-1,\tfrac{1}{2};\quad 0\} && \text{for k } =\#\nabla_j-2,\\
&\{1,-2,1;\quad 0\} && \text{for } k=\#\nabla_j-1.
\end{aligned} \tag{A.1.16}$$

The dual wavelets can be seen in Figure A.8. It was pointed out in [36] that in this case, the dual generators decay in the following fashion:

$$|\widetilde{\phi}(x)| \le C(1+|x|)^{-\alpha}, \quad \alpha > 1.2777.$$

With Definition 1.11 this directly translates to $\widetilde{\gamma} > 1.2777$. Hence, these wavelets satisfy the norm equivalences of Theorem 2.18 for the range from $(H^1(\mathrm{I}))'$ to $H^1(\mathrm{I})$.

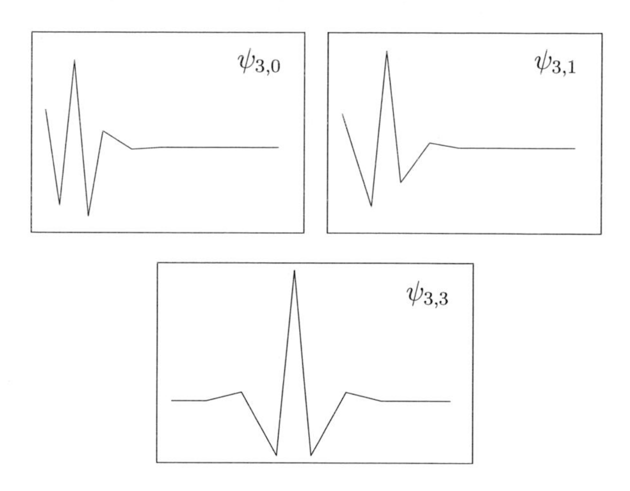

Figure A.7: Wavelets of type DKU24NB, consisting of piecewise linear polynomials. The recurring functions is $\psi_{3,4}$, the others are boundary adapted special cases.

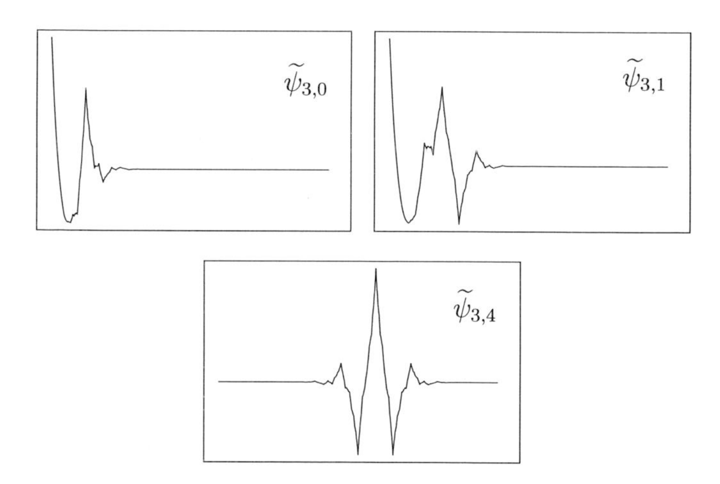

Figure A.8: Dual wavelets of type DKU24NB, projected onto $S(\Phi_{10})$. This means every function is approximated by piecewise linear functions of support $\approx 2^{-10}$. The recurring functions is $\widetilde{\psi}_{3,3}$, all others are boundary adapted special cases.

A.1.6 Boundary Adapted Wavelets $d = 2$, $\tilde{d} = 4$, $j_0 = 4$ (Primbs)

These wavelets are taken from [126]. The coefficients here shown must also be multiplied with a factor of $\frac{1}{\sqrt{2}}$.

$$
\begin{aligned}
&\left\{\frac{-385}{512}, \frac{9625}{12288}, \frac{-2651}{6144}, \frac{-583}{4096}, \frac{451}{3072}, \frac{737}{12288}, \frac{-55}{2048}, \frac{-55}{4096}; \quad 0\right\} && \text{for } k = 0, \\
&\left\{\frac{165}{512}, \frac{-495}{4096}, \frac{-1155}{2048}, \frac{3795}{4096}, \frac{-341}{1024}, \frac{-583}{4096}, \frac{99}{2048}, \frac{99}{4096}; \quad -2\right\} && \text{for } k = 1, \\
&\left\{\frac{33}{1024}, \frac{33}{512}, \frac{-11}{64}, \frac{-209}{512}, \frac{495}{512}, \frac{-209}{512}, \frac{-11}{64}, \frac{33}{512}, \frac{33}{1024}; \quad -3\right\} && \text{for } k = 2, \ldots, \#\nabla_j - 3, \\
&\left\{\frac{99}{4096}, \frac{99}{2048}, \frac{-583}{4096}, \frac{-341}{1024}, \frac{3795}{4096}, \frac{-1155}{2048}, \frac{-495}{4096}, \frac{165}{512}; \quad -3\right\} && \text{for k } = \#\nabla_j - 2, \\
&\left\{\frac{-55}{4096}, \frac{-55}{2048}, \frac{737}{12288}, \frac{451}{3072}, \frac{-583}{4096}, \frac{-2651}{6144}, \frac{9625}{12288}, \frac{-385}{512}; \quad -5\right\} && \text{for } k = \#\nabla_j - 1.
\end{aligned}
\tag{A.1.17}
$$

These wavelets offer the same smoothness and regularity as Section A.1.5, but have proven to produce very low absolute condition numbers in conjunction with the orthogonal transformation of Section 2.3.2 in previous numerical experiments, see [122].

Dual Wavelets

The dual generators are identical with the `DKU24NB` construction and can be found in Figure A.4. Although there are several different wavelets, they all use the same mask:

$$
\left\{\frac{-8}{11}, \frac{16}{11}, \frac{-8}{11}; \quad 0\right\} \quad \text{for } k = 0, \ldots, \#\nabla_j - 1, \tag{A.1.18}
$$

The difference in the dual wavelets stems from the boundary adapted dual generators. The three different types of wavelets can be seen in Figure A.9.

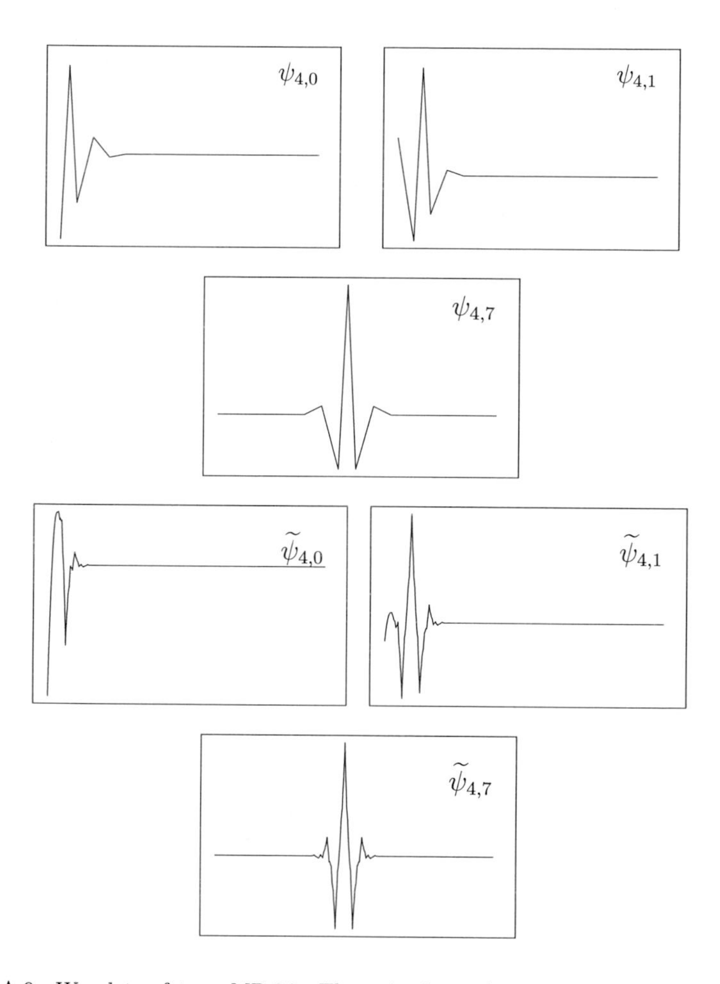

Figure A.9: Wavelets of type MP-24. The primal wavelets consist of piecewise linear polynomials. The dual wavelets are projected onto $S(\Phi_{10})$. The recurring wavelet is in each case $\psi_{4,7}$, the two others are boundary adapted special cases.

A.2 Wavelet Diagrams

We briefly explain the types of wavelet diagrams used in this work.

A.2.1 1D Diagram

The one dimensional wavelet coefficients diagrams are set up in a natural way. The horizontal axis in Figure A.2.1 represents the spatial position k, the vertical axis the level j. The lowermost row contains the wavelet indices of Ψ_{j_0-1}. Within each row, the location takes on the values $k = 0, \ldots, \#\nabla_j$ from left to right. Since the number of wavelets double and their support halves from each level to the next, the width of the cells represent roughly the size of each wavelets support.

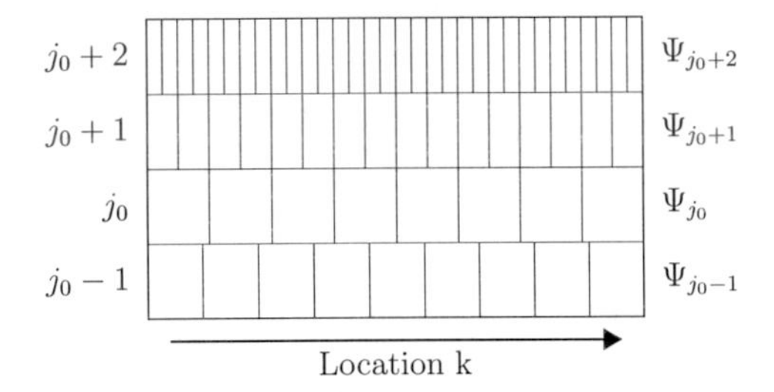

Figure A.10: Standard 1D wavelet diagram for boundary adapted wavelets on the interval $(0, 1)$.

A.2.2 2D Diagrams

Classic

The standard diagram is built according to the multi-dimensional decomposed space structure depicted in Figure 2.3 and Figure 2.4. Each rectangle or square in Figure A.11 represents the whole domain $(0, 1)^2$, with the lower left point being the origin. The side lengths are proportional to the number of functions of each dimension and type (wavelet or single scale). Thus, each wavelet index gets an equal share of the space for representation.

Level-wise Type Plot

The level-wise type is a different form of arrangement for the isotropic class diagram Figure A.11. One chooses a type $\mathbf{e} \in \{01, 10, 11\}$ and then places all squares with this type and the lowermost left above each other in a 3-D cube fashion. For this to work, the rectangles have to be stretched to conform in the picture dimensions. This diagram is particularly well suited to visualize patterns present in the space coordinates of the same wavelet types.

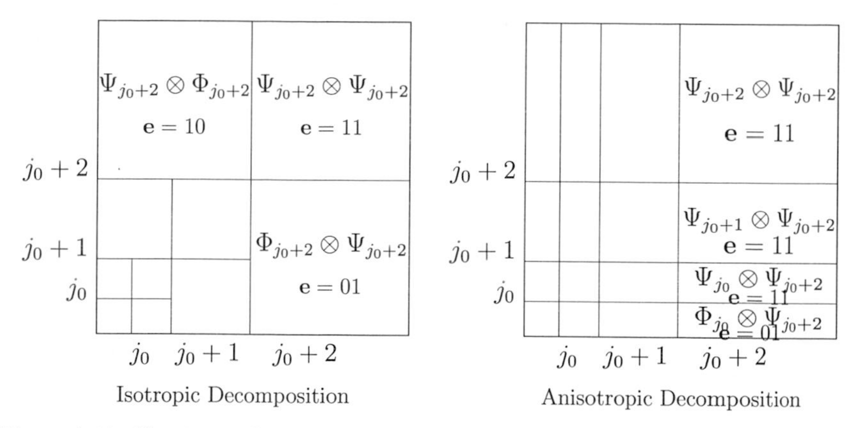

Figure A.11: Classic wavelet diagrams for isotropic and anisotropic decompositions. For the level $j_0 + 2$, some explicit bases and their types of each segment are given. The lower left squares contain the bases $\Psi_{j_0-1} \otimes \Psi_{j_0-1}$.

Scatter Plot

The aforementioned diagrams are not very well suited to display sets of wavelet indices with very high levels and many zero-valued (or non-existent) coefficients. For high levels, it is more convenient to put all coefficients into a single diagram representing the complete domain Ω. For this purpose, every wavelet index λ gets associated with a unique point in Ω. This is usually the center of the support of the function and the position of its highest value. For tensor products, the point is simply determined by calculating the unique points in each direction.
Since, for $j \to \infty$, the union of all these points in dense in $\Omega \subset \mathbb{R}^2$, the space associated to each coefficient becomes infinitesimally small. Hence, with this diagram, one cannot determine easily the properties of a single pictured coefficient, but the distribution of coefficients hints to locations of important features of the presented function.

A.2.3 3D Diagrams

To visualize data in three dimensions, one can either select specific slices of coefficients and use 2D diagrams or visualize all coefficients in a single scatter diagram.

Scatter Plot

The scatter plot for $\Omega = (0,1)^3$ is set up exactly as in two dimensions except for the "depth" information given by the unique point of the wavelet in x_3. Thankfully, most plotting libraries compute the projection onto a $2D$ canvas itself and one can directly output location and value data as three-dimensional points and color information.

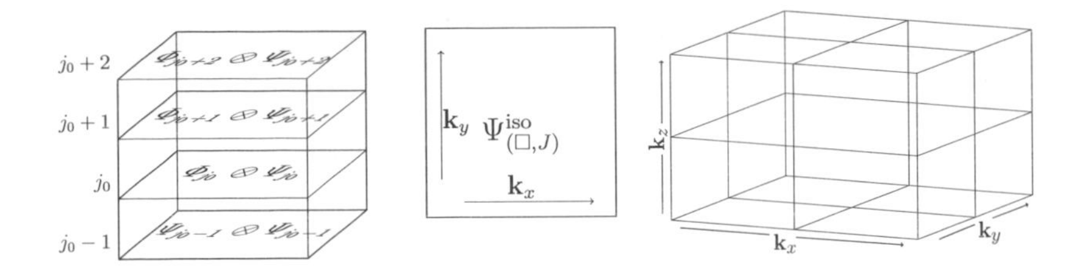

Figure A.12: **Left:** Isotropic level-wise plot for levels $j_0 - 1$ to $j_0 + 2$ and exemplified type $\mathbf{e} = 01$.
Center: The scatter plot diagram combines all coefficients into a single diagram represented by the domain $(0,1)^2$. **Right:** Scatter plot diagram (box) for the domain $(0,1)^3$.

A.3 Local Polynomial Bases

Here, we present three suitable 1D monomial bases for the local polynomial representation (3.4.24) of Section 3.4.1. For any $t \geq 0$, we denote the set of monomial basis functions on a dyadic grid of level j at location k up to order t as $\Pi^t_{j,k} := \{p^r_{j,k}(x) : r = 0, \dots, t\}$. On the standard interval $(0,1)$, the step size on level j of this dyadic grid is $h := 2^{-j}$ and the number of possible cells $n := 2^j$. For simplicity, we call any basis function of a polynomial space a "monomial", although it does not need to be a monomial in the strict sense. Also, for readability, we will omit any factors of the kind $\chi_{2^{-j}[k,k+1]}$, as the compact support of the involved functions should be obvious from the context.

A.3.1 Standard Monomials

The standard monomial basis, e.g., $p^r_{j,k}(x) := x^r$, is numerically unstable for high levels $j \gg 1$. It should therefore only be used for debugging purposes.
The basis is hierarchically constructed, i.e., $\Pi^{r+1}_{j,k} := \Pi^r_{j,k} \cup \{x^{r+1}\}$ and it holds $p^r_{j,k}\, p^s_{j,k} = p^{r+s}_{j,k}$. Therefore, the only integral values that have to be determined are

$$\begin{aligned}
\int_{2^{-j}k}^{2^{-j}(k+1)} p^r_{j,k}(x)\, dx &= \frac{1}{r+1}\left((2^{-j}(k+1))^{r+1} - (2^{-j}k)^{r+1}\right), \\
&= \frac{h^{r+1}}{r+1}\left(\sum_{i=1}^{r+1} \binom{r+1}{i} k^i\right).
\end{aligned}$$

The cause for numerical instability can now obviously be recognized here in the extremely high values of the exponent. Even for the case of linear polynomials, the nonlinear operator $F(u) = u^3$ requires the integration of fourth order polynomials and thus the value h^5 during the above evaluation. In two dimensions, integration is done in both coordinate directions and the results are multiplied. Then, values of the integral get below **machine precision** even for $j = 5$.
The polynomial refinement matrices $\mathbf{M}_{\square',\square}$ defined in (3.4.15) for the standard monomials are simply the identity matrices $\mathbf{I} \in \mathbb{R}^{(t+1)\times(t+1)}$.

A.3.2 Normalized Monomials

Another approach is to shift and scale the monomial basis in same the way wavelets are constructed from a mother wavelet, i.e.,

$$p^r_{j,k}(x) := 2^{j/2}(2^j x - k)^r. \tag{A.3.1}$$

This basis is again constructed in a hierarchical fashion. Using this basis, it is easy to represent monomials of different orders alike, e.g. constant and linear monomials. Evaluation of a polynomial in this representation can be done efficiently by transforming the point x into the **local coordinate** $\mu := \mu_{j,k}(x) := 2^j x - k$ and then using the **Horner scheme**. Lastly, the scaling factor $2^{j/2}$ must be applied.
Inner products can be evaluated by substituting $\mu_{j,k}$ to get

$$\int_{2^{-j}k}^{2^{-j}(k+1)} p^r_{j,k}(x)\, p^s_{j,k}(x)\, dx = \int_0^1 \mu^r \mu^s \, d\mu = \int_0^1 \mu^{r+s}\, d\mu = \frac{1}{r+s+1}. \tag{A.3.2}$$

The independence of the values of the integrals of the level j and the location k exemplifies the usefulness of the scaled and shifted monomial basis. For $r \geq 1$ holds

$$\frac{d}{dx} p^r_{j,k}(x) = \frac{d}{dx}\left(2^{j/2}(2^j x - k)^r\right) = 2^{j/2}\,(2^j x - k)^{r-1}\, r\, 2^j = 2^j\, r\, p^{r-1}_{j,k}(x),$$

which leads for $r, s \geq 1$ in one dimension with (A.3.2) to

$$\int_{2^{-j}k}^{2^{-j}(k+1)} \frac{d}{dx} p^r_{j,k}(x) \frac{d}{dx} p^s_{j,k}(x)\, dx = 2^{2j}\, r\, s \int_{2^{-j}k}^{2^{-j}(k+1)} p^{r-1}_{j,k}(x) p^{s-1}_{j,k}(x)\, dx = \frac{2^{2j}\, r\, s}{r+s-1}. \tag{A.3.3}$$

For either $r = 0$ or $s = 0$, the constant polynomial is differentiated, which directly result in a zero value.

Refinement/Coarsening Operators

Since the polynomial construction is now adapted to the dyadic grid cell, it follows that the polynomial refinement matrices can no longer be the identity matrices. It even follows, that there must be different matrices for the left and right subcells, $\square_1, \square_2 \in \mathcal{D}_{j+1}$, of a box $\square \in \mathcal{D}_j$. To calculate these matrices, each monomial $p^r_{j,k}$ is restricted to the subcells and the coefficients in the local polynomial basis have to be computed. For a monomial of order r, follows for the left child (level $j+1$, location $2k$),

$$p^r_{j,k}(x) = 2^{j/2}(2^j x - k)^r = \frac{1}{\sqrt{2}}\,(1/2)^r\, 2^{(j+1)/2}\,(2^{j+1}x - 2k)^r = \frac{1}{\sqrt{2}}\,(1/2)^r\, p^r_{j+1,2k}(x),$$

and for the right child (level $j+1$, location $2k+1$),

$$\begin{aligned} p^r_{j,k}(x) &= \frac{1}{\sqrt{2}}\, 2^{(j+1)/2}\,(1/2)^r\,((2^{j+1}x - (2k+1)) + 1)^r \\ &= \frac{1}{\sqrt{2}}\,(1/2)^r\, 2^{(j+1)/2} \sum_{i=0}^{r} \binom{r}{i} (2^{j+1}x - (2k+1))^i 1^{r-i} \\ &= \frac{1}{\sqrt{2}}\,(1/2)^r \sum_{i=0}^{r} \binom{r}{i} p^i_{j+1,2k+1}(x). \end{aligned}$$

The refinement matrices $\mathbf{M}_{\square_1,\square}, \mathbf{M}_{\square_2,\square} \in \mathbb{R}^{(r+1)\times(r+1)}$ for (A.3.1) are thus

$$\mathbf{M}^T_{\square_1,\square} = \frac{1}{\sqrt{2}} \begin{pmatrix} 1 & & & & \\ & \frac{1}{2} & & & \\ & & \frac{1}{4} & & \\ & & & \ddots & \\ & & & & \frac{1}{2^r} \end{pmatrix}, \quad \mathbf{M}^T_{\square_2,\square} = \frac{1}{\sqrt{2}} \begin{pmatrix} 1 & 1/2 & 1/4 & \cdots & \frac{1}{2^r} \\ & 1/2 & 2/4 & \cdots & \vdots \\ & & 1/4 & \cdots & \binom{r}{i}\frac{1}{2^r} \\ & & & \ddots & \vdots \\ & & & & \frac{1}{2^r} \end{pmatrix}. \tag{A.3.4}$$

This reveals that this basis could also suffer from numerical instabilities for high polynomial degrees as the factor $(1/2)^r$ goes to zero for increasing values of r.

Multiplication and Power Operations

Because of the normalization, the product of two monomials is not a monomial of higher order, i.e., $p^r_{j,k}\, p^s_{j,k} \neq (p^{r+s}_{j,k})$. In fact, it holds

$$p^r_{j,k}(x)\, p^s_{j,k}(x) = 2^{j/2}(2^j x - k)^r\, 2^{j/2}(2^j x - k)^s = 2^j(2^j x - k)^{r+s} = 2^{j/2}\, p^{r+s}_{j,k}(x),$$

and, repeating this rule for power operations,

$$\left(p^r_{j,k}(x)\right)^s = 2^{(s-1)j/2}\, p^{rs}_{j,k}(x).$$

Hence, except for a (on each level j and in the second case exponent s) constant factor, inner products of polynomials can yet be reduced to calculations of inner products of basis functions. The nonlinear operator $F(u) = u^3$ requires calculation of the third power of a polynomial on each cell. Because of the above identity, the operator $F(u)$ on a normalized monomial first uses the standard algorithm for the third power to calculate the coefficients of all monomials of third order and then multiplies the result with the factor 2^j.

Conversion to/from Standard Monomials

The transformation matrix exporting a polynomial in the normalized basis into the standard basis is easily calculated by expanding (A.3.1), i.e.,

$$p^r_{j,k}(x) = 2^{j/2}(2^j x - k)^r = 2^{j/2} \sum_{i=0}^{r} \binom{r}{i} (2^j x)^i (-k)^{r-i}.$$

This defines the values of the export matrix $\mathbf{C}_{\text{norm}\to\text{std}}$. It is important to note that, unlike the refinement matrices (A.3.4), this operation depends on the level and location of each cell.

Importing a polynomial given in the standard monomial basis on the whole domain $(0,1)$ or a part thereof into the normalized polynomial is done by importing the same polynomial for all applicable levels j and locations k.

$$\mathbf{C}_{\text{norm}\to\text{std}} = 2^{j/2} \begin{pmatrix} 1 & & & & \\ & h^{-1} & & & \\ & & h^{-2} & & \\ & & & h^{-3} & \\ & & & & \ddots \end{pmatrix} \begin{pmatrix} 1 & -k & k^2 & -k^3 & \cdots \\ & 1 & -2k & 3k^2 & \cdots \\ & & 1 & -3k & \cdots \\ & & & 1 & \cdots \\ & & & & \ddots \end{pmatrix}.$$

The local transformation matrix of the import, $\mathbf{C}_{\text{std}\to\text{norm}}$, is the inverse of the matrix $\mathbf{C}_{\text{norm}\to\text{std}}$, i.e.,

$$\mathbf{C}_{\text{std}\to\text{norm}} = 2^{-j/2} \begin{pmatrix} 1 & k & k^2 & k^3 & \cdots \\ & 1 & 2k & 3k^2 & \cdots \\ & & 1 & 3k & \cdots \\ & & & 1 & \cdots \\ & & & & \ddots \end{pmatrix} \begin{pmatrix} 1 & & & & \\ & h & & & \\ & & h^2 & & \\ & & & h^3 & \\ & & & & \ddots \end{pmatrix}.$$

A.3.3 Normalized Shape Functions

Since spline wavelets usually are made up only from polynomial pieces of maximal order, e.g. are all linear but not a constant on some part of their support, another typical choice are **shape functions**, see [19]. Introducing $t+1$ **control points** for order $t \in \mathbb{N}_0$,

$$\widetilde{x}_{\{s,t\}} := 2^{-j}(k + s/t) \in 2^{-j}[k, k+1], \quad s = 0, \ldots, t, \tag{A.3.5}$$

these functions are elements of Π_d, constructed such that each attains the value $2^{j/2}$ at one point $\widetilde{x}_{\{s,t\}}$ and is zero at all the others. This means the shape function bases are not hierarchically defined as the standard and normalized monomials, i.e., $\Pi^t_{j,k} \not\subseteq \Pi^{t+1}_{j,k}$. Thus, a single index r is not sufficient to identify a particular shape function. We denote the r-th shape function of $\Pi^t_{j,k}$, $r = 0, \ldots, t$, as $p^{\{r,t\}}_{j,k}(x)$. Using again the **local coordinate** $\mu := \mu_{j,k}(x) := 2^j x - k$, the first bases can be found, except for the common factor $2^{j/2}\,\chi_{2^{-j}[k,k+1]}$, in Table A.3.3.

	$r=0$	$r=1$	$r=2$	$r=3$
$t=0$	1			
$t=1$	$(-1)(\mu-1)$	μ		
$t=2$	$2(\mu-\frac{1}{2})(\mu-1)$	$(-4)\mu(\mu-1)$	$2\mu(\mu-\frac{1}{2})$	
$t=3$	$-\frac{9}{2}(\mu-\frac{1}{3})(\mu-\frac{2}{3})$ $(\mu-1)$	$\frac{27}{2}\mu(\mu-\frac{2}{3})(\mu-1)$	$-\frac{27}{2}\mu(\mu-\frac{1}{3})(\mu-1)$	$\frac{9}{2}\mu(\mu-\frac{1}{3})(\mu-\frac{2}{3})$

Table A.1: Shape Functions $r = 0, \ldots, t$ up to order $t = 3$ in the local coordinate $\mu \in [0, 1]$.

Due to the overall w.r.t. the order t quadratically growing number of basis functions, the number of combinations for the internal product grows even faster. The scaling of the shape functions again results in the values of the inner product

$$\int_{2^{-j}k}^{2^{-j}(k+1)} p_{j,k}^{\{r_1,t_1\}}(x)\, p_{j,k}^{\{r_2,t_2\}}(x)\, dx \tag{A.3.6}$$

being independent of the level j and location k. The values corresponding to the above given basis functions can be read in Table A.2.

	$p_{j,k}^{\{0,0\}}$	$p_{j,k}^{\{0,1\}}$	$p_{j,k}^{\{1,1\}}$	$p_{j,k}^{\{0,2\}}$	$p_{j,k}^{\{1,2\}}$	$p_{j,k}^{\{2,2\}}$	$p_{j,k}^{\{0,3\}}$	$p_{j,k}^{\{1,3\}}$	$p_{j,k}^{\{2,3\}}$	$p_{j,k}^{\{3,3\}}$
$p_{j,k}^{\{0,0\}}$	1	$\frac{1}{2}$	$\frac{1}{2}$	$\frac{1}{6}$	$\frac{2}{3}$	$\frac{1}{6}$	$\frac{1}{8}$	$\frac{3}{8}$	$\frac{3}{8}$	$\frac{1}{8}$
$p_{j,k}^{\{0,1\}}$		$\frac{1}{3}$	$\frac{1}{6}$	$\frac{1}{6}$	$\frac{1}{3}$	0	$\frac{13}{120}$	$\frac{3}{10}$	$\frac{3}{40}$	$\frac{1}{60}$
$p_{j,k}^{\{1,1\}}$			$\frac{1}{3}$	0	$\frac{1}{3}$	$\frac{1}{6}$	$\frac{1}{60}$	$\frac{3}{40}$	$\frac{3}{10}$	$\frac{13}{120}$
$p_{j,k}^{\{0,2\}}$				$\frac{2}{15}$	$\frac{1}{15}$	$-\frac{1}{30}$	$\frac{11}{120}$	$\frac{3}{20}$	$-\frac{3}{40}$	0
$p_{j,k}^{\{1,2\}}$					$\frac{8}{15}$	$\frac{1}{15}$	$\frac{1}{30}$	$\frac{3}{10}$	$\frac{3}{10}$	$\frac{1}{30}$
$p_{j,k}^{\{2,2\}}$						$\frac{2}{15}$	0	$-\frac{3}{40}$	$\frac{3}{20}$	$\frac{11}{120}$
$p_{j,k}^{\{0,3\}}$							$\frac{8}{105}$	$\frac{33}{560}$	$-\frac{3}{140}$	$\frac{19}{1680}$
$p_{j,k}^{\{1,3\}}$								$\frac{27}{70}$	$-\frac{27}{560}$	$-\frac{3}{140}$
$p_{j,k}^{\{2,3\}}$									$\frac{27}{70}$	$\frac{33}{560}$
$p_{j,k}^{\{3,3\}}$										$\frac{8}{105}$

Table A.2: Values of the inner product (A.3.6) for up to third order shape functions. Since the inner product is symmetric, only half of the values are presented.

Refinement/Coarsening Operators

Because of its interpolating nature, many mathematical operations with the shape function basis can often be reduced to evaluation procedures. For the linear shape function $p_{j,k}^{\{0,1\}}$, evaluation yields

$$p_{j,k}^{\{0,1\}}(2^{-j}k) = 2^{j/2}, \quad p_{j,k}^{\{0,1\}}(2^{-j}(k - \frac{1}{2})) = \frac{1}{2}2^{j/2}, \quad p_{j,k}^{\{0,1\}}(2^{-j}(k-1)) = 0.$$

Since the scaling from each level j to the next $j+1$ differs only by a factor of $\sqrt{2}$, it follows in this case

$$p_{j,k}^{\{0,1\}}(x) = \frac{1}{\sqrt{2}}\left(p_{j+1,2k}^{\{0,1\}}(x) + \frac{1}{2}p_{j+1,2k}^{\{1,1\}}(x) + \frac{1}{2}p_{j+1,2k+1}^{\{0,1\}}(x) + 0\, p_{j+1,2k+1}^{\{1,1\}}(x)\right).$$

Doing the same calculation for $p_{j,k}^{\{1,1\}}$, the refinement matrices can thus directly deduced to be

$$\mathbf{M}_{\square_1,\square}^T = \frac{1}{\sqrt{2}}\begin{pmatrix} 1 & 1/2 \\ 0 & 1/2 \end{pmatrix}, \quad \mathbf{M}_{\square_2,\square}^T = \frac{1}{\sqrt{2}}\begin{pmatrix} 1/2 & 0 \\ 1/2 & 1 \end{pmatrix}.$$

Thus, to assemble the refinement matrices, one only needs to evaluate the shape function on $\square$ at the control points of the left and right subcells $\square_1, \square_2 \in \mathcal{D}_{j+1}$, and account for the different scaling on each level. For the quadratic shape functions, this results in

$$\mathbf{M}^T_{\square_1,\square} = \frac{1}{\sqrt{2}} \begin{pmatrix} 1 & 3/8 & 0 \\ 0 & 3/4 & 1 \\ 0 & -1/8 & 0 \end{pmatrix}, \quad \mathbf{M}^T_{\square_2,\square} = \frac{1}{\sqrt{2}} \begin{pmatrix} 0 & -1/8 & 0 \\ 1 & 3/4 & 0 \\ 0 & 3/8 & 1 \end{pmatrix}.$$

Whereas it is an implementation disadvantage that the refinement matrices are not as easily calculable as in the prior cases Section A.3.1 and Section A.3.2, the interpolating nature makes other calculations particularly simple. In general, the value of the refinement matrix in row k and column i is the value of the monomial $p_{j,k}^{\{k,m\}}(\widehat{x}_{\{i,m\}})$, with $k = 0, \ldots, m$ and $\widehat{x}_{\{i,m\}}$ being the i-th control point on either the left of right subinterval. Since the actual value is, because of the same relative scaling $2^{1/2}$, independent of the level j, these values can generally be computed easily in the local coordinate μ: using the unscaled interpolating polynomial

$$q_{k,m}(x) := \prod_{i=0, i\neq k}^{m} (x - i/m),$$

the value is simply the value of the normalized polynomial, i.e.,

$$\frac{q_{k,m}(\mu)}{q_{k,m}(k/m)} = \prod_{i=0, i\neq k}^{m} \frac{\mu - i/m}{(k-i)/m} = \prod_{i=0, i\neq k}^{m} \frac{m\,\mu - i}{k - i}.$$

The relative values of the control points of the left and right subintervals are simply $\widehat{x}_{\{\cdot,m\}} := \left\{\frac{i}{2m} \,|\, i = 0, \ldots, m\right\}$ and $\widehat{x}_{\{\cdot,m\}} := \left\{\frac{i}{2m} \,|\, i = m+1, \ldots, 2m\right\}$, respectively.

Multiplication and Power Operations

The nonlinear operator $F(u) = u^3$ requires calculation of the third power of a polynomial on each cell. For piecewise linear shape functions, the result of the operator is a polynomial of third order and thus representable in the basis of $\Pi^3_{j,k}$. To calculate the coefficients, one simply evaluates the linear function at the control points of the cubic shape functions and applies the operator F to these values.

Similarly, the required multiplication of a quadratic polynomial u with a linear one v in the operator application $DF(u)\,v = 3u^2 v$ is computed by evaluating the factors at the control points for the cubic shape functions and then applying the operation.

The values of the shape functions up to $t = 3$ evaluated at the control points of same or higher order shape functions can be found in Table A.3.3, again except for the common factor $2^{j/2}$.

In higher dimensions, the above mentioned operators are computed by evaluating the higher dimensional polynomials at all control points of the cell, which is simply the tensor product of the one dimensional set of control points.

	$t=0$	$t=1$		$t=2$			$t=3$			
$\widetilde{x}_{\{s,t\}}$	any	0	1	0	1/2	1	0	1/3	2/3	1
$p_{j,k}^{\{0,0\}}$	1	1	1	1	1	1	1	1	1	1
$p_{j,k}^{\{0,1\}}$		1	0	1	1/2	0	1	2/3	1/3	0
$p_{j,k}^{\{1,1\}}$		0	1	0	1/2	1	0	1/3	2/3	1
$p_{j,k}^{\{0,2\}}$				1	0	0	1	2/9	−1/9	0
$p_{j,k}^{\{1,2\}}$				0	1	0	0	8/9	8/9	0
$p_{j,k}^{\{2,2\}}$				0	0	1	0	−1/9	2/9	1
$p_{j,k}^{\{0,3\}}$							1	0	0	0
$p_{j,k}^{\{1,3\}}$							0	1	0	0
$p_{j,k}^{\{2,3\}}$							0	0	1	0
$p_{j,k}^{\{3,3\}}$							0	0	0	1

Table A.3: Point evaluation of the shape functions up to third order on the control points of same or higher order shape functions.

Conversion to/from Standard Monomials

Conversion of a polynomial given in the standard monomial basis is simply done by evaluating the polynomial at the required control points in $2^{-j}[k, k+1]$ and then scaling these values with the factor $2^{-j/2}$. This operation can of course be written in matrix form, $\mathbf{C}^t_{\text{std}\to\text{shpf}}$, and the inverse of this matrix is then the operator $\mathbf{C}^t_{\text{shpf}\to\text{std}}$. For the linear case, this yields the matrices

$$\mathbf{C}^1_{\text{std}\to\text{shpf}} = 2^{-j/2}\begin{pmatrix} 1 & k\,h \\ 1 & (k+1)h \end{pmatrix} = 2^{-j/2}\begin{pmatrix} 1 & k \\ 1 & k+1 \end{pmatrix}\begin{pmatrix} 1 & \\ & h \end{pmatrix},$$

$$\mathbf{C}^1_{\text{shpf}\to\text{std}} = 2^{j/2}\begin{pmatrix} k+1 & -k \\ -h^{-1} & h^{-1} \end{pmatrix} = 2^{j/2}\begin{pmatrix} 1 & \\ & h^{-1} \end{pmatrix}\begin{pmatrix} k+1 & -k \\ -1 & 1 \end{pmatrix},$$

and for the quadratic case follows

$$\mathbf{C}^2_{\text{std}\to\text{shpf}} = 2^{-j/2}\begin{pmatrix} 1 & k & k^2 \\ 1 & k+1/2 & (k+1/2)^2 \\ 1 & k+1 & (k+1)^2 \end{pmatrix}\begin{pmatrix} 1 & & \\ & h & \\ & & h^2 \end{pmatrix},$$

$$\mathbf{C}^2_{\text{shpf}\to\text{std}} = 2^{j/2}\begin{pmatrix} 1 & & \\ & h^{-1} & \\ & & h^{-2} \end{pmatrix}\begin{pmatrix} 2k^2+3k+1 & -4k(k+1) & k(2k+1) \\ -(4k+3) & 8k+4 & -(4k+1) \\ 2 & -4 & 2 \end{pmatrix}.$$

Due to the nature of the construction, each order t requires a completely different transformation matrix, but the entries in the matrix $\mathbf{C}^t_{\text{std}\to\text{shpf}} \in \mathbb{R}^{(t+1)\times(t+1)}$ for order t are simply given by

$$\left(\mathbf{C}^t_{\text{std}\to\text{shpf}}\right)_{r,c} = 2^{-j/2}\left(k + \frac{r-1}{t}\right)^{c-1} h^{c-1}, \qquad \text{for } r = 1,\ldots,t+1, \text{ and } c = 1,\ldots,t+1.$$

That means the matrix is a **Vandermonde Matrix** with the abbreviation $v_r := k + \frac{r-1}{t}$, see [70], which has an explicit inverse with elements

$$\left(\mathbf{C}^t_{\text{shpf}\to\text{std}}\right)_{r,c} = 2^{j/2}(-h)^{1-r} \frac{\displaystyle\sum_{\substack{0\le m_0<\ldots<m_{t+1-r}\le t+1\\ m_0,\ldots,m_{t+1-r}\neq c}} \prod_{j=0}^{t+1-r} v_{m_j}}{\displaystyle\prod_{\substack{0\le m\le t+1\\ m\neq c}} (v_m - v_c)}$$

$$\text{for } r = 1,\ldots,t+1, \text{ and } c = 1,\ldots,t+1.$$

And efficient, i.e., $\mathcal{O}(t^2)$, algorithm to apply the matrix $\mathbf{C}^t_{\text{shpf}\to\text{std}}$ can be found in [125].

B Implementational Details

In this chapter, I will lay down and describe some of the principal design and coding details of my software.

B.1 Compilers and Computers

When writing code in numerics, many students and professionals strive to create the most optimized code, which is not necessarily the most easily understandable for another person. In fact, this thinking often creates very obfuscated code that is hard to maintain, even for the original writer and not even necessarily better or faster than "non-optimized" code. In practice, the compiler is often more adept at optimizing code than the average developer and the developer should better focus on writing maintainable and understandable code instead.

To better understand this point, it helps to know a few facts about computer and compilers.

B.1.1 A Few Remarks about CPUs

Many people in applied mathematics and numerics judge a computers' speed by only looking at the clock speed. This single number, nowadays in the range of $3\,\mathrm{GHz} = 3\,10^9\,/\mathrm{s}$, can be grossly misleading. Just noting the maximum distance that any information can travel (with the speed of light) within a single clock cycle,

$$\Delta x = c\,\Delta t = 3\,10^8\,\mathrm{m/s}\,\frac{1}{3\,10^9\,/\mathrm{s}} = 0.1\,\mathrm{m} = 10\,\mathrm{cm},$$

shows that the **physical distance** of the main memory to the CPU could be an influence to the systems' performance. Also note that electrical current only travels with a quarter of light speed and this ignores the time the logic blocks need to interpret any kind of data and other delaying hardware factors.

In reality, the architecture of a CPU is at least as important to the performance as the clock speed. A modern CPU goes to great lengths to optimize the code, even during its execution. For example, out-of-order execution and predicting the results of future commands to load possibly required data in advance, are very effective - but complex - strategies. Since the memory latency, i.e., the time it takes to read a single information from the main memory, is limited by the speed of light, the only way out was to increase the throughput by data parallelization (**SIMD** - single instruction, multiple data), i.e., applying the same (mathematical/floating point) operation to many data points simultaneously.

The floating point unit (**FPU**) was originally an optional part of a CPU which executes all commands related to floating point numbers. It can handle single-, double- and extended precision floating point numbers, i.e., `float`, `double` and `long double`, because it internally uses an 80-bit extended precision representation, but cannot employ SIMD principles. Therefore, additional floating point execution units have been included into CPUs by extensions known as **MMX**, **SSE**, **SSE2**, **AVX**, ..., which today can handle single- and double precision commands. The compiler usually chooses the most efficient execution unit on its own and the programmer can simply rely on the guarantees given by the (updated) **IEEE-754** standard, see [69, 81].

But the developer has to be aware of many pitfalls of this system. Optimizations and automatically generated interim results can have unexpected values, especially if the FPU with its internal extended precision is involved. It is therefore never a bad investment to learn to handle a **debugger** and understand **assembly code**.

Remark B.1 *It is important to recognize that* `long double` *precision can only be achieved by employing the FPU, which can slow down mathematical operations considerably because of the missing SIMD features Using* `float` *and* `double` *types in conjunction with the FPU, the results can differ from expectation because of the extended precision interim results. This is not an academic discussion: simple code examples can be given where FPU and SSE calculations differ and the results of one of them is simply false. Also, these types have to be converted into (and out of) the internal extended precision format, which is not straightforward because of denormalized values and other special cases.*

Example B.2 *Using the FPU, dividing the smallest representable* `double` *by* 2 *and immediately multiplying by* 2 *again will result in different results depending on whether the FPU or the SSE logic blocks are used. In the FPU case, the perfectly representable* `long double` *value* $2.47 \cdot 10^{-324}$ *is converted to* 0.0 *when it is converted back into a* `double`. *If the value is not read out between steps, the results can differ:*

```
                                   FPU                       SSE
                                   -----------------------   -----------------------
Smallest representable double:     4.9406564584124654e-324   4.9406564584124654e-324
Divided by 2
 -> should be zero:                0.0000000000000000e+00    0.0000000000000000e+00
Divided and multiplied by 2
 -> should still be zero:          4.9406564584124654e-324   0.0000000000000000e+00
```

One can argue for either of these behaviors to be "the right one". Another big complication of this subject-matter is that the FPU is used by default on i686 Linux installations and SSE by default on modern x86_64 Linux installations.

Since the FPU with its decade old design lacks modern SIMD features and the internal extended precision can produce unintuitive results, one should avoid using the `long double` data type.

B.1.2 About Compilers

A compiler transforms the source code of a program into machine interpretable instructions. It has to correctly implement the intent of the programmer, i.e., using the appropriate machine language instructions for the operations in the (high-level) programming language. High-level mathematical operations, e.g., adding two vectors, have to be expressed using a large number of machine code instructions. On the other hand, most high-level programming languages do not contain primitives for many machine instructions, e.g., for bit rotations:

`(x >> n) | (x << 8*sizeof(x)-n)` $\Leftrightarrow$ `ror` (Rotate Right) .

This explains why different compilers can produce highly differently behaving code, especially w.r.t. the runtime.

Letting the compiler do the optimizations does not preclude avoiding inefficient code, but it should be every programmers' motivation to write working elegant code and let the compiler do much of the optimization. An example for this can be found under the topic of "bit twiddling", see [93], which eliminates the conditional jump ("IF") from the often used `abs()` function,

```
T abs( T x ) { return (x < 0 ?  -x :  x); } ,
```

by implementing it for integer types using the highest bit (which indicates the sign) and XOR:

```
T abs( T x ) { T tmp = x >> (sizeof(x)*8-1); return (tmp ^ x) - tmp; } .
```

Here, the right shift operator (>>) must preserve the sign of the represented number, so it will fill up with 0's for a positive and 1's for a negative number.

The second version is incomprehensible even to most professional developers, but the machine code generated from both functions would be identical with most compilers, see [145]. The second version is generally preferred because conditional jumps can be very costly (w.r.t. execution performance) if the CPU incorrectly predicted (using a jump predictor in the CPU) which branch the code execution will take.

Used Compilers

I tested the program with three easily available compilers on Linux: GCC >= 4.6, CLang >= 3.2 and IntelCC 2013.1.117. Although the CLang project is the youngest, it produced the fastest program and I used it by default. For instructions on how to use a specific compiler, see the file `libwavelets/README`.

Although it should not be a problem to employ different compilers, I have not tested this and compilation might fail because of trivial code language details. In the same manner, compilation on non-Linux systems might or might not be successful.

B.2 Code Design Rationale

My software consists of several individual components:

`libwavelets` The main library. It is structured into several subdirectories:

- `Accessories` As the name suggests, this directory includes supplementary code, for example classes providing command line option parsing, logging and for measuring execution times.
- `MathStructures` This directory contains mathematical machinery used in the other parts, e.g., implementations of the polynomial bases of Section A.3, tensor product and tensor vector structures and a multi-index class.
- `Int_Full_Grid` Here, the full-grid code, first developed for [122], can be found. This code is used for the generation of the data used in Section 3.5.1 and serves to corroborate the results of the adaptive code by comparing results for linear operators. The prefix "Int" stands for "Interval".

`Int_Adaptive` The adaptive code can be found in this directory. This includes code for the adaptive data structures explained in Section B.2.4 and the implementations of all the polynomial operators. The prefix "Int" stands for "Interval".

`Implementations` The "Implementations" directories (also found under `Int_Full_Grid` and `Int_Adaptive`) contain base classes used in the other projects, in particular, the "Action" classes like `Assemble`, `Solve` and `EigenValues`.

`wavdata` Contains precomputed data. This includes the values of the bilinear forms of Section 3.5.1, i.e., computed data of (3.5.4) for different types of single-scale functions. These are loaded at runtime when needed or generated on-the-fly, which is not always feasible.

`wavtools` This contains the program `wavdiagram`, which generates the wavelet diagrams shown in Section A.2 and `polynomplot`, which plots wavelet vectors, both of them using Gnuplot, PostScript or TEX/TikZ. Apart from these, there are more than a dozen other tools for single purposes, usually to apply a specific operation to a wavelet vector in a file. This includes various types of coarsening (Section 3.3.2) and projecting a vector of indices w.r.t. dual wavelets onto a primal (piecewise polynomial) base as to plot it.

`diagassemble` Generates an adaptive version of the diagonal of the stiffness or mass matrices, see Section 2.2.5. This data can quickly be generated for any number of dimensions, levels and wavelet types by taking into account the translation of wavelets (2.1.14), so that it is not necessary to precompute this data. It is often faster to recalculate this than to load it from disk. Nevertheless, the program can write this data to disk and you can load it into a running program later, e.g., for testing purposes.

`linopassemble` Generates the data required to apply bilinear forms as described in Section 3.5.1. For positive or negative level differences up to 10 levels, the generation of the data is usually possible within a second. Higher level differences should be loaded from disk; the data is supplied in `wavdata`.

`testproblem` This project serves mostly to test individual operators of the library. It is split into an adaptive and a full-grid part, each corresponding to the code of the library used.

`boundaryvalueproblem` As the name suggests, this project implements the boundary value problem of Section 5, in both an adaptive and full-grid form.

`schurcomplement` This project was used in [122] to compute condition numbers of the **Schur complement** of a saddle point problem.

`controlproblem` This project implements the control problem from [122], currently only in full-grid form.

When installing the software, one should go through the list of projects in the order given above. The individual projects usually do not have options to configure the internals;

these details are all set at compile time of the library. After installing the library, it saves its configuration alongside the header files and `CMake` loads these when building the other projects.

B.2.1 Library CMake Options

At compile time, a few options have to be chosen which impose limits at runtime:

- `FP_TYPE` Floating point type for computations. Possible values are "`float`", "`double`", "`long_double`", "`long_double_96`", "`long_double_128`". The standard choice "`double`" should rarely be changed, performance of "`long_double(_96/128)`" types is very bad considering the miniscule possible increase in accuracy, see Section B.1.1.
- `MAX_DIM` Maximum spatial dimension n: $\{1, 2, 3, 4, \texttt{inf}\}$. A finite choice will result in the use of a data structure laid out in Section B.3. The choice `inf` leads to dynamic storage types like vectors and lists being used, which increases memory usage.
- `MAX_MONOM` Maximum polynomial order d of wavelets: $\{1, 2, 3\}$, which stands for piecewise linear, quadratic and cubic, respectively.
- `MONOM_BASE` Local polynomial basis: Possible choices are "StandardMonomials", "NormalizedMonomials" and "ShapeFunctions", see Section A.3.

The complete documentation can be found within the source directory of the software, can be found in the files `README` and `INSTALL`. Also, documentation about the software including example program runs can be found in [144].

B.2.2 Tensor Products

The guiding principle of my software is the **tensor product**. In the code, this comes in two forms: First, a template `class TensorProductContainerT<>`, which resembles a tensor product of the form (2.4.1). The purpose of the class is to emulate a high dimensional tensor product object by only saving a few elements. To achieve this, it always returns the highest available (w.r.t. to dimension) object when any dimension beyond its highest is requested. This is particularly useful if a tensor product of only one object is sought.

Secondly, the template `class TensorProductT<>` can be used to compute tensor products of the type (2.4.3). Its factors can be of arbitrary type, as long as a policy describing how to construct the product is given. For example, the factor type could be `std::vector<double>`, the product type could be `std::vector<double>` again or simply `double`. In the first case, the standard policy (here for two vectors) is to compute all permutations

$$\{a_0, a_1, \ldots, a_n\} \otimes \{b_0, b_1, \ldots, b_m\} \mapsto \{\{a_0, b_0\}, \{a_1, b_0\}, \ldots, \{a_n, b_m\}\}.$$

A possible policy when the product type is a single `double`, could be

$$\{a_0, a_1, \ldots, a_n\} \otimes \{b_0, b_1, \ldots, b_m\} \mapsto \{a_0 b_0, a_1 b_0, \ldots, a_n b_m\}.$$

The only real requirement to the factor types is a direct way of accessing consecutive elements, for which a **bidirectional** `const_iterator` is required. To save memory, the set of all resulting objects is not computed beforehand, but during traversing all possible states of the tensor product using the `class TensorProductIteratorT<>`. In this way, the memory requirement is constant instead of order $\mathcal{O}(n\,m)$ for the above example. If a complete list of all tensor products is needed, one simply has to traverse the range using the tensor product iterator type and save the resulting elements in another storage unit.

Since these techniques simplify a big portion of the complexity when coding tensor products, it was easy to design the software independent of the dimension. Another point in case is the application of trace operators explained in Section 3.6, which also requires simultaneous use of wavelet indices of different dimensions. The dimension is therefore usually specified at **runtime**, although it is possible to set a **maximum** possible value at compile time, see Section B.2.1. The point of this feature is to counter excessive memory usage for dynamic structures like `std::vector` and `std::list` by employing constant size arrays `boost::array` instead. Since constraining the dimension is not enough to use constant size arrays in all cases, e.g., for polynomials of unknown orders, it is possible to set limits on other variables as well, see Section B.2.1.

Another design rationale was to prescind the details of the wavelets described in Section A from the abstract use of operators like the mass matrix or nonlinear operator $f(u) := u^3$. Since there are too many types of variations of wavelets even for a fixed dimension, only a polymorphic approach using `virtual` methods in an **abstract base classes** could achieve this feature. The downside of increased complexity of virtual function calls can be alleviated by **caching** often requested data, e.g., mathematical description of the polynomial representation. Taking into account a translation like (2.1.9), the caches can be set up extremely quickly and memory requirements can be kept minimal, making efforts to precompute these (and save them to disk) unnecessary. The advantage of this design is the ability to choose completely different wavelet types in any spatial direction, with different boundary conditions, polynomial exactness or moment conditions.

B.2.3 One-dimensional Wavelet Implementations

The mathematical details of the wavelet types implemented in my software are given in Section A. Here, I just want to state what information is required to add the implement of another wavelet. General information of course includes the **masks** (2.1.10), (2.1.17), and their dual counterparts. If one uses the standard primal or dual single-scale functions, their implementations could already be available.

Int_Full_Grid

The wavelet masks are used to implement the one-dimensional FWT. In the full-grid setting, the application of wavelet vectors is reduced to the application of said operators in the single-scale representation. Therefore, implementations of matrices like the primal and dual mass matrices, Laplace operator and others have to programmed. Not all of these functions have to be supplied, but calling unimplemented ones will lead to a runtime error and program abort. All these algorithms must work on arbitrary random access iterators and therefore have to be supplied as templates.

Int_Adaptive

Based on the theory explained in Section 3, one needs to supply the details of the **tree structure** laid out in Section 3.2.1. Because of the construction of wavelets by translation and dilation (2.1.14), the details of the tree structure mostly depend on the boundary conditions. As long as a wavelet of the same boundary construction is already present, the types can simply be adopted. As laid out in the previous paragraph, the implementations of the single-scale bases can be reused if possible.

More importantly, to use any nonlinear operator, the polynomial description (3.4.3) has to be supplied. Because of the relation (2.1.18), only the exact representation of the single scale functions must provided, the wavelets can be constructed by the single scale basis using their masks. Also, the individual (wavelet or single-scale) functions have to be "categorized", i.e., it has to be provided for which $k \in \mathbb{Z}$ (2.1.14) holds. This is important for the quick computation of the diagonal of the stiffness matrix and other data. These algorithms recognize the repetitions of functions and use this to compute only really unique values and general data only once.

One important feature not to be forgotten here is the following: Since my code is based upon tensor products, only **one-dimensional** wavelet information need to be provided. The multidimensional wavelets, either isotropic as in Section 2.4.3 or anisotropic as in Section 2.4.2, including all operators or data, are constructed from the one-dimensional types.

B.2.4 Level-wise Storage in Unordered Containers

There are several data structures in the code using the associative containers explained in Section B.3. The prime example is of course an **adaptive vector**, implemented in the template `class AdaptiveVectorT<>`. Other use cases are the **adaptive polynomials** of Section 3.4.1 and an implementation of the **diagonal preconditioner** (2.2.33) of Section 2.2.5. All these classes share the same **storage backend**, where the data is stored **level-wise**, i.e., a separate storage unit is used per level. This has the advantage that all data on a specific level can be accessed directly, which is the very approach used in many algorithms in Section 3. Since these containers are important for the implementation, I discuss some details of these next.

B.3 Adaptive Storage Containers with Amortized $\mathcal{O}(1)$ Random Access

The C++ library **Boost** provides an unordered associative container `unordered_map` that associates unique keys ($\in \mathcal{K}$) with another value. Another container, called `unordered_set`, only saves the key without an associated value type.
Within these containers, the elements are organized into buckets. In principle, keys with the same hash code ($\in \mathcal{H}$) are stored in the same bucket. Theoretically, this structure promises (amortized) $\mathcal{O}(1)$ complexity for random accesses. After a deletion or insertion of several elements, a **rehashing** must be executed, as to continue to guarantee the constant complexity of access operations. This process increases the time of the individual function call disproportionately, but it is required so rarely that the average times is

constant.

B.3.1 Hash Functions

To employ the above structure, it is necessary to provide a **hash function** $h : \mathcal{K} \to \mathcal{H}$ which should **not** be **collision free**, i.e., it should be **injective**. On the contrary, if every bucket only contained a single element, then an abundance of buckets would be created and a lot of memory would be wasted, see Figure B.5. Since the search time inside a bucket is linear w.r.t. the number of elements contained, the buckets should not be filled up too much either. The range of the hash function thus must not be too small or too large. Conveniently, the bucket occupancy can be controlled by a **load factor** m defined as

$$m := \frac{\#\text{Number of Nodes}}{\#\text{Number of Buckets}}. \tag{B.3.1}$$

The strategy of the unordered container is to generate new buckets as soon as the actual load factor number grows bigger than a user defined **maximum load factor** $m_{\max}$.

In practice, the key type $\mathcal{K}$ is usually some kind of (an-)isotropic wavelet index, the hash type $\mathcal{H}$ is an 16, 32 or 64-bit unsigned integer. The value type, i.e., the type that is actually stored, is independent of the other two types, it could be a single value or an n-dimensional polynomial.
A hash function should be fast to evaluate, since any kind of random access into a vector will result in an execution of the hash function. Only a linear but unordered walkthrough using **iterators** is done without evaluation of the hash function.

On the other hand, the hash function should ideally depend on all properties of the wavelet index λ. If it were only to depend on the spatial location of one dimension, the range might be greatly diminished because the problem might favor indices close to one side and thus with very low location values k.

Division-Remainder Method

If the maximum number of dimensions is not known or set at compile time, wavelet indices are internally saved (per dimension) in a container of 1D indices, e.g., a list. A standard algorithm to compute hash values of input data with more bits than the word size is to successively arrange old and new data into one and then calculate the remainder modulo a constant value, e.g., a prime number. To compute a 32-bit hash value from several 64-bit variables `locations`, a step in the algorithm is performed in this way:

```
U64 key = locations[ 0 ] % prime;
for( d = 1; d < locations.size(); ++d ) {
  key <<= 32;
  key += ( locations[ d ] & 0x00000000FFFFFFFF );
  key %= prime;
}
return U32( key );
```

The computed hash value from the last step is shifted to the upper 32 bits of the 64-bit variable and the lower 32 bits are filled from the current 64 bits of data by clipping off the upper half. The modulo operation then asserts that the hash value really only uses the lower half of bits. The return value encompasses then usually only the essential lower 32 bits.

Specialization for Finite Dimensional Indices

In case the **maximum number of dimensions** is set to a fixed constant, e.g., $n = 2, 3, 4$, at compile time, the whole n-dimensional wavelet index is squeezed into a single 64-bit data structure.

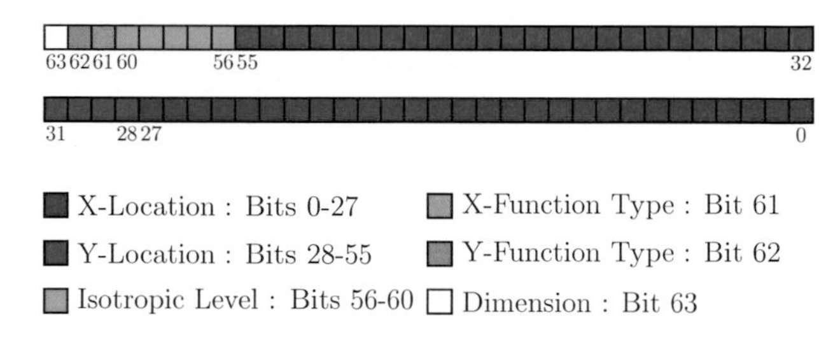

Figure B.1: Diagram for a **Packed Isotropic Wavelet Index** for up to two dimensions. The single bit for dimension encodes a one (bit not set) or two-dimensional (bit is set) value. The 5 bits for the isotropic level can represent up to 32 different levels. But the 28 location bits only allow levels up to 27, since the location index k on level J can attain values larger than 2^J, but not larger than 2^{J+1}.

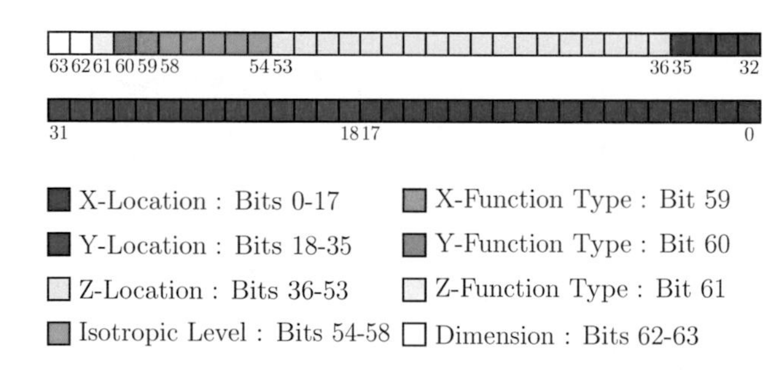

Figure B.2: Diagram for a **Packed Isotropic Wavelet Index** for up to three dimensions. For three dimensions to encode properly, two bits are needed, but these can hold four different states. Again, the 5 bits for the isotropic level could represent up to 32 different levels, but the 18 location bits only allow levels up to $J = 17$.

This is possible since the dimension, level, location and type information are all integer types and their maximum values - and thus the required bits - can be controlled. The individual field must be chosen so that the overall distributions make sense, i.e., ideally that any single bit can actually be used. For example, the level variable should not allow levels of which the maximum location index could not be represented, see Figure B.1 and Figure B.2. The layouts given in these figures are purely exemplary and the fields could be rearranged freely to fit any kind of agenda.

Remark B.3 *The implementation of such a* **bit field** *is possible with C/C++ language intrinsics; no explicit bit operations need to be employed by the programmer. This is not only much less error-prone, the compiler can often do a much better job at optimizing this construction than handwritten code.*

The advantage is that, on a modern 64-bit computer, this uses the existing memory very efficiently. But there are also two main disadvantages:

First, there is an increase in runtime since the memory controller in a computer addresses pages and bytes, nevertheless, extracting the state of a single bit is still a relatively expensive endeavor. Even on modern CPUs, comparatively simple tasks, e.g., finding the highest valued set bit, are not implemented in hardware. This has the unusual and counter intuitive effect that simple bit operations can take several clock cycles, but - at least on average - floating point operations can be executed in less than one clock cycle[2].

Second, the small space of just 64 Bits limits the possible ranges of the individual variables with increasing dimensions, thus requiring adaptations for dimensions $n > 4$, e.g., by using two 64 Bit data blocks.

Thus, it is not a good strategy to first extract the individual data from such a "packed" embodiment and then apply the hash function of the previous paragraph. Here, it is much more efficient to simply interpret the 64-bit data structure as an integer and then apply a single **modulo** operation and use the return value as the hash value:

```
return U32( bitfield % prime );
```

Using a prime number ensures that the result will depend on all bits up the highest bit set. In contrast, using a power of 2 will void the influence of any bit higher than the exponent of the 2.

B.3.2 Optimization Strategies

Although the theoretical complexity of the `unordered_map` container is optimal, the memory management can be improved upon by changing details of the memory management. There are two possible approaches for optimizations: First, one can tune the **load factor** (B.3.1). Increasing the load factor entails potentially longer lookup times since the search within a bucket is linear w.r.t. the number of nodes and thus the actual load factor. Thus, the load factor should not be too large, or execution times will increase. Choosing a very small maximum load factor, e.g., the standard value is 1.0, then for each node containing the actual user data, a separate bucket is created. The node structure

[2]This stems from the massive parallelization taking place inside the CPU by the SIMD extensions explained in Section B.1.1.

containing the user data also includes a certain amount of management data structure, enlarging the user data structure by (usually) 2 octets. The buckets also use memory, at least 1 octet. Thus, an overabundance of buckets, with a load factor of ~ 1, will have an impact on memory usage. Secondly, the memory management used in the background to allocate memory for saving buckets and nodes can be replaced, which I address in the next paragraph.

Memory Management

Here, I exchange the **standard allocator** `std::allocator` to a **pool allocator**, e.g., `boost::fast_pool_allocator` or `boost::pool_allocator`, which are optimized for fast allocation and deallocation of many small objects. A pool allocator, instead of calling **new/delete** for every object, allocates large chunks of raw memory and then constructs objects inside these memory pools by keeping track of the used and unused memory blocks itself. This way, calls to **allocate/deallocate** are as simple as returning a memory address and a little book keeping. The book keeping can be very simple, using an (unordered) **linked list**, which is the approach of the `fast_pool_allocator`. This unordered approach has the drawback that adjacent blocks are not recognized and thus an instruction to allocate a multiple of the block size needs to be taken from almost empty or new memory blocks, where contiguous blocks are most easily available. Keeping the list of free blocks in an ordered state is not trivial and especially requires work every time a block is erased. This extra step is done when using the `pool_allocator`, which is advisable to use when several contiguous blocks are requested more often than single blocks.

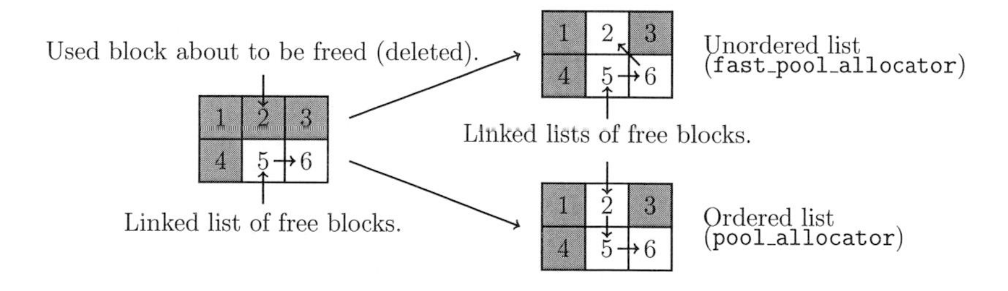

Figure B.3: Fragmentation withing a chunk (with 6 blocks) arises when blocks (here number 2) between occupied blocks of memory are freed.

The simple change to a pool allocator can make insertions and removals from the container an order of magnitude faster. There are a few pitfalls in the standard implementation of the `fast_pool_allocator` together with the `unordered_map` though, which can be comparatively easily evaded:

- In the standard implementation, **all** instantiations (objects) of any pool allocator use the **same** pool by the use of a **singleton** (see [3]). This seems to be a good idea, but can become a problem because of the next bullet points.

- To decrease the amount of chunks to allocate, the standard approach of the pool allocator is to double the block size each time, thus needing fewer calls to the real `new/delete` in the later program execution. This quickly leads to extremely large chunks of memory being allocated, the size of which will overstretch even the largest RAM installations.
- Memory chunks can only be released if all its individual blocks are unused. The larger the chunks become, the more unlikely it is that any chunks could ever be released. Thus, in the most extreme case, although only a linear amount of memory is actually used, an exponential amount of memory could be allocated in the pools.

To remedy these problems, I use the following approaches:

- An upper limit of the size of the chunks is set. To ease memory management for the operating system in the background, the size of the chunk should be multiples of the page size, usually 4096 bytes. Because of **padding** of structures in C++, it suffices to choose the chunk size as 1024 blocks or even smaller powers of 2.
- Even then, a lot of memory fragmentation could occur over time that only very few memory chunks would ever be freed. To battle fragmentation even more, the pools are no longer shared between all instances of the adaptive vectors, but each adaptive data structure uses its own pool. The pool can be shared internally for several `unordered_map/unordered_set` instances, but the memory is completely freed once the adaptive storage object is destroyed.

I call the allocator implemented according to these principles "SharedFastPool". The effects of the above mentioned optimization strategies can be seen in the Figures B.5 and B.4. Because of these results, I choose to use my SharedFastPool allocator with a maximum load factor of $m = 3$, since it only increases the runtime marginally compared to $m = 1$, but uses only slightly more memory than higher values of the load factor.

Other strategies

Another common idea in this setting is not to save actual values in the indices, e.g., j, but instead the **offset** w.r.t. the minimal implementation value, e.g., $j_0 - 1$. For the level variables, this implementation would free only 2 bits for $j_0 = 3$, but particular care must be taken to make sure that the readout of the variable is properly adjusted under all circumstances. Moreover, the necessary additions and subtractions of j_0 each time the variable is read or written would impose another runtime slowdown, in addition to all the bit set operations. For these reasons, I chose not to implement this strategy.

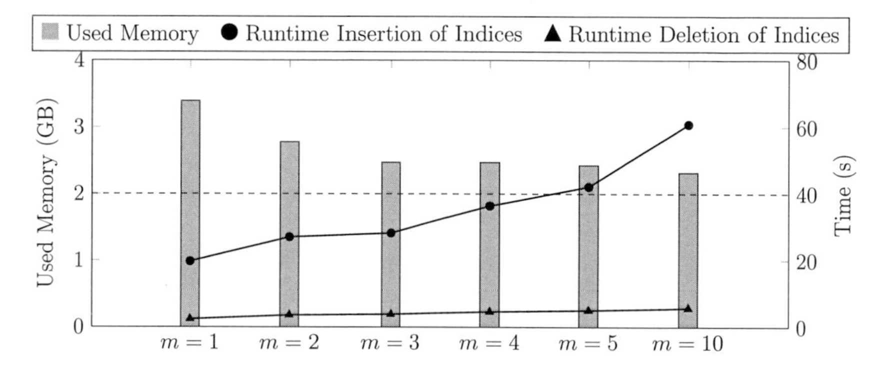

Figure B.4: Comparison of the SharedFastPool allocator with different values of the maximum load factors (B.3.1). The scenario is exactly the same as in Figure B.5, except that the `unordered_map` storage containers were "prepared" by informing them how many elements are going to be inserted. This speeds up insertions, because the required number of buckets can be preallocated instead of adaptively constructing them during the insertions. Since there was very low deviation for each run, I present the average numbers of the five runs. The decrease in memory requirements but increase in computing time for higher values of m is clearly recognizable.

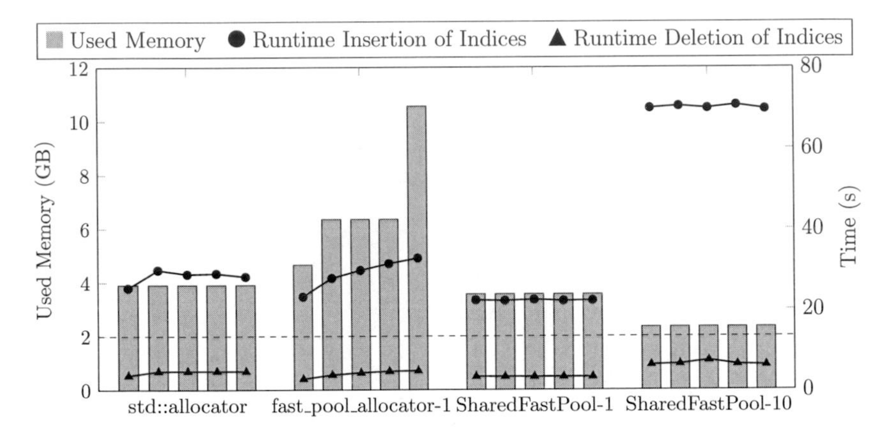

Figure B.5: Comparison of four different allocators together with `unordered_map`. Each time, a vector containing 67 million wavelet indices and `double` values is constructed 5 consecutive times. The nodes containing the wavelet coefficients and values comprise roughly 2 GB, which is marked by the blue dashed line. Any allocated memory exceeding the line is thus used internally for the storage management, foremost for the buckets. Ideally, the times and memory requirements should be constant over the course of the program run. The `fast_pool_allocator`, because of the employed **singleton** and the internal fragmentation of the pools, needs new memory over time. Since the allocation size of the pools follows an exponential law, this leads to sudden and huge increases. The last allocator (SharedFastPool) also uses one pool for all unordered storage containers holding the data level-wise, but it is completely freed when the data is deleted. The numbers after the name of the allocator denote the value of the **maximum load factor** (B.3.1). The default ($m = 1$) leads to very quick execution times, but the memory requirements for the high number of buckets is clearly visible. Also, in the last case ($m = 10$), only very few buckets were created, reducing the memory requirements significantly, but the execution time required for the insertion of all elements increases also significantly.

C Notation

The following notations and definitions should hold anywhere in this document unless explicitly specified otherwise. I have taken care to avoid using any mathematical symbol more than once for different purposes if possible.

C.1 General Notation

keywords	are emphasized
`computer`	the typewriter font indicates programming language or computer reference
a.e.	almost everywhere, i.e., valid everywhere except for a domain of measure zero
w.r.t.	with respect to
$\wedge, \vee$	logical *and*, logical *or*
$\overset{!}{<}, \overset{!}{\leq}, \overset{!}{=}, \ldots$	the exclamation mark designates statements that have to be proven
$\mathcal{H}$	a Hilbert space
$X, Y, \ldots$	spaces are denoted by capital letters
$u, v, w, \ldots$	elements of spaces
$\mathbf{u}, \mathbf{v}, \mathbf{w}, \ldots$	elements of sequence spaces, especially of ℓ_2
$\mathbf{u}_J, \mathbf{v}_J, \mathbf{w}_J, \ldots$	elements of finite sequence spaces, especially of $\ell_2(\Delta_J)$
$\mathbf{A}, \mathbf{B}, \mathbf{C}, \ldots$	discretized operators in wavelet representation, i.e. matrices of possibly infinite size
$\Vert\cdot\Vert$, $\Vert\cdot\Vert_X$	norms in some Banach space X
$(\cdot,\cdot)$, $(\cdot,\cdot)_X$	inner product of a Hilbert space X
$\vert\cdot\vert$, $\vert\cdot\vert_X$	seminorms
$\vert\cdot\vert$	the absolute value function in $\mathbb{R}$; any (equivalent) norm in $\mathbb{R}^n$
$\langle\cdot,\cdot\rangle$, $\langle\cdot,\cdot\rangle_X$	dual form on the space X, i.e. for $u \in X, v \in X' : \langle u, v\rangle_{X\times X'} := v(u)$
$\langle\cdot,\cdot\rangle$	scalar product of $\mathbb{R}^n$, i.e. for $v, w \in \mathbb{R}^n : \langle v, w\rangle := v^T w$
Ω	an open bounded set in $\mathbb{R}^n$ with boundary $\partial\Omega \supset \Gamma$
I; $\square$, $\square^n$	the interval $(0,1)$; the open unity cube in $\mathbb{R}^n$, $(0,1)^n$
$x_1, \ldots, x_d$	Euclidean coordinates in $\mathbb{R}^n$

$\delta_{(i,j)}$	Kronecker delta for numbers or multi-indices $= \begin{cases} 0, & i \neq j, \\ 1, & i = j. \end{cases}$
χ_A	characteristic function on some interval/domain A
$f\|_A$	restriction of a function $f : \Omega \to \mathbb{R}$ to a subdomain $A \subseteq \Omega$
$d\mu; ds$	the Lebesgue measure in $\mathbb{R}^n$; surface measure
$\lesssim$ ($\gtrsim$)	lesser (greater) or equal to except for a positive constant which is independent of any parameters of the arguments, see (1.1.4)
$\sim$	stands for both $\lesssim$ and $\gtrsim$, see (1.1.4)
#	cardinality of a set
∇	Gradient of a function $u : \mathbb{R}^n \to \mathbb{R}$: $\nabla u := \left(\frac{\partial u}{\partial x_1}, \dots, \frac{\partial u}{\partial x_n}\right)^T$
Δ	Laplacian of a function $u : \mathbb{R}^n \to \mathbb{R}$: $\Delta u := \sum_{i=1}^n \frac{\partial^2 u}{\partial x_1^2}$
$a \mod b$	modulus: remainder of a divided by b: $r := \arg\min_{s \in \mathbb{N}_0} \{a = m * b + s \mid m \in \mathbb{N}_0\}$
$\lfloor\cdot\rfloor$ ($\lceil\cdot\rceil$)	Gauss parentheses: the highest (smallest) non-negative integer smaller (higher) than or equal to the argument
$\kappa(\cdot), \kappa_2(\cdot)$	(spectral) condition of an operator, matrix or function set
$A^{\updownarrow}$	matrix whose rows and columns are reversed, i.e. $(A^{\updownarrow})_{i,j} := (A)_{n-i,n-j}$

$(\alpha_1, \dots, \alpha_n)$ **multi-index**:

The n-dimensional **multi-index** α is a n-tuple of non-negative integers α_i with $i = 1, \dots, n$. Two multi-indices are equal, if and only if all indices are equal. The length is denoted by $|\alpha| := \alpha_1 + \dots + \alpha_n$. For $x \in \mathbb{R}^n$ we define the shorthand expressions:

$$x^\alpha = x_1^{\alpha_1} \cdots x_n^{\alpha_n},$$

$$\partial^\alpha = D^\alpha = \frac{\partial^{|\alpha|}}{\partial x^\alpha} = \frac{\partial^{\alpha_1}}{\partial x_1^{\alpha_1}} \cdots \frac{\partial^{\alpha_n}}{\partial x_n^{\alpha_n}}.$$

$o(\cdot), \mathcal{O}(\cdot), \Theta(\cdot)$ **Landau symbols**:

Let n be an integer that tends to infinity and x be a variable that approaches some limit $x^* \geq x_0$. Also let g be a positive function and f another function. Then the Landau symbols are defined as:

$$\begin{aligned} f = \mathcal{O}(g) \quad &:\Longleftrightarrow \quad |f(x)| \leq k\, g(x), \quad \text{for all } x_0 \leq x \leq x^* \text{ and a } k > 0, \\ f = o(g) \quad &:\Longleftrightarrow \quad |f(x)|/g(x) \to 0, \quad \text{for all } x_0 \leq x \leq x^*, \\ f = \Theta(g) \quad &:\Longleftrightarrow \quad f = \mathcal{O}(g) \wedge g = \mathcal{O}(f). \end{aligned}$$

$[t_0, \ldots, t_n]\, f$ **divided difference**:
For any $f \in C^n$ and knot points $-\infty < t_0 \le t_1 \le \ldots \le t_{n-1} \le t_n < \infty$ the divided differences are recursively defined as

$$\begin{aligned} &[t_0]\, f = f(t_0), \\ &[t_0, \ldots, t_n]\, f = \tfrac{[t_1,\ldots,t_n]f - [t_0,\ldots,t_{n-1}]f}{t_n - t_0}, \quad t_n \ne t_0, \\ &[t_0, \ldots, t_n]\, f = f^{(n)}(t_0)/n!, \qquad t_0 = \cdots = t_n. \end{aligned}$$

$\phi^d(x)$ **cardinal B-Spline**:
The cardinal B-spline ϕ^d of order $d \in \mathbb{N}$ is given as

$$\phi^d(x) := d\,[0, \ldots, d]\left(\cdot - x - \left\lfloor \frac{d}{2} \right\rfloor\right)_+^{d-1},$$

where $x_+^d := \max\left\{0, x^d\right\}$.

C.2 Special Mathematical Symbols

Section 1

$\mathcal{F}(\cdot), \hat{(\cdot)}$	Fourier transform, see (1.2.10)
n	spatial dimension
$s; m+\sigma$	Sobolev smoothness indices: $s \in \mathbb{R}$; $m \in \mathbb{N}_0$, $0 < \sigma < 1$; see Section 1.2.2
γ_j	general **trace operator** of order j, see (1.2.26)
$\Delta_{\mathbf{h}}(u)(\mathbf{x})$, $\Delta_{\mathbf{h}}^d(u)$	difference operator (1.3.1), n-dimensional difference operator (1.3.2)
$\omega_d(u,t)_p$	moduli of smoothness, see (1.3.3)
$\mathcal{L}, \mathcal{A}$	linear partial differential operator of order $2m$, see (1.4.2)
$N(N_J)$	generic number of unknowns (on resolution level J), may be different in every section
h, h_j	discretization error on level j, see (1.4.28)
$N_\varphi(u)(x)$	Nemytskij operator, see (1.5.4)
DF	Fréchet derivative, see (1.5.6)

Section 2

j_0	minimum level in a **multiresolution analysis** (**MRA**), see $(\mathcal{R})$(2.1.6)
J	maximum level of resolution in a given context, see Section 2.1.3

d ($\widetilde{d}$)	primal (dual) order of **polynomial exactness**, see $(\mathcal{P})$(2.2.3) and $(\widetilde{\mathcal{P}})$(2.2.4)
γ ($\widetilde{\gamma}$)	primal (dual) **regularity**, range of smoothness for norm equivalences, (2.2.7)
S_j ($\widetilde{S}_j$)	closed subspace of primal (dual) Hilbert space, see (2.1.7) and (2.1.45)
Φ_j ($\widetilde{\Phi}_j$)	primal (dual) **single-scale basis** (also called **generators**), see (2.1.7) and (2.1.45)
Δ_j	index set for single-scale bases Φ_j and $\widetilde{\Phi}_j$, see (2.1.7) and (2.1.45)
$\phi_{j,k}$ ($\widetilde{\phi}_{j,k}$)	primal (dual) single-scale function on level j located at position $k \in \Delta_j$
W_j ($\widetilde{W}_j$)	primal (dual) detail spaces, see (2.1.13) and (2.1.49)
Ψ_j ($\widetilde{\Psi}_j$)	primal (dual) complement basis for space W_j ($\widetilde{W}_j$), see (2.1.13)
∇_j	index set for complement bases Ψ_j and $\widetilde{\Psi}_j$, see (2.1.13)
$\psi_{j,k}$ ($\widetilde{\psi}_{j,k}$)	primal (dual) **wavelet function** on level j located at position $k \in \nabla_j$, see (2.1.14)
$\Psi_{(J)}$ ($\widetilde{\Psi}_{(J)}$)	primal (dual) wavelet basis up to level J, see (2.1.29)
$\mathcal{S}$ ($\widetilde{\mathcal{S}}$)	primal (dual) **multiresolution analysis(MRA)** of $\mathcal{H}$ ($\mathcal{H}'$), see Definition 2.2
$\mathbf{M}_j$ ($\widetilde{\mathbf{M}}_j$)	primal (dual) two-level transformation matrix from level j to $j+1$, see (2.1.19)
$\mathbf{M}_{j,0}$ ($\widetilde{\mathbf{M}}_{j,0}$)	left part of $\mathbf{M}_j(\widetilde{\mathbf{M}}_j)$; matrix of dimensions $\#\Delta_{j+1} \times \#\Delta_j$, see (2.1.11)
$\mathbf{M}_{j,1}$ ($\widetilde{\mathbf{M}}_{j,1}$)	right part of $\mathbf{M}_j(\widetilde{\mathbf{M}}_j)$; matrix of dimensions $\#\Delta_{j+1} \times \#\nabla_j$, see (2.1.18)
P_j ($\widetilde{P}_j$)	primal (dual) projector onto the space Φ_j ($\widetilde{\Phi}_j$), see (2.1.46) and (2.1.47)
$\mathbf{T}_J$ ($\widetilde{\mathbf{T}}_J$)	primal (dual) **fast wavelet transform**, see (2.1.32) and (2.1.63)
$\mathbf{G}_j$	inverse of $\mathbf{M}_j$, see (2.1.21)
$\mathbf{G}_{j,0}$	upper part of $\mathbf{G}_j$; matrix of dimensions $\#\Delta_j \times \#\Delta_{j+1}$, see (2.1.25)
$\mathbf{G}_{j,1}$	lower part of $\mathbf{G}_j$; matrix of dimensions $\#\nabla_j \times \#\Delta_{j+1}$, see (2.1.25)
$\mathcal{I}$	infinite index set, see (2.1.41)
Ψ ($\widetilde{\Psi}$)	primal (dual) **wavelets** associated to the index set $\mathcal{I}$, see (2.1.41)
Ψ^A ($\widetilde{\Psi}^A$)	primal (dual) **wavelets** associated to the index set A, see (2.1.41)

$\mathbf{M}_{\mathcal{H}}$ $(\widetilde{\mathbf{M}}_{\mathcal{H}'})$ — primal (dual) **mass matrices** using the inner product $(\cdot,\cdot)$ of space $\mathcal{H}$ $(\mathcal{H}')$, see (2.1.60) and (2.1.61)

$\mathbf{D}^{\pm s}$ — any diagonal matrix that can be used for shifting wavelet coefficient vectors in the Sobolev scale by $\pm s$, see (2.2.13) and (2.2.14)

Ψ^s $(\widetilde{\Psi}^s)$ — scaled wavelet bases constituting **Riesz bases** for H^{+s} $((H^{+s})')$, see Corollary 2.23

$\mathbf{D}_1^{\pm s}$ — diagonal matrix consisting of powers of 2, see (2.2.15)

c_X, C_X — lower and upper constants used in norm equivalences, see (2.2.17)

$\mathbf{D}_{a,J}^{\pm s}$ — inverse diagonal of stiffness matrix, see (2.2.33)

$R_{\mathcal{H}}$ $(\mathbf{R}_{\mathcal{H}})$ — (wavelet discretized) Riesz operator for space $\mathcal{H} \in \{L_2, H^s, \ldots\}$, see (2.2.34)

$\widehat{\mathbf{D}}$ — diagonal matrix used in the construction of Riesz operator $\widehat{\mathbf{R}}_{H^s}$, see (2.2.38)

$\widehat{R}_{H^s}$ $(\widehat{\mathbf{R}}_{H^s})$ — specifically constructed (wavelet discretized) Riesz operator for H^s, see (2.2.43)

$\mathring{\widehat{\mathbf{R}}}_{H^s}$ — Riesz operator $\widehat{\mathbf{R}}_{H^s}$ normalized w.r.t. constant functions, see (2.2.54)

$\widetilde{\mathbf{R}}_{H^s}$ $(\mathring{\widetilde{\mathbf{R}}}_{H^s})$ — wavelet discretized (normalized) interpolating Riesz operator for H^s, see (2.2.52)

$\mathbf{C}_j$ — basis transformation working on the boundary functions only, see (2.3.7)

$\mathbf{M}'_{j,0}$; $\mathbf{M}'_{j,1}$ — refinement matrices incorporating basis transformations, see (2.3.8) and (2.3.9)

$\mathbf{T}'_J$ — fast wavelet transform incorporating basis transformations, see (2.3.10)

$\mathbf{A}$ $(\mathbf{A}_J)$ — **stiffness matrix** (discretized on level J), see (2.3.15)

O — orthogonal basis transformation, see (2.3.16)

Ψ'_{j_0-1} — transformed coarsest level wavelet basis, see (2.3.16)

$\mathbf{D}_{\{O,X\}}^{\pm s}$ — special preconditioner for basis transformed wavelet bases, see (2.3.23)

$\Phi_{(\square,j)}, \Psi_{(\square,J)}$ — **tensor product** analogons to their respective 1D objects, see Section 2.4

$\psi_\lambda^{\mathbf{ani}}(\mathbf{x})$; $\Psi_{\square,(J)}^{\mathbf{ani}}$ — **anisotropic wavelet**, see (2.4.9) and Section 2.4.2

$\mathbb{E}_n$ $(\mathbb{E}_n^\star)$ — set of isotropic types (without type $\mathbf{e} = \mathbf{0}$), see Section 2.4.3

$\psi_\lambda^{\mathbf{iso}}(\mathbf{x})$; $\Psi_{\square,(J)}^{\mathbf{iso}}$ — **isotropic wavelet**, see (2.4.19) and Section 2.4.3

$\psi^{\mathbf{iso}}_{j,k,e}(x)$; $\nabla_{j,e}$	type dependent one-dimensional wavelets and index sets, see (2.4.16) and (2.4.17)
$\mathbf{M}^{\mathbf{iso}}_{j}$ $(\mathbf{M}^{\mathbf{iso}}_{j,\mathbf{e}})$	isotropic refinement matrices (of type $\mathbf{e} \in \mathbb{E}_n$), see (2.4.23)
$\lambda := (j, \mathbf{k}, \mathbf{e})$	isotropic wavelet index in short notation; $\|\lambda\| := j$ for $n = 1$, $\|\lambda\| := \max\{j_1, \ldots, j_n\}$ for $n > 1$, see Section 2.4.3
Ψ^s_J	wavelet basis basis of all levels up to J, see (2.5.1)
$\mathcal{I}_{(J)}$	finite subset of $\mathcal{I}$ created by eliminating all indices with $\|\cdot\| > J$, see (2.5.2)

Section 3

$\mathcal{T}$ $(\mathcal{T}_j)$	**tree** of wavelet indices (on level j), see Definition 3.8 and (3.4.5)
$\mu \prec \lambda$	μ is a **descendant** of λ according to the tree structure, see (3.2.1)
$\mathcal{C}(\lambda)$; $\Pi(\lambda)$	**children** and **parents** of a node λ, see Definition 3.10
$\mathcal{N}_0(\mathcal{T})$; $\mathcal{L}(\mathcal{T})$	**roots** and **leaves** of a tree, see (3.2.2) and (3.2.3)
$S(\mathbf{v})$; $\text{supp}(\mathbf{v})$	**support** of the vector $\mathbf{v}$, see Section 3.2.2
$\sigma_N(\cdot)$ $(\sigma^{\mathbf{tree}}_N(\cdot))$	best (tree) N-term approximation, see (3.2.13) and (3.2.15)
$\mathcal{A}^s$ $(\mathcal{A}^s_{\mathbf{tree}})$	(tree) approximation classes, see (3.2.16)
$\mathcal{T}^\star$; $\mathcal{T}^\star(\eta, \mathbf{v})$	η-**best tree** of vector $\mathbf{v}$, see (3.3.3)
$\mathcal{T}'(\eta, C, \mathbf{v})$	(η, C)-**near-best tree** of vector $\mathbf{v}$, see (3.3.4)
$e(\lambda)$	ℓ_2-value of the branch at index $\lambda \in \mathcal{T} \subset \mathcal{I}$, see (3.3.5)
$E(\mathbf{v}, \mathcal{T}')$	error of restricting $\mathbf{v}$ to a **subtree** $\mathcal{T}'$, see (3.3.6)
$\widetilde{e}(\lambda)$	modified error functional of index λ, see (3.3.7)
$\mathcal{B}_j(\mathbf{v})$	the j-th **binary bin** of coefficients in $\mathbf{v}$, see (3.3.11)
$\widehat{\Delta}_j$	tree prediction layer j, see (3.3.17)
$\Lambda_\lambda(c\,j)$	**influence set** of depth $c\,j \geq 0$, see (3.3.18)
$\Lambda^{1,2,3}_\lambda(c)$	approximations to the influence sets, see (3.3.24), (3.3.25) and (3.3.27)
$\Lambda_\lambda(c)$	influence set at λ in multiple dimensions, see (3.3.29)
$\mathcal{P}(\Omega)$	**partition** of a domain $\Omega \subset \mathbb{R}^n$, see Definition 3.38
$\mathcal{D}(\mathcal{D}_j)$	**dyadic partitioning** of $\Omega = [0,1]^n$ (on level j), see (3.4.1)
$\mathcal{C}(\square)$	set of children of $\square \in \mathcal{D}$, see (3.4.2)

$\mathcal{D}_\lambda$	**support cubes** of wavelet λ, see (3.4.4)
$\mathcal{S}(\mathcal{T})$; $\mathcal{D}(\mathcal{T})$	support and dyadic support of $\mathcal{T}$, see (3.4.7) and (3.4.8)
$\mathcal{M}_d^n$	possible exponential indices of n-dimensional monomials up to order d, see (3.4.10)
$\mathbf{G}_j, \mathbf{G}_\mathcal{T}$	linear mapping wavelet coefficients to polynomial coefficients, see (3.4.14) and (3.4.18)
$\mathbf{P}_\mathcal{T}$	local polynomial representation on the leaves of $\mathcal{T}$, see (3.4.20)
$p_\square^{\mathbf{t}}(\mathbf{x})$	tensor product polynomial, see (3.4.24)
$\Upsilon_\lambda(\mathcal{T})$ ($\overline{\Upsilon}_\lambda(\mathcal{T})$)	(lower level) Dependency set, see (3.5.2) and (3.5.5)
$\overline{\Upsilon}_\lambda^a(\mathcal{T})$	Lower level Dependency set w.r.t. bilinear form $a(\cdot,\cdot)$ (3.5.6)
$\Theta(X,Y)$ ($\overline{\Theta}(X,Y)$)	set of all wavelets $\lambda \in X$ (not) intersecting with wavelets $\mu \in Y$, see (3.5.7) and (3.5.8)
$a_0(\cdot,\cdot)$	mass matrix bilinear form, see (3.5.9)
$a_1(\cdot,\cdot)$	Laplace matrix bilinear form, see (3.5.15)
$a_s(\cdot,\cdot)$	bilinear form of Riesz operator $\widetilde{\mathbf{R}}_{H^s}$ (2.2.52), see (3.5.17)
$a_z(\cdot,\cdot)$	square weighted bilinear form, see (3.5.19)
$b(v,\widetilde{q})$	trace operator bilinear form, see (3.6.1)
B (B')	(adjoint) trace operator, see (3.6.1) and (3.6.2)

Section 4

$\mathbf{R}(\mathbf{u})$	**residual** of an equation, see (4.1.1)
$\mathbf{u}^n$	**iteration variable** in step n of a solution algorithm, see (4.1.2)
$\mathbf{u}^\star$	**exact solution** of a problem, see (4.1.3)
$\mathbf{e}(\mathbf{u}^n)$	exact error of iterand w.r.t. exact solution $\mathbf{u}^\star$, see (4.1.3)
ρ	reduction factor of a convergent iteration, see (4.1.4)
$\rho_{n+1,n}(\widetilde{\rho}_{n+1,n})$	(residual) reduction factor of steps n and $n+1$, see (4.1.5) and (4.1.6)
$s^\star$	best approximation rate, see (4.1.15)

Section 5

$\mathcal{Z}(\mathcal{Z}')$	primal (dual) product Hilbert space, see (5.1.1)
L_A ($\mathbf{L_A}$)	**linear saddle point operator** (in wavelet discretization), see (5.1.11) (see (5.2.19))
L_F ($\mathbf{L_F}$)	**nonlinear saddle point operator** (in wavelet discretization), see (5.1.26) (see (5.2.18))
S ($\mathbf{S}$)	**Schur complement** (in wavelet discretization), see (5.1.16)
$\widetilde{S}(B,F,f)(\cdot)$	**Reduced equation** operator, see (5.1.32)
$a_\Omega(\cdot,\cdot)$	PDE bilinear form on domain Ω, see (5.2.2)
$a_\Box(\cdot,\cdot)$	PDE bilinear form $a_\Omega(\cdot,\cdot)$ extended on fictitious domain $\Box \supset \Omega$, see (5.2.6)
f_Ω $f_\Box$	Right hand side f on specific domains Ω and fictitious domain $\Box \supset \Omega$, see (5.2.5)
$\mathbf{n} = \mathbf{n}(x)$	outward normal at any point $x \in \partial\Omega$
$\Gamma_{i,k}$	Trace space in dimension $1 \le i \le n$ onto the faces at $k \in \{0,1\}$, see (5.2.13)
γ_W (γ_E)	Trace operator onto the west (east) face of the two dimensional square $\Box = (0,1)^2$, see (5.2.14) (TraceEast)
$\widetilde{\mathbf{A}}$	Operator of the normalized equation to $\mathbf{A}\mathbf{u} = \mathbf{f}$, see (5.3.1)
$\boldsymbol{\mu}^i$	**Defect** of the saddle point problem, see (5.3.5)
$\kappa(t)$	Mapping from $[0,1)$ to the boundary domain $\Gamma \subset \Box$

C.3 Spaces

$\mathbb{N}$	the natural numbers $= \{1,2,\ldots\}$
$\mathbb{N}_0$	the natural numbers including zero $= \{0,1,2,\ldots\}$
$\mathbb{Z}$	integers $= \{\ldots,-2,-1,0,1,2,\ldots\}$
$\mathbb{Q}$	the rational numbers
$\mathbb{R},\mathbb{R}^n$	the real numbers, the n-dimensional Euclidean Vector space
$\mathbb{R}_+,\mathbb{R}^n_+$	the positive real numbers, $\{(x_1,\ldots,x_n)\,\vert\, x_i > 0 \text{ for } 1 \le i \le n\}$

ℓ_2 — sequence space of all sequences for which the ℓ_2-norm is finite, i.e. let $\mathbf{c} \in \mathbb{R}^{\mathbb{N}} := \{x = (x_i)_{i\in\mathbb{N}} : x_i \in \mathbb{R} \text{ for } i \in \mathbb{N}\}$, then

$$\mathbf{c} \in \ell_2 \quad \Longleftrightarrow \quad \|\mathbf{c}\|_{\ell_2} := \left(\sum_{k\in\mathbb{R}} |c_k|^2\right)^{1/2} < \infty.$$

$L(X;\ Y)$ — linear operators from X to Y, see (1.1.1)

$X'; L(X;\ \mathbb{R})$ — dual space of space X, see (1.1.7)

Π_r — polynomials of order $\leq r-1$, i.e., r degrees of freedom, see (2.2.2)

C.4 Function Spaces

$C^k(\Omega)$ — $\{\phi : \Omega \to \mathbb{R} \mid \text{all derivatives } \partial^\alpha\phi \text{ of order } |\alpha| \leq k \text{ are continuous in } \Omega\}$

$C^k(\bar{\Omega})$ — $\{\phi \in C^k(\Omega) \mid$ all derivatives $\partial^\alpha\phi$ of order $|\alpha| \leq k$ have continuous extensions to $\bar{\Omega}\}$

$C^k_0(\Omega)$ — $\{\phi \in C^k(\Omega) \mid \operatorname{supp}\phi \subset\subset \Omega$, i.e. ϕ has compact support fully contained in $\Omega\}$

$C^{k,1}(\Omega)$ — Lipschitz continuous functions $= \{f \in C^k(\Omega)$ with

$$|D^s f(x) - D^s f(y)| \leq L|x-y| \text{ for all } |s| \leq k, 0 < L < \infty\}$$

$C^{k,\alpha}(\Omega)$ — Hölder continuous functions of order $0 < \alpha < 1 = \{f \in C^k(\Omega)$ with

$$\sup_{x\neq y\in\Omega} \frac{|D^s f(x) - D^s f(y)|}{|x-y|^\alpha} < \infty \ \forall\, \alpha\, \textbf{multi-index}, |s| < k\}$$

$C^\infty(\Omega)$ — space of infinitely differentiable functions on Ω with values in $\mathbb{R} = \bigcap_{k\in\mathbb{N}} \{C^k(\Omega)\}$

$\mathcal{D}(\Omega), C_0^\infty(\Omega)$ — space of infinitely differentiable functions with values in $\mathbb{R}$ and with compact support fully contained in Ω

$\mathcal{D}'(\Omega)$ — dual space of $\mathcal{D}(\Omega)$ = space of distributions on Ω

$L_2(\Omega)$ — space (equivalence class) of all real valued square Lebesgue-integrable functions on the domain Ω

$H^m(\Omega)$ — Definition 1.9: Sobolev space of order $m \in \mathbb{N} = \{\phi \in L_2(\Omega)$ with

$$\frac{\partial\phi}{\partial x_i} \in L_2(\Omega), \ldots, D^\alpha\phi \in L_2(\Omega) \ \forall\, \alpha\, \textbf{multi-index}, |\alpha| \leq m\}$$

$H^s(\Omega)$ — Sobolev space of fractional order s on Ω, see Definition 1.11

$H_0^m(\Omega)$	$\{\phi \in H^m(\Omega) \mid D^\alpha \phi = 0 \text{ on } \partial\Omega, \lvert\alpha\rvert \leq m-1\}$, see Definition 1.16
$H^{-s}(\Omega)$	Dual Space of $H_0^s(\Omega)$, see Definition 1.24
$H_{0,\Gamma}^1(\Omega)$	Space of test functions with zero values on Γ, see (5.2.3)
$\mathcal{H}_{\mathbf{mix}}^{(r_1,r_2)}(I_1 \otimes I_2)$	Sobolev space with dominating mixed derivative, see Definition 1.31
$B_q^\alpha(L_p(\Omega))$	Besov space on Ω, see Definition 1.33

References

[1] M. D. Adams and R. Ward. Wavelet Transforms in the JPEG-2000 Standard. In *In Proc. of IEEE Pacific Rim Conference on Communications, Computers and Signal Processing*, pages 160–163, 2001.

[2] R. Adams and J. Fournier. *Sobolev Spaces.* Academic press, second edition, 2003.

[3] A. Alexandrescu. *Modern C++ Design: Generic Programming and Design Patterns Applied.* Addison-Wesley, 2001.

[4] H. Alt. *Lineare Funktionalanalysis.* Springer, 2002. (In German).

[5] R. Andreev. Stability of sparse space-time finite element discretizations of linear parabolic evolution equations. *IMA J. Numer. Anal.*, 33:242–260, 2013.

[6] J.-P. Aubin. *Applied Functional Analysis.* Wiley, second edition, 2000.

[7] B. E. Baaquie. A Path Integral Approach to Option Pricing with Stochastic Volatility: Some Exact Results. *J. Phys. I France*, 7(12):1733–1753, 1997.

[8] B. E. Baaquie. Financial modeling and quantum mathematics. *Computers & Mathematics with Applications*, 65(10):1665–1673, 2013.

[9] I. Babuška. The Finite Element Method with Lagrange Multipliers. *Numer. Math.*, 20:179–192, 1973.

[10] A. Barinka. *Fast Computation Tools for Adaptive Wavelet Schemes.* Dissertation, RWTH Aachen, 2005.

[11] A. Barinka, T. Barsch, P. Charton, A. Cohen, S. Dahlke, W. Dahmen, and K. Urban. Adaptive Wavelet Schemes for Elliptic Problems – Implementation and Numerical Experiments. *SIAM J. Scient. Comput.*, 23(3):910–939, 2001.

[12] A. Barinka, T. Barsch, S. Dahlke, and M. Konik. Some Remarks on Quadrature Formulas for Refinable Functions and Wavelets. *ZAMM - Journal of Applied Mathematics and Mechanics / Zeitschrift für Angewandte Mathematik und Mechanik*, 81:839–855, 2001.

[13] A. Barinka, T. Barsch, S. Dahlke, M. Konik, and M. Mommer. Quadrature Formulas for Refinable Functions and Wavelets II: Error Analysis. *Journal of Computational Analysis and Applications*, 4(4):339–361, 2002.

[14] A. Barinka, W. Dahmen, and R. Schneider. Adaptive Application of Operators in Standard Representation. *Adv. Comp. Math.*, 24(1–4):5–34, 2006.

[15] P. Binev, W. Dahmen, and R. DeVore. Adaptive Finite Element Methods with Convergence Rates. *Numerische Mathematik*, 97(2):219–268, 2004.

[16] P. Binev and R. DeVore. Fast Computation in Adaptive Tree Approximation. *Numerische Mathematik*, 97(2):193–217, 2004.

[17] D. Boffi, M. Fortin, and F. Brezzi. *Mixed Finite Element Methods and Applications.* Springer series in computational mathematics. Springer, Berlin, Heidelberg, 2013.

[18] A. N. Borodin and P. Salminen. *Handbook of Brownian Motion : Facts and Formulae.* Probability and its applications. Birkhäuser Verlag, Basel, Boston, Berlin, 1996.

[19] D. Braess. *Finite Elements: Theory, Fast Solvers, and Applications in Solid Mechanics.* Cambridge University Press, third edition, 2007.

[20] J. H. Bramble and J. E. Pasciak. A preconditioning technique for indefinite systems resulting from mixed approximations of elliptic problems. *Math. Comp.*, 50(181):1–17, 1988.

[21] F. Brezzi. On the Existence, Uniqueness and Approximation of Saddle-Point Problems arising from Lagrangian Multipliers. *RAIRO Anal. Numér.*, 8(R-2):129–151, 1974.

[22] F. Brezzi and M. Fortin. *Mixed and Hybrid Finite Element Methods.* Springer, 1991.

[23] C. Burstedde. *Fast Optimised Wavelet Methods for Control Problems Constrained by Elliptic PDEs.* Dissertation, Institut für Angewandte Mathematik, Universität Bonn, December 2005.

[24] C. Burstedde and A. Kunoth. Fast Iterative Solution of Elliptic Control Problems in Wavelet Discretizations. *J. Comp. Appl. Maths.*, 196(1):299–319, 2006.

[25] A. Caglar, M. Griebel, M. Schweitzer, and G. Zumbusch. Dynamic Load-Balancing of Hierarchical Tree Algorithms on a Cluster of Multiprocessor PCs and on the Cray T3E. In H. Meuer, editor, *Proceedings 14th Supercomputer Conference.* Mannheim, Germany, 1999.

[26] C. Canuto, A. Tabacco, and K. Urban. The Wavelet Element Method. I. Construction and Analysis. *Applied Comput. Harmon. Analysis*, 6:1–52, 1999.

[27] J. Carnicer, W. Dahmen, and J. Peña. Local Decompositions of Refinable Spaces. *Appl. Comp. Harm. Anal.*, 3:127–153, 1996.

[28] J. Cascon, C. Kreuzer, R. Nochetto, and K. Siebert. Quasi-Optimal Convergence Rate for an Adaptive Finite Element Method. *SIAM J. Numer. Anal.*, 46(5):2524–2550, 2008.

[29] N. Chegini and R. Stevenson. Adaptive Wavelet Schemes for Parabolic Problems: Sparse Matrices and Numerical Results. *SIAM J. Numer. Anal*, 49(1):182–212, 2011.

[30] L. Chen, M. Holst, and J. Xu. Convergence and Optimality of Adaptive Mixed Finite Element Methods. *Mathematics of Computation*, 78:35–53, 2009.

[31] A. Cohen, W. Dahmen, and R. DeVore. Adaptive Wavelet Methods for Elliptic Operator Equations – Convergence Rates. *Math. Comput.*, 70(233):27–75, 2001.

[32] A. Cohen, W. Dahmen, and R. DeVore. Adaptive Wavelet Methods II – Beyond the Elliptic Case. *Found. Comput.Math.*, 2(3):203–246, 2002.

[33] A. Cohen, W. Dahmen, and R. DeVore. Adaptive Wavelet Schemes for Nonlinear Variational Problems. *SIAM J. Numerical Analysis*, 41(5):1785–1823, 2003.

[34] A. Cohen, W. Dahmen, and R. DeVore. Sparse Evaluation of Compositions of Functions using Multiscale Expansions. *SIAM J. Math. Anal.*, 35(2):279–303, 2003.

[35] A. Cohen, W. Dahmen, and R. DeVore. Adaptive Wavelet Techniques in Numerical Simulation. In *Encyclopedia of Computational Mechanics*, page 64. John Wiley & Sons, 2004.

[36] A. Cohen, I. Daubechies, and J.-C. Feaveau. Biorthogonal Bases of Compactly Supported Wavelets. *Communications on Pure and Applied Mathematics*, 45(5):485–560, 1992.

[37] A. Cohen, R. DeVore, and R. Nochetto. Convergence Rates of AFEM with H^{-1} Data. *Foundations of Computational Mathematics*, 12(5):671–718, 2012.

[38] S. Dahlke. Besov Regularity for Elliptic Boundary Value Problems in Polygonal Domains. *Applied Mathematics Letters*, 12(6):31–36, 1999.

[39] S. Dahlke, W. Dahmen, R. Hochmuth, and R. Schneider. Stable Multiscale Bases and Local Error Estimation for Elliptic Problems. *Applied Numerical Mathematics*, 23(1):21–47, 1997.

[40] S. Dahlke, W. Dahmen, and K. Urban. Adaptive Wavelet Methods for Saddle Point Problems - Optimal Convergence Rates. *SIAM J. Numerical Analysis*, 40(4):1230–1262, 2002.

[41] S. Dahlke and R. DeVore. Besov Regularity for Elliptic Boundary Value Problems. *Appl. Math. Lett*, 22:1–16, 1995.

[42] S. Dahlke, R. Hochmuth, and K. Urban. Adaptive Wavelet Methods for Saddle Point Problems. *Mathematical Modelling and Numerical Analysis (M2AN)*, 34:1003–1022, 2000.

[43] W. Dahmen. Decomposition of Refinable Spaces and Applications to Operator Equations. *Numer. Algor.*, 5:229–245, 1993.

[44] W. Dahmen. Stability of Multiscale Transformations. *J. Fourier Anal. Appl.*, 4:341–362, 1996.

[45] W. Dahmen. Wavelet and Multiscale Methods for Operator Equations. *Acta Numerica*, 6:55–228, 1997.

[46] W. Dahmen and A. Kunoth. Multilevel Preconditioning. *Numer. Math.*, 63(1):315–344, 1992.

[47] W. Dahmen and A. Kunoth. Appending Boundary Conditions by Lagrange Multipliers: Analysis of the LBB Condition. *Numer. Math.*, 88:9–42, 2001.

[48] W. Dahmen and A. Kunoth. Adaptive Wavelet Methods for Linear-Quadratic Elliptic Control Problems: Convergence Rates. *SIAM J. Control and Optimization*, 43(5):1640–1675, 2005.

[49] W. Dahmen, A. Kunoth, and K. Urban. Biorthogonal Spline-Wavelets on the Interval – Stability and Moment Conditions. http://www2.math.uni-paderborn.de/ags/kunoth/prog/databw.html. Online Resources.

[50] W. Dahmen, A. Kunoth, and K. Urban. Wavelets in Numerical Analysis and their Quantitative Properties. In A. Le Mehaute, C. Rabut, and L. Schumaker, editors, *Surface Fitting and Multiresolution Methods*, pages 93–130. Vanderbilt Univ. Press, 1997.

[51] W. Dahmen, A. Kunoth, and K. Urban. Biorthogonal Spline-Wavelets on the Interval – Stability and Moment Conditions. *Applied and Comp. Harmonic Analysis*, 6(2):132–196, 1999.

[52] W. Dahmen and C. A. Micchelli. Using the Refinement Equation for Evaluating Integrals of Wavelets. *SIAM J. Numer. Anal.*, 30:507–537, 1993.

[53] W. Dahmen and R. Schneider. Wavelets on Manifolds I: Construction and Domain Decomposition. *SIAM J. Math. Anal.*, 31:184–230, 1999.

[54] W. Dahmen, R. Schneider, and Y. Xu. Nonlinear Functions of Wavelet Expansions – Adaptive Reconstruction and Fast Evaluation. *Numer. Math.*, 86(1):49–101, 2000.

[55] W. Dahmen, K. Urban, and J. Vorloeper. Adaptive Wavelet Methods: Basic Concepts and Application to the Stokes Problem. In D.-X. Zhou, editor, *Wavelet Analysis*, pages 39–80. World Scientific, 2002.

[56] R. Dautray and J.-L. Lions. *Mathematical Analysis and Numerical Methods for Science and Technology*, volume 2: Functional and Variational Methods. Springer, second edition, 1988.

[57] C. de Boor. *A Practical Guide to Splines*. Springer, revised edition, December 2001.

[58] P. Deuflhard and A. Hohmann. *Numerical Analysis: A First Course in Scientific Computation*. de Gruyter, 1995.

[59] P. Deuflhard and A. Hohmann. *Numerische Mathematik I: Eine algorithmisch orientierte Einführung*. de Gruyter, 3rd edition, 2002.

[60] P. Deuflhard and M. Weiser. Local Inexact Newton Multilevel FEM for Nonlinear Elliptic Problems. In M.-O. Bristeau, G. Etgen, W. Fitzigibbon, J.-L. Lions, J. Periaux, and M. Wheeler, editors, *Computational Science for the 21st Century*, pages 129–138. Wiley-Interscience-Europe, Tours, France, 1997.

[61] P. Deuflhard and M. Weiser. Global Inexact Newton Multilevel FEM for Nonlinear Elliptic Problems. In W. Hackbusch and G. Wittum, editors, *Multigrid Methods, Lecture Notes in Computational Science and Engineering*, pages 71–89. Springer, 1998.

[62] R. DeVore. Nonlinear Approximation. *Acta Numerica*, 7:51–150, 1998.

[63] R. DeVore and A. Kunoth, editors. *Multiscale, Nonlinear and Adaptive Approximation.* Springer Berlin Heidelberg, 2009.

[64] R. DeVore and G. Lorentz. *Constructive Approximation.* Springer, 1993.

[65] T. Gantumur, H. Harbrecht, and R. Stevenson. An Optimal Adaptive Wavelet Method without Coarsening of the Iterands. *Math. Comp.*, 76(258):615–629, 2006.

[66] V. Girault and R. Glowinski. Error Analysis of a Fictitious Domain Method applied to a Dirichlet Problem. *Japan Journal of Industrial and Applied Mathematics*, 12(3):487–514, 1995.

[67] R. Glowinski. *Numerical Methods for Nonlinear Variational Problems.* Springer Series in Computational Physics. Springer-Verlag, 1984.

[68] R. Glowinski, T. Pan, and J. Periaux. A Fictitious Domain Method for Dirichlet Problem and Application. *Comp. Math. Appl. Mechs. Eng.*, 111:282–303, 1994.

[69] D. Goldberg. What Every Computer Scientist Should Know About Floating-Point Arithmetic. *ACM Computing Surveys*, 23:5 – 48, 1991. doi:10.1145/103162.103163.

[70] G. Golub and C. Van Loan. *Matrix Computations.* Johns Hopkins University Press, 3rd. edition, 1996.

[71] M. Griebel and S. Knapek. Optimized Tensor-Product Approximation Spaces. *Constructive Approximation*, 16(4):525–540, 2000.

[72] M. Griebel and P. Oswald. Tensor Product Type Subspace Splitting and Multilevel Iterative Methods for Anisotropic Problems. *Adv. Comput. Math.*, 4:171–206, 1995.

[73] P. Grisvard. *Elliptic Problems in Nonsmooth Domains.* Pitman, 1985.

[74] M. Gunzburger and A. Kunoth. Space-Time Adaptive Wavelet Methods for Control Problems Constrained by Parabolic Evolution Equations. *SIAM J. Control and Optimization*, 49(3):1150–1170, 2011. DOI: 10.1137/100806382.

[75] W. Hackbusch. *Elliptic Differential Equations. Theory and Numerical Treatment.* Springer, 2003.

[76] W. Hackbusch. *Tensor Spaces and Numerical Tensor Calculus.* Springer Berlin Heidelberg, 2012.

[77] H. Haug and S. Koch. *Quantum Theory of the Optical and Electronic Poperties of Semiconductors.* World Scientific Publishing, Singapore, fourth edition, 2004.

[78] S. Heston. A Closed-Form Solution for Options with Stochastic Volatility with Applications to Bond and Currency Options. *Rev. Finan. Stud.*, 6(2):327–343, 1993.

[79] R. Hochmuth, S. Knapek, and G. Zumbusch. Tensor Products of Sobolev Spaces and Applications. *Technical Report 685, SFB 256, Univ. Bonn*, 2000.

[80] M. Holst, G. Tsogtgerel, and Y. Zhu. Local Convergence of Adaptive Methods for Nonlinear Partial Differential Equations. *(submitted)*, 2010.

[81] IEEE. IEEE754-2008 – Standard for Floating-Point Arithmetic. Technical report, IEEE Standards Association, 2008. doi:10.1109/IEEESTD.2008.4610935.

[82] K. Ito and K. Kunisch. Augmented Lagrangian-SQP-Methods for Nonlinear Optimal Control Problems of Tracking Type. *SIAM J. Control and Optimization*, 34:874–891, 1996.

[83] H. Johnen and K. Scherer. On the Equivalence of the K-functional and Moduli of Continuity and some Applications. In *Constr. Theory of functions of several variables*, Proc. Conf. Oberwolfach, 1976.

[84] J. Kacur. Method of Rothe in Evolution Equations. *Teubner Texte zur Mathematik*, 80, 1985.

[85] P. Kantartzis. *Multilevel Soft-Field Tomography.* Dissertation, City University London, July 2011.

[86] P. Kantartzis, A. Kunoth, R. Pabel, and P. Liatsis. Towards Non-Invasive EIT Imaging of Domains with Deformable Boundaries. In *Engineering in Medicine and Biology Society (EMBC), 2010 Annual International Conference of the IEEE*, pages 4991–4995. IEEE, 2010.

[87] P. Kantartzis, A. Kunoth, R. Pabel, and P. Liatsis. Wavelet Preconditioning for EIT. *Journal of Physics: Conference Series*, 224:012023, 2010. DOI:10.1088/1742-6596/224/1/012023.

[88] P. Kantartzis, A. Kunoth, R. Pabel, and P. Liatsis. Multi-Scale Interior Potential Approximation for EIT. In *12th International Conference on Biomedical Applications of Electrical Impedance Tomography (University of Bath, UK, 4–6 May 2011).* IoP, 2011.

[89] P. Kantartzis, A. Kunoth, R. Pabel, and P. Liatsis. Single-Scale Interior Potential Approximation for EIT. In *12th International Conference on Biomedical Applications of Electrical Impedance Tomography (University of Bath, UK, 4–6 May 2011).* IoP, 2011.

[90] S. Kestler and R. Stevenson. Fast Evaluation of System Matrices w.r.t. Multi-Tree Collections of Tensor Product Refinable Basis Functions. *Journal of Computational and Applied Mathematics*, 260(0):103–116, 2014.

[91] H. Kim and J. Lim. New Characterizations of Riesz Bases. *Appl. Comp. Harm. Anal.*, 4:222–229, 1997.

[92] D. Knuth. *The Art of Computer Programming Volume 3, Sorting and Searching.* Addison Wesley Professional, second edition, 1998.

[93] D. Knuth. *The Art of Computer Programming Volume 4, Fascicle 1: Bitwise tricks & techniques; Binary Decision Diagrams.* Addison Wesley Professional, 2009.

[94] B. Korte and J. Vygen. *Combinatorial Optimization.* Algortihms and Combinatorics. Springer, 2005.

[95] J. Krumsdorf. Finite Element Wavelets for the Numerical Solution of Elliptic Partial Differential Equations on Polygonal Domains. Diplomarbeit, Institut für Angewandte Mathematik, Universität Bonn, 2004.

[96] A. Kunoth. *Multilevel Preconditioning.* Shaker, Aachen, 1994.

[97] A. Kunoth. On the Fast Evaluation of Integrals of Refinable Functions. *Wavelets, Images, and Surface Fitting*, pages 327–334, 1994.

[98] A. Kunoth. Computing Integrals of Refinable Functions – Documentation of the Program – Version 1.1. *Technical Report ISC-95-02-MATH*, May 1995.

[99] A. Kunoth. *Wavelet Methods – Elliptic Boundary Value Problems and Control Problems.* Advances in Numerical Mathematics. Teubner Verlag, 2001.

[100] A. Kunoth. Wavelet Techniques for the Fictitious - Domain - Lagrange - Multiplier - Approach. *Numer. Algor.*, 27:291–316, 2001.

[101] A. Kunoth and J. Sahner. Wavelets on Manifolds: An Optimized Construction. *Math. Comp.*, 75:1319–1349, 2006.

[102] A. Kunoth, C. Schneider, and K. Wiechers. Multiscale Methods for the Valuation of American Options with Stochastic Volatility. *International Journal of Computer Mathematics*, 89(9):1145–1163, June 2012.

[103] A. Kunoth and C. Schwab. Analytic Regularity and GPC Approximation for Control Problems Constrained by Linear Parametric Elliptic and Parabolic PDEs. *SIAM J. Control and Optimization*, 51(3):2442–2471, 2013.

[104] R. Leighton, M. Sands, and R. Feynman. *Feynman Lectures on Physics: The Complete and Definite Issue.* Addison Wesley, 2nd edition, 2005.

[105] J. Lions and E. Magenes. *Non-Homogeneous Boundary Value Problems and Applications*, volume I,II and III. Springer, 1972.

[106] S. Mallat. A Theory for Multiresolution Signal Decomposition: A Wavelet Representation. *IEEE Transactions on Pattern Analysis and Machine Intelligence*, 11(7), 1989.

[107] S. Mallat. *A Wavelet Tour of Signal Processing.* Academic Press, 3rd edition, 2008.

[108] R. Masson. *Wavelet Methods in Numerical Simulation for Elliptic and Saddle Point Problems.* Dissertation, University of Paris VI, January 1999.

[109] T. Meier, P. Thomas, and S. Koch. *Coherent Semiconductor Optics.* Springer, 2007.

[110] K. Mekchay and R. Nochetto. Convergence of Adaptive Finite Element Methods for general Second Order Linear Elliptic PDEs. *SIAM J. on Numerical Analysis*, 43(5):1803–1827, 2005.

[111] C. Mollet. *Excitonic Eigenstates in Disordered Semiconductor Quantum Wires: Adaptive Computation of Eigenvalues for the Electronic Schrödinger Equation Based on Wavelets.* Shaker–Verlag, 2011. DOI: 10.2370/OND000000000098.

[112] C. Mollet. Uniform Stability of Petrov-Galerkin Discretizations of Boundedly Invertible Operators: Application to the Space-Time Weak Formulation for Parabolic Evolution Problems. *Comput. Methods Appl. Math.*, 14(2):135–303, April 2014.

[113] C. Mollet. *Space-Time Adaptive Wavelet Methods for the Nonlinear Electron-Hole Schrödinger Equation with Random Disorder (working title).* Dissertation, Universität zu Köln, in preparation.

[114] C. Mollet, A. Kunoth, and T. Meier. Excitonic Eigenstates of Disordered Semiconductor Quantum Wires: Adaptive Wavelet Computation of Eigenvalues for the Electron-Hole Schrödinger Equation. *Commun. Comput. Physics*, 14(1):21–47, July 2013. DOI: 10.4208/cicp.081011.260712a.

[115] C. Mollet and R. Pabel. Efficient Application of Nonlinear Stationary Operators in Adaptive Wavelet Methods – The Isotropic Case. *Numerical Algorithms*, 63(4):615–643, July 2013. http://dx.doi.org/10.1007/s11075-012-9645-z.

[116] M. Mommer. *Towards a Fictitious Domain Method with Optimally Smooth Solutions.* Dissertation, RWTH Aachen, June 2005.

[117] M. Mommer and R. Stevenson. A goal-oriented adaptive finite element method with convergence rates. *SIAM J. Numer. Anal.*, 47(2):861–886, 2009.

[118] Message Passing Interface. http://www.mcs.anl.gov/research/projects/mpi/.

[119] R. Nochetto, K. Siebert, and A. Veeser. Theory of Adaptive Finite Element Methods: An Introduction. In DeVore and Kunoth [63], pages 409–542.

[120] OpenMP. http://openmp.org/.

[121] P. Oswald. On the Degree of Nonlinear Spline Approximation in Besov-Sobolev Spaces. *Journal of Approximation Theory*, 61(2):131–157, 1990.

[122] R. Pabel. *Wavelet Methods for PDE Constrained Control Problems with Dirichlet Boundary Control.* Shaker–Verlag, 2007. DOI:10.2370/236_232.

[123] A. Papoulis. *Probability, Random Variables, and Stochastic Processes.* McGraw-Hill series in electrical engineering. McGraw-Hill, New York, Saint Louis, Paris, 1991.

[124] P. Petrushev. Direct and Converse Theorems for Spline and Rational Approximation and Besov Spaces. In M. Cwikel, J. Peetre, Y. Sagher, and H. Wallin, editors, *Function Spaces and Applications*, volume 1302 of *Lecture Notes in Mathematics*, pages 363–377. Springer Berlin Heidelberg, 1988.

[125] W. Press, S. Teukolsky, W. Vetterling, and B. Flannery. *Numerical Recipes: The Art of Scientific Computing*. Cambridge University Press, third edition, 2007.

[126] M. Primbs. *Stabile biorthogonale Spline-Waveletbasen auf dem Intervall.* Dissertation, Universität Duisburg-Essen, Campus Duisburg, January 2006.

[127] M. Renardy and R. C. Rogers. *An Introduction to Partial Differential Equations.* Springer, New York, second edition, 1993, 2004.

[128] T. Rohwedder, R. Schneider, and A. Zeiser. Perturbed Preconditioned Inverse Iteration for Operator Eigenvalue Problems with Application to Adaptive Wavelet Discretization. *Adv. Comput. Math.*, 34(1):43–66, 2011.

[129] J. Rudi. Refinable Integrals `MATLAB` Implementation. https://github.com/johannrudi/refine. Online Resources.

[130] T. Runst and W. Sickel. *Sobolev Spaces of Fractional Order, Nemytskij Operators, and Nonlinear Partial Differential Equations.* de Gruyter Ser. Nonlinear Anal. and Appl. 3, 1996.

[131] K. Scherer. Approximation mit Wavelets I/II. Lecture courses at the Universität Bonn, Summer Semester 2001 / Winter Semester 2002.

[132] W. Schiesser. *The Numerical Method of Lines : Integration of Partial Differential Equations.* Academic Press, 1991.

[133] C. Schwab and R. Stevenson. Space-time Adaptive Wavelet Methods for Parabolic Evolution Problems. *Mathematics of Computation*, 78(267):1293–1318, 2009.

[134] C. Schwab and R. Stevenson. Fast Evaluation of Nonlinear Functionals of Tensor Product Wavelet Expansions. *Numerische Mathematik*, 119(4):765–786, July 2011. DOI: 10.1007/s00211-011-0397-9.

[135] R. Seydel. *Tools for Computational Finance.* Universitext. Springer, 2012.

[136] E. Somersalo, M. Cheney, and D. Isaacson. Existence and Uniqueness for Electrode Models for Electric Current Computed Tomography. *SIAM J. Appl. Math.*, 52(4):1023–1040, 1992.

[137] F. Stapel. *Space-Time Tree-Based Adaptive Wavelet Methods for Parabolic PDEs.* Shaker–Verlag, 2011. DOI:10.2370/OND000000000173.

[138] R. Stevenson. Locally Supported, Piecewise Polynomial Biorthogonal Wavelets on non-uniform Meshes. *Const. Approx.*, 19(4):477–508, 2003.

[139] R. Stevenson. Optimality of a Standard Adaptive Finite Element Method. *Found. Comput. Math.*, 7(2):245–269, 2007.

[140] R. Stevenson. Adaptive Wavelet Methods for Solving Operator Equations: An Overview. In DeVore and Kunoth [63], pages 543–597.

[141] B. Stroustrup. C++ FAQ. http://www.stroustrup.com/C++11FAQ.html#std-threads.

[142] F. Tröltzsch. *Optimal Control of Partial Differential Equations: Theory, Methods, and Applications.* American Mathematical Society, 2010.

[143] L. Villemoes. Sobolev Regularity of Wavelets and Stability of Iterated Filter Banks. In Y. Meyer and S. Roques, editors, *Progress in Wavelet Analysis and Applications*, pages 243–251. Editions Frontières, 1993.

[144] A. Vogel. Das LibWavelets-Programmpaket: Eine Anleitung. Bachelorarbeit, Mathematisches Institut, Universität Paderborn, 2014.

[145] F. von Leitner. Source Code Optimization. http://www.linux-kongress.org/2009/slides/compiler_survey_felix_von_leitner.pdf. Online Resource.

[146] J. Vorloeper. Multiskalenverfahren und Gebietszerlegungsmethoden. Diplomarbeit, RWTH Aachen, 1999. (In German).

[147] J. Vorloeper. *Adaptive Wavelet Methoden für Operator Gleichungen – Quantitative Analyse und Softwarekonzepte.* VDI–Verlag, 2010. (In German).

[148] J. Weidmann. *Linear Operators in Hilbert Spaces.* Springer, New York, 1980.

[149] E. Zeidler and L. Boron. *Nonlinear Functional Analysis and Its Applications: Part II/B: Nonlinear Monotone Operators.* Springer, New York, 1990.